Bone Regeneration and Repair Materials

Bone Regeneration and Repair Materials

Editors

Adalberto Luiz Rosa
Marcio Mateus Beloti

MDPI

Basel • Beijing • Wuhan • Barcelona • Belgrade • Novi Sad • Cluj • Manchester

Editors
Adalberto Luiz Rosa
University of São Paulo
Ribeirão Preto
Brazi

Marcio Mateus Beloti
University of São Paulo
Ribeirão Preto
Brazi

Editorial Office
MDPI
St. Alban-Anlage 66
4052 Basel, Switzerland

This is a reprint of articles from the Special Issue published online in the open access journal *Journal of Functional Biomaterials* (ISSN 2079-4983) (available at: https://www.mdpi.com/journal/jfb/special_issues/EH711Q9383).

For citation purposes, cite each article independently as indicated on the article page online and as indicated below:

Lastname, A.A.; Lastname, B.B. Article Title. *Journal Name* **Year**, *Volume Number*, Page Range.

ISBN 978-3-7258-1189-2 (Hbk)
ISBN 978-3-7258-1190-8 (PDF)
doi.org/10.3390/books978-3-7258-1190-8

Cover image courtesy of Adalberto Luiz Rosa

Contents

About the Editors

Adalberto Luiz Rosa

Adalberto Luiz Rosa is the full Professor at Department of Oral & Maxillofacial Surgery, and Periodontology of the Ribeirão Preto School of Dentistry, University of São Paulo. His projects, scientific interests and publications cover the following topics: bone biology, bone regeneration, cell culture, cell therapy, bone biomaterials and molecular biology.

Marcio Mateus Beloti

Marcio Mateus Beloti is the full Professor at Department of Basic and Oral Biology of the Ribeirão Preto School of Dentistry, University of São Paulo. His projects, scientific interests and publications cover the following topics: bone, biomaterials, cell therapy, regenerative medicine and cell signalling.

Journal of *Functional Biomaterials*

Editorial

Bone Regeneration and Repair Materials

Marcio Mateus Beloti and Adalberto Luiz Rosa *

Bone Research Lab, Ribeiraão Preto School of Dentistry, University of São Paulo, Ribeiraão Preto 14040-904, SP, Brazil; mmbeloti@usp.br
* Correspondence: adalrosa@forp.usp.br; Tel.: +55-16-3315-4106

1. Introduction

Bone tissue has a remarkable ability to regenerate following injury and trauma [1–3]. However, the extent of bone loss or the presence of concurrent diseases can often surpass the regenerative ability, leading to the failure of conventional procedures and, consequently, the need for additional treatments [4,5]. In the field of bone repair, regenerative medicine encompasses all currently available treatments, including biological and material approaches and their combination, which are being evaluated by researchers and clinicians [6–8]. This Special Issue, entitled "Bone Regeneration and Repair Materials", is composed of 14 original and 2 review articles that can be grouped into the following three categories: (1) physicochemical and mechanical characterizations of biomaterials for bone regeneration and implants, (2) strategies to induce bone repair using biomaterials and/or cells and (3) titanium (Ti) implants. We believe that this collection of information is of great interest to researchers and clinicians dealing with bone tissue and offers new insights into the interactions between bones and materials. We are appreciative to the *Journal of Functional Biomaterials* team for inviting us to Guest Edit this Special Issue and to the authors from eight countries who helped us build this impressive collection of scientific knowledge.

2. Overview of Published Articles

One of the main challenges in developing biomaterials as substitutes for bone tissue is the emulation of the physicochemical and mechanical properties of bones. Using nacre from mollusk shells with layered structures as a natural model for bio-inspired materials, Trabizian et al. (contribution 1) fabricated nacre-like composites of hydroxyapatite (HA) and polymers using a bidirectional freeze-casting technique. The mechanical characterization of the composites indicated that increasing the HA fraction enhanced the mineral bridge density, resulting in composites with higher flexural and compressive strengths, making them potential candidates for use in orthopedics, such as spinal fusion and bone fracture fixation implants. Based on the use of goose bone as a traditional medicine in Malay culture, Abdul Rahman et al. (contribution 2) described a method for preparing goose bone ash via bone calcination. They observed that sintering bones at 900 °C generated HA in the mineralogical phase with a calcium/phosphate atomic ratio of 1.64, which is very close to the ideal stoichiometric ratio of 1.67, creating possibilities for further investigations into therapeutic approaches using goose bone ash to repair bone tissue. Marine sponges have highly porous bodies and inorganic (biosilica) and collagen-like (spongin) organic contents, making them potential structures for use as natural scaffolds in bone tissue engineering [9]. Santos et al. (contribution 3) demonstrated that scaffolds produced from two species of marine sponges, despite exhibiting similar chemical compositions and porosities, presented distinct osteogenic potential when implanted in noncritical bone defects created in rat tibiae, with the scaffold with a higher degradation rate inducing more bone formation. As mentioned by the authors, the development of sustainable mariculture techniques is crucial for the generation of large-scale biomaterials derived from marine sponges for clinical therapeutic applications. With a focus on materials for bone tissue engineering, synthetic

Citation: Beloti, M.M.; Rosa, A.L. Bone Regeneration and Repair Materials. *J. Funct. Biomater.* **2024**, *15*, 78. https://doi.org/10.3390/jfb15030078

Received: 11 March 2024
Revised: 18 March 2024
Accepted: 19 March 2024
Published: 21 March 2024

polymers with adequate printability and mechanical properties have been employed to fabricate scaffolds using several printing processing methods [10]. Gao et al. (contribution 4) applied digital light processing printing technology to fabricate a scaffold in which the addition of gelatin methacrylate enhanced the osteoblast differentiation of mesenchymal stem cells (MSCs) derived from rabbit bone marrow and increased bone formation in bone defects created in rabbit femurs. Among the biomaterials with potential to substitute bone tissue, 45S5 Bioglass® is of great relevance as a synthetic glass that was found to chemically bond to bone [11,12]. Considering the repair of demanding bone defects such as the ones in osteoporotic bones, Araújo et al. (contribution 5) incorporated teriparatide, a recombinant fragment of the human parathyroid hormone, into 45S5 Bioglass® and observed a promising result in terms of the bone repair of critical size defects created in ovariectomized rat calvariae. The association between biomaterials and cells is a smart approach that has been extensively investigated in bone tissue engineering [13,14]. For the first time, Adolpho et al. (contribution 6) combined photobiomodulation therapy, which is known to enhance bone repair, with a ceramic/polymer scaffold and MSCs and demonstrated that the association of these three tools increased the bone formation in rat calvarial defects, highlighting the need for innovative approaches and the combination of different techniques to regenerate large bone defects. Another potential therapeutical application of stem cells is in the prevention of the progressive degeneration of cartilage and subchondral bone triggered by temporomandibular disorders, a subject deeply explored in a bibliometric study performed by da Silva et al. (contribution 7). Oral health is directly linked to the integrity of mineralized tissues of the stomatognathic system, including bone, enamel and dentin, and biomaterials may contribute to the preservation of these tissues [15,16]. Dotta et al. (contribution 8) synthesized strontium-containing nanoparticles that formed a mineral layer and penetrated dentin tubules, which resisted an acidic environment and induced mineral deposition by human dental pulp stem cells. These nanoparticles combine the abrasive properties of calcium carbonate with the ability of strontium to induce mineralization and can be added to dentifrice formulations to treat dentin hypersensitivity. In the dental setting, implant placement in the posterior region of the maxilla often requires prior procedures, and sinus floor elevation is the most common surgical approach for oral implant-based rehabilitation [17]. Miyauchi et al. (contribution 9) used a rabbit model to compare the healing pattern after sinus floor lifting with either non-collagenated bovine or collagenated porcine xenografts, and they observed that despite both materials allowing bone formation, the collagenated xenograft underwent higher resorption, resulting in a greater amount of new bone. Despite the benefits of sinus floor elevation, Omori et al. (contribution 10) used a rabbit model to demonstrate that contact with grafts induced the thinning and possible perforation of the sinus mucosa, which have implications for clinical outcomes and need to be further investigated. Ti is a powerful tool for promoting oral rehabilitation because of its ability to osseointegrate, which might be affected by both bone quality and quantity and implant surface features [18,19]. Santiago et al. (contribution 11) developed and characterized a fluorapatite coating prepared using a hydrothermal method and deposited it on commercial Ti implants. In a rabbit tibia model, this surface promoted more bone formation, with increased bone-to-implant contact, compared to HA-coated implants, making fluorapatite coatings an interesting approach for the enhancement of implant osseointegration under challenging clinical conditions. Systemic diseases such as osteoporosis and hypertension may disrupt the process of Ti osseointegration [20,21]. Indeed, Mulinari-Santos et al. (contribution 12) worked with spontaneously hypertensive rats under antihypertensive therapy and showed that osteoporotic conditions induced by estrogen deficiency impaired Ti osseointegration, even when the implant surface was coated with an antiresorptive agent used to treat osteoporosis. Angiogenesis–osteogenesis coupling is crucial for Ti osseointegration, and obesity may disturb this circuit, disrupting bone–implant interactions [22]. Pinto et al. (contribution 13) established an in vitro experimental model of high adipogenesis and demonstrated that the proinflammatory environment created by obesity interferes with endothelial cell responses to a Ti-enriched medium, which could explain the high implant failure ratio in the obese population. To better understand bone response to Ti implants, it is important to investigate the cell signaling pathways involved in osseointegration [23,24]. Teixeira et al. (contribution 14) used a conditioned medium approach to show that a laser-modified Ti surface enhanced the osteoblast differentiation of MSCs by downregulating the Wnt signaling inhibitor, Dickkopf 1. Souza et al. (contribution 15) stated that the high osteogenic potential of a nanostructured Ti surface generated by chemical conditioning with H_2SO_4/H_2O_2 may be related to its capacity to regulate the Hedgehog and Notch signaling pathways, and that the activation of Hedgehog and the inhibition of Notch might synergistically affect osteoblast differentiation, especially in cells grown on nanotopography. Although Ti, ceramics, glasses, and polymers are the most commonly studied and used bone substitutes and implants, alternative materials have been investigated such as wood, which is a sustainable and renewable source suitable for the production of biomaterials using more environmentally friendly processes, a topic that is explored in a review prepared by Nefjodovs et al. (contribution 16).

3. Conclusions

This Special Issue demonstrates that a diverse range of materials and approaches have emerged as promising and powerful tools in regenerative medicine to promote bone repair, regeneration, and implant osseointegration. Additionally, this collection of studies sheds light on the need to understand the cellular mechanisms involved in the interactions between bones and materials in the search for optimized and smart therapies. We know that there is a long way to go before most of the strategies presented here can be implemented in clinical practice. Furthermore, we believe that the efforts of scientists and clinicians in both basic and translational fields will accelerate this ride and uncover innovative therapies to treat damaged bone tissue in a plethora of clinical situations, always seeking optimal patient well-being.

Author Contributions: Conceptualization, M.M.B. and A.L.R.; writing—original draft preparation, M.M.B. and A.L.R.; writing—review and editing, M.M.B. and A.L.R. All authors have read and agreed to the published version of the manuscript.

Funding: This work was supported by the São Paulo Research Foundation (FAPESP; grant: 2017/12622-7) and the National Council for Scientific and Technological Development (CNPq; grants: 305033/2022-0 and 307698/2021-1).

Conflicts of Interest: The authors declare no conflicts of interest.

List of Contributors:

1. Tabrizian, P.; Sun, H.; Jargalsaikhan, U.; Sui, T.; Davis, S.; Su, B. Biomimetic Nacre-like Hydroxyapatite/Polymer Composites for Bone Implants. *J. Funct. Biomater.* **2023**, *14*, 393. https://doi.org/10.3390/jfb14080393
2. Abdul Rahman, F.S.; Abdullah, A.M.; Radhi, A.; Shahidan, W.N.S.; Abdullah, J.Y. Physicochemical Characterization of Thermally Processed Goose Bone Ash for Bone Regeneration. *J. Funct. Biomater.* **2023**, *14*, 351. https://doi.org/10.3390/jfb14070351
3. Santos, C.P.G.; Prado, J.P.S.; Fernandes, K.R.; Kido, H.W.; Dorileo, B.P.; Parisi, J.R.; Silva, J.A.; Cruz, M.A.; Custódio, M.R.; Rennó, A.C.M.; et al. Different Species of Marine Sponges Diverge in Osteogenic Potential When Therapeutically Applied as Natural Scaffolds for Bone Regeneration in Rats. *J. Funct. Biomater.* **2023**, *14*, 122. https://doi.org/10.3390/jfb14030122
4. Gao, J.; Li, M.; Cheng, J.; Liu, X.; Liu, Z.; Liu, J.; Tang, P. 3D-Printed GelMA/PEGDA/F127DA Scaffolds for Bone Regeneration. *J. Funct. Biomater.* **2023**, *14*, 96. https://doi.org/10.3390/jfb14020096
5. de Araújo, J.C.R.; Sobral Silva, L.A.; de Barros Lima, V.A.; Bastos Campos, T.M.; Lisboa Filho, P.N.; Okamoto, R.; de Vasconcellos, L.M.R. The Local Release of Teriparatide Incorporated in 45S5 Bioglass Promotes a Beneficial Effect on Osteogenic Cells and Bone Repair in Calvarial Defects in Ovariectomized Rats. *J. Funct. Biomater.* **2023**, *14*, 93. https://doi.org/10.3390/jfb14020093
6. Adolpho, L.F.; Ribeiro, L.M.S.; Freitas, G.P.; Lopes, H.B.; Gomes, M.P.O.; Ferraz, E.P.; Gimenes, R.; Beloti, M.M.; Rosa, A.L. Mesenchymal Stem Cells Combined with a P(VDF-TrFE)/$BaTiO_3$ Scaffold and Photobiomodulation Therapy Enhance Bone Repair in Rat Calvarial Defects. *J. Funct. Biomater.* **2023**, *14*, 306. https://doi.org/10.3390/jfb14060306

7. da Silva, Z.A.; Melo, W.W.P.; Ferreira, H.H.N.; Lima, R.R.; Souza-Rodrigues, R.D. Global Trends and Future Research Directions for Temporomandibular Disorders and Stem Cells. *J. Funct. Biomater.* **2023**, *14*, 103. https://doi.org/10.3390/jfb14020103
8. Dotta, T.C.; Hayann, L.; de Padua Andrade Almeida, L.; Nogueira, L.F.B.; Arnez, M.M.; Castelo, R.; Cassiano, A.F.B.; Faria, G.; Martelli-Tosi, M.; Bottini, M.; et al. Strontium Carbonate and Strontium-Substituted Calcium Carbonate Nanoparticles Form Protective Deposits on Dentin Surface and Enhance Human Dental Pulp Stem Cells Mineralization. *J. Funct. Biomater.* **2022**, *13*, 250. https://doi.org/10.3390/jfb13040250
9. Miyauchi, Y.; Izutani, T.; Teranishi, Y.; Iida, T.; Nakajima, Y.; Xavier, S.P.; Baba, S. Healing Patterns of Non-Collagenated Bovine and Collagenated Porcine Xenografts Used for Sinus Floor Elevation: A Histological Study in Rabbits. *J. Funct. Biomater.* **2022**, *13*, 276. https://doi.org/10.3390/jfb13040276
10. Omori, Y.; Botticelli, D.; Migani, S.; Ferreira Balan, V.; Pires Godoy, E.; Xavier, S.P. Sinus Mucosal Damage Triggered by Synthetic or Xenogeneic Bone Substitutes: A Histological Analysis in Rabbits. *J. Funct. Biomater.* **2022**, *13*, 257. https://doi.org/10.3390/jfb13040257
11. Santiago, E.; Martin, V.; Colaço, B.; Fernandes, M.H.; Santos, C.; Gomes, P.S. Hydrothermal Synthesis of Fluorapatite Coatings over Titanium Implants for Enhanced Osseointegration—An In Vivo Study in the Rabbit. *J. Funct. Biomater.* **2022**, *13*, 241. https://doi.org/10.3390/jfb13040241
12. Mulinari-Santos, G.; dos Santos, J.S.; Kitagawa, I.L.; de Souza Batista, F.R.; Botacin, P.R.; Antoniali, C.; Lisboa-Filho, P.N.; Okamoto, R. Estrogen Deficiency Impairs Osseointegration in Hypertensive Rats Even Treated with Alendronate Coated on the Implant Surface. *J. Funct. Biomater.* **2023**, *14*, 471. https://doi.org/10.3390/jfb14090471
13. Pinto, T.S.; Gomes, A.M.; de Morais, P.B.; Zambuzzi, W.F. Adipogenesis-Related Metabolic Condition Affects Shear-Stressed Endothelial Cells Activity Responding to Titanium. *J. Funct. Biomater.* **2023**, *14*, 162. https://doi.org/10.3390/jfb14030162
14. Teixeira, J.F.L.; de Souza, J.A.C.; Magalhães, F.A.C.; de Oliveira, G.J.P.L.; de Santis, J.B.; de Souza Costa, C.A.; de Souza, P.P.C. Laser-Modified Ti Surface Improves Paracrine Osteogenesis by Modulating the Expression of DKK1 in Osteoblasts. *J. Funct. Biomater.* **2023**, *14*, 224. https://doi.org/10.3390/jfb14040224
15. Souza, P.G.; Adolpho, L.F.; Lopes, H.B.; Weffort, D.; Souza, A.T.P.; Oliveira, F.S.; Rosa, A.L.; Beloti, M.M. Effects of Modulation of the Hedgehog and Notch Signaling Pathways on Osteoblast Differentiation Induced by Titanium with Nanotopography. *J. Funct. Biomater.* **2023**, *14*, 79. https://doi.org/10.3390/jfb14020079
16. Nefjodovs, V.; Andze, L.; Andzs, M.; Filipova, I.; Tupciauskas, R.; Vecbiskena, L.; Kapickis, M. Wood as Possible Renewable Material for Bone Implants—Literature Review. *J. Funct. Biomater.* **2023**, *14*, 266. https://doi.org/10.3390/jfb14050266

References

1. Schelling, S.H. Secondary (Classical) Bone Healing. *Semin. Vet. Med. Surg. Small Anim.* **1991**, *6*, 16. [PubMed]
2. Tseng, S.S.; Lee, M.A.; Reddi, A.H. Nonunions and the Potential of Stem Cells in Fracture-Healing. *J. Bone Joint Surg. Am.* **2008**, *90* (Suppl. 1), 98. [CrossRef]
3. Bosch-Rué, È.; Díez-Tercero, L.; Buitrago, J.O.; Castro, E.; Pérez, R.A. Angiogenic and Immunomodulation Role of Ions for Initial Stages of Bone Tissue Regeneration. *Acta Biomater.* **2023**, *166*, 14. [CrossRef]
4. Ring, D.; Jupiter, J.B.; Sanders, R. A, Quintero, J.; Santoro, V.M.; Ganz, R.; Marti R.K. Complex Nonunion of Fractures of the Femoral Shaft Treated by Wave-Plate Osteosynthesis. *J. Bone Joint. Surg. Br.* **1997**, *79*, 289. [CrossRef] [PubMed]
5. Seebach, C.; Nau, C.; Henrich, D.; Verboket, R.; Bellen, M.; Frischknecht, N.; Moeck, V.; Eichler, K.; Horlohé, K.H.S.; Hoffmann, R.; et al. Cell-Based Therapy by Autologous Bone Marrow-Derived Mononuclear Cells for Bone Augmentation of Plate-Stabilized Proximal Humeral Fractures: A Multicentric, Randomized, Open Phase IIa study. *Stem Cells Transl. Med.* **2024**, *13*, 3. [CrossRef] [PubMed]
6. Arnold, A.M.; Holt, B.D.; Daneshmandi, L.; Laurencin, C.T.; Sydlik, S.A. Phosphate Graphene as an Intrinsically Osteoinductive Scaffold for Stem Cell-Driven Bone Regeneration. *Proc. Natl. Acad. Sci. USA* **2019**, *116*, 4855. [CrossRef] [PubMed]
7. Freitas, G.P.; Lopes, H.B.; Souza, A.T.P.; Gomes, M.P.O.; Quiles, G.K.; Gordon, J.; Tye, C.; Stein, J.L.; Stein, G.S.; Lian, J.B.; et al. Mesenchymal Stem Cells Overexpressing BMP-9 by CRISPR-Cas9 Present High In Vitro Osteogenic Potential and Enhance In Vivo Bone Formation. *Gene Ther.* **2021**, *28*, 748. [CrossRef]
8. Zhang, Y.; Dai, J.; Hang, R.; Yao, X.; Bai, L.; Huang, D.; Hang, R. Impact of Surface Biofunctionalization Strategies on Key Effector Cells Response in Polyacrylamide Hydrogels for Bone Regeneration. *Biomater. Adv.* **2024**, *158*, 213768. [CrossRef] [PubMed]

9. Martins, E.; Diogo, G.S.; Pires, R.; Reis, R.L.; Silva, T.H. 3D Biocomposites Comprising Marine Collagen and Silica-Based Materials Inspired on the Composition of Marine Sponge Skeletons Envisaging Bone Tissue Regeneration. *Mar. Drugs* **2022**, *20*, 718. [CrossRef]
10. Tabatabaei, F.; Gelin, A.; Rasoulianboroujeni, M.; Tayebi, L. Coating of 3D Printed PCL/TCP Scaffolds Using Homogenized-Fibrillated Collagen. *Colloids Surf. B Biointerfaces* **2022**, *217*, 112670. [CrossRef]
11. Xynos, I.D.; Hukkanen, M.V.; Batten, J.J.; Buttery, L.D.; Hench, L.L.; Polak, J.M. Bioglass 45S5 Stimulates Osteoblast Turnover and Enhances Bone Formation In Vitro: Implications and Applications for Bone Tissue Engineering. *Calcif. Tissue Int.* **2000**, *67*, 321. [CrossRef]
12. Jones, J.R. Review of Bioactive Glass: From Hench to Hybrids. *Acta Biomater.* **2013**, *9*, 4457. [CrossRef] [PubMed]
13. Freitas, G.P.; Lopes, H.B.; Almeida, A.L.G.; Abuna, R.P.F.; Gimenes, R.; Souza, L.E.B.; Covas, D.T.; Beloti, M.M.; Rosa, A.L. Potential of Osteoblastic Cells Derived from Bone Marrow and Adipose Tissue Associated with a Polymer/Ceramic Composite to Repair Bone Tissue. *Calcif. Tissue Int.* **2017**, *101*, 312. [CrossRef] [PubMed]
14. Mazzoni, E.; Iaquinta, M.R.; Mosaico, M.; De Pace, R.; D'Agostino, A.; Tognon, M.; Martini, F. Human Mesenchymal Stem Cells and Innovative Scaffolds for Bone Tissue Engineering Applications. *Tissue Eng. Part. B Rev.* **2023**, *29*, 514. [CrossRef]
15. Degli Esposti, L.; Ionescu, A.C.; Gandolfi, S.; Ilie, N.; Adamiano, A.; Brambilla, E.; Iafisco, M. Natural, Biphasic Calcium Phosphate from Fish Bones for Enamel Remineralization and Dentin Tubules Occlusion. *Dent. Mater.* 2024. *in press*. [CrossRef] [PubMed]
16. Qian, Y.; Zhao, Y.; Lu, Y.; Bao, S.; Zhu, W.; Chen, Q.; Si, M. Clinical and Radiographic Outcomes of Lateral Sinus Floor Elevation with Simultaneous Hydrophilic Implants Placement: A Retrospective Study of 2–5 Years. *Clin. Oral Implants Res.* **2024**, 1–13. [CrossRef]
17. Listl, S.; Faggion, C.M., Jr. An Economic Evaluation of Different Sinus Lift Techniques. *J. Clin. Periodontol.* **2010**, *37*, 777. [CrossRef]
18. Brånemark, P.I.; Hansson, B.O.; Adell, R.; Breine, U.; Lindström, J.; Hallén, O.; Ohman, A. Osseointegrated Implants in the Treatment of the Edentulous Jaw. Experience from a 10-Year Period. *Scand. J. Plast. Reconstr. Surg. Suppl.* **1977**, *16*, 1.
19. Freitas, G.P.; Lopes, H.B.; Martins-Neto, E.C.; de Oliveira, P.T.; Beloti, M.M.; Rosa, A.L. Effect of Surface Nanotopography on Bone Response to Titanium Implant. *J. Oral Implantol.* **2016**, *42*, 240. [CrossRef]
20. Wu, X.; Al-Abedalla, K.; Eimar, H.; Arekunnath Madathil, S.; Abi-Nader, S.; Daniel, N.G.; Nicolau, B.; Tamimi, F. Antihypertensive Medications and the Survival Rate of Osseointegrated Dental Implants: A Cohort Study. *Clin. Implant. Dent. Relat. Res.* **2016**, *18*, 1171. [CrossRef]
21. Siverino, C.; Tirkkonen-Rajasalo, L.; Freitag, L.; Günther, C.; Thompson, K.; Styger, U.; Zeiter, S.; Eglin, D.; Stadelmann, V.A. Restoring Implant Fixation Strength in Osteoporotic Bone With a Hydrogel Locally Delivering Zoledronic Acid and Bone Morphogenetic Protein 2. A Longitudinal In Vivo MicroCT Study in Rats. *Bone* **2024**, *180*, 117011. [CrossRef] [PubMed]
22. Rőszer, T.; Józsa, T.; Kiss-Tóth, E.D.; De Clerck, N.; Balogh, L. Leptin Receptor Deficient Diabetic (Db/Db) Mice are Compromised in Postnatal Bone Regeneration. *Cell Tissue Res.* **2014**, *356*, 195. [CrossRef]
23. Bighetti-Trevisan, R.L.; Almeida, L.O.; Castro-Raucci, L.M.S.; Gordon, J.A.R.; Tye, C.E.; Stein, G.S.; Lian, J.B.; Stein, J.L.; Rosa, A.L.; Beloti, M.M. Titanium with Nanotopography Attenuates the Osteoclast-Induced Disruption of Osteoblast Differentiation by Regulating Histone Methylation. *Biomater. Adv.* **2022**, *134*, 112548. [CrossRef] [PubMed]
24. Shirazi, S.; Ravindran, S.; Cooper, L.F. Topography-Mediated Immunomodulation in Osseointegration; Ally or Enemy. *Biomaterials* **2022**, *291*, 121903. [CrossRef] [PubMed]

Journal of *Functional Biomaterials*

Article

Biomimetic Nacre-like Hydroxyapatite/Polymer Composites for Bone Implants

Parinaz Tabrizian [1], Huijun Sun [1], Urangua Jargalsaikhan [2], Tan Sui [2], Sean Davis [3] and Bo Su [1,*]

1 Biomaterials Engineering Group, Bristol Dental School, University of Bristol, Bristol BS1 2LY, UK; yg20581@bristol.ac.uk (P.T.)
2 School of Mechanical Engineering Sciences, University of Surrey, Guildford GU2 7XH, UK
3 School of Chemistry, University of Bristol, Bristol BS8 1TS, UK
* Correspondence: b.su@bristol.ac.uk

Abstract: One of the most ambitious goals for bone implants is to improve bioactivity, incapability, and mechanical properties; to reduce the need for further surgery; and increase efficiency. Hydroxyapatite (HA), the main inorganic component of bones and teeth, has high biocompatibility but is weak and brittle material. Cortical bone is composed of 70% calcium phosphate (CaP) and 30% collagen and forms a complex hierarchical structure with anisotropic and lamellar microstructure (osteons) which makes bone a light, strong, tough, and durable material that can support large loads. However, imitation of concentric lamellar structure of osteons is difficult to achieve in fabrication. Nacre from mollusk shells with layered structures has now become the archetype of the natural "model" for bio-inspired materials. Incorporating a nacre-like layered structure into bone implants can enhance their mechanical strength, toughness, and durability, reducing the risk of implant catastrophic failure or fracture. The layered structure of nacre-like HA/polymer composites possess high strength, toughness, and tunable stiffness which matches that of bone. The nacre-like HA/polymer composites should also possess excellent biocompatibility and bioactivity which facilitate the bonding of the implant with the surrounding bone, leading to improved implant stability and long-term success. To achieve this, a bi-directional freeze-casting technique was used to produce elongated lamellar HA were further densified and infiltrated with polymer to produce nacre-like HA/polymer composites with high strength and fracture toughness. Mechanical characterization shows that increasing the ceramic fractions in the composite increases the density of the mineral bridges, resulting in higher flexural and compressive strength. The nacre-like HA/(methyl methacrylate (MMA) + 5 wt.% acrylic acid (AA)) composites with a ceramic fraction of 80 vol.% showed a flexural strength of 158 ± 7.02 MPa and a Young's modulus of 24 ± 4.34 GPa, compared with 130 ± 5.82 MPa and 19.75 ± 2.38 GPa, in the composite of HA/PMMA, due to the higher strength of the polymer and the interface of the composite. The fracture toughness in the composition of 5 wt.% PAA to PMMA improves from 3.023 ± 0.98 MPa·m$^{1/2}$ to 5.27 ± 1.033 MPa·m$^{1/2}$ by increasing the ceramic fraction from 70 vol.% to 80 vol.%, respectively.

Keywords: biomimetic; bioactive; hydroxyapatite; nacre; bi-directional freeze-casting; nacre-like composite; bone implants; mechanical properties; fracture toughness

Citation: Tabrizian, P.; Sun, H.; Jargalsaikhan, U.; Sui, T.; Davis, S.; Su, B. Biomimetic Nacre-like Hydroxyapatite/Polymer Composites for Bone Implants. *J. Funct. Biomater.* **2023**, *14*, 393. https://doi.org/10.3390/jfb14080393

Academic Editors: Adalberto Luiz Rosa, Marcio Mateus Beloti and Adriana Bigi

Received: 28 May 2023
Revised: 21 July 2023
Accepted: 22 July 2023
Published: 25 July 2023

1. Introduction

After decades of research on bioactive implants for bone repair and regeneration, replicating the mechanical strength and toughness of cortical bone is still a challenge for engineers and clinicians [1–3]. A perfect candidate biomaterial has biocompatibility and mechanical properties similar to the bone in which it is implanted, such as Young's modulus, high tensile strength, stiffness, and fatigue resistance [4,5]. The most commonly used materials for bone implants such as rods, screws, spinal fusion cages, and plates are stainless steel (SS), titanium (Ti), and polyetheretherketone (PEEK). However, each of these materials also has its drawbacks, such as problems with stress-shielding in Ti and SS implants and poor osteoconductivity in PEEK [3,6]. Therefore, there is a tremendous need to find bioactive and biocompatible materials with bone-like stiffness, strength, and toughness, that can prevent further postoperative complications, like limitation of range of motion, reduce pain, and minimize the need for additional surgery [1,6].

Bone tissue can be divided into two types: cortical (or dense bone) and cancellous (or spongy bone). Cortical bone has a porosity of 5–15%, while the porosity of cancellous bone ranges from 40–95%. Cortical bone is composed of 70% calcium phosphate (CaP) and 30% collagen, which forms a complex hierarchical structure with anisotropic and lamellar microstructures. Osteons have a lamellar structure with a thickness of 3–7 µm surrounded by blood vessels and nerves. Osteons are highly mineralized concentric bone layers composed of aligned collagen fibrils, which can be considered as reinforcing and toughening microelements. This structure makes bone a light, strong, tough, and flexible material that can support large loads. The toughness of bone is the result of mutual competition between extrinsic (crack-tip shielding) toughening mechanisms and intrinsic (plastic deformation) toughening mechanisms. Intrinsic toughening mechanisms, such as sliding of collagen fibrils and nucleation of micro- and nano-scale damages, are defined as those that confer resistance to microstructural perturbations upstream of the crack tip. Extrinsic toughening mechanisms, such as crack bridging and crack deflection reduce the driving force contributing to crack propagation. The preferential orientation of osteons in cortical bone provides an effective extrinsic toughness that is anisotropic. A crack that propagates perpendicular to the osteons is more likely to be deflected and twisted than a crack that propagate parallel to the osteons and thus may explain the anisotropy of fracture toughness [1,3,6].

To develop synthetic biomaterial composites that can be used as bone implants, it is necessary to imitate bone-like structures to reach the mechanical properties and other biological functionalities of real bone. Biological bone tissues indeed tend to have whisker-like structures at nanoscales which consist of mineralized collagen nanofibers. The preparation of such bone-like HA as a scaffold is essential to enhance the bone repair biologically. In addition to bioactivity and biocompatibility, mimicking the hierarchical structure of natural bone provides load-bearing capacity in the implant and minimizes stress concentration. So, the composite exhibits similar mechanical properties to natural bone with long-term stability, enhancing the lifespan of the implant and functionality for the patient [7–9].

In this regard, the comparison between bone and nacre structure shows good similarity in terms of hierarchical microstructure and the toughening mechanisms through which they operate [8].

So, nacre-like ceramic composites have opened new horizons for the fabrication of biomimetic bone implants with high strength and toughness [5,6]. In the nacre structure of seashells, a mechanism of toughness occurs based on the creation of weak interfaces where cracks can be deflected and energy is dissipated by opening surfaces, resulting in higher fracture toughness in ceramic constituents [10]. So, a good example of tough composites are nacre-like composites, that mimic the brick-and-mortar microstructure of seashells. Based on the design of the microstructure, the composite exhibits a rising resistance curve (R-curve) indicating a higher amount of energy is required to propagate a crack. Various methods have been used to develop nacre-like composites with hierarchical brick-and-mortar microstructure, of which bi-directional freeze-casting has proven to be a powerful method for producing bulk composites with excellent mechanical properties [11,12]. The bi-directional freeze-casting method is a modification of the unidirectional freeze-casting method, a polydimethylsiloxane (PDMS) wedge with different slopes is placed between the suspension and the cold finger [10,13]. Due to the low thermal conductivity of the PDMS wedge, the thinner side cools faster during freezing, resulting in a temperature gradient not only in the vertical but also in the horizontal direction. The ice crystals nucleate only at the bottom of the wedge and continue to grow preferentially in two directions. Bi-directional freeze-casting is an effective method for generating long-range aligned lamellar structures, and has already been used to build various functional building blocks into nacre-like composites [11,13,14].

In terms of materials, hydroxyapatite (HA) is the most stable calcium phosphate phase with a Ca/P ratio of 1.67. It is formed in nature and can be extracted, but various ions and vacancies form defective structures. Its solubility is low in physiological environments determined by temperature, pH, body fluids, etc. Hydroxyapatite is considered a bioactive material and is a mineral form of calcium apatite that closely resembles the composition and structure of natural bone minerals. It is the primary inorganic component of human bones and teeth. When used in medical and dental applications, hydroxyapatite demonstrates bioactive properties, meaning it can interact with living tissues and promote biological activity. This bioactivity is mainly due to the chemical similarity between hydroxyapatite and the mineral component of bones and teeth. In terms of contacting living tissue, such as bone or tooth surfaces, it can bond to the surrounding tissues through osseointegration. This integration enables the formation of a direct chemical and mechanical bond between the hydroxyapatite material and the surrounding bone, facilitating bone growth and regeneration [15,16].

Hydroxyapatite is commonly used in various biomedical applications, including bone grafts, dental implants, and coatings for orthopedic implants. Its bioactive nature makes it a desirable material for promoting tissue regeneration and improving the long-term success and stability of implantable medical devices [17–19].

The surface of HA can serve as a nucleus for bone minerals in body fluid, which do not cause inflammatory reactions when used clinically. They are osteoconductive but not osteoinductive. However, they are mechanically weak and brittle and exhibit weaknesses in strength and toughness, making it difficult to achieve the mechanical properties of cortical bone [16,20].

Methyl methacrylate (MMA) is an important monomer, which is widely used for producing polymethyl methacrylate in different biomedical applications, such bone cements, screw fixation in bone, and filler for bone cavities and skull defects [21,22]. To fabricate nacre-like composite, PMMA with low viscosity and high strength makes it an appropriate polymer phase compared to synthetic polymer like PLA, which can infiltrate the ceramic scaffold to make the final brick-and-mortar microstructure. Based on the literature, by infiltrating freeze-cast alumina scaffolds with poly (methyl methacrylate) (PMMA) as a compliant layer, bending strengths of 210 MPa with remarkable fracture toughness have been achieved that exceed a stress intensity of 30 MPa [23]. PMMA can replace ceramic materials in areas where higher strength and toughness are preferred. Resulting in a nacre-like SiC/PMMA composite, it reveals a rising crack resistance (R-curve) behavior where the toughness increases with crack extension [24]. Investigations into the role of the polymer phase in the mechanics of nacre-like composites show that the polymer phase has a resounding impact on the mechanical performance of nacre-like composites. The composite strength significantly increases with stiffer polymers like polyether urethane diacrylate-co-poly(2-hydroxyethyl methacrylate) (PUA-PHEMA) and poly(methyl methacrylate) (PMMA) by avoiding the stress concentrations at the mineral bridges [25]. Research shows that adding acrylic acid (AA) to MMA can modify the properties of PMMA and may also imply the onset of new characteristics. Thus, the copolymerization of MMA with acrylic acid (AA) has been intensively studied, and a variety of materials with useful performance features have been obtained with potential medical applications for regeneration of soft tissues [21,26]. A copolymer of polymethyl methacrylate (PMMA) and polyacrylic acid (PAA) can exhibit varying degrees of biocompatibility and biodegradability. PMMA is generally considered to be biocompatible and has been used in various medical applications. PAA, on the other hand, is also generally biocompatible. When PMMA and PAA are combined in a copolymer, resulting in a biocompatible material. The incorporation of PAA into a copolymer can introduce biodegradability to some extent. PAA is known to be biodegradable under certain conditions, and by combining with PMMA, it can improve the biological activity of the final composite as well [27,28].

In this study, the authors investigate and fabricate HA/polymer composites to improve the brittleness and weaknesses of bioactive HA materials by mimicking the brick-and-mortar structure of nacre through a bi-directional freeze-casting method. The effective processing parameters for the mechanical properties of the composites were studied, and the effects of the ceramic, interface, and polymer phases were cited as the most important parameters for producing bone-like strength and tough, nacre-like HA/polymer composites that could potentially be used in orthopedics such as spinal fusion and bone fracture fixation implants.

2. Materials and Methods

2.1. Slurry Preparation

Hydroxyapatite (Captal S, Plasma biotal, Buxton, UK) with an average particle size of 2.5 μm was used as the ceramic material. Darvan 821A (R. T. Vanderbilt Co., Norwalk, CT, USA) served as a dispersant that aided solute colloid formation, and polyvinyl alcohol (PVA powder, MW: 30,000–70,000, Sigma Aldrich, Dorset, UK) was used as a binder. The slurry contained 15 vol.% HA with 1.5 wt.% dispersant and 6 wt.% binder based on solids loading were dispersed in deionized water and transferred to a polyethylene (PE) bottle containing zirconia grinding balls and then milled (1600-VS-A, Pascal engineering, Crawley, UK) for 24–48 h at ambient temperature and high speed (~200 rpm). The amount of solid loading, dispersant, and binder were optimized based on the rheological behavior and processability of the sample after freeze-drying. After ball milling, the slurries were degassed by adding 0.1 mL octanol for 30 min to remove the air bubbles generated during slow ball-milling.

2.2. Fabrication of Nacre-Like HA/Polymer Composites

The slurry was poured into the PDMS molds with a slope angle of 10° and covered with silicone to obtain a flat surface at the bottom of the sample. Freezing started at the top line of the copper wedge and ended at the other side of the mold. The technical details of bi-directional freeze-casting were reported in a previous work [29]. After freezing, the ice formed in the frozen samples was sublimated in a freeze dryer (Lyotrap, Lte Scientific Ltd., Oldham, UK) at −60 °C for at least 48 h for each sample under a vacuum pressure of 0.03 mbar. To control the ceramic fraction in the final composites, highly aligned lamellar ceramic scaffolds were densified by a hydraulic press (PerkinElmer, Waltham, MA, USA). Finally, a furnace with an oxidation atmosphere was used to sinter the green bodies; in the first stage, the binder was burned out at 600 °C for 2 h, followed by dwelling at 1300 °C for 4 h. The sintering procedure was determined based on the literature [5,30,31]. The as-prepared scaffold with its long-range aligned lamellar structure was first grafted with γ–methacryloxypropyltrimethoxysilane (γ–MPS, Sigma Aldrich, Dorset, UK) in ethanol at different concentrations and for different time periods. Characterization of the grafted ceramic surface was performed by using an attenuated total reflectance (ATR) FTIR spectrometer (PerkinElmer, spectrum-one FTIR, Waltham, MA, USA). In addition, the hydrophobicity of the grafted surface was measured using a drop shape analyzer (DSA100, KRUSS, Hamburg, Germany). Subsequently, the grafted scaffolds were infiltrated with a solution of methyl methacrylate (MMA, Sigma Aldrich, Dorset, UK) and 0.5–1 wt.% azobisisobutyronitrile (AIBN, Sigma Aldrich, UK) as an initiator, and acrylic acid (AA, Sigma Aldrich, Dorset, UK) as a second polymer phase in Cast' N Vac for 2 h, then heated at 45 °C for 24 h to complete polymerization, and annealed at 90 °C for 2 h.

2.3. Characterization

A Zwick Roell universal testing machine (Z020, Zwick Roell, Ulm, Germany) was used to determine compressive and flexural strength (ASTM Standard C1424-15 and D790-15) [32]. Five specimens were tested for each condition. Porosity was measured by the Archimedes method according to the standard (ASTM B962-17) [33]. Scanning electron microscope (EM, Quanta 400-FEI Scanning Electron Microscope, San Diego, CA, USA) was used to study the morphology of the different specimens. X-ray diffraction

(XRD, Bruker D8 Advance, Coventry, UK), with Cu-Ka (λ = 1.54 Å) was conducted to investigate any potential changes in the crystalline phase. The fracture toughness of the specimens was measured by three-point bending on single-edged notched bending (SENB) with ASTM E1820 [34]. The prepared SENB specimens were tested in situ at SEM using a Deben Micro-Test 150 N bending stage (Deben, Bury Saint Edmunds, UK) with a support span of 20 mm and a displacement rate for loading and/unloading of 0.1 mm/min. In situ SEM was used to monitor crack propagation in the real-time and record high resolution images after each loading and/unloading cycle.

Based on standard fracture mechanics, the post-fracture dissipated energy (J-integral) was calculated as a function of crack extension with an overall straight trajectory of the crack over the specimen depth (mode I). Fracture toughness, K_J, was determined by back-calculation from the equivalence of mode I, J–K:

$$K_J = \sqrt{J\,E'} \quad (1)$$

The J-integral was formed from the elastic and plastic components as follows:

$$J = J_{el} + J_{pl} \quad (2)$$

The elastic contribution was calculated using the theory of linear-elastic fracture mechanics:

$$J_{el} = K^2/\,E' \quad (3)$$

where E′ is the elastic modulus of the composite under pure strain conditions and is expressed as follows:

$$E' = E/(1-\nu)^2 \quad (4)$$

Here, E is the modulus of elasticity of the composite, and ν is the Poisson's ratio. E was determined using the rule of mixtures, and ν = 0.3 was used for all composites. The mode I stress-intensity K was calculated as:

$$K = \frac{PS}{BW^{\frac{3}{2}}} f\left(\frac{a}{w}\right) \quad (5)$$

$$f\left(\frac{a}{w}\right) = \frac{3\left(\frac{a}{w}\right)^{\frac{1}{2}}\left[1.99 - \left(\frac{a}{w}\right)\left(1-\frac{a}{w}\right)\left(2.15 - 3.93\left(\frac{a}{w}\right) + 2.7\left(\frac{a}{w}\right)^2\right)\right]}{2\left(1+2\frac{a}{w}\right)\left(1-\frac{a}{w}\right)^{\frac{3}{2}}} \quad (6)$$

where P is the maximum load (N), S is the span (mm), B is the specimen thickness (mm), W is the specimen width (mm), and a is the crack length (mm).

The plastic component J_{pl} was defined as:

$$J_{pl} = 1.9A_{pl}/Bb \quad (7)$$

where b is the uncracked ligament, and A_{pl} is the plastic area under the curve between loading and plastic displacement. According to ASTM, the maximum crack extension for a specimen is given by $\Delta a_{max} = 0.25b_0$, where b_0 is the initial uncracked ligament determined as $b_0 = W - a_0$. Only the crack expansions within the range valid according to ASTM are included in the linear fit [34–36].

3. Results

3.1. *Effect of Ceramic Phase*

After sublimation and sintering, an HA scaffold with a long-range lamellar structure with 70% porosity was obtained. The scaffold was further densified by uniaxial pressing to ≈20–40% porosity. Figure 1a illustrates the effects of ceramic fraction on wall thickness, density of ceramic bridges, and compressive strength. The average value of five specimens for each ceramic fraction was given. Increasing the ceramic fraction from 60 vol.% to 80 vol.% increases the wall thickness from 17.89 ± 1.08 to 36.01 ± 1.89 μm with higher bridges density and compressive strength of 167.5 ± 2.87, which was 57 47 ± 2.60 MPa. Figure 1b–d show the microstructure of the composites based on ceramic fractions of 60 vol.%, 70 vol.%, and 80 vol.%, respectively. By increasing the ceramic fractions, the wall becomes longer and thicker; moreover, the density of ceramic bridges increases from 5.89 ± 1.56 to 23.03 ± 2.07 (%), which is crucial for improving the strength and toughness of the final composite because they can transfer and redistribute stresses and enhance the frictional sliding between the ceramic layers, leading to improved mechanical properties. (d. The yellow arrows show the increase in ceramic bridges due to the increase in the ceramic fractions) [37,38]. The ceramic bridges are formed due to the conflict between forced and preferential ice growth during bi-directional freeze-casting, resulting in an oblique ice growth direction [39,40].

a

Ceramic fraction (vol%)	60	70	80
Ceramic wall thickness (μm)	17.89±1.08	25.41±1.31	36.015±1.89
Density of bridges (%)	5.89±1.56	11.56±1.23	23.03±2.07
Compressive strength (MPa)	57.47±2.60	84.03±1.98	167.5±2.87

e

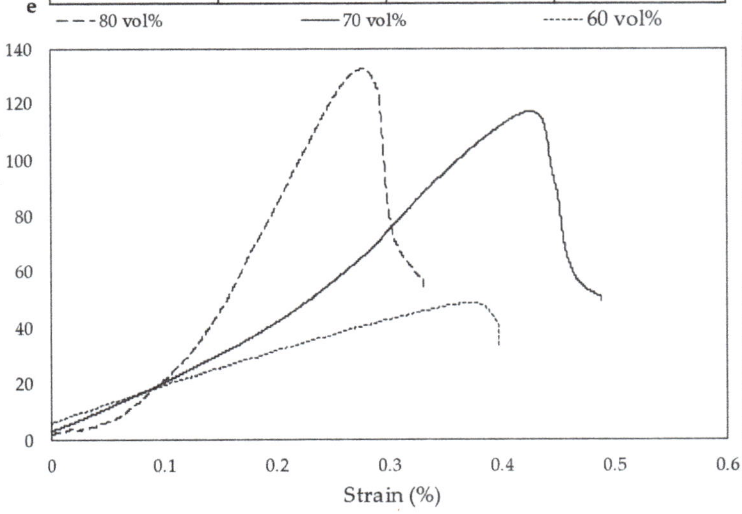

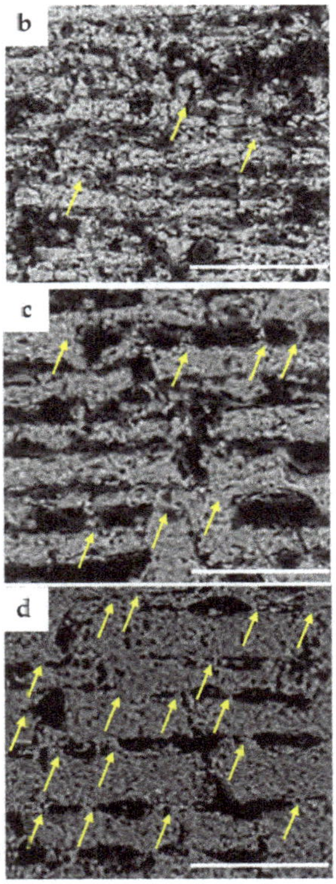

Figure 1. (**a**) Effect of ceramic fraction on ceramic wall thicknesses, ceramic bridges density, and compressive strength. Increasing the ceramic fraction from 60 vol.% to 80 vol.%, results in an increase in wall thickness to 36.01 ± 1.89 μm with a higher ceramic bridges density of 23.03 ± 2.07 (%) and a compressive strength of 167.5 ± 2.87 MPa. (**b**–**d**) show the microstructure of the composite at ceramic fractions of 60 vol.%, 70 vol.%, and 80 vol.%, respectively (Scale bars are 500 μm). (**d**) The yellow arrows show the ceramic bridges increase by increasing the ceramic fraction. (**e**) Stress–strain diagram of different composites with different ceramic fractions. Flexural strength increases to 130 ± 5.82 MPa and the Young's modulus is 19.75 ± 2.38 GPa for composites with 80 vol.%.

Increasing the ceramic fraction to more than 80 vol.% destroyed the brick-and-mortar as the ceramic walls fused together and the distinct layered structure was lost in the subsequent sintering stage. Figure 1e shows the stress–strain curve of HA/PMMA composites at different ceramic fractions. The composites with different ceramic fractions showed similar failure mechanisms but different strength values, the composite with 80 vol.% having a flexural strength of 130 ± 5.82 MPa and Young's modulus of 19.75 ± 2.38 GPa. However, the composite with 70 vol.% ceramic fractions had a flexural strength of 115 ± 2.67 MPa and Young's modulus of 14.36 ± 2.38 GPa. These values were 52.68 ± 3.78 MPa and 10.11 ± 1.23 GPa for 60 vol.% ceramic fractions. Thus, higher ceramic fractions increase the density of the ceramic walls and the mineral bridges, resulting in a stronger composite.

3.2. Effect of Interface

To produce a nacre-like ceramic composite, it is important to improve the interface between the ceramic and the polymer. To this end, the silanization process is an important procedure to establish a strong interface between the ceramic and polymer phases. HA itself has O–H groups in the surface chemistry, which can establish a bond with the silane group. Therefore, it is important to optimize the grafting of the HA scaffold to find the best bonding and flexural strength based on Fourier-transform infrared spectroscopy (FTIR) and contact angle measurement, which can confirm the silanization of the ceramic surface.

Figure 2a shows the results of Fourier-transform infrared spectroscopy (FTIR) in the region of 4000–1000 cm^{-1} and contact angle analysis after silanization in different ratios of 25 wt.% and 50 wt.% γ–MPS/ethanol solutions at both grafting times of 12 h and 24 h. The silane coupling agent usually acts as a kind of mediator that connects organic materials with inorganic materials. The silane molecule consists of a silicon-based head and an organic tail that is linked to inorganic and organic phases, respectively. In particular, the silicon-based head can firmly bond to reactive groups on a substrate surface (e.g., glass, metal, and ceramic) via a covalent bond. In other words, the O–H (hydroxyl group) on the HA surface reacts with the –Si–OR (R is $-CH_3$ for γ–MPS) on the head of the silane molecule and then generates H–OR and the silane-grafted HA scaffold.

As shown in all ratios and grafting times compared to the non-grafted scaffold, there is no peak and no change on non-grafted scaffold. The peaks at 1630 cm^{-1} were assigned to the stretching bonds of carbon–oxygen double bonds (C=O), which were seen in all spectra at all concentrations or times, and at 1720 cm^{-1}, the carbon–carbon double bonds (C=C) were assigned with increasing the time from 12 h to 24 h. Based on FTIR analysis, this strong bond can occur in all configurations, but it was developed with time, which can possibly show reaction progress through the conversion of a different functional group into a double bond. The C–H stretching bonds was localized at 2980 cm^{-1} and relates to the $-CH_2CH_2-$ on the silane molecule determined after 24 h grafting for both concentrations of 25 wt.% and 50 wt.% γ-MPS in ethanol. Figure 2b shows the schematic of silanization process and effect of silanization conditions. The C=O and C=C found in FTIR spectra show the organic tail on hydroxyapatite surface after grafting with γ-MPS. Based on FTIR results, the silanization of the HA scaffold depends on the time and concentration of γ–MPS. Longer immersion time and higher concentration of γ–MPS results in more complete coverage of hydrophobic organic tails, leading to more covalent bond formation and strong interface.

The analysis of the contact angle of the grafted scaffold can confirm the hydrophobicity of the surface which can lead to the optimization of the degree of silanization to achieve a stronger bond between the ceramic and the polymer. Figure 2c–e show the effect of silanization conditions on the hydrophobicity behavior of the scaffold. As shown in Figure 2d the contact angles increase to the 83.5° due to silanization at the concentration of 25 wt.% γ–MPS grafted for 24 h; however, when the concentration of γ–MPS is increased to 50 wt.%, the angle reaches 107°, which shows hydrophobicity of the surface, which is due to the grafting with a higher concentration of γ–MPS with functional carbon double bonds. Thus, the ceramic surface forms covalent bonds with the polymer phase, resulting in stronger interface between the polymer and ceramic phases and improving the strength of the final composite.

Figure 3 shows how silanization affects the flexural strength of the final composite at a constant ceramic fraction of 70 vol.%. Based on the FTIR and contact angle analysis, the maximum flexural strength of 117 ± 1.89 MPa belongs to the composite where the scaffold was silanized in 50 wt.% γ–MPS for 24 h which shows stretching bonds of carbon–carbon double bonds (C=C), carbon–oxygen double bonds (C=O), and carbon–hydrogen stretching bonds (C–H) with a hydrophobic surface, resulting in stronger interface bonds between the ceramic and the polymer.

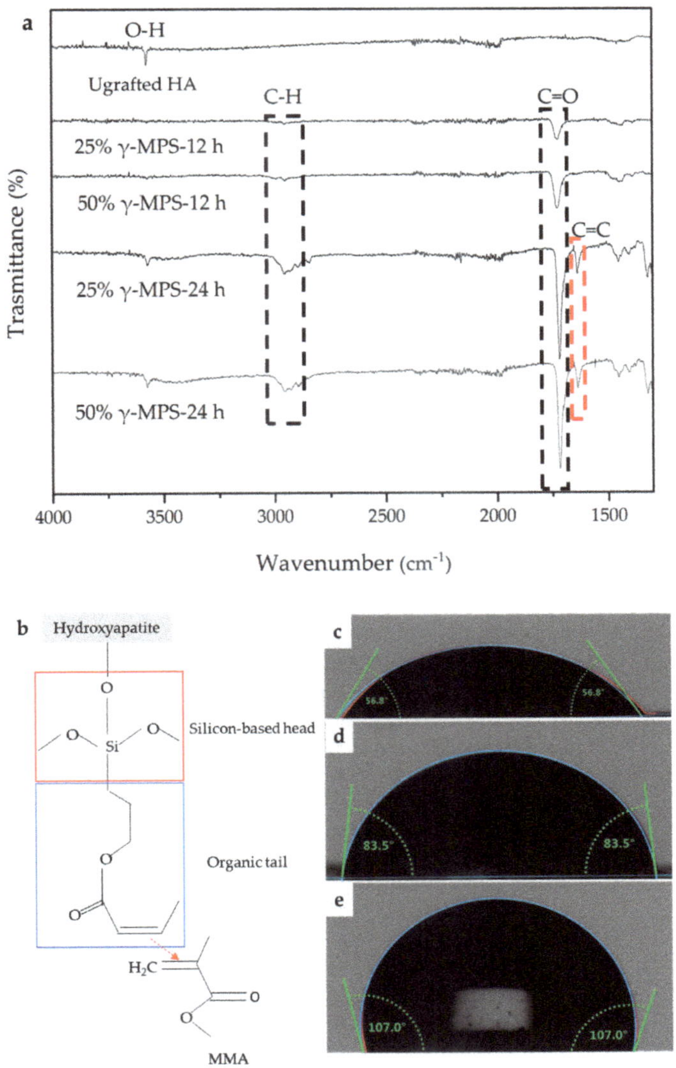

Figure 2. (**a**) FTIR spectra of silane-grafted hydroxyapatite with γ–MPS in ethanol at different concentrations and soaking times. (**b**) The schematic of silanization process and effect of silanization conditions. The C=O and C=C found in FTIR spectra show the organic tail on hydroxyapatite surface after grafting with γ–MPS. The contact angle images of the different samples show, (**c**) non-grafted HA scaffold, (**d**) ethanol with 25 wt.% γ–MPS after 24 h, and (**e**) ethanol with 50 wt.% γ–MPS after 24 h. It can be seen that increasing the ratio of γ-MPS improves the hydrophobicity of the scaffold surface, which leads to stronger adhesion of the ceramic and polymer phases and thus better mechanical properties.

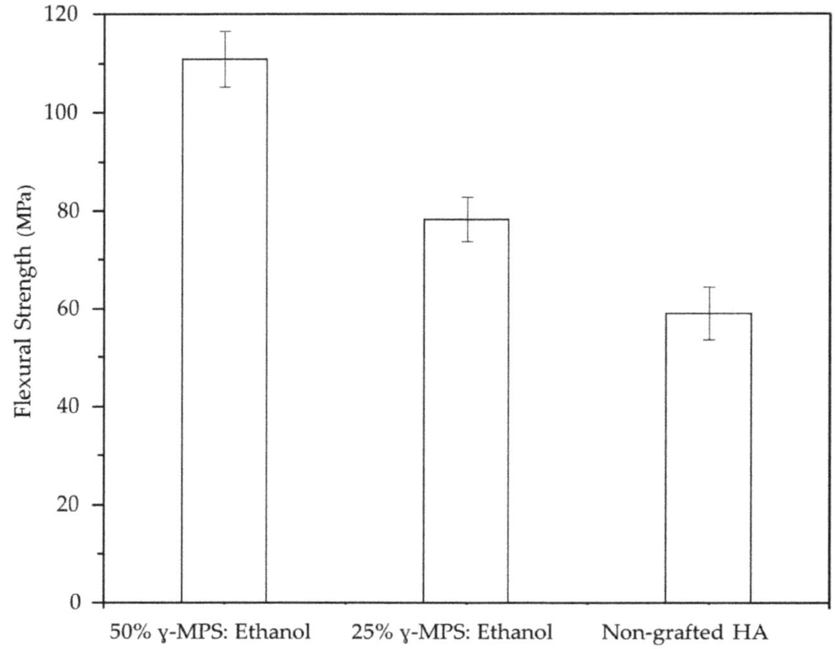

Figure 3. Flexural strength is based on different situations of silanization process. The maximum value of the strength reaches 117 ± 1.89 MPa for the sample silanized with the highest concentration of salinization solution and soaked for 24 h.

3.3. Effect of Polymer Phase

The polymer phase has a resounding influence on the mechanical properties of a nacre-like composite. Figure 4a shows the effects of adding PAA to PMMA on the flexural strength of the pure polymer. As shown by the addition of 5, 10, and 15 wt.% acrylic acid to the polymer composition, the flexural strength increases compared to pure PMMA; however, the results show that the optimum value in terms of highest flexural strength is 5 wt.% PAA. The presence of AA can form a covalent bond with the free radical of MMA, which increases flexibility and allow stronger chain and elongation. Figure 4b shows the FTIR results of HA/PMMA, the different polymer systems and the HA/PMMA + PAA. Based on the FTIR, the combination of PMMA and PAA shows the presence of hydroxyl group (O–H) around 3500 cm^{-1}. In the composite spectra, there is no hydroxyl peak (O–H), which shows that O-H groups introduce new bonds in the interface of ceramic and polymer, resulting in stronger interfacial connection. Figure 4c shows the X-ray diffraction (XRD) spectra of the ungrafted HA, grafted HA, HA/PMMA + PAA, and HA/PMMA exhibit consistent patterns, indicating that hydroxyapatite remains the predominant crystalline phase. The results from the XRD analyses indicate that neither the silanization process nor the subsequent polymerization processes exert any discernible influence on peak shifting or phase transformation, so the crystallinity of the samples remains un-changed throughout the silanization and polymerization procedures. Therefore, the addition of PAA to the polymer phase not only makes strong copolymer, but also with introducing new bonds in the interface improves interfacial strength, which can increase the mechanical properties of final composite shown in Figure 5.

From Figure 4a it can be seen that the addition of 5 wt.% PAA to PMMA is an optimum level for the use of PAA in the polymer phase. Too high addition of PAA would lead to an increase in stresses at the interfaces, resulting in lower flexural strength. At 5 wt.% PAA, the mechanical properties improved, for instance the flexural strength increases from

130 ± 5.82 to 158 ± 7.02 MPa shown in Figure 5, compared to the cortical bone values of 160 MPa which can be considered a good candidate for future bone implants.

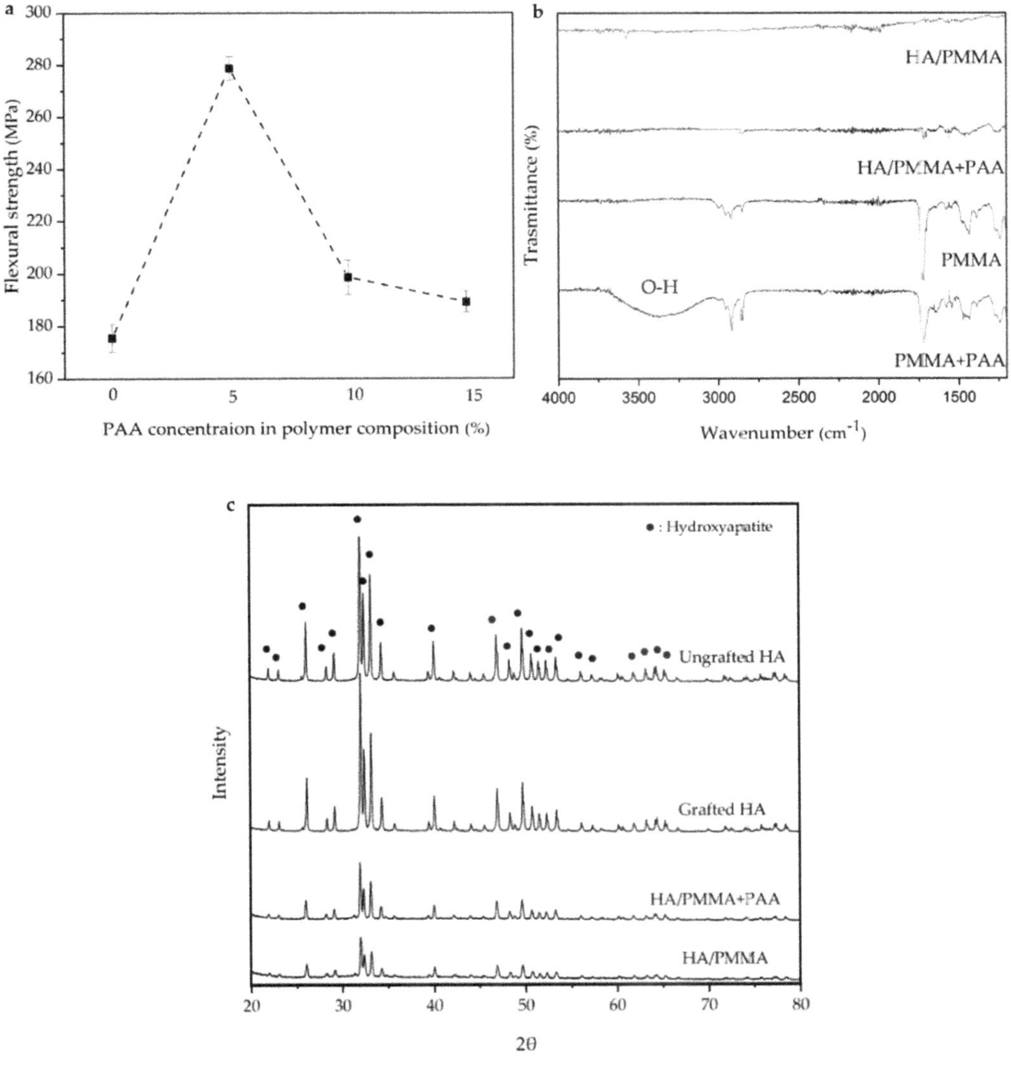

Figure 4. (**a**) The effect of adding different amounts of PAA to PMMA on the flexural strength of the polymer phase. (**b**) The FTIR results of the HA/PMMA, the different polymer phases, and HA/PMMA + PAA. (**c**) The XRD spectra of the ungrafted HA, grafted HA, HA/PMMA + PAA, HA/PMMA, show that hydroxyapatite is the main crystalline phase and silanization and polymerization do not have effect on peak shifting or phase changing and the crystallinity is identical.

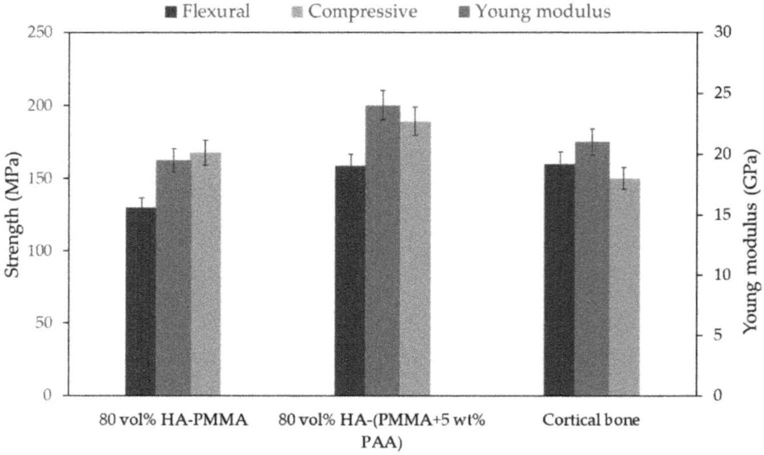

Figure 5. The effect of the addition of PAA on the mechanical properties of HA/PMMA with the same ceramic fraction composites. The flexural, compressive strength, and Young's modulus increase to 158 ± 7.02, 189.09 ± 6.45 MPa, and 24 ± 4.34 GPa, respectively, compared with the value of cortical bone of 160, 150 MPa, and 21 GPa, respectively. The mechanical data of the cortical bone were taken from the literature [41–43].

Figure 6a shows the crack-propagation behavior of different composites with different ceramic fractions and polymer systems, observed in situ in a SEM; this allowed real-time observation of the crack propagation and its interaction with the microstructure during toughness measurement. Cracks are deflected at the ceramic/polymer. Stretching and tearing in the polymeric (mortar) can serve as a ligament bridge spanning the crack to carry the load for crack propagation, resulting in additional crack bridging and subsequent pull-out in ceramic (bricks). According to Section 3.1, ceramic bridges can improve the strength and fracture toughness of nacre-like composites by transferring and redistributing stresses to control sliding of individual ceramic layers and prevent delamination. Ceramic bridges increase the roughness of the ceramics, which leads to sliding interference between the ceramic walls during crack propagation, which can improve fracture toughness by energy dissipation of the interference [23,44]. In addition, the polymer phase in a nacre-like composite plays the role of a lubricant phase through which the sliding of the ceramic walls is controlled, leading to high stress release and energy dissipation (Figure 6 indicated with yellow circles). Therefore, as discussed in Section 3.2, silanization can provide stronger interfaces between the ceramic and the polymer and improve the effectiveness of polymer as a viscoelastic adhesive. All of these intrinsic toughening mechanisms are similar to those operating in natural nacre and cortical bone [8,45].

Figure 6b shows the K_J curve based on crack extension of three composites groups with different ceramic fractions and different polymer phases. The fracture toughness of the composites was measured by determining the area under the load–displacement curve and dividing by the area of the fracture surface. All composites show the resulting R-curve behavior, with the average value of fracture toughness increasing to 3.023 ± 0.98 MPa·m$^{1/2}$ by adding 5 wt.% PAA at 70 vol.% ceramic fraction compared to the composite with the same ceramic fraction containing PMMA. By increasing the ceramic fraction to 80 vol.% in the composite of PMMA + 5 wt.% PAA, the fracture toughness increases to 5.27 ± 1.033 MPa·m$^{1/2}$, which is quite close to toughness behavior of the cortical bone.

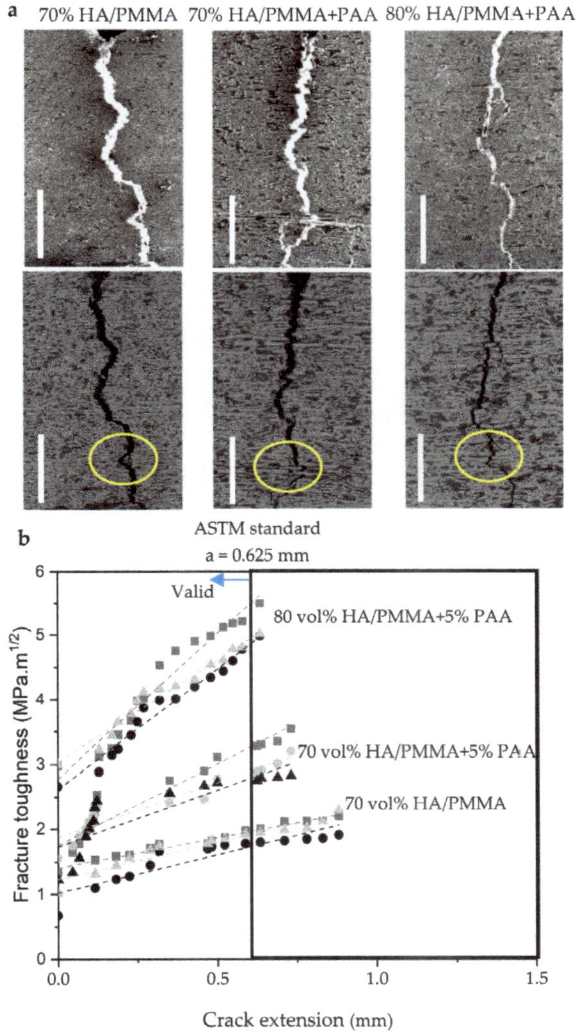

Figure 6. (**a**) The crack-propagation behavior of various samples with different ceramic fractions and polymer systems, observed in situ in a scanning electron microscope (SEM). The yellow circles indicated the crack deflections and ceramic pull-out. Scale bars are 500 μm. (**b**) R–curve of fracture toughness (K_J) of composites, addition of 5 wt.% PAA to PMMA increased the fracture toughness to 3.023 ± 0.98 MPa·m$^{1/2}$ at 70 vol.% ceramic fraction and increases to the value of 5.27 ± 1.033 MPa·m$^{1/2}$ at higher ceramic fraction of 80 vol.%.

4. Discussion

The main goal of this research is to study and fabricate newly designed bone implants that are bioactive and biomimetic in addition to having high strength and fracture toughness. These new implants can improve the efficiency of bone implants and save patients from further problems and surgeries. Taking a cue from nature, a nacre-like composite in which brittle materials such as ceramics can form a tough and strong composite by designing the microstructure. As a result, a hydroxyapatite (HA)/polymer composite with brick-and-mortar structure was developed based on bi-directional freeze-casting method.

These HA/polymer composites are bioactive and, because they mimic the microstructure and mechanical properties of cortical bone, can be considered a good candidate for a new bone implant.

The effective parameters in composite materials can be divided into three main groups, ceramic, interfacial, and polymer. The higher the percentage of ceramic fractions, the thicker and denser the ceramic wall, resulting in higher density of mineral bridges, and higher flexural and compressive strength. The microstructural characterization of composites with different ceramic fractions shows that by increasing the ceramic fractions from 60 vol.% to 80 vol.%, the ceramic walls become thicker and longer, and the density of ceramic bridges increases from 5.89 ± 1.56 to 23.03 ± 2.07 (%) with increasing the ceramic fractions from 60 vol.% to 80 vol.% (see yellow arrows in Figure 1d). The strain-stress curve shows that all the composites without considering the ceramic components exhibit ductile failure caused by the brick-and-mortar microstructure of the composites. The flexural strength and Young's modulus increase from 115 ± 2.67 MPa and 14.36 ± 2.38 GPa to 130 ± 5.82 MPa and 19.75 ± 2.38 GPa, for ceramic fraction of 70 vol.% and 80 vol.%, respectively. These results are consistent with other nacre-like composites. Sana et al. reported that the flexural strength of Alumina/polymer nacre-like composite increases from 145.77 to 172.65 MPa when the ceramic fraction is increased from 70 vol.% to 76 vol.% [29]. Tan et al. show that increasing the ceramic fractions of zirconia in a nacre-like 3Y–TZP/polymer composite increases the density of mineral bridges and Young's modulus [6].

In addition to the ceramic phase, the interface is also an effective parameter for the development of a tough and strong nacre-like composite. In order to obtain a stronger interface between ceramic and polymer, ceramic surfaces must be grafted to establish the covalent bonds with polymer. The silane-grafted surfaces of HA grafted in 50 wt.% γ–MPS for 24 h, as shown in Figure 2a,e, exhibit stretch bonds of carbon–carbon double bonds (C=C), carbon–oxygen double bonds (C=O), and carbon–hydrogen stretch bonds (C–H) with hydrophobic surfaces. This results in covalent bonds of the ceramic with the monomer infiltrated into the scaffolds, giving the flexural strength of 117 ± 1.89 MPa shown in Figure 3, which is higher than that of the composite non-grafted and the composite grafted at the concentration of 25 wt.% for 24 h. Thus, the flexural strength of the composite materials correlates with the interface, and the interface can be improved by grafting or silanization of ceramic surfaces at a suitable level. Olga et al. reported that silanization affects the mechanical properties of the composite HA/PMMA. The hydroxyapatite particles were treated with a silane coupling agent, so that the adhesion between HA and PMMA was improved and the compressive strength increased [46]. Launey et al. show that grafted alumina/PMMA has higher mechanical properties than non-grafted. For a composite with a constant ceramic wall thickness of 20 μm, the flexural strength was 90 MPa in non-grafted composite and 112 MPa for grafted composite [23].

The polymer composition affects the mechanical properties of the final composite. As can be seen in Figure 4a, mixing PAA with PMMA increases the flexural strength of polymer compared to PMMA; however, this improvement has an optimum value of 5 wt.%; beyond that, there is a downward trend. The FTIR analysis shown in Figure 4b shows that mixing PAA with PMMA forms hydroxyl groups on the surface of the polymer phase, which can also be introduce new bonds in interface. As shown in Figure 4b, O-H group on a surface of PAA can interact with free or residue groups on a surface of ceramic, which enhances the strength of interface, as well. The X-ray diffraction (XRD) spectra reveal that the crystalline phase remains unchanged subsequent to both silanization and polymerization processes, as illustrated in Figure 4c. Figure 5 shows how the mechanical properties of the composites change when PAA is mixed with PMMA. The flexural strength increases from 130 ± 5.82 to 158 ± 7.02 MPa, the compressive strength improves from 167.5 ± 2.87 to 189.09 ± 6.45 MPa, and the Young's modulus increases from 19.75 ± 2.38 GPa to 24 ± 4.34 GPa. All this is due to the improvement in the flexural strength of the neat polymer, as shown in Figure 4a, and improvement in the interface due to the introduction of hydroxyl groups into the interface. The results of other work show that the addition of PAA to PMMA increases the elongation of the polymer and improves the mechanical properties as a copolymer. In addition, the number of polymer chain anchors was increased, and the interface interaction becomes stronger. Therefore, PAA molecules can act as bridges between polymer and ceramic, resulting in higher interfacial bonding force, and significant improvement in mechanical strength. In the absence of PAA, the interfacial strength is relatively weak, resulting in relatively low flexural strength [26].

Fracture toughness is an important feature for mimicking bone implant development because osteons play a role of toughening microelement in cortical bone. The toughness of cortical bone is the result of competition between intrinsic and extrinsic mechanisms operating in this tissue [2]. A nacre-like composite as a result of brick-and-mortar microstructure shows an intrinsic toughening mechanism like cortical bone. Hydroxyapatite itself, like other ceramics, is a brittle material that exhibits low toughness and has catastrophic failure. The fabricating nacre-like HA/polymer composites resulted in extrinsic toughness mechanisms in the composite, where the brick-and-mortar microstructure reduces the crack-driving force behind the crack tip. In addition, the polymer phase can contribute to plastic deformation and act as a lubricant.

Figure 6a shows the in situ images SEM for composites with ceramic fraction of 70 vol.% infiltrated with PMMA and PMMA + 5 wt.% PAA; also, the composites with higher ceramic fractions of 80 vol.% in the same polymer compositions of PMMA + 5 wt.% PAA, all composites show tortuous cracking. In addition, crack deformation, pull-out of ceramic bricks, and frictional sliding between ceramic walls were also observed, all of which are extrinsic toughening mechanisms. Fracture toughness characterization of HA/polymer composites showed a rising R–curve behavior in different ceramic fractions of 70 vol.%, and 80 vol.% and polymer composition approved the hypothesis of a rising R–curve behavior of nacre-like composites as shown in Figure 6b. As shown, the addition of PAA to PMMA increases the fracture toughness of the composite from 1.99 $\pm$ 0.78 to 3.023 $\pm$ 0.98 MPa$\cdot$m$^{1/2}$ which is caused by higher strength in the composites consist of copolymer of PMMA + PAA. In turn, increasing the ceramic fraction from 70 vol.% to 80 vol.% in the same polymer compositions, the fracture toughness reaches 5.27 $\pm$ 1.033 MPa$\cdot$m$^{1/2}$. Hao et al. showed the same results for HA/PMMA nacre-like composite, where the fracture toughness improved with increasing ceramic fraction. The value of fracture toughness increases with the increase in ceramic fractions [5]. Therefore, it can be concluded that the fracture toughness correlates with the ceramic fraction. It could be explained by the extrinsic toughness mechanisms mentioned above that when the ceramic fraction is increased from 60 vol.% to 80 vol.%, the density of the ceramic bridges and wall thickness increase, resulting in thinner polymer layers and increasing the stress-relieving lubricating effect. Hongbo et al. show the same results by decreasing the thickness of the polymer layers and thus increasing the fracture toughness in a nacre-like composite material [34].

5. Conclusions

In summary, a good combination of high strength and toughness was achieved in a nacre-like, bioactive HA-polymer composite fabricated by bi-directional freeze-casting. Ceramic, interface, and polymer are the main effective parameters leading to potential composites for future bone implants. Increasing the ceramic fractions in the brick-and-mortar microstructure increases the density of the ceramic bridges, resulting in better mechanical properties. The composites HA/PMMA with ceramic fractions of 60 vol.%, 70 vol.%, and 80 vol.% exhibit flexural strength of 52.68 $\pm$ 3.78, 115 $\pm$ 2.67, and 130 $\pm$ 5.82 MPa, respectively. The nacre-like HA/polymer composite exhibits a-rising R–curve behavior caused by the brick-and-mortar microstructure. The addition of 5 wt.% PAA to PMMA increases the number of polymer chains and the chains are more elongated, resulting in higher mechanical properties of the composites compared to pure PMMA. The flexural strength and fracture toughness were improved by about 20% to 158 $\pm$ 7.02 MPa and 5.27 $\pm$ 1.033 MPa$\cdot$m$^{1/2}$, respectively, at 80 vol.% ceramic fraction. The strong, tough, bone-matching and bioactive materials are expected to out-perform the materials currently used in orthopedics such as spinal fusion and bone fracture fixation implants.

Author Contributions: Conceptualization, P.T., B.S. and S.D.; methodology, P.T.; software, not applicable; validation, P.T. and B.S.; formal analysis, P.T. and B.S.; investigation, P.T.; resources, B.S., S.D. and T.S.; data curation, P.T.; writing—original draft preparation, P.T.; writing—review and editing, P.T., B.S., H.S., T.S. and U.J.; visualization, P.T.; supervision, B.S. and S.D.; project administration, P.T. and B.S.; funding acquisition, not applicable. All authors have read and agreed to the published version of the manuscript.

Funding: This research received no external funding.

Data Availability Statement: Data are available upon request from the corresponding author.

Conflicts of Interest: The authors declare no conflict of interest.

References

1. Morgan, E.; Unnikrisnan, G.; Hussein, A. Bone Mechanical Properties in Healthy and Diseased States. *Annu. Rev. Biomed. Eng.* **2018**, *20*, 119–143. [CrossRef] [PubMed]
2. Huang, W.; Restrepo, D.; Jung, J.; Su, F.Y.; Liu, Z.; Ritchie, R.O.; McKittrick, J.; Zavattieri, P.; Kisailus, D. Multiscale Toughening Mechanisms in Biological Materials and Bioinspired Designs. *Adv. Mater.* **2019**, *31*, e1901561. [CrossRef] [PubMed]
3. Nair, A.K.; Gautieri, A.; Chang, S.-W.; Buehler, M.J. Molecular mechanics of mineralized collagen fibrils in bone. *Nat. Commun.* **2013**, *4*, 1724–1729. [CrossRef]
4. Mirzaali, M.J.; Schwiedrzik, J.J.; Thaiwichai, S.; Best, J.P.; Michler, J.; Zysset, P.K.; Wolfram, U. Mechanical properties of cortical bone and their relationships with age, gender, composition and microindentation properties in the elderly. *Bone* **2016**, *93*, 196–211. [CrossRef]
5. Bai, H.; Walsh, F.; Gludovatz, B.; Delattre, B.; Huang, C.; Chen, Y.; Tomsia, A.P.; Ritchie, R.O. Bioinspired Hydroxyapatite/Poly(methyl methacrylate) Composite with a Nacre-Mimetic Architecture by a Bidirectional Freezing Method. *Adv. Mater.* **2016**, *28*, 50–56. [CrossRef]
6. Tan, G.; Zhang, J.; Zheng, L.; Jiao, D.; Liu, Z.; Zhang, Z.; Ritchie, R.O. Nature-Inspired Nacre-Like Composites Combining Human Tooth-Matching Elasticity and Hardness with Exceptional Damage Tolerance. *Adv. Mater.* **2019**, *31*, e1904603. [CrossRef]
7. Tang, M.; Xu, K.; Shang, H.; Li, X.; He, X.; Ke, L.; Xie, M.; Zhou, Z.; Liu, C.; Du, S.; et al. Biomineralization of bone-like hydroxyapatite to upgrade the mechanical and osteoblastic performances of poly(lactic acid) scaffolds. *Int. J. Biol. Macromol.* **2023**, *226*, 1273–1283. [CrossRef] [PubMed]
8. Ferrand, H.; Athanasiou, C. A Materials Perspective on the Design of Damage-Resilient Bone Implants Through Additive/Advanced Manufacturing. *JOM* **2020**, *72*, 1195–1210. [CrossRef]
9. Tatli, Z.; Bretcanu, O.; Çalışkan, F.; Dalgarno, K. Fabrication of porous apatite-wollastonite glass ceramics using a two steps sintering process. *Mater. Today Commun.* **2022**, *30*, 103216. [CrossRef]
10. Su, F.Y.; Mok, J.R.; McKittrick, J. Radial-Concentric Freeze Casting Inspired by Porcupine Fish Spines. *Ceramics* **2019**, *2*, 161–179. [CrossRef]
11. Bai, H.; Chen, Y.; Delattre, B.; Tomsia, A.P.; Ritchie, R.O. Bioinspired large-scale aligned porous materials assembled with dual temperature gradients. *Sci. Adv.* **2015**, *1*, e1500849. [CrossRef] [PubMed]
12. Algharaibeh, S.; Wan, H.; Al-Fodeh, R.; Ireland, A.J.; Zhang, D.; Su, B. Fabrication and mechanical properties of biomimetic nacre-like ceramic/polymer composites for chairside CAD/CAM dental restorations. *Dent. Mater.* **2022**, *38*, 121–132. [CrossRef] [PubMed]
13. Wu, Z.; Zhou, Z.; Hong, Y. Isotropic freeze casting of through-porous hydroxyapatite ceramics. *J. Adv. Ceram.* **2019**, *8*, 256–264. [CrossRef]
14. Gao, W.; Wang, M.; Bai, H. A review of multifunctional nacre-mimetic materials based on bidirectional freeze casting. *J. Mech. Behav. Biomed. Mater.* **2020**, *109*, 103820. [CrossRef] [PubMed]
15. Kumar, R.; Wang, M. Functionally graded bioactive coatings of hydroxyapatite/titanium oxide composite system. *Mater. Lett.* **2002**, *55*, 133–137. [CrossRef]
16. Eliaz, N.; Metoki, N. Calcium phosphate bioceramics: A review of their history, structure, properties, coating technologies and biomedical applications. *Materials* **2017**, *10*, 334. [CrossRef] [PubMed]
17. Canillas, M.; Pena, P.; de Aza, A.H.; Rodríguez, M.A. Calcium phosphates for biomedical applications. *Bol. Soc. Esp. Ceram. Vidr.* **2017**, *56*, 91–112. [CrossRef]
18. Habraken, W.; Habibovic, P.; Epple, M.; Bohner, M. Calcium phosphates in biomedical applications: Materials for the future? *Mater. Today* **2016**, *19*, 69–87. [CrossRef]

19. Singh, J.; Chath, S. Microstructural and in-vitro characteristics of functional calcium silicate topcoat on hydroxyapatite coating for bio-implant applications. *Progree Biomater.* **2022**, *11*, 95–108. [CrossRef]
20. Jaiswal, S.; Dubey, A.; Lahiri, D. The influence of bioactive hydroxyapatite shape and size on the mechanical and biodegradation behaviour of magnesium based composite. *Ceram. Int.* **2020**, *46*, 27205–27218. [CrossRef]
21. Frazer, R.Q.; Byron, R.T.; Osborne, P.B.; West, K.P. PMMA: An essential material in medicine and dentistry. *J. Long Term Eff. Med. Implant.* **2005**, *15*, 629–639. [CrossRef] [PubMed]
22. Kozłowski, M.; Bula, A.; Hulimka, J. Determination of Mechanical Properties of Methyl Methacrylate Adhesive (MMA). *Archit. Civ. Eng. Environ.* **2018**, *11*, 87–96. [CrossRef]
23. Launey, M.; Munch, E.; Alsem, D.; Barth, H.; Saiz, E.; Tomsia, A.; Ritchie, R. Designing highly toughened hybrid composites through nature-inspired hierarchical complexity. *Acta Mater.* **2009**, *57*, 2919–2932. [CrossRef]
24. Zhang, M.; Jiao, D.; Tan, G.; Zhang, J.; Wang, S.; Wang, J.; Liu, Z.; Zhang, Z.; Ritchie, R.O. Strong, Fracture-Resistant Biomimetic Silicon Carbide Composites with Laminated Interwoven Nanoarchitectures Inspired by the Crustacean Exoskeleton. *ACS Appl. Nano Mater.* **2019**, *2*, 1111–1119. [CrossRef]
25. Niebel, T.P.; Bouville, F.; Kokkinis, D.; Studart, A.R. Role of the polymer phase in the mechanics of nacre-like composites. *J. Mech. Phys. Solids* **2016**, *96*, 133–146. [CrossRef]
26. Yu, H.-P.; Zhu, Y.-J.; Xiong, Z.-C.; Lu, B.-Q. Bioinspired fiberboard-and-mortar structural nanocomposite based on ultralong hydroxyapatite nanowires with high mechanical performance. *Chem. Eng. J.* **2020**, *399*, 125666. [CrossRef]
27. Li, T.; Hu, X.; Zhang, Q.; Zhao, Y.; Wang, P.; Wang, X.; Qin, B.; Lu, W. Poly(acrylic acid)-chitosan @ tannic acid double-network self-healing hydrogel based on ionic coordination. *Polym. Adv. Technol.* **2020**, *31*, 1648–1660. [CrossRef]
28. Cui, C.; Shao, C.; Meng, L.; Yang, J. High-Strength, Self-Adhesive, and Strain-Sensitive Chitosan/Poly(acrylic acid) Double-Network Nanocomposite Hydrogels Fabricated by Salt-Soaking Strategy for Flexible Sensors. *ACS Appl. Mater. Interfaces* **2019**, *11*, 39228–39237. [CrossRef]
29. Algharaibeh, S.; Ireland, A.J.; Su, B. Bi-directional freeze casting of porous alumina ceramics: A study of the effects of different processing parameters on microstructure. *J. Euro. Ceram. Soc.* **2018**, *39*, 514–521. [CrossRef]
30. Ramesh, S.; Aw, K.; Tolouei, R.; Amiriyan, M.; Tan, C.; Hamdi, M.; Purbolaksono, J.; Hassan, M.; Teng, W. Sintering properties of hydroxyapatite powders prepared using different methods. *Ceram. Int.* **2013**, *39*, 111–119. [CrossRef]
31. Tang, Y.; Zhao, K.; Hu, L.; Wu, Z. Two-step freeze casting fabrication of hydroxyapatite porous scaffolds with bionic bone graded structure. *Ceram. Int.* **2013**, *39*, 9703–9707. [CrossRef]
32. *ASTM-D790-15*; Standard Test Methods for Flexural Properties of Unreinforced and Reinforced Plastics and Electrical Insulating Materials 1. ASTM International: West Conshohocken, PA, USA, 2016.
33. *ASTM B962-17*; Standard Test Methods for Density of Compacted or Sintered Powder Metallurgy (PM) Products Using Archimedes' Principle. ASTM International: West Conshohocken, PA, USA, 2017.
34. Wan, H.; Leung, N.; Algharaibeh, S.; Sui, T.; Liu, Q.; Peng, H.-X.; Su, B. Cost-effective fabrication of bio-inspired nacre-like composite materials with high strength and toughness. *Compos. Part B Eng.* **2020**, *202*, 108414. [CrossRef]
35. Algharaibeh, S. *Fabrication and Characterization of Biomimetic Nacre-Like Ceramic/Polymer Composite a Potential CAD/CAM Dental Material*; University of Bristol: Bristol, UK, 2019.
36. Magrini, T.; Senol, A.; Style, R.; Bouville, F.; Studart, A.R. Fracture of hierarchical multi-layered bioinspired composites. *J. Mech. Phys. Solids* **2022**, *159*, 104750. [CrossRef]
37. Grossman, M.; Bouville, F.; Masania, K.; Studart, A.R. Quantifying the role of mineral bridges on the fracture resistance of nacre-like composites. *Proc. Natl. Acad. Sci. USA* **2018**, *115*, 12698–12703. [CrossRef] [PubMed]
38. Gim, J.; Schnitzer, N.; Otter, L.M.; Cui, Y.; Motreuil, S.; Marin, F.; Wolf, S.E.; Jacob, D.E.; Misra, A.; Hovden, R. Nanoscale deformation mechanics reveal resilience in nacre of Pinna nobilis shell. *Nat. Commun.* **2019**, *10*, 4822. [CrossRef]
39. Deville, S. *Ice-Templating: Processing Routes, Architectures, and Microstructures*; Springer: Cham, Switzerland, 2017. [CrossRef]
40. Zhang, H. *Ice Templating and Freeze-Drying for Porous Materials and Their Applications*; Wiley-VCH Verlag GmbH & Co. KGaA: Weinheim, Germany, 2018. [CrossRef]
41. Iyo, T.; Maki, Y.; Sasaki, N.; Nakata, M. Anisotropic viscoelastic properties of cortical bone. *J. Biomech.* **2004**, *37*, 1433–1437. [CrossRef]
42. Boruah, S.; Subit, D.L.; Paskoff, G.R.; Shender, B.S.; Crandall, J.R.; Salzar, R.S. Influence of bone microstructure on the mechanical properties of skull cortical bone—A combined experimental and computational approach. *J. Mech. Behav. Biomed. Mater.* **2017**, *65*, 688–704. [CrossRef]
43. Larsson, B.A.M.; Sundh, D.; Mellström, D.; Axelsson, K.F.; Nilsson, A.G.; Lorentzon, M. Association between Cortical Bone Microstructure and Statin Use in Older Women. *J. Clin. Endocrinol. Metab.* **2018**, *104*, 250–257. [CrossRef]
44. Petrini, M.; Ferrante, M.; Su, B. Fabrication and characterization of biomimetic ceramic/polymer composite materials for dental restoration. *Dent. Mater.* **2013**, *29*, 375–381. [CrossRef]

45. Launey, M.E.; Buehler, M.J.; Ritchie, R.O. On the mechanistic origins of toughness in bone. *Annu. Rev. Mater. Res.* **2010**, *40*, 25–53. [CrossRef]
46. Cisneros-Pineda, O.G.; Kao, W.H.; Loría-Bastarrachea, M.I.; Veranes-Pantoja, Y.; Cauich-Rodríguez, J.V.; Cervantes-Uc, J.M. Towards optimization of the silanization process of hydroxyapatite for its use in bone cement formulations. *Mater. Sci. Eng. C* **2014**, *40*, 157–163. [CrossRef] [PubMed]

Journal of
Functional Biomaterials

MDPI

Article

Physicochemical Characterization of Thermally Processed Goose Bone Ash for Bone Regeneration

Fatimah Suhaily Abdul Rahman [1], Abdul Manaf Abdullah [2], Asanah Radhi [3,*], Wan Nazatul Shima Shahidan [1] and Johari Yap Abdullah [1,*]

1 School of Dental Sciences, Health Campus, Universiti Sains Malaysia, Kubang Kerian 16150, Kelantan, Malaysia; fsuhaily@gmail.com (F.S.A.R.); shima@usm.my (W.N.S.S.)
2 School of Mechanical Engineering, College of Engineering, Universiti Teknologi MARA, Shah Alam 40450, Selangor, Malaysia; abdulmanaf@uitm.edu.my
3 Faculty of Bioengineering and Technology, Universiti Malaysia Kelantan, Jeli 17600, Kelantan, Malaysia
* Correspondence: asanah@umk.edu.my (A.R.); johariyap@usm.my (J.Y.A.)

Abstract: Goose bone is traditionally applied for many ailments including bone fractures. Goose bone that consists of calcium phosphate plays a major role in bone regeneration. In this study, the production of goose bone ash (GBA) was translated from a traditional process into one of a laboratory scale via thermal and mechanical methods. The GBA was thermally processed via calcination at 300 °C and 900 °C. The differences in physicochemical properties between studied GBA (SGBA) and commercial GBA (CGBA) were elucidated via Fourier transform infrared (FT-IR), X-ray fluorescence (XRF), X-ray diffraction (XRD) and electron diffraction X-Ray (EDX). The morphological properties of SGBA and CGBA were characterized using field emission scanning electron microscopy (FESEM) in which nano-sized particles were detected. The results showed that the SGBA of 300 °C had comparable physicochemical properties to those of CGBA. A high processing temperature was associated with decreasing organic compounds and increasing crystallinity. The finding from EDX suggests that sintering at 900 °C (SGBA 900) demonstrated the presence of hydroxyapatite in the mineralogical phase and had a Ca/P atomic ratio of 1.64 which is comparable to the ideal stoichiometric ratio of 1.67. Findings from this study could be used for the further exploration of GBA as a potential material for bone regeneration via the elucidation of their biological properties in the next experimental setting.

Keywords: goose bone; bioceramic; calcium phosphate; bone regeneration; physicochemical property; hydroxyapatite

Citation: Abdul Rahman, F.S.; Abdullah, A.M.; Radhi, A.; Shahidan, W.N.S.; Abdullah, J.Y. Physicochemical Characterization of Thermally Processed Goose Bone Ash for Bone Regeneration. *J. Funct. Biomater.* **2023**, *14*, 351. https://doi.org/10.3390/jfb14070351

Academic Editors: Adalberto Luiz Rosa and Marcio Mateus Beloti

Received: 25 May 2023
Revised: 22 June 2023
Accepted: 27 June 2023
Published: 30 June 2023

1. Introduction

Bone is one of the natural bioceramics that contains calcium phosphate as a potential material for improving the treatment of bone disease. An injured bone can be easily healed due to the high capacity of bone regeneration, but if the injuries exceed a critical size, a scaffold or template is required to induce the growth of bone tissue [1]. For this purpose, goose bone was selected as one alternative scaffold with which to regenerate defective bone tissue. The utilization of goose bone as a traditional remedy is widely used in Malay culture as it can help in the healing of bone fractures due to its mineral content. Other than that, it was believed that a consistent consumption of goose bone in daily life could boost the immune system [2].

Avian species such as goose have a unique type of bone that is mainly found in their females, known as medullary bone. Medullary bone is a layer of material found inside a regular bone in birds. It forms a porous and spongy layer inside the bones of females when they are going to lay eggs. In terms of mineral content, the skeleton consists of a calcium-rich medullary bone. The formation of medullary bone was induced via the use of estrogenic and androgenic hormones during the egg laying cycle [3]. During the oviposition cycle, the demand for calcium is higher and calcium metabolism is largely derived from dietary sources and skeletal reserves [4]. To accommodate the increased calcium demand from the bone, the medullary bone is made without a mechanical function [5]. This spongy bone material supplies a labile calcium source for eggshell formation and is located within the medullary cavity of the mid-diaphysis of long bones, particularly those of the legs [4]. From previous reports, the source of calcium is one of the crucial ingredients that must be in the bioceramic scaffold [6,7]. Therefore, the utilization of unique avian bones such as goose bone may present a great opportunity to develop beneficial calcium phosphate-based material to fill the gap of the current issue in bone regeneration treatment.

Generally, normal bone tissue is made up of type 1 collagen and hydroxyapatite (HAp), which are highly compatible with bone functions throughout the body [8]. These structured, mineralized and elastic natural materials support and protect other tissues in the body system. The ideal bone repair material must possess good biocompatibility to maximize the osteogenesis process and good degradability to accelerate the regeneration of bone. These characteristics are the requirements for the implantation site of the scaffold [8–10].

Calcium phosphate-based material is biocompatible, osteoconductive, and osteoinductive, properties which are favored in medical industry, particularly in orthopedics, dentistry, and the coating application of metallic implants [11]. This bioceramic material is densely structured with textured granulated fillers suitable for bone reconstruction due to its similarity with the natural mineral composition of bone tissue. HAp, Ca_{10} $(PO_4)_6$ $(OH)_2$, is the main mineral component of bones and teeth; it occurs naturally in calcium phosphate-based material. The physical and chemical properties of natural hydroxyapatite and bone are found to be identical as both of them have a porous structure, that makes them biocompatible [12]. HAp can be obtained naturally or synthetically, and the need for this material depends on the medical applications due to its different physicochemical properties. Nowadays, the production of nanohydroxyapatite (nano-HA) has drawn significant interest in the field of bone regeneration. The utilization of nanoparticles in many applications including bone regeneration therapy is an added advantage. Surface properties behave differently at the nanoscale and due to these properties, nanoparticles can improve the surface functionalization that is useful in medical treatments.

Natural HAp can be derived from natural sources such as animal bones, scales, shells, and mineral materials, which are basically composed of calcium, as the major constituents. The production of natural HAp is non-stoichiometric since trace elements such as Na, Zn, Mg, K, Si, Ba, Fe, and F, and ions such as $CO_2{}^{-3}$, were found in the organic materials [13]. Basically, these impurities affect the content of calcium and hence produce a higher HAp ratio (>1.67) compared to that of stoichiometric HAp. The chemical composition of natural HAp resembles that of the inorganic part of structured bone, which is beneficial for rapid bone regeneration due to its richness of anions and cations [14]. The presence of ion substitution in the apatite structure affects the crystallinity phase, which influences the dissolution rate, significantly enhances the cell proliferation of human osteoblast cells and eventually promotes osseointegration [15].

It has also been reported that the existence of amino acid sequences in the collagen of structured bone helps cell growth by binding the cells and initiating signal transduction [16,17]. Other organic sources in the bone skeleton, including fibronectin and laminin, have a biological property that regulates cell growth upon contact with cells [16]. Collagen type I in the bone matrix is the most abundant extracellular matrix (ECM) protein and is a crucial material for bone strength. A previous study demonstrated that the interaction between bone marrow cells and collagen type I led to their differentiation into osteoblast cells via an increase in osteogenic gene expression and an elevation alkaline phosphatase activity [18].

Previously, several studies have shown the usage of different animal bone ashes for bone regeneration therapies [19–21]. The efficacy of bone ashes as scaffolds depend on the bone composition ratios. These ratios determine the mineral content of thermally processed bone ash. The processing temperature plays an important role, resulting in the mineralogical structure of the calcified bone. As reported, the crystallinity of the bone ash increases as the temperature increases. In mammals such as porcine animals, cancellous bone particles demonstrated a larger surface area and greater pore volume at a lower burning temperature (below 400 °C) than at a high temperature (above 1200 °C) [20]. A denser structure achieved at a high temperature may offer a good property for bone scaffold, biologically. In other species such as ovine species, cortical bone revealed an increase in crystallinity with increased temperature and a longer processing time [21]. Similar characteristics were observed for goose bone ash at a high burning temperature, except the structure of goose bone is more porous than that of the other bones. This structure provides a greater surface area with a larger pore size that is potentially good for osteoblast and bone regeneration [22]. The purpose of this research is to provide in-depth observations of the differences in physicochemical and morphological properties between CGBA and SGBA via XRF, FTIR, XRD and FESEM. The SGBA was subjected to two different heating temperatures, i.e., 300 °C and 900 °C. The heating temperature of 300 °C was selected based on a common approach for the deproteinization of bone which involves a low processing temperature [23]. Meanwhile, the selection of 900 °C is based on a previous study which mostly suggested the ideal sintering temperature for HAp production from natural bone to be in the range of 900–1200 °C [24]. These results will pave a way for the investigation of goose bone ash potential as an alternative material with which to promote bone regeneration.

2. Materials and Methods

2.1. Preparation of Goose Bone Ash (GBA)

Goose bones which were poultry waste were collected from a farmhouse located in Pasir Mas, Kelantan, Malaysia. The goose bones were collected from the waste of adult geese slaughtered for commercial purposes. In this study, two types of bone samples were used, which were commercial GBA (control group) from Roxhana Ventures Sdn. Bhd. and the studied GBA (study group). The studied GBA (SGBA) is a bone ash that is produced at the laboratory scale. The SGBA was prepared by washing the raw bones, cutting them into small pieces and finally drying them in the oven for 4 h. Eventually, the ash bone sample was obtained via the calcination of the bones at 300 °C and 900 °C using a furnace (Daihan Wise Therm, Seoul, Republic of Korea) for 3 h. There were two types of SGBA samples (300 °C and 900 °C) and the physicochemical characterizations of these materials were executed and were compared to those of CGBA.

2.2. Physicochemical Characterizations

2.2.1. X-ray Fluorescence (XRF) Measurement

The CGBA and SGBA elemental analyses were conducted using Rigaku Supermini 200 Wavelength Disperse X-ray F Spectrometer (WDXRF). The elemental oxide composition of the ash bone samples was determined. The value of each element was quantified as a percentage and the value was calculated based on mass

2.2.2. Fourier Transform Infrared (FTIR) Measurement

The functional groups in the GBA were detected via FTIR spectroscopy. The ash samples were mixed with potassium bromide (KBr) powder and crushed in a mortar before pellets were made. The samples were analyzed with Perkin Elmer. Data spectra were collected in the absorption mode between 4000 and 400 cm^{-1} with a resolution of 1 cm^{-1}.

2.2.3. X-ray Diffraction (XRD) Measurement

The X-ray diffraction patterns were obtained using Rigaku Smartlab X-Ray Diffractometer (Tokyo, Japan) with Cu Kα radiation (λ = 1.5406 Å), at an accelerating voltage of 30 kV and a 20 mA current, and the measurement was performed at room temperature. Diffractograms were recorded from 10–80° on a 2θ scale with a rate of 10° per minute. Prior to the analysis, the GBAs samples were oven-dried at 60 °C overnight. Then, the powder samples were ground with a pestle and pressed into a disc pellet with a 30 mm diameter.

The samples were inserted in the vacuum and analyzed for 20 min. The XRD analysis determined the changes in the mineral phase of bone ash during heating or burning. The amorphous and crystalline phase of the samples generalized the pattern of mineral phases.

2.2.4. Field Emission Scanning Electron Microscopy (FESEM) Measurement with Electron Diffraction X-ray (EDX)

The morphology, shape, and texture of the GBAs were analyzed via field emission scanning electron microscopy (FESEM) (Quanta 450 FEG, Fei, Eindhoven, The Netherlands). The GBA powders were stored in a desiccator prior to the analysis. Then, the samples were mounted on the SEM specimen stubs and coated with a thin layer of gold using an ion sputtering device (Leica EM SCD 005, Prague, Czech Republic) at current and vacuum values of 20 mA and 0.05 mbar, respectively. The morphology of GBAs was observed using FESEM at magnifications of 500 and 100,000.

3. Results and Discussion

3.1. XRF Analysis

The elemental comparison between GBAs is outlined in Table 1. The differences in the elemental content were calculated using mass (%). In the X-ray fluorescence analysis, it was revealed that calcium (Ca) was the main constituent in GBAs with an average of 75% to 81%. The SGBA 300 demonstrated the highest value of 80.6% compared to those of SGBA 900 and CGBA which were 77.7% and 75.7%, respectively. In addition, other major elements that were generally found in bone such as phosphorus (P), potassium (K), chlorine (Cl), ferrous (Fe), magnesium (Mg) and sulfur (S) were detected in all GBA samples except for SGBA 300 in which no Mg was detected.

Table 1. Results of elemental analysis of GBA via XRF (% mass).

	Elemental Content (% Mass)							
Element	Mg	Si	P	S	Cl	K	Ca	Fe
CGBA	0.316	0.407	18.5	0.379	0.913	3.48	75.7	0.381
SGBA 300	-	0.487	16.7	0.363	0.495	1.25	80.6	0.107
SGBA 900	0.641	0.445	20.2	0.0913	0.314	0.328	77.7	0.277

Based on the results, there were differences in the element contents of the GBA samples, such as those of Ca, P, and Mg. This could be because of the type of goose bone that was utilized as the raw material in producing bone ash. There was no information provided about the breed and the exact age of the obtained goose bones from the local slaughterhouse. These goose attributes can influence the hydroxyapatite composition of calcified tissues [25]. A similar study by Bahrololoom et al. successfully demonstrated the detection of numerous elements in cortical bone from cattle [26]. The variations of elements in bone noticeably rely on some biological factors such as nutrition [27].

In addition to having Ca and P as its main constituents, GBA contains small amounts of inorganic elements such as Mg, Na, K, Zn, and Sr in the HA structure. Apatite replacement with the aforementioned inorganic ions is known to have a significant impact to such an extent that it affects biological HAp functions. The introduction of Mg impurities into HAp structures has been demonstrated in earlier research to help promote osteoblast proliferation, which is essential for the generation of bone [28].

Ca is one of the ions that are mostly found in the bone matrix, and is mainly found in the form of calcium phosphate [7]. From Table 1, the element phosphorus (P) shows a range from 16% to 21% across all samples in which SGBA 900 exhibited the highest value (20.2%). A previous study showed that a high concentration of calcium ions from bone can induce the cell proliferation of osteoblasts and help osteoblastic activity for bone regeneration [29]. Meanwhile, phosphorus is an essential element to maintain biological structure and also bone mineralization [30].

The phase composition of GBA (Table 2) shows the content of calcium, phosphorus and magnesium (in oxide forms) of natural HAp extracted from goose bone. The range of CaO in all GBA samples lies between 63% and 67%, and this result is somewhat higher than that for the CaO found in bovine bone that was only around 53.4% [26]. The higher composition of Ca in GBA might be due to the presence of medullary bone in female avian species that is rich with Ca content as a labile source. Based on the phase composition of CaO and P_2O_5, the calculated values of Ca/P for CGBA, SGBA 300 and SGBA 900 were 1.99, 2.30 and 1.87, respectively.

Table 2. Phase composition of GBA under different heating processes determined via XRF (% mass).

	Phase Composition (% Mass)							
Element	MgO_2	SiO_2	P_2O_5	SO_3	Cl	K_2O	CaO	Fe_2O_3
CGBA	0.413	0.667	31.6	0.652	0.623	2.76	63.0	0.276
SGBA 300	0.385	0.820	29.0	0.677	0.369	1.54	67.0	0.0833
SGBA 900	0.826	0.717	33.9	0.152	0.208	0 256	63.7	0.194

3.2. FTIR Analysis

Basically, calcium phosphate-based materials such as GBA were determined according to their Ca/P molar ratio, but from a chemistry perspective, they are formed of three main elements: calcium, phosphorus, and oxygen. In Figure 1, the FTIR analysis of GBA shows the characteristic bands associated with functional groups PO_4^{3-}, OH^- and CO_3^{2-}. In general, the highest intensity and sharpness of the peaks were due to the phosphate and hydroxyl group which was the highest in SGBA 900 compared to that in SGBA 300 and CGBA, indicating the increased crystallinity of HA. The spectrum also shows that multiple peaks of phosphate ions were detected which were 1052 cm^{-1} (asymmetric), 604 cm^{-1} and 548 cm^{-1} (bending). The absorption bands of CGBA specifically appeared at 1037 cm^{-1} (asymmetric stretching) and at 603 cm^{-1} and 564 cm^{-1} (bending). Meanwhile, the presence of phosphate ions for SGBA 300 were detected at 1031 cm^{-1} (asymmetric) and 604 cm^{-1} and 562 cm^{-1} (bending).

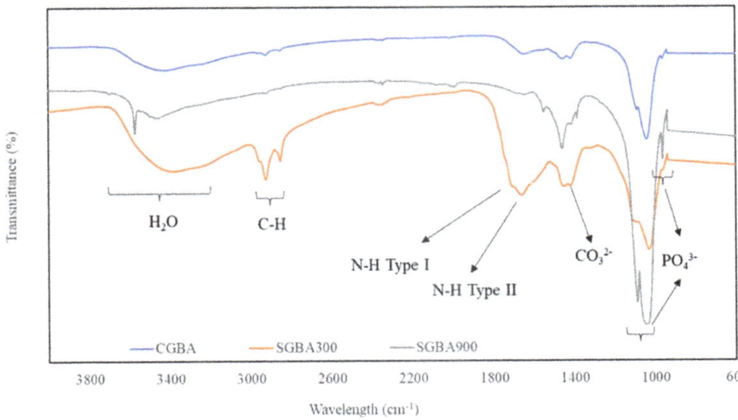

Figure 1. FTIR spectrum of CGBA and SGBA.

The displayed bands at approximately 2854 cm^{-1} and 2924 cm^{-1} (Figure 1) were obviously detected for SGBA 300, but the bands were visible for the rest of the bone ashes. These bands could be attributed to asymmetric C-H bonds in the aliphatic chains of collagen, particularly those of the amide group [31]. A previous study reported that three types of amide absorption bands were identified from the processed natural bones: amide I (1600–1700 cm^{-1}; C=O stretching vibrations), amide II (1500–1550 cm^{-1}; N-H deformation and amide III (1200–1300 cm^{-1}; N-H deformation and other complex modes resulting from a mixture of several coordinate displacements) [31].

The detection of the absorption band of amide type I at 1706 cm^{-1} and amide type II at 1508 cm^{-1} of SGBA 300 was observed but, not for SGBA 900, and this result was similarly found in a previous study [32]. Meanwhile, the absorption band of amide (type II) for CGBA was slightly detected. However, all GBA samples did not reveal any absorption band for amide (type III) formation. The previous study reported that when the amide group was associated with a water molecule, the formation of a hydrogen bridge occurred; therefore, the absorption band in a range of 3330 cm^{-1}–3425 cm^{-1} were very noticeable for SGBA 300 [33]. This result may indicate the co-existence of an inorganic part (calcium phosphate) and organics traces in goose bone ashes [32]. As reported, bone tissue is composed of 60% inorganic components (hydroxyapatite), 30% inorganic components (bone matrix protein) and 10% water [34].

Functional analysis for SGBA 300 revealed that the thermal processing of the sample did not fully destroy the organic part and that it was considered hybrid organic–inorganic material. Additionally, the FTIR analysis of CGBA demonstrated the reduction property of the organic component compared to that in goose bone calcified at 300 °C. This occurrence of this phenomenon might be due to the unknown traditional processing method of CGBA, particularly in the thermal aspect which might be due to a secrecy issue of the industry. However, it can be speculated that CGBA was processed either at a higher temperature or with several more stages of the heating process than those used for processing SGBA 300 based on the current analysis. A small peak at 1414 cm^{-1} which corresponds to that of the carbonate groups (CO_3^{2-}) was detected for CGBA and SGBA 300 but not for SGBA 900, confirming the isolation of carbonated HA from the goose bone at 900 °C. The amount of the remaining organic material is crucial as it determines the ability to safely use grafting materials without provoking any immunological reactions [35]. In addition, the absorption bands of amide groups are also no longer visible for SGBA 900, indicating the degradation of the collagen structure [36]. Regardless of the intensity of the peaks, both SGBA samples demonstrated a similar fingerprint region of the IR spectra to that of CGBA where the absorption bands below 1500 cm^{-1} indicated the presence of molecules in the mineral part of bone.

3.3. XRD Analysis

The phase identification of GBA samples was performed using X-ray diffraction. Figure 2 demonstrated the XRD patterns of commercial goose bone ash and calcinated goose bone at temperatures of 300 °C and 900 °C. The crystalline nature and phase composition were confirmed via the XRD analysis. The obtained XRD spectra were compared to those in the COD 9,002,214 standard HAp data. As shown in Figure 2, all the crystalline peaks in the XRD spectra closely matched with the peaks in standard HAp. This means that the thermal process produced natural HAp. The XRD results also suggest that the HAp present in the goose bone matrix was not disrupted by calcination at up to 900 °C.

It is evident that when the calcination temperature was increased to 900 °C, the intensity of the diffraction peaks increased, and the peaks sharpened and narrowed. This may have been due to the crystal size and crystalline nature becoming more prevalent at a higher temperature. The CGBA and SGBA 300 that was calcined at 300 °C displayed larger peaks with a low intensity, indicating that the organic material remaining in the bone matrix had not completely been removed.

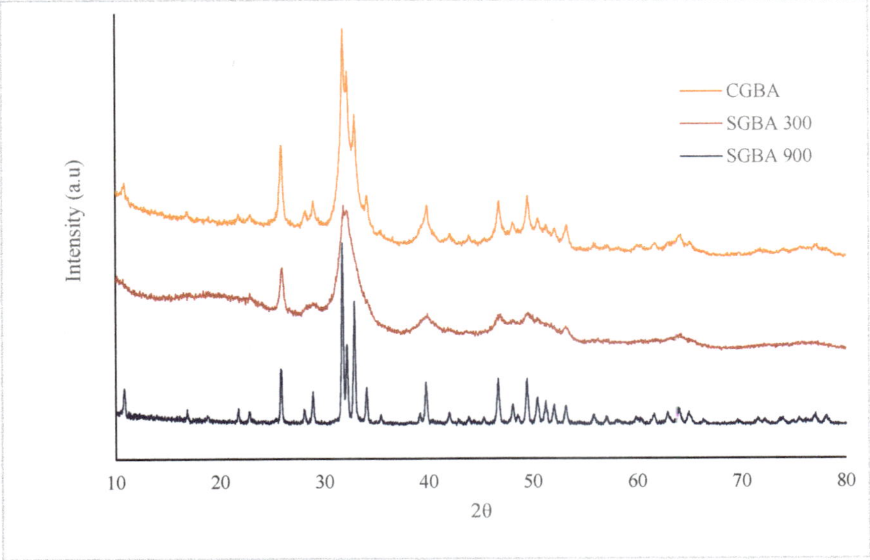

Figure 2. XRD spectra for commercial (CGBA) and calcinated (SGBA 300 and SGBA 900) goose bone ash.

Heating at a lower temperature, such as 300 °C, produces broad diffraction peaks that correlate to poor-crystallinity apatite, possibly due to the presence of a low concentration of carbonated groups in the sample. This is indicated by the FTIR spectrum of SGBA 300 (Figure 1) that exhibited that the band of carbonate ions is at 1000 which is stronger and broader, indicating that an increase in carbonate ions causes a decrease in the crystallinity of the structure [37]. Spence et al. reported that carbonate hydroxyapatite structured material accelerates osteogenesis by enhancing bioresorption [38]. By increasing the heating temperature to 900 °C, the diffraction peaks become more intense, sharper, and narrower, suggesting an increase in crystallinity and crystal size. This suggests that as the temperature increased to 900 °C, the raw, amorphous goose bone transformed into a crystalline phase with a diminishing organic phase and carbonates [39]. This is supported by Haberko et al. that found the concentration of carbonate groups to decrease when the calcination temperature is above 700 °C [40].

The percentage of crystallinity and the crystallite size of the samples were calculated using Scherrer's equation. Table 3 shows the percentage of crystallinity and the crystallite size of commercial goose bone ash and calcinated goose bone. The calcination process being conducted at a higher temperature causes changes in the crystallite size. Broader diffraction peaks reflect a smaller crystallite size. The crystallite size was bigger for SGBA 900 compared to that for SGBA 300 and CGBA, which might be due to particle coarsening during sintering.

Table 3. Percentage of crystallinity and crystallite size of commercial and calcinated GBA.

Samples	Crystallinity	Crystallite Size
CGBA	71.6%	20.0 nm
SGBA 300	62.7%	23.6 nm
SGBA 900	81.4%	38.3 nm

A previous study by Stastny et al. suggested that HA with higher crystallinity tends to have lower solubility and slower degradation rates [41]. Fulmer et al. also investigated the impact of crystallinity on the solubility of the studied ceramics. Among the investigated apatite materials with reduced crystallinity and carbonate substitutions, sintered hydroxyapatite characterized by high crystallinity was the least soluble [42].

3.4. FESEM Analysis

The investigation of the morphology and size of GBA powders was carried out via FESEM analysis as shown in Figure 3.

Figure 3. FESEM image of (**a**) CGBA, (**b**) SGBA 300 and (**c**) SGBA 900 at different magnifications.

Figure 3a–c demonstrate the images of the GBA samples under a resolution/magnification of 80,000 and 50,000, respectively. The result shows the changes in the surface morphology of the commercial bone powder and the studied bone powder at different temperatures. As shown in Figure 3a, the particles of CGBA exhibit the formation of highly agglomerated particles. This bone powder had no well-defined shape with an irregular size. The cotton-shaped nanoparticles of the GBA samples were observed via FESEM imaging. The range of nanoparticles of CGBA was detected to be between 23 nm and 67 nm through FESEM imaging. The broader size distribution of CGBA may have contributed to the random aggregation of nanoparticles, resulting in space formation between nanometer-sized particles [43]. Consequently, an irregular porous structure was created, and this corresponded with the semi-crystalline phase of the electron diffraction pattern (Figure 2). A previous study also reported that there were changes in the surface morphology of the bone powder during the calcination process. A high temperature disintegrated the organic and inorganic phase of the sample powder; thus, multiple pores with a more compact microstructure were able to form [44]. This finding correlates with that of the morphological property of SGBA 300 (Figure 3b) which indicates the dense and thick structure of the sample. These denser nanoparticles (>10 nm) had a smooth texture with a sharp edge appearance.

Meanwhile, nanoparticles of SGBA 900 (Figure 3c) were measured (>100 nm) and structured as crystallite particles. The decomposed bone powder was also observed to be aggregated as coral reefs and appeared as distinct nanoparticles. Hoque et al. and Odusote et al. found out that the temperature of calcination affects the transformation of particle sizes [44,45].

At a high temperature (>650 °C), ab enlargement of particle sizes was discovered, and this may be related to the absorption of heat energy by the particles [45]. Scaffolding materials such as nanohydroxyapatite can have potential in strengthening the mechanical properties of bone substitutes as well as promoting the process of osteogenesis of the osteoblast cell [46].

Through the EDX spectrum shown in Figure 4, the existence of two main peaks corresponded to the elements Ca and P for every GBA. The Ca/P ratios of CGBA and both SGBA samples were obtained through FESEM/EDX semiquantitative chemical analysis. The calculated ratios for CGBA, SGBA 300 and SGBA 900 were approximately 2.05, 2.50 and 1.64, respectively. These values follow a similar trend and are close to those values from the XRF data. The SGBA 900 that was thermally treated and sintered at 900 °C showed a Ca/P ratio that appeared to be very close to the stoichiometric value, which indicates an optimum thermal temperature. This value is comparable to the ratio of calcified goose beak bone studied by Kim et al. which was reported to be 1.63 [19]. In contrast, CGBA and SGBA 300 had a higher Ca/P ratio compared to the stoichiometric HAp value of 1.67. This might have been a result of the CaO produced during the calcinations [47]

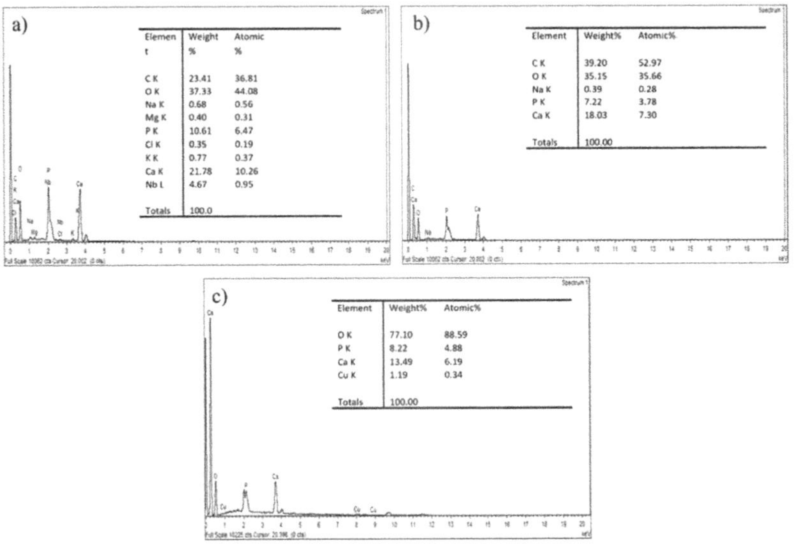

Figure 4. EDX spectrum of (**a**) CGBA, (**b**) SGBA 300 and (**c**) SGBA 900.

This result is consistent with that of the XRD analysis (Figure 2), which demonstrated the level of the crystalline phase in the GBA samples' structures. SGBA 900 is considered pure HAp due to its smaller variation in calcium/phosphorus atomic ratios (Ca/P). The range of the Ca/P ratio of 1.650 to 1.667 plays an important role in the mechanical characteristics of the bioceramic materials because different values will affect the treatment behavior [48]. The physicochemical and biological properties of HAp significantly changed when the Ca/P ratio exceeded 1.67. The strength of HAp decreased, and the defects of the HAp crystal structure were inhibited by ionic nuclei such as F^-, Cl^-, CO_3^{2-}, Mg^{2+}, and Sr^{2+} [49]. However, the greater number of these apatite nuclei enhanced biological activity, compared to the case with pure HAp. Overall, natural HAp has high chemical stability, but has flaws in its mechanical properties which will influence the scaffold's property.

4. Conclusions

In this study, we produced SGBA 300 and SGBA 900 and compared their physicochemical properties to those of commercial goose bone ash (CGBA). The experimental results show that the different temperatures used during calcination revealed the distinguished physicochemical properties of decomposed bone powder. FTIR analysis indicated the existence of an organic compound (an amide group) in CGBA and SGBA 300, but at a higher temperature of calcination (SGBA 900), the absorbance peak of the amide group disappeared. SGBA 900 was found to be highly crystalline via the XRD spectrum, showing a small variation in the Ca/P ratio (1.64) compared to that of other GBA samples. The different variations in the Ca/P ratio obtained through semiquantitative (FESEM/EDX) analysis indicated the different structures of the mineral phase in every GBA powder. In addition, the morphological property of the decomposed bone powder revealed the existence of nanoparticles. CGBA was indicated to have an amorphous structure with greater porosity, whilst SGBA 300 had a compact and dense particle structure. Meanwhile, SGBA 900 presented crystallite particles with a minimally agglomerate structure. In conclusion, the physicochemical properties of SGBA 300 are comparable with those of CGBA, but SGBA 900 shows a significant improvement in mineralogical properties, and the molar ratio of this natural HAp is in the range of that of stoichiometric HAp. Therefore, SGBA 900 is a promising substitute material for bone regeneration. However, the efficacy of the materials in terms of regenerative properties needs to be assessed to warrant their function. Additional in vitro research is needed to evaluate cytotoxicity, osteogenic cell attachment and proliferation to measure immunogenicity, graft resorption, and new bone formation.

Author Contributions: Conceptualization, F.S.A.R., A.M.A. and A.R.; methodology, F.S.A.R. and A.M.A.; software, F.S.A.R. and J.Y.A.; validation, F.S.A.R., A.M.A. and W.N.S.S.; formal analysis, F.S.A.R.; investigation, F.S.A.R.; resources, A.R. and J.Y.A.; writing—original draft preparation, F.S.A.R., A.M.A., A.R., J.Y.A. and W.N.S.S.; writing—review and editing, F.S.A.R., A.M.A., A.R., J.Y.A. and W.N.S.S.; visualization, F.S.A.R.; supervision, J.Y.A. and W.N.S.S.; project administration, J.Y.A.; funding acquisition, J.Y.A. All authors have read and agreed to the published version of the manuscript.

Funding: This research was funded by the industrial grant from Roxhana Ventures Sdn. Bhd via Universiti Sains Malaysia, grant number 304/PPSG/6150197/R124.

Institutional Review Board Statement: Not applicable.

Informed Consent Statement: Not applicable.

Data Availability Statement: Not applicable.

Acknowledgments: The authors thank Roxhana Omar, the Managing Director of Roxhana Ventures Sdn. Bhd. for providing patented goose bone samples (patent number MY-169206-A) in this study. Special thanks also to Nora from Multidisciplinary Dental Laboratory (MDL), School of Dental Sciences, Universiti Sains Malaysia, for the support in the FTIR analysis. The authors also appreciate the team of SIRIM-Industry Scientific Industry Sharing (SISS) for their support in the XRD and FESEM analysis.

Conflicts of Interest: The authors declare no conflict of interest.

References

1. Barba, A.; Maazouz, Y.; Diez-Escudero, A.; Rappe, K.; Espanol, M.; Montufar, E.B.; Öhman-Mägi, C.; Persson, C.; Fontecha, P.; Manzanares, M.-C.; et al. Osteogenesis by foamed and 3D-printed nanostructured calcium phosphate scaffolds: Effect of pore architecture. *Acta Biomater.* **2018**, *79*, 135–147. [CrossRef] [PubMed]
2. Ali, M.Y.; Kabir, A.; Khandker, S.S.; Hossan, T.; Shakil, M.S.; Islam, M.A.; Hasan, M.A.; Hossen, M.S.; Gan, S.H.; Khalil, M.I. Assessment of toxicity and therapeutic effects of goose bone in a rat model. *J. Chem.* **2019**, *2019*, 1943601. [CrossRef]
3. Prondvai, E. and Stein, K.H.W. Medullary bone-like tissue in the mandibular symphyses of a pterosaur suggests non-reproductive significance. *Sci. Rep.* **2014**, *4*, 6253. [CrossRef]
4. Kerschnitzki, M.; Zander, T.; Zaslansky, P.; Fratzl, P.; Shahar, R.; Wagermaier, W. Rapid alterations of avian medullary bone material during the daily egg-laying cycle. *Bone* **2014**, *69*, 109–117. [CrossRef]

5. Shipov, A.; Sharir, A.; Zelzer, E.; Milgram, J.; Monsonego-Ornan, E.; Shahar, R. The influence of severe prolonged exercise restriction on the mechanical and structural properties of bone in an avian model. *Veter.-J.* **2010**, *183*, 153–160. [CrossRef]
6. Plimpton, C.H. Calcium metabolism in health and disease. *Med. Clin. N. Am.* **1995**, *39*, 881–898. [CrossRef]
7. Jeong, J.; Kim, J.H.; Shim, J.H.; Hwang, N.S.; Heo, C.Y. Bioactive calcium phosphate materials and applications in bone regeneration. *Biomater. Res.* **2019**, *23*, 4. [CrossRef]
8. Khalaf, A.T.; Wei, Y.; Wan, J.; Zhu, J.; Peng, Y.; Abdul Kadir, S.Y.; Zainol, J.; Oglah, Z.; Cheng, L.; Shi, Z. Bone tissue engineering through 3d bioprinting of bioceramic scaffolds: A review and update. *Life* **2022**, *12*, 903. [CrossRef]
9. Chen, X.; Gleeson, S.E.; Yu, T.; Khan, N.; Yucha, R.W.; Marcolongo, M.; Li, C.Y. Hierarchically ordered polymer nanofiber shish kebabs as a bone scaffold material. *J. Biomed. Mater. Res. A* **2015**, *105*, 1786–1798. [CrossRef]
10. Prado, G.D.C.; Weinand, W.R.; Volnistem, E.A.; Baesso, M.L.; Noronha, J.N.G.; Truite, C., de Souza, B.M.; Bonadio, T.G.M.; José dos Reis, P.; Hernandes, L. Physicochemical and bone regeneration studies using scaffoldings of pure natural hydroxyapatite or associated with Nb2O. *Mater. Chem. Phys.* **2021**, *272*, 124922. [CrossRef]
11. Pokhrel, S. Hydroxyapatite: Preparation, Properties and Its Biomedical Applications. *Adv. Chem. Eng.* **2018**, *8*, 225–240. [CrossRef]
12. Mohd Pu'Ad, N.A.S.; Koshy, P.; Abdullah, H.Z.; Idris, M.I.; Lee, T.C. Syntheses of hydroxyapatite from natural sources. *Heliyon* **2019**, *5*, e01588. [CrossRef]
13. Akram, M.; Ahmed, R.; Shakir, I.; Ibrahim, W.A.W.; Hussain, R. Extracting hydroxyapatite and its precursors from natural resources. *J. Mater. Sci.* **2013**, *49*, 1461–1475. [CrossRef]
14. Cacciotti, I. Cationic and anionic substitutions in hydroxyapatite. In *Handbook of Bioceramics and Biocomposites*; Springer: Berlin/Heidelberg, Germany, 2016; pp. 145–211.
15. Tsai, K.-S.; Kao, S.-Y.; Wang, C.-Y.; Wang, Y.-J.; Wang, J.-P.; Hung, S.-C. Type I collagen promotes proliferation and osteogenesis of human mesenchymal stem cells via activation of ERK and Akt pathways. *J. Biomed. Mater. Res. Part A* **2010**, *94*, 673–682. [CrossRef]
16. Teti, A. Regulation of cellular functions by extracellular matrix. *J. Am. Soc. Nephrol.* **1992**, *2*, S83. [CrossRef]
17. Mizuno, M.; Fujisawa, R.; Kuboki, Y. Type I collagen-induced osteoblastic differentiation of bone-marrow cells mediated by collagen-a2β1 integrin interaction. *J. Cell. Physiol.* **2000**, *184*, 207–213. [CrossRef]
18. Tan, C.Y.; Singh, R.; Teh, Y.C.; Tan, Y.M.; Yap, B.K. The effects of calcium-to-phosphorus ratio on the densification and mechanical properties of hydroxyapatite ceramic. *Int. J. Appl. Ceram. Technol.* **2015**, *12*, 223–227. [CrossRef]
19. Kim, J.M.; Choi, S.H. Osteogenic evaluation of goose-beak bones according to processing temperature. *Vivo* **2019**, *33*, 419–424. [CrossRef]
20. Go, A.; Kim, S.E.; Shim, K.M.; Lee, S.-M.; Choi, S.H.; Son, J.S.; Kang, S.S. Osteogenic effect of low-temperature-heated porcine bone particles in a rat calvarial defect model. *J. Biomed. Mater. Res. Part A* **2014**, *102*, 3609–3617. [CrossRef]
21. Hiller, J.; Thompson, T.; Evison, M.; Chamberlain, A.; Wess, T. Bone mineral change during experimental heating: An X-ray scattering investigation. *Biomaterials* **2003**, *24*, 5091–5097. [CrossRef]
22. Kim, J.M.; Han, T.S.; Kim, M.H.; Oh, D.S.; Kang, S.S.; Kim, G.; Kwon, T.-Y.; Kim, K.-H.; Lee, K.-B.; Son, J.S.; et al. Osteogenic evaluation of calcium phosphate scaffold with drug-loaded poly (lactic-co-glycolic acid) microspheres in beagle dogs. *Tissue Eng. Regen. Med.* **2012**, *9*, 175–183. [CrossRef]
23. Abdelmoneim, D.; Porter, G.C.; Coates, D.E.; Duncan, W.J.; Waddell, J.N.; Hammer, N.; Li, K.C. The effect of low-processing temperature on the physicochemical and mechanical properties of bovine hydroxyapatite bone substitutes. *Materials* **2022**, *15*, 2798. [CrossRef] [PubMed]
24. Alshemary, A.Z.; Akram, M.; Taha, A.; Tezcaner, A.; Evis, Z.; Hussain, R. Physico-chemical and biological properties of hydroxyapatite extracted from chicken beaks. *Mater. Lett.* **2018**, *215*, 169–172. [CrossRef]
25. Asaduzzaman, S.M. Extraction of hydroxyapatite from bovine and human cortical bone by thermal decomposition and effect of gamma radiation: A comparative study. *J. Altern. Complement. Med.* **2017**, *8*, 1–10.
26. Bahrololoom, M.E.; Javidi, M.; Javadpour, S.; Ma, J. Characterisation of natural hydroxyapatite extracted from bovine cortical bone. *Ash. J. Ceram. Process. Res.* **2009**, *10*, 129–138.
27. Rey, C. Calcium phosphate biomaterials and bone mineral. Differences in composition, structures and properties. *Biomaterials* **1990**, *11*, 13–15.
28. Zhou, H.; Liang, B.; Jiang, H.; Deng, Z.; Yu, K. Magnesium-based biomaterials as emerging agents for bone repair and regeneration: From mechanism to application. *J. Magnes. Alloys* **2021**, *9*, 779–804. [CrossRef]
29. Lee, M.N.; Hwang, H.-S.; Oh, S.-H.; Roshanzadeh, A.; Kim, J.-W.; Song, J.H.; Kim, E.-S.; Koh, J.-T. Elevated extracellular calcium ions promote proliferation and migration of mesenchymal stem cells via increasing osteopontin expression. *Exp Mol. Med.* **2018**, *50*, 1–16. [CrossRef]
30. Penido, M.G.M.G.; Alon, U.S. Phosphate homeostasis and its role in bone health. *Pediatr. Nephrol.* **2012**, *27*, 2039–2048. [CrossRef]
31. Miculescu, F.; Stan, G.E.; Ciocan, L.T.; Miculescu, M.; Berbecaru, A.; Antoniac, I. Cortical bone as resource for producing biomimetic materials for clinical use. *Dig. J. Nanomater. Biostructures* **2012**, *7*, 1667–1677.
32. Haque, S.T.; Islam, R.A.; Gan, S.H.; Chowdhury, E.H. Characterization and evaluation of bone-derived nanoparticles as a novel ph-responsive carrier for delivery of doxorubicin into breast cancer cells. *Int. J. Mol. Sci.* **2020**, *21*, 6721. [CrossRef] [PubMed]
33. De Campos Vidal, B.; Mello, M.L.S. Collagen type I amide I band infrared spectroscopy. *Micron* **2011**, *42*, 283–289. [CrossRef]
34. Feng, X. Chemical and biochemical basis of cell-bone matrix interaction in health and disease. *Curr. Chem. Biol.* **2009**, *3*, 189–196.

35. Perić Kačarević, Z.; Kavehei, F.; Houshmand, A.; Franke, J.; Smeets, R.; Rimashevskiy, D.; Wenisch, S.; Schnettler, R.; Jung, O.; Barbeck, M. Purification processes of xenogeneic bone substitutes and their impact on tissue reactions and regeneration. *Int. J. Artif. Organs* **2018**, *41*, 789–800. [CrossRef]
36. Bozec, L.; Odlyha, M. Thermal denaturation studies of collagen by microthermal analysis and atomic force microscopy. *Biophys. J.* **2011**, *101*, 228–236. [CrossRef]
37. Wang, M.; Qian, R.; Bao, M.; Gu, C.; Zhu, P. Raman, FT-IR and XRD study of bovine bone mineral and carbonated apatites with different carbonate levels. *Mater. Lett.* **2018**, *210*, 203–206. [CrossRef]
38. Spence, G.; Patel, N.; Brooks, R.; Rushton, N. Carbonate substituted hydroxyapatite: Resorption by osteoclasts modifies the osteoblastic response. *J. Biomed. Mater. Res. A* **2009**, *90*, 217–224. [CrossRef]
39. Venkatesan, J.; Kim, S.K. Effect of Temperature on isolation and characterization of hydroxyapatite from tuna (thunnus obesus) bone. *Materials* **2010**, *3*, 4761–4772. [CrossRef]
40. Haberko, K.; Bućko, M.M.; Brzezińska-Miecznik, J.; Haberko, M.; Mozgawa, W.; Panz, T.; Pyda, A.; Zarębski, J. Natural hydroxyapatite—Its behaviour during heat treatment. *J. Eur. Ceram. Soc.* **2006**, *26*, 537–542. [CrossRef]
41. Stastny, P.; Sedlacek, R.; Suchy, T.; Lukasova, V.; Rampichova, M.; Trunec, M. Structure degradation and strength changes of sintered calcium phosphate bone scaffolds with different phase structures during simulated biodegradation in vitro. *Mater. Sci. Eng. C Mater. Biol. Appl.* **2019**, *100*, 544–553. [CrossRef]
42. Fulmer, M.T.; Ison, I.C.; Hankermayer, C.R.; Constantz, B.R.; Ross, J. Measurements of the solubilities and dissolution rates of several hydroxyapatites. *Biomaterials* **2002**, *23*, 751–755. [CrossRef] [PubMed]
43. Sun, R.; Åhlén, M.; Tai, C.-W.; Bajnóczi, E.G.; de Kleijne, F.; Ferraz, N.; Persson, I.; Strømme, M.; Cheung, O. Highly porous amorphous calcium phosphate for drug delivery and bio-medical applications. *Nanomaterials* **2019**, *10*, 20. [CrossRef] [PubMed]
44. Odusote, J.K.; Danyuo, Y.; Baruwa, A.D.; Azeez, A.A. Synthesis and characterization of hydroxyapatite from bovine bone for production of dental implants. *J. Appl. Biomater. Funct. Mater.* **2019**, *17*, 2280800019836829. [CrossRef]
45. Hoque, M.E.; Sakinah, N.; Chuan, Y.L.; Ansari, M.N.M. Synthesis and characterization of hydroxyapatite bioceramic. *Int. J. Sci. Res. Eng. Technol.* **2014**, *3*, 458–462.
46. Tanaka, M.; Izumiya, M.; Haniu, H.; Ueda, K.; Ma, C.; Ueshiba, K.; Ideta, H.; Sobajima, A.; Uchiyama, S.; Takahashi, J.; et al. Current methods in the study of nanomaterials for bone regeneration. *Nanomaterials* **2022**, *12*, 1195. [CrossRef]
47. Joschek, S.; Nies, B.; Krotz, R.; Göpferich, A. Chemical and physicochemical characterization of porous hydroxyapatite ceramics made of natural bone. *Biomaterials* **2000**, *21*, 1645–1658. [CrossRef]
48. Tariq, U.; Haider, Z.; Chaudhary, K.; Hussain, R.; Ali, J. Calcium to phosphate ratio measurements in calcium phosphates using LIBS. *J. Physics: Conf. Ser.* **2018**, *1027*, 012015. [CrossRef]
49. Hou, X.; Zhang, L.; Zhou, Z.; Luo, X.; Wang, T.; Zhao, X.; Lu, B.; Chen, F.; Zheng, L. Calcium phosphate-based biomaterials for bone repair. *J. Funct. Biomater.* **2022**, *13*, 187. [CrossRef]

Journal of
Functional
Biomaterials

MDPI

Article

Different Species of Marine Sponges Diverge in Osteogenic Potential When Therapeutically Applied as Natural Scaffolds for Bone Regeneration in Rats

Cíntia P. G. Santos [1], João P. S. Prado [1], Kelly R. Fernandes [1], Hueliton W. Kido [1], Bianca P. Dorileo [1], Julia R. Parisi [2], Jonas A. Silva [1], Matheus A. Cruz [1], Márcio R. Custódio [3], Ana C. M. Rennó [1] and Renata N. Granito [1,*]

[1] Department of Biosciences, Federal University of São Paulo (UNIFESP), Santos 11015-020, SP, Brazil
[2] Department of Physiotherapy, Federal University of São Carlos (UFSCar), São Carlos 13565-905, SP, Brazil
[3] Institute of Biosciences, University of São Paulo (USP), São Paulo 05508-000, SP, Brazil
* Correspondence: rn.granito@unifesp.br

Abstract: A highly porous structure, and an inorganic (biosilica) and collagen-like organic content (spongin) makes marine sponges potential candidates to be used as natural scaffolds in bone tissue engineering. The aim of this study was to characterize (through SEM, FTIR, EDS, XRD, pH, mass degradation and porosity tests) scaffolds produced from two species of marine sponges, *Dragmacidon reticulatum* (DR) and *Amphimedon viridis* (AV), and to evaluate the osteogenic potential of these scaffolds by using a bone defect model in rats. First, it was shown that the same chemical composition and porosity (84 ± 5% for DR and 90 ± 2% for AV) occurs among scaffolds from the two species. Higher material degradation was observed in the scaffolds of the DR group, with a greater loss of organic matter after incubation. Later, scaffolds from both species were surgically introduced in rat tibial defects, and histopathological analysis after 15 days showed the presence of neo-formed bone and osteoid tissue within the bone defect in DR, always around the silica spicules. In turn, AV exhibited a fibrous capsule around the lesion (19.9 ± 17.1%), no formation of bone tissue and only a small amount of osteoid tissue. The results showed that scaffolds manufactured from *Dragmacidon reticulatum* presented a more suitable structure for stimulation of osteoid tissue formation when compared to *Amphimedon viridis* marine sponge species.

Keywords: bone tissue engineering; demospongiae; natural biomaterial; marine sponges; scaffolds

Citation: Santos, C.P.G.; Prado, J.P.S.; Fernandes, K.R.; Kido, H.W.; Dorileo, B.P.; Parisi, J.R.; Silva, J.A.; Cruz, M.A.; Custódio, M.R.; Rennó, A.C.M.; et al. Different Species of Marine Sponges Diverge in Osteogenic Potential When Therapeutically Applied as Natural Scaffolds for Bone Regeneration in Rats. *J. Funct. Biomater.* **2023**, *14*, 122. https://doi.org/10.3390/jfb14030122

Academic Editors: Adriana Bigi and Marco Tatullo

Received: 14 December 2022
Revised: 2 February 2023
Accepted: 21 February 2023
Published: 24 February 2023

1. Introduction

Biomaterial-based therapy has increasingly become a viable strategy for treating bone fractures. Synthetic or naturally occurring, every type of biomaterial has a unique set of characteristics. Polyurethanes, polyesters, metals such as titanium and other synthetic polymers have advantageous properties over natural ones especially because of their abundance, limitless designs and customizable attributes. They require, however, chemical alterations since they lack cell adhesion sites and are, therefore, considered to be less biocompatible [1,2].

As a vast source of natural biomaterials, marine biodiversity has lately been gaining prominence in scientific research around the world. Among the invertebrates that stand out in different studies, marine sponges (Phylum Porifera) are primitive sessile animals characterized by being multicellular, filter-feeding and structurally porous [3].

It is indeed a fact that these animals archaically filter water for feeding with microorganisms that confer them unique characteristics for their use as biomaterials in the engineering of bone tissue. Their porous architecture with interconnected pores would favor, once implanted in patients, the passage of cells and blood vessels and, consequently, tissue regeneration [3]. Thus, adequate structural characteristics of the implant play a fundamental role in the efficiency of the treatment since the formation of the new tissue depends on an adequate migration and proliferation of the cells responsible for this process and also on the invasion of new blood vessels within the implanted scaffold. This infiltration will only properly occur if the scaffold is structurally favorable, especially in terms of porosity, pore size and pore interconnectivity. In addition to being a source of precursor cells that will be later responsible for the new tissue formation, vascularization may also have its relevance since it provides the nutrients needed for cellular metabolism and the removal of its residual products [4,5].

Besides their structural peculiarities, marine sponges have, like the bone itself, a skeleton composed of an organic component, named spongin, and inorganic components such as biosilica [6,7]. Spongin is a protein similar to vertebrate collagen that has also been used in the production of scaffolds for tissue bioregeneration [8] because it allows cell attachment, proliferation and migration through the biomaterial [9]. Based on this, spongin is an excellent alternative to animal collagen, due to the low risk of transmission of infection-causing agents and good biocompatibility [10,11]. Biosilica is an inorganic element that is known to drive bone cell differentiation and increase mineralization [6,11,12].

The ideal scaffold for bone substitution, regardless of whether it is natural or synthetic, needs to be biocompatible, porous, have osteoconductive and osteoinductive capacities, mechanical properties similar to those of bone and a rate of degradation compatible with the process of bone remodeling [3,13].

Previous in vitro studies showed that osteoprogenitor cells were able to grow and attach onto the sponge skeleton. Green et al. [14] demonstrated that the skeleton of *Spongia* sp. enabled human osteoprogenitor cells' adhesion, expansion and invasion. Alkaline phosphatase and type I collagen histochemical stains showed that the bone matrix could be formed. Lin et al. [15], by also evaluating a sponge skeleton with a collagenous fibrous network (*Callyspongiidae* sp.), showed that mouse primary osteoblasts were able to anchor onto the surface of collagen fibers, express osteoblast markers (osteocalcin and osteopontin) and form mineralization nodules. Likewise, SaOS-2 cells were able to grow and colonize the bioceramic structure of *Petrosia ficiformis* sponges after calcination [16], addionally demonstrating the advantages conferred by sponge scaffold architecture as a template for bone cell growth, differentiation and mineralization [17].

These ancient multicellular organisms boast a diverse array of skeletal structures that have evolutionary-approved 3D-scaffold-like qualities and appear to be highly suitable for use across a range of fields within modern bioinspired materials science, biomimetics and regenerative medicine [16]. Even though several in vitro studies speculate its potential for use as a scaffold, only one study has been conducted with the aim of analyzing the in vivo potential of using marine sponges in their natural state until now. Nandi et al. [18] carried out an investigation to identify and characterize marine sponges as potential bone scaffolds. For this purpose, samples of marine sponge *Biemna fortis* (class Demospongiae), collected from the intertidal region of Anjuna, India, were implanted in femoral defects in rabbits. After 90 days, the results showed that marine sponges of the mentioned species, combined or not with growth factors (IGF-1 and BMP-2), were biocompatible and biodegradable, and could be considered as a new natural biomaterial for bone tissue engineering purposes.

Based on the above, sea sponges, with their porosity, biosilica and spongin, can be potentially considered as innovative bone substitutes, which would allow the development of new therapeutic resources for the improvement of patients' quality of life. Additionally, since they a natural, abundantly found material, their use would contribute to a reduction in the treatment costs in public health systems. However, this potential is still largely unexplored. Despite several studies evaluating the use of compounds extracted from marine sponges, only one in vivo study was found in the literature evaluating the marine sponge skeleton as a bone-mimicking biomaterial. Moreover, this single study did not perform quantitative evaluations of the bone healing process. Finally, since sponges structurally differ among species, we hypothesize that their osteogenic potential would also be different. In this way, the present study performed a physical-chemical characterization and an evaluation of the in vivo osteogenic potential of scaffolds manufactured from two species of marine sponges with the aim of developing efficient substitutes for guided bone regeneration. The species of Porifera used in this study were *Dragmacidon reticulatum* and *Amphimedon viridis*, which are abundantly present on the Brazilian coast.

2. Materials and Methods

It is important to note that this study is registered in the National Management System of Genetic Patrimony (Sistema Nacional de Gestão do Patrimônio Genético, SisGen, registration number A56D034).

The marine sponge species *Dragmacidon reticulatum* and *Amphimedon viridis*, both belonging to the class Demospongiae, were used in this study. These two species (Figure 1) were specifically chosen for being abundant and easily found on the Brazilian coast. Both have porous skeletons composed of spongin and biosilica, which are interesting elements in the context of bone repair. The sponges were collected in high hydrodynamic coasts, in the intertidal zone, in the area of São Sebastião, Brazil (23°49′23.76″ S, 45°25′01.79″ W) and in the area of Enseada do Araçá (23 No. 81′73.78″ S, 45° 40′66.39″ W, São Sebastião, Brazil). An amount of 200 g of each marine sponge species was collected for this study.

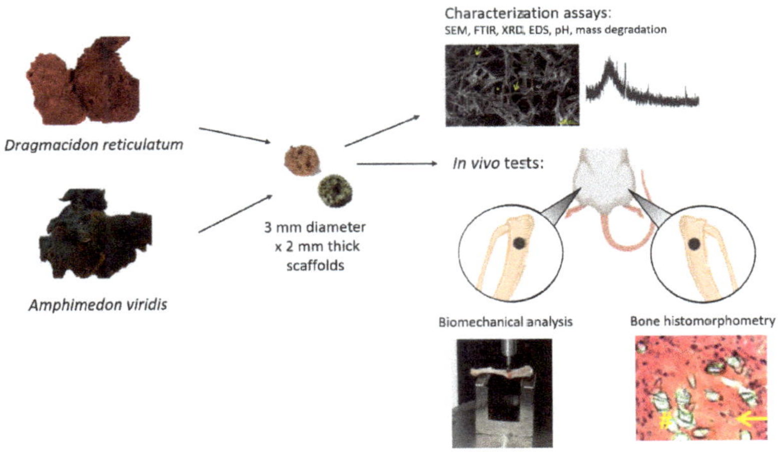

Figure 1. Summary of research methodology showing samples of marine sponges of the species *Dragmacidon reticulatum* and *Amphimedon viridis*, scaffolds manufactured from these sponges and the experiments performed.

After the collection, the sponges were washed with running water, classified according to their species' characteristics and kept in the freezer until use. To produce the scaffolds, the sponges were cut with a trephine-type dental drill (3i Implant Innovations Inc., Palm Beach Gardens, Florida, USA) and a scalpel blade for the manufacture of 3 mm diameter × 2 mm thick scaffolds (Figure 1). The produced scaffolds were freeze-dried and sterilized by ethylene oxide (Acecil Central de Esterilização Comércio e Indústria Ltd.a—Campinas/SP, Brazil). The parameters used for the freeze drying were −40 °C and 600 uHG (DIM Liofilizador LT X.X00—Terroni Equipamentos Cientificos Ltd.a—São Carlos/SP, Brazil).

Characterization analyses of the scaffolds were performed before and after an incubation period, the conditions of which were determined according to the Kokubo protocol [19–21]. For the SEM, FTIR, XRD and EDS aanlyses, the scaffolds were incubated in simulated body fluid (SBF; pH 7.4) in a ratio of 1:10 (mass of the material (g): volume of the SBF (mL)) during a period of 21 days. Samples were evaluated before (day zero) and after incubation at periods 1, 7 and 21 days. For pH and mass loss evaluations, the scaffolds were placed in 3 mL of phosphate-buffered saline (PBS, pH 7.4) and incubated at 37 °C in a water bath on a shaker table (70 rpm) for 21 days. These analyses were performed before (day zero) and after incubation of scaffolds at periods 1, 3, 7, 14 and 21 days.

2.1. Surface Morphology Analysis (SEM)

A scanning electron microscope (SEM, Le0 440, Carl Zeiss, Jena, Germany), operating at a 10 keV electron beam, was used to morphologically analyze the scaffolds surface. This technique consists of obtaining the enlarged image of the sample from the interaction of an electron beam with the material. For this, the samples were fixed on an aluminum base using a carbon tape. Next, due to the non-conductive properties of the samples (organic origin), they were covered with a thin layer of conductive material; in this case, the material used was gold. Thus, images were obtained with magnification of 500×, in the above-mentioned periods, in order to assess the morphology of the initial surface and the degradation behavior of the biomaterials (n = 5).

2.2. Fourier Transform Infrared Spectroscopy (FTIR)

To identify the chemical bonds present in the material, Fourier-transform infrared spectroscopy (IRAffinity-1S-FTIR Shimadzu spectrophotometer, São Paulo, Brazil) was used. The specters were obtained in the range of 4000–400 cm^{-1}, with a resolution of 4 cm^1 (n = 3).

2.3. X-ray Diffraction (XRD)

The crystalline phase of the material was evaluated by X-ray diffraction (XRD) with Philips X'Pert MPD diffractometer, Cu-Ka (l = 0.154 nm), 45 kV, 30 mA. Data were collected at angles between 20° and 60° at the IQSC(USP), São Paulo (n = 3).

2.4. Energy Dispersive X-ray Spectroscopy (EDS)

This analysis was carried out by means of equipment IXRF Systems 500 coupled to a scanning electron microscope (SEM) that allowed a qualitative evaluation of the chemical elements present in the samples, with 0.5% mass detection [22] (n = 5).

2.5. pH Evaluation

Directly after removing samples from the incubation medium (n = 5), pH of incubation medium was measured with a pH meter (Orion Star A211, Thermo Scientific, Waltham, MA, USA).

2.6. Degradation Analyses

The mass loss evaluation was performed to determine the degradation of biomaterials in liquid medium. After the experimental periods of incubation in PBS, the samples were oven-dried overnight at 37 °C and weighed on a precision balance. The relation between the final weight obtained and the initial weight was calculated to quantify the mass loss in percentage, according to the following equation: % Mass loss = ((fm − im)/im) × 100%, where fm is the sample mass after immersion in PBS and im is the sample mass before immersion in the same solution (n = 5).

2.7. Porosity Evaluation

The porosity of *Dragmacidon reticulatum* and *Amphimedon viridis* scaffolds were evaluated according to the Archimedes' principle (Equation (1)). First, scaffolds (1 cm × 0.4 cm) were obtained by cutting the sponges with a cutter. The scaffolds were dried overnight at 37 °C. The scaffolds were then weighed on a precision balance. The density of water was also calculated (mass/volume). Later, the scaffolds were placed in a glass container with 5 mL of water and the values of mass (grams) and volume (mL) were recorded. The scaffolds were removed from the glass container and the weight of the water was again recorded. The scaffolds' density was calculated using the values of water + scaffolds and water weight differences. For the porosity calculation, the following formula was used:

$$\text{Porosity (\%)} = \text{m1} - \text{m3.WD}[(\text{m1.WD}) + (\text{SM.SD})] - (\text{m3/WD}) \times 100 \quad (1)$$

where m1 is initial mass of water, m3 is water mass after removing the scaffold, WD is the water density, SM is the scaffold mass and SD is the scaffold density ($n = 5$).

2.8. Biological Evaluation

2.8.1. In Vivo Study

Thirty male Wistar rats were used (12 weeks old, weight 300–350 g) in this study. All rats were submitted to a surgical procedure in which a unicortical non-critical bone defect was performed in both tibias. The animals were randomly divided into 3 groups: control group (CG)—defects were left unfilled; *Dragmacidon reticulatum* group (DR)—defects were implanted with scaffolds of marine sponge *Dragmacidon reticulatum*; *Amphimedon viridis* group (AV)—scaffolds that were implanted belonged to the species *Amphimedon viridis*. After surgical procedure, the animals were kept at a controlled temperature, 12 h light–dark period, with free access to water and standard food. This study was approved by the Ethics Committee on the Use of Animals (CEUA) of the Federal University of São Paulo (2017/3011170417).

2.8.2. Surgical Procedures

A non-critical-sized bone defect, 3 mm in diameter, was performed in the upper third of each rat tibia (10 mm below the knee joint) by using a motorized trephine drill (3i Implant Innovations Inc., Palm Beach Gardens, Florida, USA) irrigated with saline solution (Figure 2A). Then, the wounds were closed with resorbable Vicryl® 5-0 (Johnson & Johnson, Sint-Stevens-Woluwe, Belgium). Surgeries were performed according to the ethical principles of animal instrumentation, at standard conditions of asepsis and general anesthesia. Initially, the animals were anesthetized with intraperitoneal injection of ketamine (80 mg/kg), xylazine (8 mg/kg), acepromazine (1 mg/kg) and fentanyl (0.05 mg/kg) in a single syringe. In addition, a single dose of cephalothin antibiotic (60 mg/kg) was given preoperatively. Next, trichotomy and antisepsis were performed with the aid of a shearing machine and sterile gauzes containing 2% degermant iodine, which was followed by three steps of 70% ethanol application in the surgical focus.

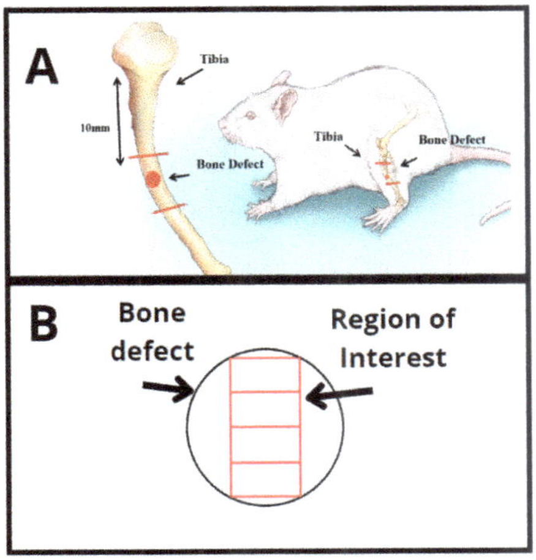

Figure 2. (**A**) Scheme showing the procedure for creating the tibial bone defect in rats. (**B**) Total region of interest analyzed for the histomorphometric parameters.

All animals were then submitted to the surgical creation of bone defects bilaterally in the tibias, but only the animals of DR and AV groups received implants (scaffolds) as treatment. After surgery, anti-inflammatory meloxicam was administered subcutaneously at a dose of 2 mg/kg and, after 24 and 48 h, at a dose of 1 mg/kg. Finally, the animals were placed in individual boxes with free access to water and food and were monitored until anesthesia was completely over. Additionally, postoperative animals were monitored daily throughout the whole treatment period, with pain parameters being constantly evaluated. The animals were euthanized by drug overdose (intraperitoneal injection of ketamine 240 mg/kg and xylazine 24 mg/kg) 15 days after surgery.

2.8.3. Histological Procedures

After sample collection, the left tibias were dehydrated with 70% ethanol for three days. They were then dehydrated in absolute ethanol (100%) for a further three days, diaphanized in toluene for one day and included in the methyl methacrylate resin (Merck acrylic resin). The obtained blocks were sanded and cut using a microtome (Leica Microsystems SP 1600, Nussloch, Germany). Five-micrometer-thick sections were perpendicularly obtained, considering the medial–lateral axis of the implants. Histological sections were stained with Goldner's tri-chromium (Merck).

2.8.4. Histological Analysis

The qualitative analysis of the slides was performed by means of the morphological description of bone defects, according to the following criteria: presence of newly formed bone tissue (primary and secondary bone), granulation tissue, presence of fibrosis and biomaterial. Analyses were performed blindly on the 10× objective.

2.8.5. Histomorphometric Analysis

A microscope (Labophot 2ª, Nikon, Minato City, Tokyo) coupled to the OsteoMeasure software (OsteoMetrics, Atlanta, GA, USA) was used for the quantitative analyses. Measurements were performed in the fields located in the medial region of the bone defect, from the upper border until the bottom of the defect, using the 10× objective. The total region of interest (ROI) was 1.85 ± 0.38 mm^2 (Figure 2B). The following histomorphometric parameters were obtained: bone volume as percentage of tissue volume (BV/TV%), osteoid volume as percentage of tissue volume (OV/TV%), number of osteoblasts per unit area of tissue analyzed (N.Ob/T.Ar mm^2), osteoblastic surface as a percentage of bone surface (Ob.S/BS%) and percentage of fibrous tissue volume as a percentage of tissue volume (Fb.V/TV%), according to international standardized nomenclature [23]. A parameter was additionally included in order to analyze the biomaterial present in the ROI: biomaterial volume as a percentage of tissue volume (Bm.V/TV%) (n = 5). Active (mature) bone-forming osteoblasts were identified by their cylinder-like shape and their arrangement in rows over an area of osteoid tissue.

2.8.6. Biomechanical Test

Biomechanical analysis was performed using the three-point bending test on the right tibia of animals of all groups. Biomechanical assays were performed on the Instron universal testing machine (model 4444, 825 University Ave Norwood, MA, 02062-2643, US) at room temperature. For the test, a load cell with a maximum capacity of 1 N and a preload of 5 N and a constant speed of 0.5 cm/min was used. Both tibia ends were supported by two metal supports, with the defect region facing downwards. The force was then perpendicularly applied to the longitudinal axis of the bone by a cylindrical rush until the moment of fracture. The force applied and the indentator displacement were monitored and recorded using the equipment's own software. From the force–displacement curve, fracture energy (J) (ability to absorb energy to breakage), elastic deformation energy (J) (ability to absorb and return energy without apparent deformation) and maximum load (N) (maximum load that the material can support) were obtained [23].

2.8.7. Statistical Analysis

Initially, the variable distribution was tested using the Shapiro–Wilk's normality test. For variables that exhibited normal distribution (BV/TV%; OV/TV%; N.Ob/T. Ar mm^2; Ob.S/BS%; Fb.V/TV%; Bm.V/TV%; fracture energy (J); elastic deformation energy (J); maximum load (N)), comparisons among groups were carried out by analysis of variance (ANOVA), followed by Tukey post hoc. The Mann–Whitney test was used for variables not exhibiting normal distribution (characterization tests: pH and mass degradation). The statistical program used was GraphPad Prism version 7.0 and the adopted significance level was 5% ($p \leq 0.05$).

3. Results

3.1. Characterization of Scaffolds

3.1.1. SEM

The qualitative analysis of the surface morphology of scaffolds showed the presence of silica spicules and pores in both materials (scaffolds of *Dragmacidon reticulatum* and *Amphimedon viridis* marine sponges) (Figure 1). *Dragmacidon reticulatum* sponge scaffolds were structurally more porous than *Amphimedon viridis* sponge scaffolds. Additionally, it was observed that the scaffolds of the *Dragmacidon reticulatum* species show a greater degradation after incubation when compared to the *Amphimedon viridis* species (Figure 3).

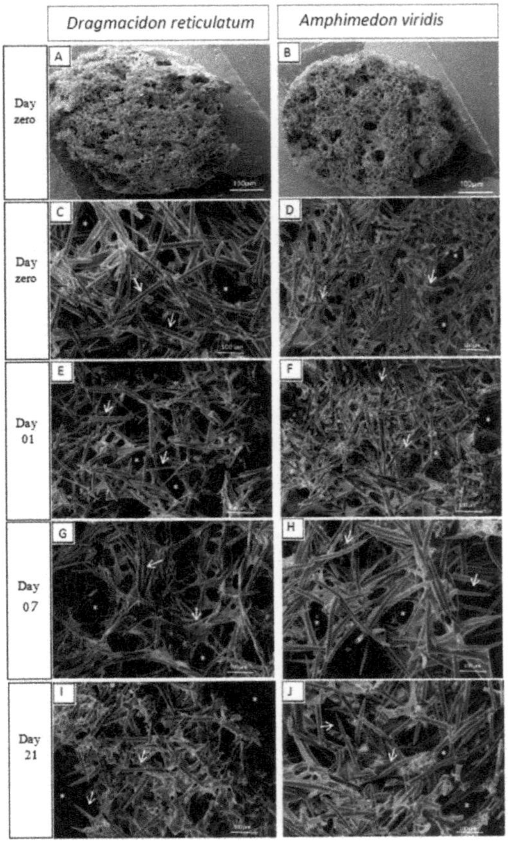

Figure 3. SEM micrographs showing scaffold morphology and its surface degradation in the periods before incubation (day zero (**A–D**)) and after incubation in SBF (days 1(**E**,**F**), 7(**G**,**H**) and 21(**I**,**J**))—increase of 500×. The signs indicate the presence of silica spicules (→) and pores (*). (n = 5).

3.1.2. FTIR

The scaffolds manufactured from the two species of marine sponges studied (*Dragmacidon reticulatum* and *Amphimedon viridis*) exhibited the same functional groups: at 3460 cm^{-1}, a broad peak consistent with the intermolecular OH bond; weak peaks at 2950 cm^{-1} and 2875 cm^{-1} for asymmetric and symmetrical -CH2- (OH-linked), respectively; weak signal between 800 cm^{-1} and 960 cm^{-1} of rotary motion -CH2-; presence of medium to strong signal between 1260 cm^{-1} and 1440 cm^{-1} being positive for primary alcohol; average signal at 770 cm^{-1} of the silicon–carbon bond; SiOH peaks at 3700 to 3200 cm^{-1}, 1030 cm^{-1} and 890 cm^{-1}; presence of signals at 1650 cm^{-1} and 1535 cm^{-1} consistent with amide I and amide II functions, respectively. However, from the seventh day of incubation, there was loss of the characteristic point of amide I, amide II and primary alcohol (CH2OH)—related to organic matter—only in *Dragmacidon reticulatum* species, as shown in Figure 4.

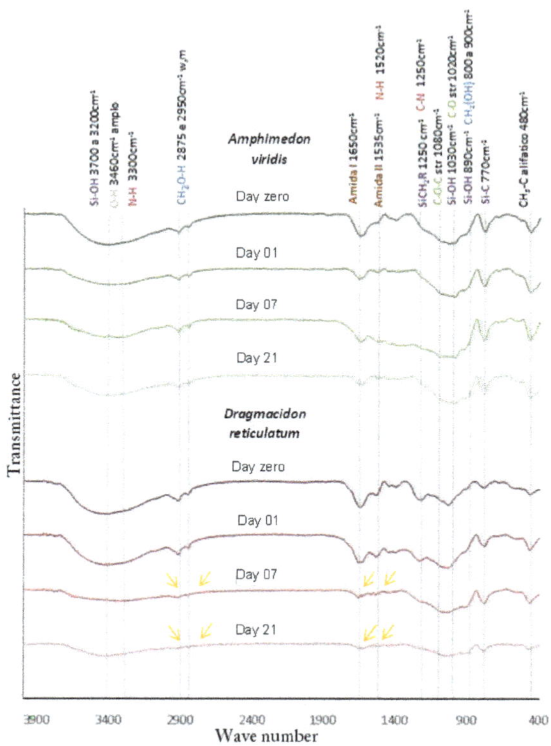

Figure 4. FTIR spectra of the scaffolds manufactured from the marine sponge species *Dragmacidon reticulatum* and *Amphimedon viridis*, before (day zero) and after incubation in SBF (day 1, 7 and 21). The yellow arrows show the loss of organic matter after incubation for *Dragmacidon reticulatum* (n = 3).

3.1.3. XRD

XRD analysis of the samples revealed that the two species of marine sponges studied are composed of amorphous content, as shown in Figure S1.

3.1.4. EDS

In the EDS analysis, the scaffolds manufactured from the marine sponge species *Dragmacidon reticulatum* and *Amphimedon viridis* showed the same chemical composition as shown in Table S1, and the three most proportionally present chemical elements were carbon, oxygen and silicon, which together represent more than 80% of the sample.

3.1.5. pH and Mass Degradation

The results of the pH measurements during the incubation period are shown in Figure 5A. The scaffolds of the *Dragmacidon reticulatum* species showed an initial drop in pH in the periods referring to the 1st and 3rd days of immersion in PBS and subsequent increase from the 14th day, reaching pH similar to the initial value (pH = 7.6) at day 21. The pH of the medium incubated with scaffolds from the species *Amphimedon viridis* did not show significant variation during the analysis period, remaining practically stable from day 1 to 21. A statistical difference was observed in the comparison between the two species, in the first ($p = 0.0158$), third ($p < 0.0001$) and seventh ($p = 0.0037$) days of incubation.

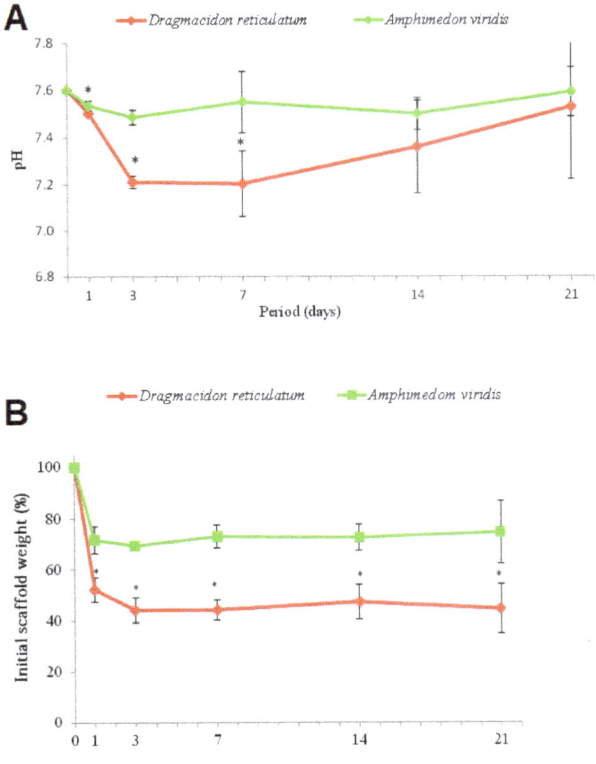

Figure 5. (**A**) pH values after incubation of the scaffolds in PBS. Means ± SD * $p < 0.05$ vs. *Dragmacidon reticulatum* (Mann–Whitney test). (**B**) Mass degradation of the scaffolds after incubation in PBS. Means ± SD * $p < 0.05$ vs. *Dragmacidon reticulatum* (Mann–Whitney test) ($n = 5$).

Figure 5B shows the results of the degradation assays of the scaffolds' mass after immersion in PBS at different periods. An initial loss of mass in both groups (day 1) was observed, being more pronounced in the species *Dragmacidon reticulatum* when compared to the species *Amphimedon viridis*. After the third day, the values remained stable. Statistical difference was verified between species in the first ($p = 0.0003$), third ($p = 0.0286$), seventh ($p < 0.0001$), fourteenth ($p = 0.0002$) and twenty-first ($p = 0.0027$) periods.

3.1.6. Porosity

The results of the porosity tests (Figure 6) demonstrate an average of 84 ± 5% porosity for *Dragmacidon reticulatum* scaffolds and 90 ± 2% for *Amphimedon viridis*. No statistical differences were found between *Dragmacidon reticulatum* and *Amphimedon viridis* scaffolds.

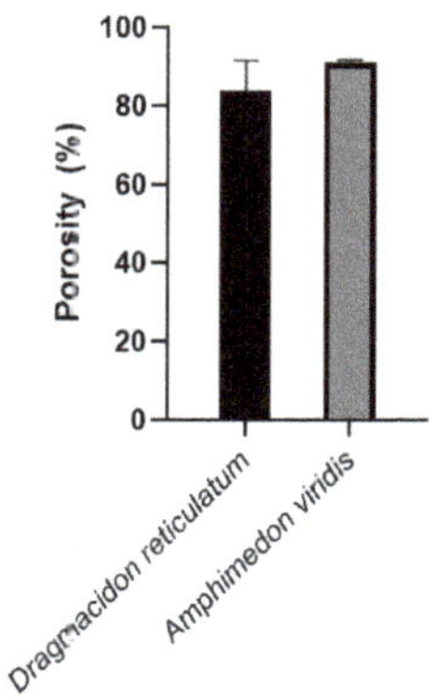

Figure 6. Porosity test of the sponges' scaffolds. No difference was found between *Dragmacidon reticulatum* scaffolds and *Amphimedon viridis* scaffolds (n = 5).

3.2. In Vivo Tests

3.2.1. Qualitative Histological Analysis

An overview of representative histological sections for all the experimental groups is shown in Figure 7. For CG, bone formation was observed at the border of the entire defect, with osteoid areas around the newly formed bone. The DR, when compared to AV, presented new bone tissue points and a greater presence of osteoid tissue, mainly around the spicules of silica and newly formed bone tissue. In AV, a fibrous capsule was formed around the lesion area where the scaffold was implanted. Neoformed bone tissue was absent in this group and the areas of osteoid tissue were smaller, with only poorly isolated portions inside the defect.

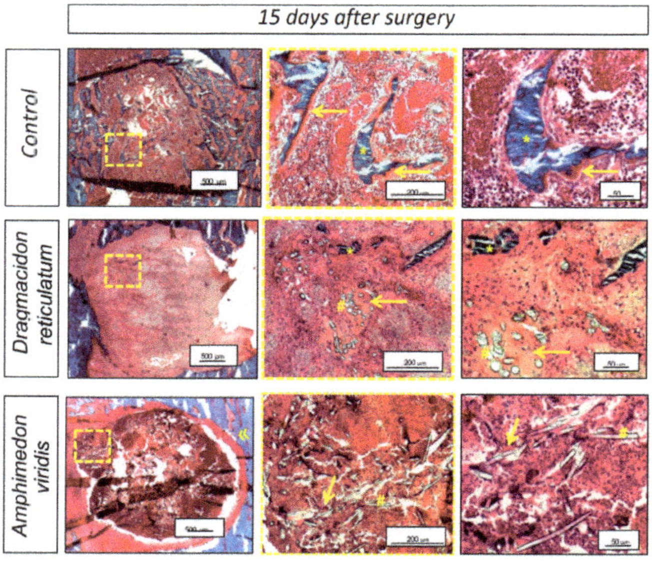

Figure 7. Bone defect photomicrography 15 days after the surgical procedure for the untreated experimental group (CG), for the group in which scaffolds of the marine sponge *Dragmacidon reticulatum* were implanted in the bone defects (DR) or for the group in which implanted scaffold belonged to the *Amphimedon viridis* species (AV). Neoformed bone tissue (*); osteoid (→); silica spike (#); and fibrous capsule (»). Bar scale = 500 µm (2.5× image), bar scale = 200 µm (20× image) and bar scale = 50 µm (40× image). Used stain: Goldner's tri-chromium (Merck).

3.2.2. Histomorphometric Analysis

Figure 8 shows the mean and standard deviation (SD) for the quantitative histomorphometric variables: BV/TV (%), OV/TV (%), N.Ob/T. Ar (mm^2), Ob.S/BS (%), Bm.V/TV (%) and Fb.V/TV (%).

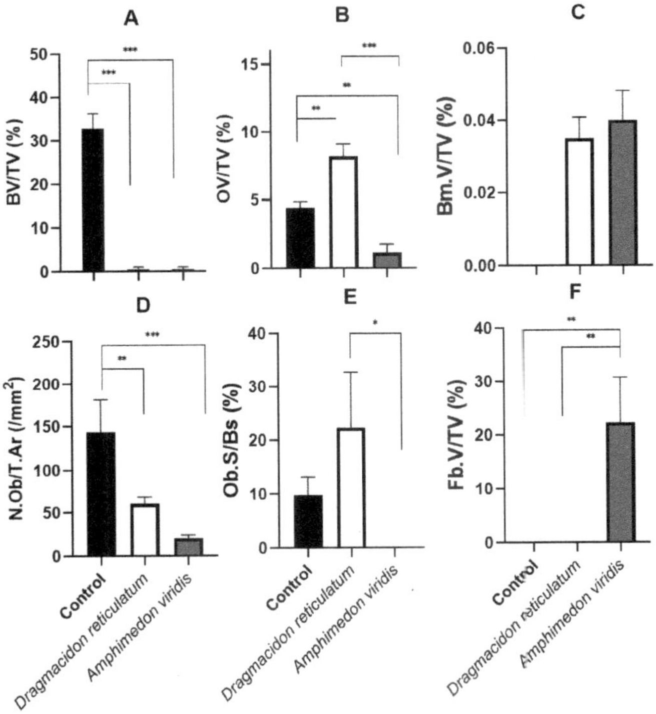

Figure 8. Bone histomorphometric analysis of bone defects 15 days after the surgical procedure for the experimental groups: no treatment (control—CG); submitted to the scaffold implant of the marine sponge *Dragmacidon reticulatum* (DR); submitted to the implant of the species *Amphimedon viridis* (AV). (**A**) Bone volume as percentage of tissue volume (BV/TV%), means ± SD *** $p < 0.0001$ (Anova/Tukey test). (**B**) Osteoid volume as percentage of tissue volume (OV/TV%), means ± SD ** $p < 0.01$; *** $p < 0.0001$ (Anova/Tukey test). (**C**) Biomaterial volume as percentage of tissue volume (Bm.V/TV%), means± SD $p > 0.05$ (Mann–Whitney test). (**D**) Number of osteoblasts per unit tissue area analyzed (N.Ob/T.Ar (mm^2)), means ± SD ** $p = 0.0039$; *** $p < 0.0001$ (Anova/Tukey test). (**E**) Osteoblastic surface as a percentage of the bone surface (Ob.S/Bs%), means ± SD * $p = 0.0237$ (Anova/Tukey test). (**F**) Percentage of fibrous tissue in the analyzed tissue (Fb.V/TV%), means ± SD ** $p = 0.0068$; $p = 0.0022$ (Anova/Tukey test).

For the BV/TV parameter, it was verified that the bone volume that was formed 15 days after the surgical procedure was proportionally higher in the control group compared to the other two groups that included scaffold implants. On average, the percentage of bone volume in the analyzed area of the defect was 32.6 ± 8.5% for CG, 0.05 ± 0.04% for DR and 0.00% for AV. Statistical differences were found between the groups: CG vs. DR ($p < 0.0001$) and CG vs. AV ($p < 0.0001$), in the ANOVA/Tukey test, as shown in Figure 8A.

On the other hand, when evaluating the percentage of osteoid in the analyzed area of the defect (OV/TV%), it was observed that the formation of this tissue was significantly superior in the group with scaffold implantation of the marine sponge *Dragmacidon reticulatum* (DR; 8.3 ± 2.6%) in comparison to the control group (CG, 4.3 ± 1.9%) and *Amphimedon viridis* (AV, 0.7 ± 0.3%). A statistically significant difference was observed in the comparisons among all groups: CG vs. DR (p = 0.0042), CG vs. AV (0.0077), DR vs. AV (p < 0.0001) (Figure 8B).

Additionally, in the evaluation of the parameter referring to the percentage of biomaterial in the analyzed tissue (Bm.V/TV%), a significant statistical difference was observed between the groups CG vs. DR and CG vs. AV, where the CG 0%, DR 0.03 ± 0.01% and AV 0.04 ± 0.01%. Thus, there was no difference between the two types of scaffolds implanted (Figure 8C).

The number of osteoblasts per unit area of tissue analyzed (N.Ob/T.Ar mm^2) of 132.0 ± 65.3 mm^2 for the CG, 56.7 ± 11.3 mm^2 for the DR and 13.2 ± 5.0 mm^2 for AV was observed. There were statistical differences between the CG vs. DR and CG vs. AV groups (Figure 8D).

On the other hand, the osteoblastic surface as a percentage of the bone surface (Ob. S/ES%) was 8.4 ± 5.6% for the CG, 17.7 ± 14.6% for DR and 0.0% for AV. A statistical difference was observed between the DR vs. AV (*) groups (Figure 8E).

In the evaluation of parameter Fb.V/TV (%), regarding the percentage of fibrous tissue in the analyzed tissue, statistical differences were observed between the CG vs. AV (p = 0.0068) and DR vs. AV (p = 0.0022) groups, as fibrosis was only verified in AV (19.9 ± 17.1%) (Figure 8F).

3.2.3. Biomechanical Analysis

In the evaluation of the fracture energy (J), the respective mean values of 0.04 ± 0.02 J, 0.03 ± 0.02 J and 0.04 ± 0.02 J were observed for CG, DR and AV. The evaluation of the variable elastic deformation energy (J) showed the mean values 0.02 ± 0.01 J, 0.02 ± 0.01 J and 0.03 ± 0.01 J for the CG, DR and AV groups, respectively. Finally, in the evaluation of the maximum load (N), an average value of 49.2 ± 5.6 N was observed for CG, 47.6 ± 9.8 N for DR and 46.7 ± 10.9 N for AV. No statistical differences were observed among groups (p > 0.005) for any of the parameters (Figure 9).

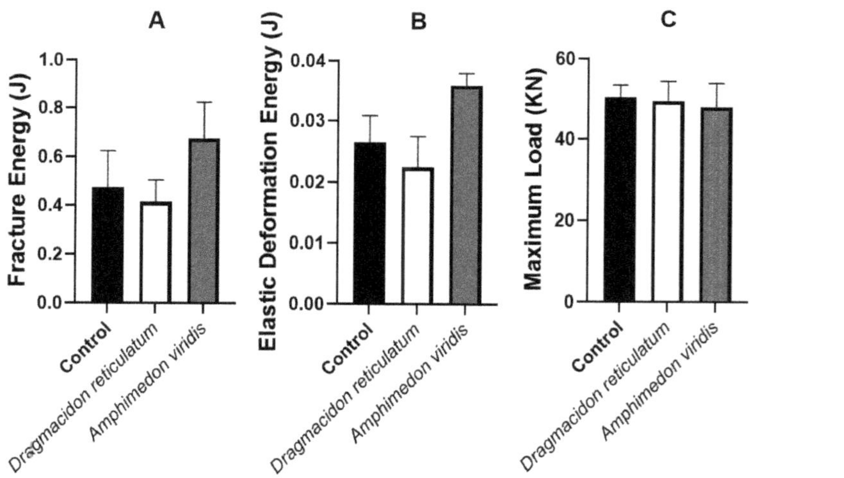

Figure 9. Mean and standard deviation of the variables of the biomechanical test: (**A**) fracture energy (J), (**B**) elastic deformation energy (J) and (**C**), and maximum load (N). Means ± SD p > 0.05. (Anova/Tukey test).

4. Discussion

This study aimed to evaluate the physical-chemical properties of scaffolds manufactured from two species of marine sponges of the Demospongiae class and to investigate the in vivo biological response of these natural biomaterials once implanted in bone defects in rat tibia. The hypothesis was that, since different species could be physically and morphologically different, the osteogenic potential of these scaffolds would also be distinct. The main results showed that scaffolds of *Dragmacidon reticulatum* presented better performance in bone repair when compared to scaffolds of the species *Amphimedon viridis*.

Initially, SEM micrographs showed that the scaffold structures of both species are constituted by a porous material, with interconnected pores and presence of silica spicules, degrading after immersion in PBS. These characteristics are very important, since previous studies prove that the presence of interconnected pores and silica spicules are important characteristics in the scaffolds destined to favor the process of bone repair, as well as the gradual degradation of the scaffold allowing the replacement of the material by bone tissue [24,25]. Interestingly, scaffolds from both species exhibited a great porosity (84 $\pm$ 5% for DR and 90 $\pm$ 2% for AV), which would be adequate for their use as bone substitutes since the scientific literature establishes a minimum of 80% for an ideal scaffold [26]. In fact, higher porosity is expected to be reflected as an increase in bone formation, as observed in several studies involving different biomaterials, used as raw material for scaffolds' manufacture [27]. In the study of Nandi et al. [15], SEM analysis also showed that the *Biemna fortis* skeleton has a collagenous fibrous network with highly networked porosity in the size range of 10–220 µm. However, the fibrous material was burnt at 725 °C and the fired material was mixed with naphthalene to induce porosity. After naphthane removal, apparent porosity was determined to be about 52%, which is substantially lower than the porosity found for *Dragmacidon reticulatum* and *Amphimedon viridis* scaffolds used in the present study (84 $\pm$ 5% for DR and 90 $\pm$ 2% for AV). This can be explained not only by the species-related difference, but also by the manufacturing method of the scaffolds themselves.

In the FTIR analysis, the scaffolds of both species had the same functional groups, with organic amide functions at 1650 cm^{-1} and 1535 cm^{-1}, a primary alcohol at 2950 cm^{-1} and 2875 cm^{-1}, inorganic silicon–carbon bonds at 1250 cm^{-1} and 770 cm^{-1} and silicon-hydroxyl at 3700 cm^{-1}, 1030 cm^{-1} and 890 cm^{-1} [28]. It was observed that from the seventh day of incubation, for the *Dragmacidon reticulatum* species, there was a reduction in the points referring to the amide and primary alcohol groups, characteristic of oxidation of the organic matter, evidencing the degradation of the sample during the incubation time. This finding confirms our previous results of electron microscopy, which indicated a greater presence of pores and greater degradation in the scaffolds of this same species. For the species *Amphimedon viridis* the points remained stable during the incubation period evaluated. This difference between the species studied also suggests a better potential of the species *Dragmacidon reticulatum* when compared to the species *Amphimedon viridis*, since the study of Kido et al. [25] has shown that, in addition to porosity, proper degradation of the scaffold is essential for the bone repair process to occur. It is known that tissue formation depends on space to occur, and that the presence of a biomaterial may be a barrier to the growth of new tissue, especially if the implanted biomaterial does not have adequate porosity or a degradation rate compatible with the rate of bone formation. In the study of Nandi et al. [18], FTIR spectra revealed the presence of Si-O-Si groups at about 800 cm^{-1} and massive carbonate groups at about 1500 cm^{-1}, which is a sign that the freeze-dried *Biemna fortis* material has a high organic content.

The XRD revealed that the scaffolds of the two species of marine sponges studied are composed of amorphous organic content, which coincides with the findings of Nandi et al. [18], who also characterized the marine sponge (*Biemna fortis*) belonging to the same class of species of this study (class Demospongiae). In addition, Schröder et al. [29], when studying the structure, biochemical composition and formation mechanism of biosilica produced by living organisms (marine sponges, diatoms and higher plants), reported it to be amorphous silica without evidence of crystalline silica. Gabbai-Armelin et al. [12] also found the mostly amorphous nature of biosilica extracted from *Dragmacidon reticulatum* species, with only a few crystalline peaks being observed. Therefore, these studies agree with the amorphous nature of sponges and their components, and the implication of this finding for bone tissue engineering would be the greatest biodegradation of these natural biomaterials once implanted, considering that the crystalline arrangement would imply greater stability due to the structural configuration of their monomers [3]. This may be beneficial for the replacement of the implant by the natural tissue, although potentially promoting, on the other hand, a significant decrease in the mechanical properties of the biomaterial [3].

For the EDS analysis, the scaffolds of both species had the same chemical composition, the three main components being carbon, oxygen and silicon, and a lower percentage of aluminum, calcium, chlorine, iron, potassium, magnesium, sodium, phosphorus and sulfur. These findings are in line with those obtained in the Sandford [30] study, which carried out a physical-chemical analysis of the siliceous skeletons of six species of marine sponges of two groups (Demospongiae and Hexactinellida), which concluded that the average chemical composition of the spicules of both groups was 85% silica (SiO_2) and presented small amounts of the other elements mentioned earlier in this study. Additionally, Schröder et al. [29] reinforce that silica spicules, besides silicon and oxygen, present trace amounts of various other elements (mainly aluminum, calcium, chlorine, iron, potassium, sodium and sulfur), where approximately 75% of the body mass of these animals is silica. Interestingly, marine organisms process approximately seven gigatons of silicon to make their silica skeletons, an interesting feature for regenerative medicine in view of the osteoinductive potential (i.e., ability to stimulate differentiation of precursor cells in functional osteoblasts) of this component [31].

For the analysis of pH and mass degradation, a more marked variation of the values was observed, in the sense of an acidification of the medium with the loss of the material, during the first week of immersion of the scaffolds in PBS, reaching stability in the subsequent periods. It is interesting to note that only the DR scaffolds showed the initial drop in pH in the periods referring to the 1st and 3rd days of immersion in PBS and following. This could be explained by the fact that the loss of the scaffolds' mass after immersion in PBS was remarkably greater in DR according to the degradation assays, with the products of this lixiviation most likely being responsible for the pH alterations. Further, this acidification would be interesting in the context of bone healing since Hazehara-Kunitomo et al. [32], who were able to measure the pH in vivo during the healing of a bone fracture, showed that the pH decreased to 6.8 during the inflammatory period (initial 2 days) and that this short-term acidification could help stem cell differentiation towards bone-forming osteoblasts [33].

In the histological analysis of the present study, bone formation was observed around the border of the whole defect in the CG, and histomorphometry (parameter BV/TV%) revealed that the percentage of bone formed was higher in this group than in the other groups that used marine sponge grafts (DR and AV). This can be explained by the adoption of a non-critical bone defect model that allows spontaneous bone repair with complete closure of the defect. In addition, other studies have shown that, initially, the presence of the biomaterial at the lesion site constitutes a barrier to bone growth, since the biomaterial needs to be degraded for the subsequent occurrence of bone tissue formation [34–39]. In the study of Nandi et al. [18], scaffolds prepared from *Biemna fortis* sponges (belonging to the same class of sponges of the present study—Demospongiae) were implanted into non-critical femoral bone defects in rabbits and tracked for up to 90 days. No quantifications were performed, but radiological, histological and scanning electron microscopy have demonstrated that bare sponge scaffolds (without any growth factor loading) promoted better osseous tissue formation, with invasion of this new bone across the porous scaffold's matrix, in comparison to bone defects without any implant, which mostly showed soft tissue formation, with the defect gap still present even after 3 months.

In addition to mineralized bone, the formation of osteoid tissue was also evaluated in this study. It was present mostly underlying the newly formed bone tissue and around the sponge silica spicules, being much more evident in the scaffolds created from *Dragmacidon reticulatum* species. In fact, the analysis of the histomorphometric parameter OV/TV (%) showed a higher osteoid percentage in DR. Therefore, we can indicate more advanced bone repair in the CG due to the greater presence of mineralized bone tissue, followed by DR, with a more important presence of bone matrix that is not yet mineralized and, finally, AV, in which the neoformation of osseous and osteoid bone tissue was lower than the others. Nandi et al. [18] also assessed newly formed osteoid tissue in implanted *Biemna fortis* scaffolds through fluorochrome labeling—oxytetracycline. They showed a greater presence of newly formed bone in sponge scaffolds in comparison to control defects, which were moderately filled with osteoid tissue and where consolidation was under process.

It is important to note that the evidence discussed above is consistent with the cellular parameters analysis. Several in vitro studies have shown that the inorganic part of the sponge skeleton (biosilica) induces bone neoformation by attracting osteoprogenitor cells and stimulating their differentiation in osteoblasts [3,26,31,34]. In the present study, the number of osteoblasts per unit area of tissue analyzed (N.Ob/T.Ar/mm^2) was higher in the CG compared to the other groups, which is compatible with the presence of a greater amount of neoformed bone tissue. Additionally, analyzing the osteoblastic surface (Ob.S/BS%), which refers to the percentage of bone surface covered by active osteoblasts, a statistical difference was found between the two groups with sponge implants, DR (17.7 $\pm$ 14.6%) and AV (0.0%), being superior in the *Dragmacidon reticulatum* species, which again indicates superior osteogenic properties in the scaffolds manufactured from this species. In the study of Nandi et al. [18], despite the absence of quantitative parameters, well-formed osteons inside sponge-filled defects exhibited osteoblastic and osteoclastic activities, whereas osteoclasts were prominently observed at the cortical region of non-filled control defects. The authors state that sponge scaffolds most likely serve as a bioactive stimulant for cell maturation, in agreement with other studies that have previously show that, at least in in vitro situations, osteoblast attachment, proliferation, migration and differentiation can be stimulated by marine sponge components [10,11,38].

However, in the present study, in addition to the lower bone and osteoid formation and the lower number of osteoblasts in AV, a fibrous capsule formation was observed around the implant area only in this group, which may indicate an attempt by the organism to isolate the biomaterial, i.e., a rejection of *Amphimedon viridis* scaffolds [39]. Urabayashi [19] has demonstrated, through the in vitro chemical analysis of the crude sponge extract of this species, a cytotoxic effect in human cell lines from retinal pigment epithelium and breast carcinoma, as well as a hemolytic action in rat cells, which may explain the results of the present study, since severe local and systemic inflammatory and cytotoxic responses caused by the implants may result in delayed or non-healing of the bone [25]. Therefore, despite previous research demonstrating the osteogenic potential of marine sponges in terms of both bone formation and cell stimulation, the current study shows that the benefits may also depend on the species of sponge studied and that the integration of the material with the original tissue can be significantly inferior or even harmful depending on the species, as shown here.

As the last histomorphometric parameter, the percentage of biomaterial inside the defect (Bm.V/TV%) was not different between DR and AV, although the characterization assays indicated a more important degradation in scaffolds manufactured from the *Dragmacidon reticulatum* species. Therefore, studies with longer experimental periods are necessary in order to allow a longer time for the biodegradation of the implanted materials and for the concomitant process of bone consolidation. In the study of Nandi et al. [18], evidence of sponge scaffold degradation was found radiologically, mostly from day 60 when the scaffold altered its shape from cylindrical to oval, with the edges of the filling material reducing its size, a sign that newly formed bone tissue was beginning to replace the sponge material. Moreover, the authors found that the enrichment of the scaffolds with the IGF-1 and BMP-2 growth factors promoted bone formation, with the presence of neoformed bone tissue and active osteoblastic cells throughout the defect region, as well as increased biodegradation of the implanted biomaterial.

Finally, for all the parameters evaluated in the biomechanical test, no statistical differences were found among the groups, although the CG had a higher bone formation than the others. This may be due to the presence of the biomaterial at the site of the defect in the two other groups, which could imply a greater resistance to fractures. In addition, interestingly, it may indicate a good integration of the implanted materials with the pre-existing living tissue, a hypothesis also reported in the study by Granito et al. [3] involving bone defects in tibiae of rats filled or not with bioactive synthetic materials.

Together, the characterization analyses and the in vivo study with the scaffolds manufactured from the species *Dragmacidon reticulatum* and *Amphimedon viridis* confirmed our hypothesis that different species, because they exhibit different chemical and structural properties, also have different osteogenic properties. Here, the *Dragmacidon reticulatum* species proved to be a more interesting option as a scaffold in bone tissue engineering and, therefore, the so-called good osteogenic potential is species-dependent, as demonstrated herein in a pioneering way.

In a pioneering way, this study compared two abundant species on the Brazilian coast and performed an extensive quantitative analysis of the bone repair process after in vivo implantation of these natural biomaterials. However, there are some limitations that still need to be addressed herein, such as the use of a non-critical bone defect model, which was chosen to allow biomechanical evaluations on a weight-bearing bone, and the short treatment period after scaffold implantion. Therefore, the critical-sized calvarial defect could be the next step of this study, together with longer follow-up periods and the inclusion of other experimental groups for the evaluation of scaffolds manufactured from marine sponges belonging to other species and even other families. Actually, a larger number of different species and families could have their potential evaluated if in silico methods had been employed. Faster outcomes and reduced costs are some of the benefits that computer simulations could bring for the prediction of the implant failure values, as in the study of Putra et al. [40]. Finally, another limitation of this study is inherent to the use of a natural biomaterial itself. Structural and morphological variations are expected among specimens collected from the environment. Subsequently, biomaterial availability would be an additional challenge, especially considering the clinical application as the final purpose of this study. Both obstacles could be at least partially overcome with the production of sponge biomass by in situ aquaculture. Mariculture could possibly satisfy the biomaterial demand without compromising natural sponge beds, constituting, therefore, a sustainable approach for biomaterial production [41].

In addition, the biological performance of scaffolds should also be assessed in other experimental animal models, as well as including the evaluation of bone repair in clinical situations in which it is compromised, such as in diabetes, in order to validate the use of these natural biomaterials as promising therapeutic alternatives.

5. Conclusions

It can be concluded that scaffolds manufactured from the marine sponge species *Dragmacidon reticulatum*, when compared to *Amphimedon viridis* species, were more effective as bone substitutes, since their structure was more porous and biodegradable, with a compatible increase in osteoid tissue formation and presence of osteoblastic cells, despite the absence of fibrous tissue formation in the studied period. Thus, sponges of this species may constitute a promising alternative source of biomaterials for use in bone tissue engineering. Further research is still needed for a better understanding of these marine sponge scaffolds. A critical bone defect model in rats should be the next step of this study, as well as other experimental models to evaluate the bone repair when it is compromised, such as an osteoporosis model. In addition, another crucial perspective for the future would be the improvement of mariculture techniques in substitution of sponge collection in the environment. Thereby, the sustainable obtainment of sufficient amounts of biomaterials would be achievable in view of their envisaged therapeutic application.

Supplementary Materials: The following supporting information can be downloaded at: https://www.mdpi.com/article/10.3390/jfb14030122/s1, Figure S1: XRD analysis of the samples; Table S1: EDS analysis of the scaffolds manufactured from the marine sponge species *Dragmacidon reticulatum* and *Amphimedon viridis*.

Author Contributions: Conceptualization, R.N.G., C.P.G.S., M.R.C. and A.C.M.R.; methodology, C.P.G.S., K.R.F., H.W.K., B.P.D., J.R.P., J.A.S., M.A.C., M.R.C., A.C.M.R. and R.N.G.; software, C.P.G.S., J.A.S. and J.P.S.P.; validation, C.P.G.S. and R.N.G.; formal analysis, C.P.G.S. and J.P.S.P.; investigation, C.P.G.S., K.R.F., H.W.K., B.P.D., J.R.P., J.A.S., M.A.C.; resources, A.C.M.R., M.R.C. and R.N.G.; data curation, C.P.G.S. and J.P.S.P.; writing—original draft preparation, C.P.G.S., J.P.S.P. and R.N.G.; writing—review and editing, K.R.F., H.W.K., B.P.D., J.R.P., J.A.S., M.A.C., M.R.C. and A.C.M.R.; visualization, C.P.G.S., M.R.C., A.C.M.R. and R.N.G.; supervision, R.N.G. and A.C.M.R.; project administration, C.P.G.S. and R.N.G.; funding acquisition, A.C.M.R. and R.N.G. All authors have read and agreed to the published version of the manuscript.

Funding: This research was funded by FAPESP (Fundação de Amparo à Pesquisa do Estado de São Paulo), grant numbers 2019/10228-5 and 2021/04957-4.

Institutional Review Board Statement: The authors confirm that the ethical policies of the journal, as noted on the journal's author guidelines page, have been adhered to and the appropriate ethical review committee approval has been received. This study was approved by the Ethics Committee on the Use of Animals (CEUA) of the Federal University of São Paulo (2017/3011170417).

Data Availability Statement: The data presented in this study are available on request from the corresponding author.

Acknowledgments: The authors are grateful to the Centre for Marine Biology of University of São Paulo (CEBIMar/USP) for the precious support in the specimen's collection (joint project No. 938).

Conflicts of Interest: The authors declare no conflict of interest. The funders had no role in the design of the study; in the collection, analyses, or interpretation of data; in the writing of the manuscript, or in the decision to publish the results.

References

1. Reddy, M.S.B.; Ponnamma, D.; Choudhary, R.; Sadasivuni, K.K. A Comparative Review of Natural and Synthetic Biopolymer Composite Scaffolds. *Polymers* **2021**, *13*, 1105. [CrossRef] [PubMed]
2. Alvarez Echazú, M.I.; Perna, O.; Olivetti, C.E.; Antezana, P.E.; Municoy, S.; Tuttolomondo, M.V.; Galdopórpora, J.M.; Alvarez, G.S.; Olmedo, D.G.; Desimone, M.F. Recent Advances in Synthetic and Natural Biomaterials-Based Therapy for Bone Defects. *Macromol. Biosci.* **2022**, *22*, e2100383. [CrossRef] [PubMed]
3. Granito, R.N.; Ribeiro, D.A.; Rennó, A.C.M.; Ravagnani, C.; Bossini, P.S.; Peitl-Filho, O.; Zanotto, E.D.; Parizotto, N.A.; Oishi, J. Effects of biosilicate and bioglass 45S5 on tibial bone consolidation on rats: A biomechanical and a histological study. *J. Mater. Sci. Mater. Med.* **2009**, *20*, 2521–2526. [CrossRef] [PubMed]
4. Mastrogiacomo, M.; Scaglione, S.; Martinetti, R.; Dolcini, L.; Beltrame, F.; Cancedda, R.; Quarto, R. Role of scaffold internal structure on in vivo bone formation in macroporous calcium phosphate bioceramics. *Biomaterials* **2006**, *27*, 3230–3237. [CrossRef] [PubMed]
5. Romano, G.; Almeida, M.; Varela Coelho, A.; Cutignano, A.; Gonçalves, L.G.; Hansen, E.; Khnykin, D.; Mass, T.; Ramšak, A.; Rocha, M.S.; et al. Biomaterials and Bioactive Natural Products from Marine Invertebrates: From Basic Research to Innovative Applications. *Mar. Drugs* **2022**, *20*, 219. [CrossRef]
6. Wang, X.; Schröder, H.C.; Müller, W.E.G. Biocalcite, a multifunctional inorganic polymer: Building block for calcareous sponge spicules and bioseed for the synthesis of calcium phosphate-based bone. *Beilstein J. Nanotechnol.* **2014**, *5*, 610–621. [CrossRef] [PubMed]
7. Jones, J.R. Reprint of: Review of bioactive glass: From Hench to hybrids. *Acta Biomater.* **2015**, *23*, S53–S82. [CrossRef]
8. Cen, L.; Liu, W.; Cui, L.; Zhang, W.; Cao, Y. Collagen tissue engineering: Development of novel biomaterials and applications. *Pediatr. Res.* **2008**, *63*, 492–496. [CrossRef]
9. Cruz, M.A.; Fernandes, K.R.; Parisi, J.R.; Vale, G.C.A.; Junior, S.R.A.; Freitas, F.R.; Sales, A.F.S.; Fortulan, C.A.; Peitl, O.; Zanotto, E.; et al. Marine collagen scaffolds and photobiomodulation on bone healing process in a model of calvaria defects. *J. Bone Miner. Metab.* **2020**, *38*, 639–647. [CrossRef]
10. Silva, T.H.; Moreira-Silva, J.; Marques, A.L.; Domingues, A.; Bayon, Y.; Reis, R.L. Marine origin collagens and its potential applications. *Mar. Drugs* **2014**, *12*, 5881–5901. [CrossRef]
11. Wiens, M. The role of biosilica in the osteoprotegerin/RANKL ratio in human osteoblast-like cells. *Biomaterials* **2010**, *31*, 7716–7725. [CrossRef] [PubMed]

12. Gabbai-Armelin, P.R.; Kido, H.W.; Cruz, M.A.; Prado, J.P.S.; Avanzi, I.R.; Custódio, M.R.; Renno, A.C.M.; Granito, R.N. Characterization and Cytotoxicity Evaluation of a Marine Sponge Biosilica. *Mar. Biotechnol.* **2019**, *21*, 65–75. [CrossRef] [PubMed]
13. Jones, J.R.; Gentleman, E.; Polak, J. Bioactive glass scaffolds for bone regeneration. *Elements* **2007**, *3*, 393–399. [CrossRef]
14. Green, D.; Howard, D.; Yang, X.; Kelly, M.; Oreffo, R.O. Natural marine sponge fiber skeleton: A biomimetic scaffold for human osteoprogenitor cell attachment, growth, and differentiation. *Tissue Eng.* **2003**, *9*, 1159–1166. [CrossRef] [PubMed]
15. Lin, Z.; Solomon, K.L.; Zhang, X.; Pavlos, N.J.; Abel, T.; Willers, C.; Dai, K.; Xu, J.; Zheng, Q.; Zheng, M. In vitro evaluation of natural marine sponge collagen as a scaffold for bone tissue engineering. *Int. J. Biol. Sci.* **2011**, *7*, 968–977. [CrossRef] [PubMed]
16. Barros, A.A.; Aroso, I.M.; Silva, T.H.; Mano, J.F.; Duarte, A.R.C.; Reis, R.L. Surface modification of silica-based marine sponge bioceramics induce hydroxyapatite formation. *Cryst. Growth Des.* **2014**, *14*, 4545–4552. [CrossRef]
17. Tsurkan, D.; Wysokowski, M.; Petrenko, I.; Voronkina, A.; Khrunyk, Y.; Fursov, A.; Ehrlich, H. Modern scaffolding strategies based on naturally pre-fabricated 3D biomaterials of poriferan origin. *Appl. Phys. A* **2020**, *126*, 382. [CrossRef]
18. Nandi, S.K.; Kundu, B.; Mahato, A.; Thakur, N.L.; Joardar, S.N.; Mandal, B.B. In vitro and in vivo evaluation of the marine sponge skeleton as a bone mimicking biomaterial. *Integr. Biol.* **2015**, *7*, 250–262. [CrossRef]
19. Urabayashi, M.S. Prospecção de Moléculas Bioativas em Esponjas Marinhas da Espécie *Amphimedon viridis*: Estudos Celulares e Moleculares. Ph.D. Thesis, Universidade de São Paulo, São Paulo, Brazil, 2015.
20. Kokubo, T.; Takadama, H. How useful is SBF in predicting in vivo bone bioactivity? *Biomaterials* **2006**, *27*, 2907–2915. [CrossRef]
21. Gabbai-Armelin, P.R.; Souza, M.T.; Kido, H.W.; Tim, C.; Bossini, P.S.; Magri, A.M.P.; Fernandes, K.R.; Pastor, F.A.C.; Zanotto, E.; Parizotto, N.; et al. Effect of a new bioactive fibrous glassy scaffold on bone repair. *J. Mater. Sci. Mater. Med.* **2015**, *26*, 177. [CrossRef]
22. Rodrigues, A.C.; Furtado, A.C.S.; Magalhaes, M.M.; Souza, F.I.B. Análise química por espectroscopia de raios-X por dispersão energia (EDS): Adaptação aplicada à raízes de dendezeiro (*Eleais guineensis* Jacq.). In Proceedings of the Seminário de iniciação científica da embrapa amazônia oriental, Belém, Brazil, 20–22 September 2017.
23. Parfitt, A.M. Bone histomorphometry: Standardization of nomenclature, symbols and units (summary of proposed system). *Bone* **1988**, *9*, 67–69. [CrossRef] [PubMed]
24. Schepers, E.J.; Ducheyne, P. Bioactive glass particles of narrow size range for the treatment of oral bone defects: A 1–24 month experiment with several materials and particle sizes and size ranges. *J. Oral Rehabil.* **1997**, *24*, 171–181. [CrossRef] [PubMed]
25. Kido, H.W.; Tim, C.; Bossini, P.S.; Parizotto, N.; de Castro, C.A.; Crovace, M.C.; Rodrigues, A.C.M.; Zanotto, E.D.; Filho, O.P.; Anibal, F.D.F.; et al. Porous bioactive scaffolds: Characterization and biological performance in a model of tibial bone defect in rats. *J. Mater. Sci. Mater. Med.* **2015**, *26*, 74. [CrossRef] [PubMed]
26. Tripathi, Y.; Shukla, M.; Bhatt, A.D. Idealization through interactive modeling and experimental assessment of 3D-printed gyroid for trabecular bone scaffold. *Inst. Mech. Eng. Part H J. Eng. Med.* **2021**, *235*, 1025–1034. [CrossRef] [PubMed]
27. Karageorgiou, V.; Kaplan, D. Porosity of 3D biomaterial scaffolds and osteogenesis. *Biomaterials* **2005**, *26*, 5474–5491. [CrossRef]
28. Sócrates, G. *Infrared and Raman Characteristic Group Frequencies*, 3rd ed.; Tables and Charts; John Wiley and Sons, Ltd.: Chichester, UK, 2001.
29. Schröder, H.C.; Wang, X.; Tremel, W.; Ushijima, H.; Müller, W.E.G. Biofabrication of biosilica-glass by living organisms. *Nat. Prod. Rep.* **2008**, *25*, 455–474. [CrossRef]
30. Sandford, F. Physical and chemical analysis of the siliceous skeletons in six sponges of two groups (demospongiae and hexactinellida). *Microsc. Res. Tech.* **2003**, *62*, 336–355. [CrossRef]
31. Granito, R.N.; Custódio, M.R.; Rennó, A.C.M. Natural marine sponges for bone tissue engineering: The state of art and future perspectives. *J. Biomed. Mater. Res. B Appl. Biomater.* **2017**, *105*, 1717–1727. [CrossRef]
32. Yuri, H.-K.; Emilio, S.H.; Mitsuaki, O.; Kyaw, A.; Keiko, K.; Hai, P.; Kentaro, A.; Masahiro, O.; Toshitaka, O.; Takuya, M.; et al. Acidic Pre-Conditioning Enhances the Stem Cell Phenotype of Human Bone Marrow Stem/Progenitor Cells. *Int. J. Mol. Sci.* **2019**, *20*, 1097. [CrossRef]
33. Boccardi, E.; Philippart, A.; Melli, V.; Altomare, L.; De Nardo, L.; Novajra, G.; Vitale-Brovarone, C.; Fey, T.; Boccaccini, A.R. Bioactivity and Mechanical Stability of 45S5 Bioactive Glass Scaffolds Based on Natural Marine Sponges. *Ann. Biomed. Eng.* **2016**, *44*, 1881–1893. [CrossRef]
34. Kido, H.W.; Bossini, P.S.; Tim, C.R.; Parizotto, N.A.; Da Cunha, A.F.; Malavazi, I.; Rennó, A.C.M. Evaluation of the bone healing process in an experimental tibial bone defect model in ovariectomized rats. *Aging Clin. Exp. Res.* **2014**, *26*, 473–481. [CrossRef] [PubMed]
35. Magri, A.M.P.; Fernandes, K.R.; Assis, L.; Mendes, N.A.; Santos, A.L.Y.D.S.; Dantas, E.D.O.; Renno, A.C. Photobiomodulation and bone healing in diabetic rats: Evaluation of bone response using a tibial defect experimental model. *Lasers Med. Sci.* **2015**, *30*, 1949–1957. [CrossRef] [PubMed]
36. Hench, L.L. The future of bioactive ceramics. *J. Mater. Sci. Mater. Med.* **2015**, *26*, 86. [CrossRef] [PubMed]
37. Martins, E.; Diogo, G.S.; Pires, R.; Reis, R.L.; Silva, T.H. 3D Biocomposites Comprising Marine Collagen and Silica-Based Materials Inspired on the Composition of Marine Sponge Skeletons Envisaging Bone Tissue Regeneration. *Mar. Drugs* **2022**, *20*, 718. [CrossRef]
38. Pallela, R.; Venkatesan, J.; Janapala, V.R.; Kim, S.K. Biophysicochemical evaluation of chitosan-hydroxyapatite-marine sponge collagen composite for bone tissue engineering. *J. Biomed. Mater. Res. A* **2012**, *100*, 486–495. [CrossRef]

39. Klopfleisch, R.; Jung, F. The pathology of the foreign body reaction against biomaterials. *J. Biomed. Mater. Res. A* **2017**, *105*, 927–940. [CrossRef]
40. Putra, R.U.; Basri, H.; Prakoso, A.T.; Chandra, H.; Ammarullah, M.I.; Akbar, I.; Syahrom, A.; Kamarul, T. Level of Activity Changes Increases the Fatigue Life of the Porous Magnesium Scaffold, as Observed in Dynamic Immersion Tests, over Time. *Sustainability* **2023**, *15*, 823. [CrossRef]
41. Brümmer, F.; Nickel, M. Sustainable Use of Marine Resources: Cultivation of Sponges. *Prog. Mol. Subcell. Biol.* **2003**, *37*, 143–162. [CrossRef]

Journal of *Functional Biomaterials*

MDPI

Article

3D-Printed GelMA/PEGDA/F127DA Scaffolds for Bone Regeneration

Jianpeng Gao [1,2,†], Ming Li [1,3,†], Junyao Cheng [1,2,†], Xiao Liu [1,2], Zhongyang Liu [1,3], Jianheng Liu [1,3,*] and Peifu Tang [1,3,*]

1 Department of Orthopaedics, Chinese PLA General Hospital, Beijing 100039, China
2 Medical School of Chinese PLA, Beijing 100039, China
3 National Clinical Research Center for Orthopedics, Sports Medicine & Rehabilitation, Beijing 100853, China
* Correspondence: jianhengliu@126.com (J.L.); pftang301@163.com (P.T.)
† These authors contributed equally to this work.

Abstract: Tissue-engineered scaffolds are an effective method for the treatment of bone defects, and their structure and function are essential for bone regeneration. Digital light processing (DLP) printing technology has been widely used in bone tissue engineering (BTE) due to its high printing resolution and gentle printing process. As commonly used bioinks, synthetic polymers such as polyethylene glycol diacrylate (PEGDA) and Pluronic F127 diacrylate (F127DA) have satisfactory printability and mechanical properties but usually lack sufficient adhesion to cells and tissues. Here, a compound BTE scaffold based on PEGDA, F127DA, and gelatin methacrylate (GelMA) was successfully prepared using DLP printing technology. The scaffold not only facilitated the adhesion and proliferation of cells, but also effectively promoted the osteogenic differentiation of mesenchymal stem cells in an osteoinductive environment. Moreover, the bone tissue volume/total tissue volume (BV/TV) of the GelMA/PEGDA/F127DA (GPF) scaffold in vivo was $49.75 \pm 8.50\%$, higher than the value of $37.10 \pm 7.27\%$ for the PEGDA/F127DA (PF) scaffold and $20.43 \pm 2.08\%$ for the blank group. Therefore, the GPF scaffold prepared using DLP printing technology provides a new approach to the treatment of bone defects.

Keywords: biomaterials; bone regeneration; bone tissue engineering; digital light processing

Citation: Gao, J.; Li, M.; Cheng, J.; Liu, X.; Liu, Z.; Liu, J.; Tang, P. 3D-Printed GelMA/PEGDA/F127DA Scaffolds for Bone Regeneration. *J. Funct. Biomater.* **2023**, *14*, 96. https://doi.org/10.3390/jfb14020096

Academic Editors: Adalberto Luiz Rosa, Marcio Mateus Beloti and Adriana Bigi

Received: 28 December 2022
Revised: 1 February 2023
Accepted: 2 February 2023
Published: 9 February 2023

1. Introduction

Bone defects caused by tumors, infections, and trauma are difficult and critical to treat clinically and thus require effective therapeutic initiatives [1–3]. Bone tissue engineering (BTE) scaffolds have been widely used in bone repair because they are capable of filling the defective area, providing mechanical support, and guiding the growth of new tissues in the early treatment of bone defects [4–6].

With recent developments in 3D printing technology, it has become possible to prepare scaffolds according to a predesigned (computer aided design, CAD) structure and achieve more precise control over the macroscopic structure of the scaffold [7–9]. Digital light processing (DLP) technology is a representative lithography-based 3D bioprinting technology characterized by a layer-by-layer based printing pattern, the core of which is the digital micromirror device (DMD), which provides superior image stability, fidelity, and reliability. Visible or ultraviolet light can be used to cross-link bioinks and complete the liquid–solid conversion in DLP printing [10–14]. Thus, some traditional natural and synthetic materials, which previously could only be printed by extrusion printers, are now available for printing via DLP, a gentler printing method, as long as the bioinks can be endowed with light-curing properties (e.g., Gel-GelMA, PEG-PEGDA, and F127-F127DA) [15–19].

Polyethylene glycol (PEG), which exhibits high biocompatibility and almost no immunogenicity, can be chemically modified to form polyethylene glycol diacrylate (PEGDA), which possesses photo-crosslinking properties, low viscosity, and high solubility, making it an ideal biomaterial for DLP bioprinting [18,20,21]. Despite its many advantages, PEGDA is generally inelastic and brittle, which makes it more likely to be used in combination with other materials rather than alone for bone tissue engineering [22]. It has been shown that PEGDA can be mixed with materials such as nanohydroxyapatite [23–25], nanoclay [26], and extracellular matrix [27] to form bioinks for the preparation of BTE scaffolds, which play an important role in the treatment of bone defects. However, the mechanical properties, printability, and bioactivity of these scaffolds still need to be improved. Pluronic F127 diacrylate (F127DA), modified with Pluronic F127 (F127), exhibits low swelling properties, fatigue resistance, and proper elastic modulus, which complement the shortcomings of PEGDA [28–30]. Bioinks combined with F127DA and PEGDA may possess excellent printability and mechanical properties. However, this combination has not been widely used in BTE perhaps due to the lack of cell adhesion [31–33].

Gelatin methacrylate (GelMA), as one of the most commonly used photosensitive hydrogel materials in bone tissue engineering, has good biocompatibility [34]. Unlike PEGDA and F127DA, GelMA can significantly promote cell adhesion and proliferation [35]. Thus, based on this, a new GelMA/PEGDA/F127DA bioink was developed in this study and a GelMA/PEGDA/F127DA scaffold was prepared using DLP printing technology; the scaffold not only possessed good mechanical properties similar to synthetic materials and played a supporting role in early implantation, but the porous structure created by 3D printing also actively promoted the growth of bone tissue. Moreover, the addition of GelMA greatly increased the cell adhesion of the scaffold, which is crucial for osteoconduction and bone regeneration. Therefore, the composite scaffold consisting of natural (GelMA) and synthetic materials (PEGDA-F127DA) prepared using DLP technology represents a promising approach for the treatment of bone defects (Figure 1).

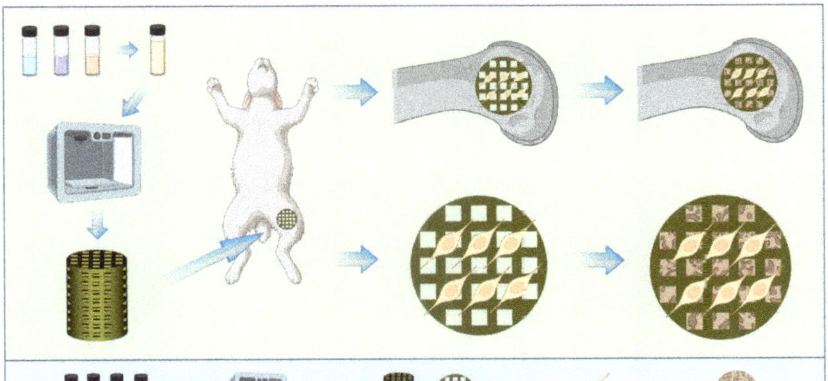

Figure 1. Schematic of the GelMA/PEGDA/F127DA scaffold for bone regeneration.

2. Materials and Methods

2.1. Materials

Gelatin from porcine skin was purchased from Sigma-Aldrich (St. Louis, MO, USA). Methacrylic anhydride (MA, 97%) and lithium phenyl-2,4,6-trimethylbenzoyl phosphinate (LAP) were purchased from J&K (Beijing, China). PEGDA and F127DA were purchased from Engineering for Life (EFL) (Suzhou, China). Cell counting kit-8 (CCK-8), a live/dead viability assay kit, phalloidin, Alizarin Red S, and an alkaline phosphatase (ALP) assay kit were purchased from Beyotime (Shanghai, China).

2.2. Preparation of Bioinks

The preparation of GelMA was carried out as described previously [36]. In brief, gelatin was dissolved in phosphate buffered saline (PBS) at 40 °C to prepare a 10% gelatin solution. After adding methacrylic anhydride (MA) dropwise into the gelatin solution, the solution was stirred with magnetic force for three hours at 40 °C and 300 rpm; then, a white porous foam was prepared after dialysis against distilled water for 5 days at 40 °C and lyophilized.

GelMA (5% (w/v)), PEGDA (10% (w/v)), and F127DA (5% (w/v)) were dissolved in PBS supplemented with lithium phenyl (2,4,6-trimethylbenzoyl) phosphinate (LAP, 0.25% (w/v)) and tartrazine (0.05% (w/v)). Then, a bioink composed of 5% GelMA/10% PEGDA/5% F127DA (GPF) was prepared. The preparation method of another bioink containing 10% PEGDA/5% F127DA (PF) was the same.

2.3. Fabrication of Scaffolds

The CAD model was designed as a cylinder with interconnected pores, with a diameter of 6 mm, height of 8 mm, and pore size of 600 μm. Then, a DLP printer (BP8601 Pro, EFL, Suzhou, China) was used to prepare the scaffolds and the parameters were adjusted for printing. Then, the scaffolds were strengthened under ultraviolet light for 3 min (kernel parameters: layer height, 100 μm; light intensity, 20 mW/cm^2; exposure time, 4 s; temperature, 29 °C).

2.4. Characterization

2.4.1. Microstructure of the 3D-Printed Scaffolds

The scaffolds were observed using a scanning electron microscope (SEM, SU8100, HITACHI, Hitachi, Japan) after lyophilization (K850, Quorum, East Sussex, UK) and gold/palladium sputter-coating (MC1000, HITACHI, Hitachi, Japan). The pore size of the printed scaffolds was calculated using the ImageJ software (V1.8.0, NIH, Bethesda, MD, USA); three images were selected for each sample, and five pores were measured for each image.

2.4.2. Compressive Tests

A compression test was performed using a universal tensile machine (3365, Instron, Boston, MA, USA) at room temperature. The compression modulus was defined as the initial slope of the linear region of the stress–strain curve. The mechanical indexes (compressive stress and modulus) were acquired according to the software (n = 3).

2.4.3. Swelling

The different scaffolds were placed into PBS and soaked for 24 h at 37 °C, and their weights (Ws) were measured after sufficient swelling. Then, the scaffolds were freeze-dried to obtain their dry weight (Wd). The swelling ratio was calculated as

$$\text{Swelling ratio} = \frac{\text{Ws} - \text{Wd}}{\text{Wd}} \quad (1)$$

2.4.4. Degradation

The different scaffolds were lyophilized, and their weights (W0) were measured. Then, the lyophilized scaffolds were placed in PBS solution and soaked at 37 °C. The PBS was changed every two days and the samples were removed on the 3rd, 6th, 9th, 12th, 15th, 20th, 25th, 30th, 40th, and 50th day. After rinsing twice with deionized water, the samples were lyophilized and weighed (Wt). The remaining weight was calculated as

$$\text{Remaining weight } (\%) = \frac{\text{Wt}}{\text{W0}} \times 100\% \quad (2)$$

2.5. Cell Culture

The mouse embryo osteoblast precursor cells (MC3T3-E1 subclone 14) and rabbit bone marrow mesenchymal stem cells (rBMSCs) used in this experiment were obtained from the Orthopedic Laboratory of the PLA General Hospital. The original generation of cells was expanded to the 3rd generation with medium containing α-MEM, fetal bovine serum (FBS, 10%) and penicillin–streptomycin (1%) for experiments. Both types of cells were cultured in this medium—refreshed every 2 days—in a 37 °C and 5% CO_2 environment.

2.6. Cell Viability

2.6.1. Extracts of Different Scaffolds

The scaffolds were soaked in medium for 48 h at 37 °C. For different scaffolds, 100% extracts were prepared according to the standard of 1.25 cm^2/mL and diluted to different concentrations of 75%, 50%, and 25%.

2.6.2. CCK-8

MC3T3-E1 cells were cultured in extracts with different concentrations of different scaffolds for 1–5 days. After the color deepened for 1 h with the addition of CCK-8 (10%), the cell viability was analyzed using a microplate reader (Thermo Fisher, Waltham, MA, USA).

2.6.3. Live/Dead Staining of Cells Cultured with Extracts

MC3T3-E1 cells were cultured for 48 h in the extracts at the optimal concentration obtained via the CCK-8 assay. The cells were incubated with live/dead dye for 15 min, and then observed under a fluorescence microscope (Ni-U, Nikon, Tokyo, Japan), where green represented living cells and red represented dead cells.

2.6.4. Phalloidin Staining of Cells Cultured with Extracts

MC3T3-E1 cells cultured for 48 h in the extracts were fixed in 4% paraformaldehyde solution for 30 min. After three washes with PBS, the cells were stained with phalloidin for 30 min and DAPI for 5 min. The morphology of the cytoskeleton was observed with a confocal microscope (FV3000, Olympus, Tokyo, Japan).

2.6.5. Live/Dead Staining of Cells Cultured on Scaffolds

Sterile scaffolds were soaked in medium for 15 min in 24-well plates; following their removal from the medium, 1 mL of cell suspension (5×10^4 cells) was added. MC3T3-E1 cells were seeded on different scaffolds and cultured in medium for 48 h. The cells were incubated with live/dead dye for 15 min, and then observed under a confocal microscope (Olympus, FV3000, Tokyo, Japan), where green represented living cells and red represented dead cells.

2.7. Effect of the Scaffold on Osteogenic Differentiation In Vitro

2.7.1. Extracts of Different Scaffolds

To distinguish the promotive effect of the scaffolds on osteogenesis under osteoinductive conditions (OIC) and non-osteoinductive conditions (non-OIC), osteoinductive extracts were prepared in a similar way to the normal extracts.

rBMSCs cultured for 5 (non-OIC) and 7 (OIC) days were fixed in 4% paraformaldehyde solution for 30 min. After three washes with PBS, they were stained with an ALP assay kit for one hour and observed with a stereomicroscope (SMZ25, Nikon, Tokyo, Japan).

2.7.2. Alizarin Red S

rBMSCs cultured for 5 (non-OIC) and 21 (OIC) days were fixed in 4% paraformaldehyde solution for 30 min. After three washes with PBS, they were stained with Alizarin Red S for ten minutes and observed with a stereomicroscope (SMZ25, Nikon, Tokyo, Japan).

2.7.3. Quantitative Real-Time PCR

The scaffolds were soaked in medium for half an hour and then removed. A total of 10^5 cells were seeded on the surface of the scaffolds and cultured for 24 h in normal medium, which was then replaced with osteoinductive medium (50 μg/mL ascorbic acid, 10 mM β-phosphoglycerol, and 10 nM dexamethasone) or osteoinductive extracts. After 5 days of culture, total RNA was extracted from the cells for real-time PCR using Trizol reagent (G3013, Servicebio, Wuhan, China); each sample was repeated three times.

2.8. Effect of the Scaffold on Bone Regeneration In Vivo

2.8.1. Ethics Statement

All animals used in this study were obtained from the Animal Experiment Center of the PLA General Hospital and approved by the Ethics Committee (2022-x18-51).

2.8.2. Implantation in Rabbit Femoral Condyle Defects

New Zealand white rabbits (2.5 kg ± 0.5 kg, male, 3 in each group) were used in this experiment. Briefly, the rabbits were anesthetized, and the distal femur was shaved and disinfected. After cutting the skin and subcutaneous tissue, a cylindrical defect 6 mm in diameter was created, without penetrating the contralateral cortex in the distal femur, using a surgical drill. The sterile scaffolds were inserted into the defect site and the subcutaneous tissue and skin were sutured layer by layer. The rabbits were sacrificed at week 4 and week 12 postoperatively for the next step of treatment.

2.8.3. Micro-CT Analysis

The Inveron MM System (Siemens, Munich, Germany) was used to evaluate the amount of new bone in each group of rabbits via micro-CT scans. The scanning parameters were an effective pixel size of 17.34 μm, a current of 500 μA, a voltage of 80 kV, and an exposure time of 1500 ms. The 2D images were reconstructed into 3D images using Inveron Research Workplace (Siemens) to calculate the bone regeneration parameters: BMD, bone volume/total volume (BV/TV), trabecular thickness (Tb.Th), and trabecular spacing (Tb.SP).

2.8.4. Histology Analysis

Samples were decalcified in 10% EDTA, dehydrated in a stepped concentration of ethanol, and cleared using xylene. The samples were then embedded in paraffin and cut into 10 mm slices using a microtome for staining.

2.9. Statistical Analysis

The results between two groups were analyzed using a paired t test. The results among three groups were analyzed using a one-way analysis of variance (ANOVA) with a Tukey–Kramer multiple comparison analysis using the GraphPad Prism software (version 8, GraphPad, San Diego, CA, USA). The data are expressed as the mean ± standard deviation (SD) and all experiments were performed at least three times. A value of $p < 0.05$ was regarded as statistically significant (* $p < 0.05$, ** $p < 0.01$, *** $p < 0.001$).

3. Results

3.1 Characterization

Two scaffolds with different compositions, 10% PEGDA/5% F127DA (PF) and 5% GelMA/10% PEGDA/5% F127DA (GPF), were successfully prepared. As predicted, the printed scaffolds possessed a favorable porous structure, with a pore size of 508.13 ± 21.28 μm for the GPF scaffold, which was lower than the 564.04 ± 17.56 μm found for the PF scaffold and 600 μm for the CAD (Figure 2A–E). However, as shown in Figure 2H–J, the addition of GelMA did not significantly change the compressive strength of the scaffolds, but the modulus decreased from 127.4 ± 12 kPa to 92.34 ± 6.80 kPa as the scaffold became more elastic. In addition, when the scaffolds were soaked in PBS for 24 h to reach swelling equilibrium, the swelling ratios of PF and GPF were 3.35 ± 0.88 and 3.90 ± 0 62, respectively. The remaining weight of GPF at day 50 was 57.81 ± 3.64%, which may be more suitable for bone regeneration than the weight of 67.10 ± 4.30% observed for PF (Figure 2F,G).

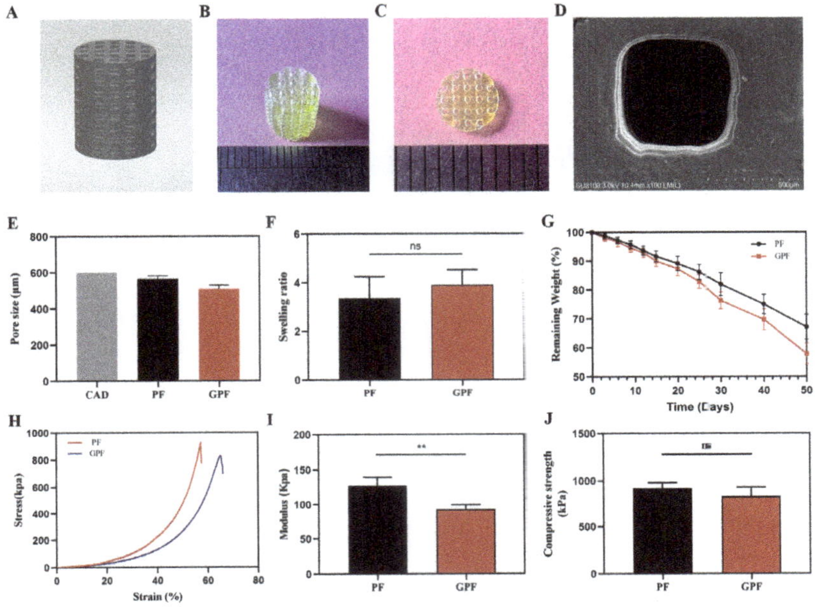

Figure 2. Characterization of scaffolds. (**A**) Image of 3D modeling. (**B**,**C**) Images of the GPF scaffold. (**D**) SEM image of the GPF scaffold. (**E**) Pore size of the CAD and scaffolds. (**F**) Swelling ratio of different scaffolds. (**G**) In vitro degradation behavior of the two scaffolds in PBS (37 °C, pH = 7.4). (**H**) Compressive stress–strain curves of the scaffolds. (**I**) Compressive modulus of the scaffolds. (**J**) Compressive strength of the scaffolds. Data were analyzed using a paired t test and are shown as the mean ± standard deviation (** $p < 0.01$, n = 3).

3.2. Biocompatibility

Good biocompatibility is the basis for the clinical application of bone tissue engineering. To investigate the cytocompatibility of PF and GPF, the extracts were prepared at different concentrations (100%, 75%, 50%, and 25%) for cell culture according to the standards of extract preparation. As shown in Figure 3A, there was no obvious effect of either scaffold on cell proliferation when the extract concentrations were 100%,75%, or 50%, while GPF significantly improved cell proliferation after day 4 compared to PF when the extract concentration was 25%. Moreover, cells were cultured in the 25% extract for 48 h and stained with a live/dead viability assay kit, and phalloidin, and similar cell numbers and morphologies were observed for PF, GPF, and normal medium (Figure 3B,C and Figure 4).

For further observation of the growth condition of the cells on the scaffold surface, MC3T3-E1 cells were inoculated on the scaffolds and cultured for 48 h. After staining with a live/dead viability assay kit, it was found that the number of cells on the surface of the PF scaffolds was low and most were dead, and the cells exhibited a spherical shape. In contrast, cells on the surface of the GPF scaffolds were observed to be well proliferated, with a low number of dead cells; in addition, MC3T3-E1 cells could extend their tentacles on the surface of the GPF scaffolds (Figure 5).

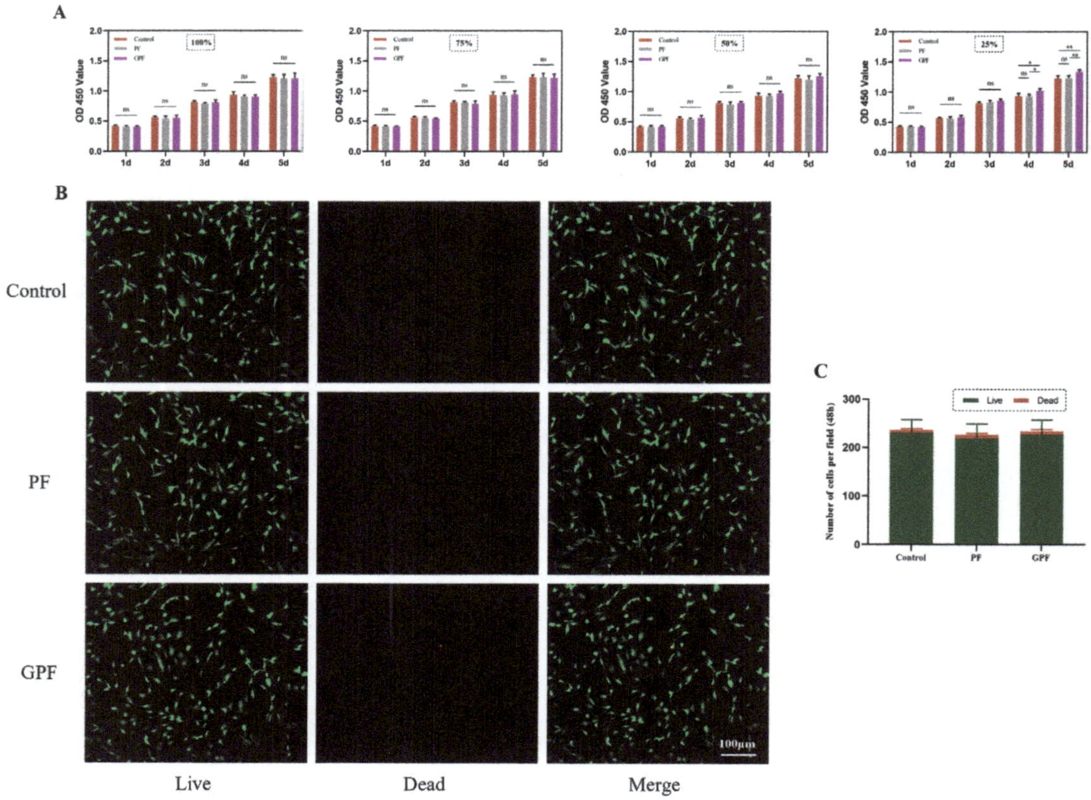

Figure 3. Biocompatibility of scaffolds as extracts in vitro. (**A**) CCK-8 assay showing the proliferation of MC3T3-E1 cells co-cultured with different extracts of different scaffolds for 1–5 days. (**B**) Live/dead assay of MC3T3-E1 cells co-cultured with 25% extract for 2 days. Green represents living cells and red represents dead cells. (**C**) Number of cells in the live/dead assay. Data were analyzed via a one-way ANOVA and are shown as the mean ± standard deviation (* $p < 0.05$, ** $p < 0.01$, n = 3).

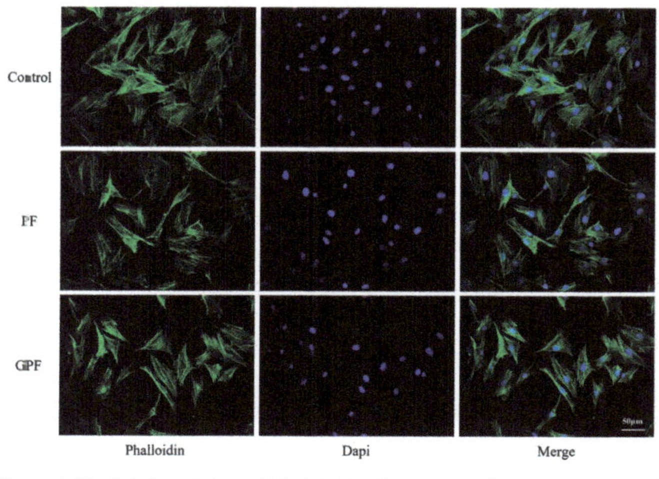

Figure 4. Phalloidin staining of MC3T3-E1 cells cocultured with 25% extract for 2 days.

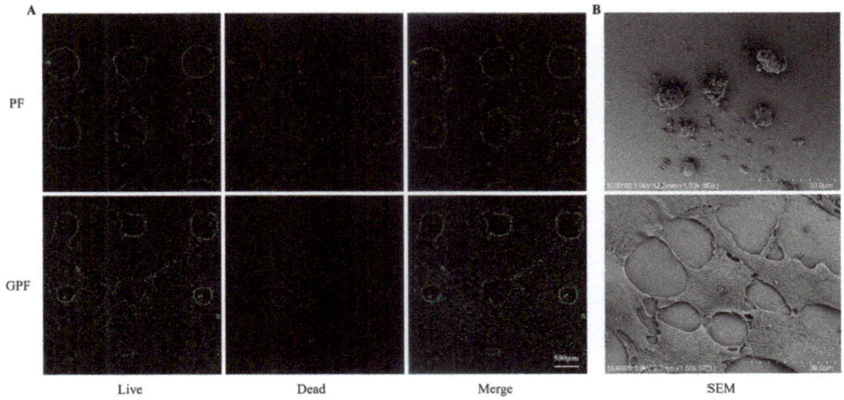

Figure 5. Biocompatibility and SEM images of scaffolds in vitro. (**A**) Live/dead assay of MC3T3-E1 cells cultured on scaffolds for 2 days. Green represents living cells and red represents dead cells. (**B**) SEM images of cells cultured on the surface of the scaffolds for 2 days.

3.3. Capacity for Osteogenic Differentiation In Vitro

To distinguish the osteogenic promotion of the scaffolds under OIC and non-OIC, we performed an in vitro validation of osteogenic differentiation in each of the two conditions.

The rBMSCs were stained with an ALP assay kit and Alizarin Red S after 5 days of culture in normal extracts. The osteogenic differentiation of rBMSCs was not promoted by either the PF or GPF scaffolds (Figure 6A,B), and the expression of osteogenic genes such as Col-1, OPN, OCN, and Runx2 was similar in each group after seeding the cells on the scaffolds for 5 days (Figure 6C).

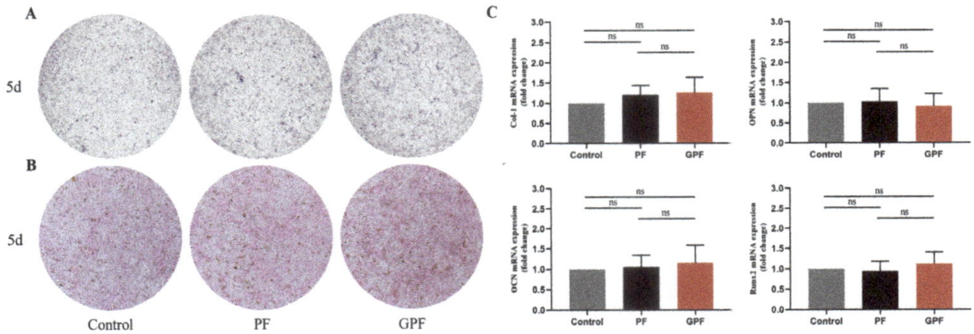

Figure 6. Effect of scaffolds on osteogenic differentiation under non-osteoinductive conditions. (**A**) ALP staining after 5 days of culture in normal extracts. (**B**) Alizarin Red S staining after 5 days of culture in normal extracts. (**C**) Expression of osteogenic-related genes determined using quantitative real-time PCR. Data were analyzed via a one-way ANOVA and are shown as the mean ± standard deviation (n = 3).

However, when rBMSCs were cultured in osteogenic extracts, a clear difference was observed between ALP staining on day 7 and Alizarin Red S staining on day 14, where GPF possessed a greater ability to promote osteogenic differentiation than PF (Figure 7A,B). Similarly, the expression of osteogenic-related genes was higher in cells seeded on GPF rather than PF scaffolds (Figure 7C).

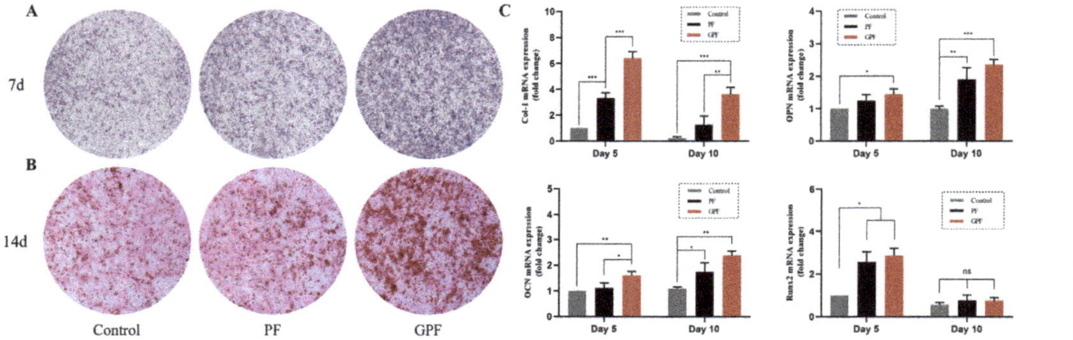

Figure 7. Effect of scaffolds on osteogenic differentiation under osteoinductive conditions. (**A**) ALP staining after 7 days of culture in osteoinductive extracts. (**B**) Alizarin Red S staining after 14 days of culture in osteoinductive extracts. (**C**) Expression of osteogenic-related genes determined using quantitative real-time PCR. Data were analyzed via a one-way ANOVA and are shown as the mean ± standard deviation (* $p < 0.05$, ** $p < 0.01$, *** $p < 0.001$, n = 3).

3.4. Bone Regeneration In Vivo

To explore the effect of the different scaffolds on the treatment of bone defects, PF and GPF scaffolds were implanted at the distal femoral defect site and a micro-CT was performed at week 4 and week 12 postoperatively. The results suggested that without intervention, there was only a small amount of new bone at the defect site at week 12. However, greater regeneration of bone tissue was observed with both the PF and GPF scaffolds, mostly from cancellous bone toward cortical bone, and the new bone took on a scaffold-like meshed shape. At week 12, the new bone latticed off, which could be related to degradation inside the scaffold (Figure 8A). Similar results to the CT images can be observed in Figure 8B, with the GPF scaffold achieving better efficacy in BMD, BV/TV, Th. Tb, and Th. Sp.

Subsequently, a histological analysis of the samples was performed at week 12. It was observed from HE and Masson staining that a gap existed between the PF scaffolds and the new bone organization, which was consistent with the in vitro study where the PF scaffolds lacked cell and tissue adhesion. In contrast, the new bone could adhere to the surface of the GPF scaffold and grow into the pores, exhibiting good osteoconductivity (Figure 9).

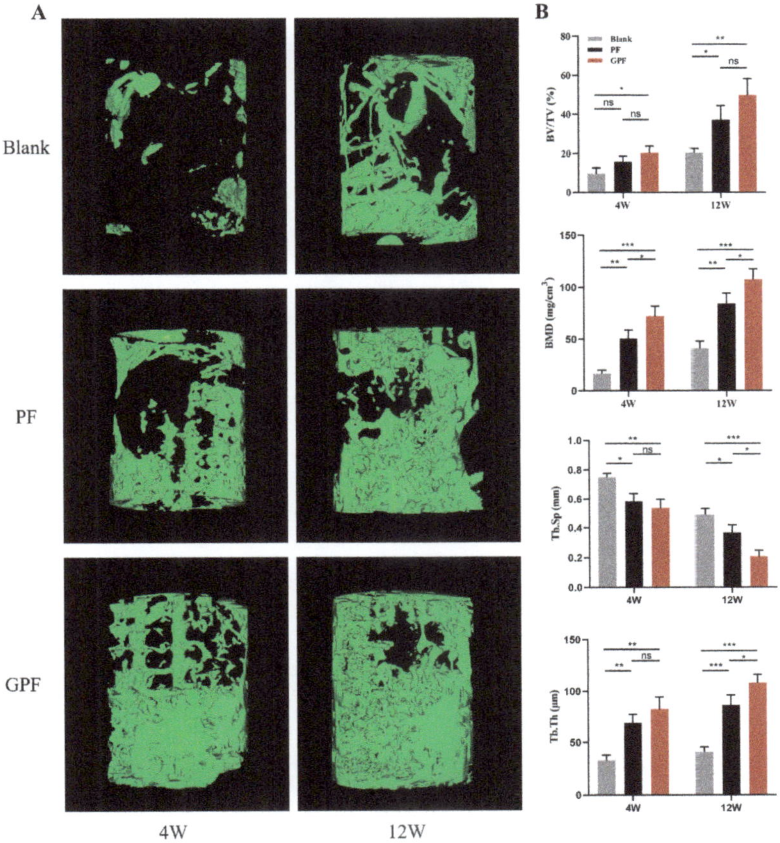

Figure 8. Micro-CT analysis of new bone formation at 4 weeks and 12 weeks. (**A**) Reconstructed 3D patterns from micro-CT images of femur defects at 4 weeks and 12 weeks. (**B**) Micro-architectural parameters of the newly formed bone. BMD-bone mineral density; BV/TV-bone tissue volume/total tissue volume; Tb.Th-trabecular thickness; Tb.Sp-trabecular separation. Data were analyzed via a one-way ANOVA and are shown as the mean ± standard deviation (* $p < 0.05$, ** $p < 0.01$, *** $p < 0.001$, n = 3).

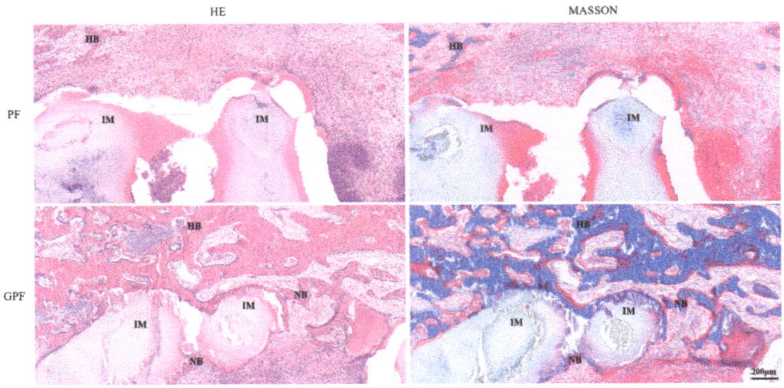

Figure 9. Histological evaluation of newly formed bone at 12 weeks. NB-new bone; HB-host bone; IM-implanted materials.

4. Discussion

DLP printing technology is widely used in bone tissue engineering owing to its good printing accuracy and gentle printing process, which requires bioink with photosensitive properties [37–39]. At this stage, the materials used mainly consist of natural materials (GelMA, HAMA, etc.), synthetic materials (PEGDA, F127DA, PPF, etc.), and inorganic materials (TCP, HA, metals, etc.). Among them, natural and synthetic materials can be chemically modified to endow them with photosensitive properties for direct DLP printing [40,41], while inorganic materials need to be mixed into a photosensitive resin for printing and sintered (1200 °C) to remove organics [42]. Greeshma et al., prepared GelMA-based bioink from autologous bone particles (BPs) and determined the appropriate printing parameters, revealing that 3D-printed GelMA/BP-based composite scaffolds could effectively promote bone regeneration by improving the proliferation, migration, and osteogenic differentiation capacity of cells [16]. Zhang et al., prepared Haversian bone-mimicking scaffolds using the DLP printing technique with bioceramics; the scaffolds were found to induce osteogenesis, angiogenesis, and neural differentiation in vitro and accelerate the growth of blood vessels and new bone formation in vivo [43].

Natural materials with good bioactivity have some disadvantages such as poor mechanical properties and rapid degradation [44]. Inorganic ceramic and some metallic materials (e.g., titanium, steel, etc.) with strong mechanical properties possess slow degradation rates, and are generally brittle [45]. In addition, some metals with weak rigidity, such as magnesium, can promote vascular and bone regeneration; however, some studies have demonstrated that the dynamic in vivo environment can lead to the accelerated fatigue of magnesium materials, making the magnesium scaffold lose its abilities as described earlier [44]. In addition, high-temperature sintering during the preparation of scaffolds using inorganic materials can lead to the inactivation of active substances in bioink; these disadvantages limit the application of these materials in BTE. Therefore, synthetic polymers need to be further explored for use in DLP-printed bone tissue engineering. PEGDA (brittle material) and F127DA (elastic material) are the most common synthetic polymers used for modification, possessing photo-crosslinking and complementary mechanical properties. Shen et al., prepared a tissue adhesive with good histocompatibility using PEGDA and F127DA; the adhesive was expected to repair wounded tissues without suturing [45]. However, neither material has been proposed for the preparation of BTE scaffolds via DLP.

In this study, PEGDA and F127DA were added to bioink and showed satisfactory printing and mechanical properties, while no obvious cytotoxicity of the printed scaffolds was observed. However, cells exhibited difficulty in adhering after seeding on the surface of the PF scaffold, which may be a hindrance for its application in BTE. Therefore, the addition of materials with stronger adhesion properties is needed to increase the bioactivity of synthetic polymers. Wang et al., prepared an injectable hydrogel by adding GelMA to a PEGDA-based bioink. Although the compressive strength of the scaffold (approximately 300 kPa) still needs to be improved and its injectable properties imply the abandonment of the macroscopic porous structure, the addition of GelMA improved the bioadhesion of the bioink [25,46]. Therefore, GelMA, with its good biocompatibility and adhesion support, was added to the PEGDA/F127DA bioink in this study to promote cell adhesion and proliferation. It can be clearly observed in Figure 5 that cells on the surface of the GPF scaffolds can extend their tentacles for better adhesion and proliferation than the spherical cells on the surface of the PF scaffolds. In addition, while retaining the good mechanical properties of the PF scaffold, the GPF scaffold had a faster degradation rate, which could help new bone to better replace the material, facilitating the regeneration of bone tissue.

In addition to the choice of material, pore size is also crucial for cell proliferation and differentiation. From the literature, pore sizes larger than 300 μm show better vascularization and osseointegration in BTE [47,48]. Zhang et al., showed that the optimal pore size for osteogenic capacity is approximately 600–700 μm, and pore sizes that are too small or too large affect cell behavior and bone regeneration [49]. Chen et al., demonstrated that a pore size of 500 μm showed the best cell proliferation and differentiation and inward bone growth [50]. Luo et al., concluded that porous scaffolds with a pore size of 400–600 μm better promote osteogenesis and osseointegration [51]. Although some scholars believe that a small pore size (188 μm) is more favorable for the osteogenic differentiation of cells in vitro [52], more studies have demonstrated that 400–700 μm is a good choice for the preparation of bone tissue engineering scaffolds [49–51,53]. In this study, the standard pore size of the scaffold was 600 μm, while the printed PF scaffold had a pore size of approximately 564 μm and the pore size of the GPF scaffold was approximately 508 μm. The addition of GelMA increased the bioadhesion of the scaffold but reduced the printability of the bioink; nevertheless, both scaffolds had a good pore size structure and exhibited satisfactory bone regeneration.

The conditions under which GPF promotes bone regeneration were also investigated in this study. There are various ways for materials to enhance bone regeneration, one of which is to induce the osteogenic differentiation of rBMSCs when there is no exogenous induction; another is to use a scaffold to accelerate the osteogenic differentiation of rBMSCs when an inductive environment exists. The GPF scaffolds in this study represent the second option, which can be observed from the in vitro experiments. When rBMSCs were cultured in normal medium, the GPF scaffolds and PF scaffolds did not exhibit osteogenesis-promoting effects, whereas when cultured in osteogenesis-inducing medium, the effect of the GPF scaffolds in promoting osteogenesis was significantly greater than that of the PF scaffolds. Combined with the 3D images of the defect site from the CT reconstruction, these results indicated that GPF can play a facilitating role in promoting bone regeneration when an osteogenesis-inducing environment exists in vivo. In addition, the lattice-like new bone showed that the bone was growing and crawling along the pore structure. At 12 weeks, the reduction of the lattice-like structure indicated the degradation of the scaffold, leading to the loss of its original aperture which was replaced by new bone tissue. Our histological observations, shown in Figure 9, were consistent with the in vitro structure. A lack of adhesion of the PF scaffold to the cells and tissues resulted in a significant gap between them in vivo. In contrast, the tightly adherent growth of bone tissue could be observed around the GPF scaffold containing GelMA, indicating that the GPF scaffold could guide the adhesion and regeneration of new bone.

Our results clearly indicate that GPF possesses a satisfactory porous structure (508.13 ± 21.28 μm) and mechanical properties (829.59 ± 89.21 kPa) to promote osteogenic differentiation under osteoinductive conditions and guide bone growth in vivo. However, the following limitations may exist. First, the accuracy of the universal tensile machine is 0.5%, which means the mechanical results may have an error of 0.5%. Second, the differential effects of GPF scaffolds on osteogenic differentiation in diverse environments still require further investigation. Third, the establishment of a 3D finite element model to simulate 3D physiological loading has made significant progress in the design of implants [54], which has given us great insight to predict the state of BTE scaffolds in vivo through computational simulations for developing an optimal structure.

5. Conclusions

GPF scaffolds prepared using DLP printing technology not only possess satisfactory mechanical properties, but also an appropriate degradation rate that is more compatible with the time course of bone regeneration. By improving the disadvantages of traditional synthetic polymers that are not conducive to cell adhesion, these scaffolds exhibit excellent histocompatibility, guiding the new bone tissue to grow inside the defect when implanted in vivo. Furthermore, GPF scaffolds can effectively promote the osteogenic differentiation of rBMSCs in an osteoinductive but not a non-osteoinductive environment. The reasons for such different results will be the focus of further research. In summary, GPF-based composite scaffolds prepared using DLP printing technology provide a new approach to the clinical treatment of bone defects.

Author Contributions: Conceptualization, J.G. and P.T.; methodology, J.G. and J.L.; software, M.L.; validation, J.L., M.L. and Z.L.; formal analysis, J.C.; investigation, X.L.; resources, P.T.; data curation, J.G.; writing—original draft preparation, J.G.; writing—review and editing, P.T.; visualization, P.T.; supervision, J.L.; project administration, P.T.; funding acquisition, J.L. and Z.L. All authors have read and agreed to the published version of the manuscript.

Funding: This research was funded by the Key Program of the National Natural Science Foundation of China (grant number 21935011), the Beijing Municipal Natural Science Foundation (grant number L202033 and L212050), and the Military Medical Science and Technology Youth Training Program (grant number 19QNP052).

Institutional Review Board Statement: The Institutional Animal Care and Use Committee of the Chinese PLA General Hospital approved the design of this study (2022-x18-51).

Informed Consent Statement: Not applicable.

Data Availability Statement: The data supporting the reported results can be provided by the authors on request.

Conflicts of Interest: The authors declare no conflict of interest.

References

1. Gillman, C.; Jayasuriya, A. FDA-approved bone grafts and bone graft substitute devices in bone regeneration. *Mater. Sci. Engineering. C Mater. Biol. Appl.* **2021**, *130*, 112466. [CrossRef] [PubMed]
2. Bose, S.; Sarkar, N. Natural Medicinal Compounds in Bone Tissue Engineering. *Trends Biotechnol.* **2020**, *38*, 404–417. [CrossRef] [PubMed]
3. Ding, Z.; Cheng, W.; Mia, M.; Lu, Q. Silk Biomaterials for Bone Tissue Engineering. *Macromol. Biosci.* **2021**, *21*, e2100153. [CrossRef] [PubMed]
4. Wan, Z.; Zhang, P.; Liu, Y.; Lv, L.; Zhou, Y. Four-dimensional bioprinting: Current developments and applications in bone tissue engineering. *Acta Biomater.* **2020**, *101*, 26–42. [CrossRef]
5. Kačarević, Ž.P.; Rider, P.; Alkildani, S.; Retnasingh, S.; Pejakić, M.; Schnettler, R.; Gosau, M.; Smeets, R.; Jung, O.; Barbeck, M. An introduction to bone tissue engineering. *Int. J. Artif. Organs* **2020**, *43*, 69–86. [CrossRef] [PubMed]
6. Peng, Z.; Zhao, T.; Zhou, Y.; Li, S.; Li, J.; Leblanc, R. Bone Tissue Engineering via Carbon-Based Nanomaterials. *Adv. Healthc. Mater.* **2020**, *9*, e1901495. [CrossRef]
7. Feng, Y.; Zhu, S.; Mei, D.; Li, J.; Zhang, J.; Yang, S.; Guan, S. Application of 3D Printing Technology in Bone Tissue Engineering: A Review. *Curr. Drug Deliv.* **2021**, *18*, 847–861. [CrossRef]
8. Matai, I.; Kaur, G.; Seyedsalehi, A.; McClinton, A.; Laurencin, C. Progress in 3D bioprinting technology for tissue/organ regenerative engineering. *Biomaterials* **2020**, *226*, 119536. [CrossRef]
9. Rahimnejad, M.; Rezvaninejad, R.; Rezvaninejad, R.; França, R. Biomaterials in bone and mineralized tissue engineering using 3D printing and bioprinting technologies. *Biomed. Phys. Eng. Express* **2021**, *7*, 062001. [CrossRef]
10. Gong, J.; Qian, Y.; Lu, K.; Zhu, Z.; Siow, L.; Zhang, C.; Zhou, S.; Gu, T.; Yin, J.; Yu, M.; et al. Digital light processing (DLP) in tissue engineering: From promise to reality, and perspectives. *Biomed. Mater.* **2022**, *17*, 062004. [CrossRef]
11. Etemad-Shahidi, Y.; Qallandar, O.B.; Evenden, J.; Alifui-Segbaya, F.; Ahmed, K.E. Accuracy of 3-Dimensionally Printed Full-Arch Dental Models: A Systematic Review. *J. Clin. Med.* **2020**, *9*, 3357. [CrossRef]
12. Zhang, J.; Hu, Q.; Wang, S.; Tao, J.; Gou, M. Digital Light Processing Based Three-dimensional Printing for Medical Applications. *Int. J. Bioprint.* **2020**, *6*, 242. [CrossRef] [PubMed]
13. Hosseinabadi, H.G.; Dogan, E.; Miri, A.; Ionov, L. Digital Light Processing Bioprinting Advances for Microtissue Models. *ACS Biomater. Sci. Eng.* **2022**, *8*, 1381–1395. [CrossRef]
14. Wu, Y.; Su, H.; Li, M.; Xing, H. Digital light processing-based multi-material bioprinting: Processes, applications, and perspectives. *J. Biomed. Mater. Res. Part A* **2022**, *111*, 527–542. [CrossRef] [PubMed]
15. Gao, Q.; Niu, X.; Shao, L.; Zhou, L.; Lin, Z.; Sun, A.; Fu, J.; Chen, Z.; Hu, J.; Liu, Y.; et al. 3D printing of complex GelMA-based scaffolds with nanoclay. *Biofabrication* **2019**, *11*, 035006. [CrossRef] [PubMed]
16. Ratheesh, G.; Vaquette, C.; Xiao, Y. Patient-Specific Bone Particles Bioprinting for Bone Tissue Engineering. *Adv. Healthc. Mater.* **2020**, *9*, e2001323. [CrossRef]
17. Gaharwar, A.K.; Dammu, S.A.; Canter, J.M.; Wu, C.J.; Schmidt, G. Highly extensible, tough, and elastomeric nanocomposite hydrogels from poly(ethylene glycol) and hydroxyapatite nanoparticles. *Biomacromolecules* **2011**, *12*, 1641–1650. [CrossRef]
18. Kim, H.D.; Lee, E.A.; An, Y.H.; Kim, S.L.; Lee, S.S.; Yu, S.J.; Jang, H.L.; Nam, K.T.; Im, S.G.; Hwang, N.S. Chondroitin Sulfate-Based Biomineralizing Surface Hydrogels for Bone Tissue Engineering. *ACS Appl. Mater. Interfaces* **2017**, *9*, 21639–21650. [CrossRef]
19. Wang, Y.; Miao, Y.; Zhang, J.; Wu, J.P.; Kirk, T.B.; Xu, J.; Ma, D.; Xue, W. Three-dimensional printing of shape memory hydrogels with internal structure for drug delivery. *Mater. Sci. Eng. C Mater. Biol. Appl.* **2018**, *84*, 44–51. [CrossRef]

20. Li, H.; Ma, T.; Zhang, M.; Zhu, J.; Liu, J.; Tan, F. Fabrication of sulphonated poly(ethylene glycol)-diacrylate hydrogel as a bone grafting scaffold. *J. Mater. Science. Mater. Med.* **2018**, *29*, 187. [CrossRef]
21. Escudero-Castellanos, A.; Ocampo-García, B.E.; Domínguez-García, M.V.; Flores-Estrada, J.; Flores-Merino, M.V. Hydrogels based on poly(ethylene glycol) as scaffolds for tissue engineering application: Biocompatibility assessment and effect of the sterilization process. *J. Mater. Science. Mater. Med.* **2016**, *27*, 176. [CrossRef] [PubMed]
22. Spotnitz, W.D.; Burks, S. Hemostats, sealants, and adhesives: Components of the surgical toolbox. *Transfusion* **2008**, *48*, 1502–1516. [CrossRef] [PubMed]
23. Sousa, A.C.; Biscaia, S.; Alvites, R.; Branquinho, M.; Lopes, B.; Sousa, P.; Valente, J.; Franco, M.; Santos, J.D.; Mendonça, C.; et al. Assessment of 3D-Printed Polycaprolactone, Hydroxyapatite Nanoparticles and Diacrylate Poly(ethylene glycol) Scaffolds for Bone Regeneration. *Pharmaceutics* **2022**, *14*, 2643. [CrossRef]
24. Mondal, D.; Haghpanah, Z.; Huxman, C.; Tanter, S.; Sun, D.; Gorbet, M.; Willett, T.L. mSLA-based 3D printing of acrylated epoxidized soybean oil—Nano-hydroxyapatite composites for bone repair. *Mater. Sci. Eng. C Mater. Biol. Appl.* **2021**, *130*, 112456. [CrossRef] [PubMed]
25. Wang, Y.; Cao, X.; Ma, M.; Lu, W.; Zhang, B.; Guo, Y. A GelMA-PEGDA-nHA Composite Hydrogel for Bone Tissue Engineering. *Materials* **2020**, *13*, 3735. [CrossRef] [PubMed]
26. Zhai, X.; Ruan, C.; Ma, Y.; Cheng, D.; Wu, M.; Liu, W.; Zhao, X.; Pan, H.; Lu, W.W. 3D-Bioprinted Osteoblast-Laden Nanocomposite Hydrogel Constructs with Induced Microenvironments Promote Cell Viability, Differentiation, and Osteogenesis both In Vitro and In Vivo. *Adv. Sci.* **2018**, *5*, 1700550. [CrossRef]
27. Luo, Y.; Pan, H.; Jiang, J.; Zhao, C.; Zhang, J.; Chen, P.; Lin, X.; Fan, S. Desktop-Stereolithography 3D Printing of a Polyporous Extracellular Matrix Bioink for Bone Defect Regeneration. *Front. Bioeng. Biotechnol.* **2020**, *8*, 589094. [CrossRef]
28. Shen, C.; Li, Y.; Wang, Y.; Meng, Q. Non-swelling hydrogel-based microfluidic chips. *Lab Chip* **2019**, *19*, 3962–3973. [CrossRef]
29. Li, Y.; Wang, Y.; Shen, C.; Meng, Q. Non-swellable F127-DA hydrogel with concave microwells for formation of uniform-sized vascular spheroids. *RSC Adv.* **2020**, *10*, 44494–44502. [CrossRef]
30. Sun, Y.N.; Gao, G.R.; Du, G.L.; Cheng, Y.J.; Fu, J. Super Tough, Ultrastretchable, and Thermoresponsive Hydrogels with Functionalized Triblock Copolymer Micelles as Macro-Cross-Linkers. *ACS Macro Lett.* **2014**, *3*, 496–500. [CrossRef]
31. Jiang, T.; Zhao, J.; Yu, S.; Mao, Z.; Gao, C.; Zhu, Y.; Mao, C.; Zheng, L. Untangling the response of bone tumor cells and bone forming cells to matrix stiffness and adhesion ligand density by means of hydrogels. *Biomaterials* **2019**, *188*, 130–143. [CrossRef] [PubMed]
32. Wang, Y.; Shang, L.; Chen, G.; Sun, L.; Zhang, X.; Zhao, Y. Bioinspired structural color patch with anisotropic surface adhesion. *Sci. Adv.* **2020**, *6*, eaax8258. [CrossRef] [PubMed]
33. Li, S.; Zhou, H.; Li, Y.; Jin, X.; Liu, H.; Lai, J.; Wu, Y.; Chen, W.; Ma, A. Mussel-inspired self-adhesive hydrogels by conducting free radical polymerization in both aqueous phase and micelle phase and their applications in flexible sensors. *J. Colloid Interface Sci.* **2022**, *607*, 431–439. [CrossRef] [PubMed]
34. Yue, K.; Santiago, G.T.-d.; Alvarez, M.M.; Tamayol, A.; Annabi, N.; Khademhosseini, A. Synthesis, properties, and biomedical applications of gelatin methacryloyl (GelMA) hydrogels. *Biomaterials* **2015**, *73*, 254–271. [CrossRef]
35. Hong, Y.; Zhou, F.; Hua, Y.; Zhang, X.; Ni, C.; Pan, D.; Zhang, Y.; Jiang, D.; Yang, L.; Lin, Q.; et al. A strongly adhesive hemostatic hydrogel for the repair of arterial and heart bleeds. *Nat. Commun.* **2019**, *10*, 2060. [CrossRef] [PubMed]
36. Gao, J.; Ding, X.; Yu, X.; Chen, X.; Zhang, X.; Cui, S.; Shi, J.; Chen, J.; Yu, L.; Chen, S.; et al. Cell-Free Bilayered Porous Scaffolds for Osteochondral Regeneration Fabricated by Continuous 3D-Printing Using Nascent Physical Hydrogel as Ink. *Adv. Healthc. Mater.* **2021**, *10*, e2001404. [CrossRef] [PubMed]
37. Feng, B.; Zhang, M.; Qin, C.; Zhai, D.; Wang, Y.; Zhou, Y.; Chang, J.; Zhu, Y.; Wu, C. 3D printing of conch-like scaffolds for guiding cell migration and directional bone growth. *Bioact. Mater.* **2023**, *22*, 127–140. [CrossRef] [PubMed]
38. Tao, J.; Zhu, S.; Liao, X.; Wang, Y.; Zhou, N.; Li, Z.; Wan, H.; Tang, Y.; Yang, S.; Du, T.; et al. DLP-based bioprinting of void-forming hydrogels for enhanced stem-cell-mediated bone regeneration. *Mater. Today Bio* **2022**, *17*, 100487. [CrossRef]
39. Zhang, B.; Gui, X.; Song, P.; Xu, X.; Guo, L.; Han, Y.; Wang, L.; Zhou, C.; Fan, Y.; Zhang, X. Three-Dimensional Printing of Large-Scale, High-Resolution Bioceramics with Micronano Inner Porosity and Customized Surface Characterization Design for Bone Regeneration. *ACS Appl. Mater. Interfaces* **2022**, *14*, 8804–8815. [CrossRef]
40. Liang, R.; Gu, Y.; Wu, Y.; Bunpetch, V.; Zhang, S. Lithography-Based 3D Bioprinting and Bioinks for Bone Repair and Regeneration. *ACS Biomater. Sci. Eng.* **2021**, *7*, 806–816. [CrossRef]
41. Mei, Q.; Rao, J.; Bei, H.; Liu, Y.; Zhao, X. 3D Bioprinting Photo-Crosslinkable Hydrogels for Bone and Cartilage Repair. *Int. J. Bioprint.* **2021**, *7*, 367. [CrossRef] [PubMed]
42. Zhang, J.; Huang, D.; Liu, S.; Dong, X.; Li, Y.; Zhang, H.; Yang, Z.; Su, Q.; Huang, W.; Zheng, W.; et al. Zirconia toughened hydroxyapatite biocomposite formed by a DLP 3D printing process for potential bone tissue engineering. *Mater. Sci. Engineering. C Mater. Biol. Appl.* **2019**, *105*, 110054. [CrossRef] [PubMed]
43. Zhang, M.; Lin, R.; Wang, X.; Xue, J.; Deng, C.; Feng, C.; Zhuang, H.; Ma, J.; Qin, C.; Wan, L.; et al. 3D printing of Haversian bone-mimicking scaffolds for multicellular delivery in bone regeneration. *Sci. Adv.* **2020**, *6*, eaaz6725. [CrossRef] [PubMed]
44. Putra, R.U.; Basri, H.; Prakoso, A.T.; Chandra, H.; Ammarullah, M.I.; Akbar, I.; Syahrom, A.; Kamarul, T. Level of Activity Changes Increases the Fatigue Life of the Porous Magnesium Scaffold, as Observed in Dynamic Immersion Tests, over Time. *Sustainability* **2023**, *15*, 823. [CrossRef]

45. Shen, C.; Li, Y.; Meng, Q. Adhesive polyethylene glycol-based hydrogel patch for tissue repair. *Colloids Surf. B Biointerfaces* **2022**, *218*, 112751. [CrossRef]
46. Wang, Y.; Ma, M.; Wang, J.; Zhang, W.; Lu, W.; Gao, Y.; Zhang, B.; Guo, Y. Development of a Photo-Crosslinking, Biodegradable GelMA/PEGDA Hydrogel for Guided Bone Regeneration Materials. *Materials* **2018**, *11*, 1345. [CrossRef]
47. Karageorgiou, V.; Kaplan, D. Porosity of 3D biomaterial scaffolds and osteogenesis. *Biomaterials* **2005**, *26*, 5474–5491. [CrossRef]
48. Roosa, S.M.; Kemppainen, J.M.; Moffitt, E.N.; Krebsbach, P.H.; Hollister, S.J. The pore size of polycaprolactone scaffolds has limited influence on bone regeneration in an in vivo model. *J. Biomed. Mater. Res. Part A* **2010**, *92*, 359–368. [CrossRef]
49. Zhang, Y.; Sun, N.; Zhu, M.; Qiu, Q.; Zhao, P.; Zheng, C.; Bai, Q.; Zeng, Q.; Lu, T. The contribution of pore size and porosity of 3D printed porous titanium scaffolds to osteogenesis. *Biomater. Adv.* **2022**, *133*, 112651. [CrossRef]
50. Chen, Z.; Yan, X.; Yin, S.; Liu, L.; Liu, X.; Zhao, G.; Ma, W.; Qi, W.; Ren, Z.; Liao, H.; et al. Fang Influence of the pore size and porosity of selective laser melted Ti6Al4V ELI porous scaffold on cell proliferation, osteogenesis and bone ingrowth. *Mater. Sci. Eng. C Mater. Biol. Appl.* **2020**, *106*, 110289. [CrossRef]
51. Luo, C.; Wang, C.; Wu, X.; Xie, X.; Wang, C.; Zhao, C.; Zou, C.; Lv, F.; Huang, W.; Liao, J. Influence of porous tantalum scaffold pore size on osteogenesis and osteointegration: A comprehensive study based on 3D-printing technology. *Mater. Sci. Eng. C Mater. Biol. Appl.* **2021**, *129*, 112382. [CrossRef] [PubMed]
52. Chang, B.; Song, W.; Han, T.; Yan, J.; Li, F.; Zhao, L.; Kou, H.; Zhang, Y. Influence of pore size of porous titanium fabricated by vacuum diffusion bonding of titanium meshes on cell penetration and bone ingrowth. *Acta Biomater.* **2016**, *33*, 311–321. [CrossRef] [PubMed]
53. Li, W.; Dai, F.; Zhang, S.; Xu, F.; Xu, Z.; Liao, S.; Zeng, L.; Song, L.; Ai, F. Pore Size of 3D-Printed Polycaprolactone/Polyethylene Glycol/Hydroxyapatite Scaffolds Affects Bone Regeneration by Modulating Macrophage Polarization and the Foreign Body Response. *ACS Appl. Mater. Interfaces* **2022**, *14*, 20693–20707. [CrossRef] [PubMed]
54. Jamari, J.; Ammarullah, M.I.; Saad, A.P.M.; Syahrom, A.; Uddin, M.; van der Heide, E.; Basri, H. The Effect of Bottom Profile Dimples on the Femoral Head on Wear in Metal-on-Metal Total Hip Arthroplasty. *J. Funct. Biomater.* **2021**, *12*, 38. [CrossRef] [PubMed]

Journal of *Functional Biomaterials*

MDPI

Article

The Local Release of Teriparatide Incorporated in 45S5 Bioglass Promotes a Beneficial Effect on Osteogenic Cells and Bone Repair in Calvarial Defects in Ovariectomized Rats

Juliani Caroline Ribeiro de Araújo [1,*], Leonardo Alvares Sobral Silva [1], Vinicius Almeida de Barros Lima [1], Tiago Moreira Bastos Campos [2], Paulo Noronha Lisboa Filho [3], Roberta Okamoto [4] and Luana Marotta Reis de Vasconcellos [1]

1 Department of Bioscience and Buccal Diagnose, Institute of Science and Technology, Sao Paulo State University, UNESP, Sao Paulo 12245-700, SP, Brazil
2 Department of Prosthodontics and Periodontology, Bauru School of Dentistry, University of Sao Paulo, Bauru 17012-230, SP, Brazil
3 Department of Physics, São Paulo State University, UNESP, Bauru 17033-360, SP, Brazil
4 Department of Surgery and Integrated Clinic, Araçatuba Dental School, São Paulo State University, UNESP, Araçatuba 16066-840, SP, Brazil
* Correspondence: juliani.ribeiro@unesp.br

Abstract: With the increase in the population's life expectancy, there has also been an increase in the rate of osteoporosis, which has expanded the search for strategies to regenerate bone tissue. The ultrasonic sonochemical technique was chosen for the functionalization of the 45S5 bioglass. The samples after the sonochemical process were divided into (a) functionalized bioglass (BG) and (b) functionalized bioglass with 10% teriparatide (BGT). Isolated mesenchymal cells (hMSC) from femurs of ovariectomized rats were differentiated into osteoblasts and submitted to in vitro tests. Bilateral ovariectomy (OVX) and sham ovariectomy (Sham) surgeries were performed in fifty-five female Wistar rats. After a period of 60 days, critical bone defects of 5.0 mm were created in the calvaria of these animals. For biomechanical evaluation, critical bone defects of 3.0 mm were performed in the tibias of some of these rats. The groups were divided into the clot (control) group, the BG group, and the BGT group. After the sonochemical process, the samples showed modified chemical topographic and morphological characteristics, indicating that the surface was chemically altered by the functionalization of the particles. The cell environment was conducive to cell adhesion and differentiation, and the BG and BGT groups did not show cytotoxicity. In addition, the experimental groups exhibited characteristics of new bone formation with the presence of bone tissue in both periods, with the BGT group and the OVX group statistically differing from the other groups ($p < 0.05$) in both periods. Local treatment with the drug teriparatide in ovariectomized animals promoted positive effects on bone tissue, and longitudinal studies should be carried out to provide additional information on the biological performance of the mutual action between the bioglass and the release of the drug teriparatide.

Keywords: biocompatible materials; bone regeneration; cell differentiation; osteoporosis; teriparatide

Citation: de Araújo, J.C.R.; Sobral Silva, L.A.; de Barros Lima, V.A.; Bastos Campos, T.M.; Lisboa Filho, P.N.; Okamoto, R.; de Vasconcellos, L.M.R. The Local Release of Teriparatide Incorporated in 45S5 Bioglass Promotes a Beneficial Effect on Osteogenic Cells and Bone Repair in Calvarial Defects in Ovariectomized Rats. *J. Funct. Biomater.* **2023**, *14*, 93. https://doi.org/10.3390/jfb14020093

Academic Editor: Marco Tatullo

Received: 6 November 2022
Revised: 24 January 2023
Accepted: 1 February 2023
Published: 9 February 2023

1. Introduction

Osteoporosis, considered a public health problem, is one of the most prevalent post-menopausal diseases in the world and is commonly associated with estrogen deficiency [1]. According to the International Osteoporosis Foundation, this disease is very common in the elderly, and one in three women over 50 will experience a fracture due to bone fragility during their lifetime [2]. This is because estrogen production is attenuated after the menopausal period [3], and this deficiency inhibits, mainly in late phases, the differentiation and maturation of osteoblasts, exerting a direct effect on these cells [4].

In this context, in patients suffering from osteoporosis, bone healing is a negatively affected process [5], making the treatment of bone defects or bone fractures in these compromised organisms challenging [6]. Therefore, due to changes in the normal bone remodeling process, it is sometimes necessary to implant biomaterials that can assist in the surgical treatment of patients with osteoporotic fractures. In these cases, one must consider not only the properties of the biomaterial but also the characteristics and regenerative capacity of the host bone [7].

Currently, autogenous bone is considered the gold standard in the replacement of bone defects [8], but the disadvantages associated with autogenous bone, such as the need for more than one surgical site, limited availability, and the possibility of the post-complications [9–11], have motivated the development of new biomaterials used in bone repair [12,13], particularly bioglass, which has excellent physical–chemical properties and a long history of applications as biomedical fillers, which has stimulated several researchers to test the use of these materials in tissue engineering and regeneration strategies [14–21]. Furthermore, the discovery of bioglass as the first artificial material with a clear ability to form an integrated bond with bones has stimulated the interest of scientists and clinicians for a long time [22].

Previous studies have shown that the formation of a surface layer of apatite hydroxy-carbonate (HCA) after the ionic exchange between glass and body fluid is responsible for the bone binding mechanism of these materials [23,24], which also makes the role of the surface of bioglass important in bioactivity studies [25]. In tissue regeneration strategies, other materials such as metals, bioceramics, and biodegradable polymers can be associated with 45S5 Bioglass® [26] increase the potential for application in tissue engineering [17], making this material attractive to be evaluated as a bone substitute in adverse situations.

The 45S5 Bioglass® has great versatility in the fields of biomedicine, orthopedics, and even dentistry, being used for bone repairs in the fields of periodontics [27], maxillofacial surgery [28], and even for soft tissue repair [29]. Furthermore, the ability of bioglasses to incorporate hydrophilic and hydrophobic groups in their structures resulted in their development in association with therapeutic agents [30–33].

One of the new perspectives in the field of tissue engineering research is the use of biomaterials associated with drugs that promote a joint action to act on bone repair [31,34,35]. Teriparatide (PTH 1-34) is a drug analogous to parathyroid hormone (PTH 1-84), which, when administered systemically, proved to be effective in reducing the osteonecrotic area in the jaws of animals receiving doses of zoledronic acid, a drug from the group of bisphosphonates, and undergoing tooth extraction compared to the group that was not treated with teriparatide [36]. In addition, teriparatide stimulates cortical and trabecular bone formation, as well as increases bone strength and volume [37]; this is because this drug has a different mechanism of action than medications a currently available for the treatment of osteoporosis, with improved bone quantity and quality in osteopenic animals [38–40]. PTH-34 facilitates cell proliferation and differentiation, acting on cell angiogenesis and endothelial cell function as demonstrated by Jiang et al. [41].

Because of the association between bioglass particles and local medicines for osteoporosis in the in vivo response, the choice of this drug was due to its favorable effects on bone tissue metabolism [39,40,42,43] and evidence in the literature on its effect on cell cultures [44,45].

The drug incorporation process on the surface of biomaterials can occur in different ways; among them, sonochemical processing stands out [31] because in this technique, the ultrasound equipment generates the formation of bullous cavitations through radiation contact with the material [46], promoting changes in the physical and structural characteristics of the biomaterial, such as reduction, homogenization of particles, and occasionally the formation of a superficial amorphous layer [46,47], which is necessary for the incorporation of the drug. Gonzalo-Juan et al. [48] used the sonochemical technique to produce silver nanoparticles and incorporate them into the surface of the bioglass and proved that this technique does not interfere with the bioactivity mechanisms of glass, as it does not significantly change the network structure of the glass, considering this easy and fast route.

From this context, the investigation of new biomaterials that can accelerate or promote bone regeneration in patients who present osteoporosis and require surgical procedures is of paramount importance and has been addressed in current studies [15–21]. Therefore, the development of a biomaterial that favors the bone regeneration process and can positively contribute to the field of tissue bioengineering, promoting a combination of the advantages of teriparatide medication, which proves to be an innovative bone anabolic drug, with bioglass that has properties positive in relation to bone repair becomes of extreme necessity.

2. Materials and Methods

2.1. 45S5 Synthesis

We used the method described by Spirandeli et al. [49] to obtain 45S5 bioglass, which is based on the processes of preparing glass ceramics by sol-gel followed by melting–quenching. The fusion of the 45S5 bioglass was carried out in an oven at 1350 °C/15 min, in a ZAS crucible (zirconia-alumina-silicon). Subsequently, the molten glass was poured into water. The glass, in the shape of fries, was collected, dried, and then ground in a mechanical mortar for 3 h and later sieved through a 200 mesh sieve. The composition of the 45S5 bioglass, in % mol, was 46.1% SiO_2, 24.4% Na_2O, 26.9% CaO e 2.6% P_2O_5.

2.2. Sonochemical Technique

All samples were subjected to ultrasonic processing carried out at the Advanced Materials Laboratory of the Center for Research and Development of Functional Materials—Health Division—CEPID/FAPESP in a Sonics VCX-750 (SONICS Vibra Cell™, Newtown, EUA) brand model, with 750 W power and 20 kHz frequency, with 5 min pulses and variable amplitude up to 70% of the Sonics equipment nominal amplitude (450 W/cm^2). The 45S5 bioglass samples were divided into equal parts, one part consisting of functionalized bioglass in the absence of the drug, which was called group BG, and the other part consisting of bioglass subjected to functionalization processing associated with the drug teriparatide, which was identified as group BGT. The relative concentration in mass was 55 g/L determined in the initial stages of the process, using the hormone PTH 1-34 (Lilly France S.A.S, Fegersheim, France), 45S5 bioglass in the form of powder, and Milli-Q water.

2.3. Characterization Biomaterial

The surface topography of the samples was analyzed by the scanning electron microscope (SEM) (JEOL/JSM-5310, Tokyo, Japan) with "Field Emission Gun" (FEG) was used (Tescan/Vega 3, Brno, Czech Republic) before and after the functionalization of the samples. The images were obtained on the SEM with a secondary electron (ES) detector, projected on the sample surface. The original magnifications used were 10,000×, 20,000×, 50,000×. The functional groups on the surface of the particles were identified by analyzing them using the Fourier transform infrared spectrophotometry (FTIR) (Parkin Elmer, model Spectrum GX) in UATR mode, installed at the Associated Laboratory of Sensors and Materials of the National Institute for Space Research (LAS/INPE), in the middle region 500–4000 cm^{-1}, 32 scans and 4 cm^{-1} resolutions using the Spectrom Search Plus program (PerkinElmer, Waltham, MA, USA). Energy-dispersive analysis was performed to map the elements using Bruker Nano GmbH 410, (Berlin, Germany) and Espirit 1.9 software (Bruker, Berlin, Germany) associated with SEM (Inspect S50, FEI Company, Brno, Czech Republic). The values of the Zeta potential of the bioglass were obtained using the dynamic light scattering equipment, Stabino Control 2.00.23, (Particle Metrix GmbH, Meerbusch, Germany) installed at the Research and Development Institute of the University of Vale do Paraíba (UNIVAP). The bioglass powders were dispersed in deionized water (pH 5.0), and the measurements were performed at 25 °C at an angle of 15°. To calculate the zeta potential from the mobility values, the Smoluchowski equation was used, with values of the refractive index, dielectric constant, and water viscosity at 25 °C.

2.4. In Vitro Experiment

The mesenquimals cells (hMSC) were obtained from the femurs of nine ovariectomized female Wistar rats (Rattus norvegicus) as previously described by Zhang et al. [50]. After cleaning the femurs, in the laminar flow, the bone marrow cells were isolated and inserted in cell culture flasks of 250 mL and 75 cm^2 (TPP, Biosystems, Curitiba, Brazil) with essential alpha MEM medium culture (Gibco) supplemented with 10% Bovine Fetal Serum (SBF) (LGC Tenchology, Campinas, Brazil) and gentamicin (500 µg/mL) (Gibco) and were incubated in an oven at 37 °C, with atmospheric humidity containing 5% carbon dioxide (CO_2). The culture medium was changed every three days, and the progression of the culture was evaluated by inverted phase microscopy (Microscope Carl Zeiss—Axiovert 40C, Oberkochen, Germany). After confluence (approximately seven days), cells were released enzymatically [51] and plated at a density of 1×10^4 viable cells in each well of the 96-well microplate (Transwell, Corning/Costar, New York, NY, USA).

Before plating, samples were weighed on a semi-analytical precision scale (BEL Engin Mark, model 210A) at 0.022 g. Subsequently, these samples were sterilized in UV light and placed inside the wells. In the wells of the control group, only cells were plated. Then, osteogenic culture medium was added to the plate. The final volume of the osteogenic medium was 250 µL per well, which was changed every 48 h.

After these procedures, all plates were incubated at 37 °C with 5% CO_2 and kept until the tests. All tests were performed according to ISO 10993-5 [52] and in triplicate, with each isolation being a pool of cells from the femurs of three animals.

2.4.1. Cell Adhesion

After 3 and 5 days of cultivation, cell morphology was evaluated by FE-MEV (Field Emission Scanning Electron Microscopy) (Zeiss—EVO MA10, São Paulo, Brazil) at the Dental Materials and Prosthesis Research Laboratory of ICT/Unesp. The samples were coated with a thin layer of gold using a sputter-coating system. The samples were placed on the stub (aluminum platform), aided by a double-sided carbon tape (3M, Sumaré, Brazil), and metallized with a thin layer of gold by sputtering in the metallizing machine (EMITECH K550X, Sputter Coater, Quorum Technologies, Lewes, UK) for a period of 130 s. The images were obtained by SEM with a secondary electron detector in several magnifications.

2.4.2. Cell Viability (MTT)

After the periods of 3 and 7 days, a quantitative assessment of live cells was performed, after exposure to the toxic agent by incubation with the MTT [3-4,5-dimethylthiazole bromide] (Sigma Aldrich, St. Louis, MO, USA), and the formazan crystals were dissolved by adding 100 µL of dimethyl sulfoxide solvent (Sigma-Aldrich) to each well. The plates were shaken at room temperature to dissolve the crystals, and the absorbance was measured spectrophotometrically at 570 nm (Micronal AJX 1900). Results are expressed accepting the absorbance of the negative control as indicating 100% viability.

2.4.3. Protein Content Determination and ALP Assays during hMSC Differentiation

The total protein content was calculated in two different periods, 3 and 7 days, to assess whether the biomaterial accelerates the production of cellular proteins or not. This measurement was performed according to the modified method of Lowry [53]. The absorbance was measured spectrophotometrically at 680 nm (Micronal AJX 1900). The alkaline phosphatase (ALP) activity was determined by releasing thymolphthalein by hydrolysis of the thymolphthalein substrate, using a commercial kit according to the manufacturer's instructions (Labtest Diagnostic), in the same periods of the total protein, using the same lysates. The absorbance was measured in a spectrophotometer (Micronal AJX 1900) at 590 nm.

2.4.4. Formation of Mineralization Nodules

After 12 days of culture by staining Alizarin S2% red (Sigma-Aldrich, St. Louis, Brazil), pH was measured to be 4.2. Specifically for this test, microplates with the presence of a Transwell net (Transwell, Corning/Costar, New York, NY, USA) were used. The red dye from Alizarin S was used to stain areas that are rich in calcium. This test was used to assess whether there was an acceleration in the production and calcification process of the cellular matrix and to characterize mesenchymal stem cells as osteoblasts due to the production of mineralized matrix. The quantification of calcium in mineralized formations was performed according to the method described by Gregory et al. [54]. The absorbance was measured in a spectrophotometer (Micronal AJX 1900) at 405 nm.

2.5. Experimental Design In Vivo Study

This work was executed according to the ethical principles of experimentation (CONCEA) and approved by the local ethic committee (Processes numbers 10/2019) adopted by the National Council of Control of Animal, and ARRIVE [55] was respected. Forty-five female Wistar rats (Rattus norvegicus) (Central Vivarium of Unesp Botucatu, São Paulo, Brazil) that were approximately 90 days old and 350 g in weight were used and kept in cages with water ad libitum and diet. The animal was randomly divided into two groups, with 30 rats having undergone bilateral ovariectomy (OVX) and 15 rats having undergone simulated ovariectomy surgery (Sham). Thus, the experimental model adopted in the present project consisted of estrogen-deficient animals in the OVX group [18,56–58]. In each OVX and Sham group, critical 5.0 mm bone defects were made in the calvaries. For biomechanical evaluation, OVX rats (n = 10) and Sham rats (n = 5) underwent 3.0 mm bone defects on the right and left tibiae. All defects on the left side in OVX group were filled with biomaterial—(a) functionalized bioglass (BG) and (b) functionalized bioglass with 10% teriparatide (BGT)—and the right sides were filled with clot. All defects on the sham group were filled with clot. This distribution aimed to avoid the possible influence of one material on the other resulting in 5 rats for each subgroup according to the period. After 2 and 6 weeks, the animals were euthanized and the bone repair area was evaluated through histological, histomorphometry, immunohistochemical, and biomechanical tests.

2.5.1. Bilateral Ovariectomy

For the induction of osteoporosis, forty rats were subjected to bilateral ovariectomy (OVX). The rats were anesthetized with xylazine hydrochloride (Xylazine—Coopers, Brasil, Ltd., Osasco, Brazil) and ketamine hydrochloride (Injectable Ketamine Hydrochloride, Fort Dodge, Health Animal Ltd., Osaka, Japan) Additionally, they were then placed in lateral decubitus to perform an incision of 1 cm (cm) on the flanks and the subcutaneous tissue, and then the peritoneum was divided into planes to access the abdominal cavity. Subsequently, the ovaries and uterine horns were located, which were cauterized using the cauterizer (Cautermax-Fabinject®). Subsequently, the sutures were made in layers, all using silk thread no3 (Ethicon/Johnson & Johnson, Blue Ash, OH, USA). Twenty rats in the sham surgery group (Sham) underwent the same procedure, but only the surgical exposure of the uterine horns and ovaries was performed without their respective cauterizations.

2.5.2. Surgical Procedure at Calvaria

After 60 days of the ovariectomy (OVX) and simulated ovariectomy (Sham) procedures, the same rats were subjected to the manufacture of critical bone defects in the calvaria. The animals were weighed and anesthetized with a solution of xylazine hydrochloride 2.3 g/100 mL (Anasedan®, Vetbrands, Jacareí, Brazil) and ketamine hydrochloride 1.16 g/10 mL (Dopalen®, Vetbrands, Jacareí, Brazil).

After the anesthesia of the animals, the surgical sites were subjected to trichotomy and antisepsis with an iodized alcohol solution. A linear incision of approximately 3.5 mm was performed with a scalpel blade number 15 in the region corresponding to the medial face of the calvaria. The tissues were divulsed to expose the calvaria cortices, in which the 5.0 mm defects were made under abundant and continuous irrigation with physiological solution to prevent heating due to the friction of the drill with the bone. To obtain the critical defect, a 5.0 mm diameter drill was used. The defects on the left side in the OVX group were made with (a) bioglass (BG) or (b) bioglass associated with 10% teriparatide (BGT), and on right side, they were filled with clot. In the sham group, the defect was filled only by clot. For the stabilization of the material in the surgical site of the defect, a larger-diameter biological membrane (GenDerm®) was positioned at the bone defect site. In all defects, the tissues were repositioned, and the layers were sutured with no3 silk thread (Ethicon/Johnson & Johnson). Again, iodized alcohol antisepsis was performed. After surgery, the rats were placed in mini-isolate of the ventilated rack and monitored until the euthanasia deadline.

2.5.3. Macroscopic Evaluation of the Uterus and Exfoliative Cytology

Euthanasia was performed within 2 and 6 weeks with an overdose of the drugs used for anesthesia, administered intramuscularly. Overdoses 4 times greater were used to ensure that the animal did not return consciousness. In addition, to prove the effect of ovariectomy, the macroscopic aspect of the uterine horns of the rats of the Sham group and of the rats of the OVX group was observed, and the estrous cycle was assessed.

2.5.4. Histological and Histomorphometric Analysis

The calvaries were removed, cleaned, and placed in a 10% formaldehyde solution for 48 h. Then, the samples were placed in a solution of ethylenediaminetetraacetic acid (EDTA Titriplex III, EMD Millipore, Burlington, MA, USA) for demineralization for about 60 days. The parts were checked regularly, and when demineralization was found, the pieces were sectioned longitudinally in the center of the bone defect of the calvaria, inserted in plastic cassettes, and placed in a tissue processor (Leica TP1020, Wetzlar, Germany) for later inclusion in paraffin. Then, semi-serial sections 5.0 μm (μm) thick were obtained, which were stained with hematoxylin and eosin (HE), and 3.5 μm sections that were extended on silanized slides for immunohistochemistry analyses. In the descriptive histological analysis, aspects of the development of bone repair were observed to evaluate the formation of granulation tissue, bone neoformation, the arrangement of immature bone trabeculae, and bone maturation until final remodeling. The morphometric analysis of the bone neoformation was realized with five slices from each animal stained with hematoxylin and eosin. Stained sections were examined by light microscopy under 5× objective lenses, and images were obtained with a Microscope Zeiss Axioskop 40 (Carl Zeiss) and analyzed with software Image J (Image Processing and Analysis in Java, NIH, Bethesda, MD, USA) by two calibrated and blinded examiners using a light microscope.

2.5.5. Immunohistochemistry Analysis

The slides were prepared for immunohistochemical reactions by blocking endogenous tissue peroxidase by incubation with 6% hydrogen peroxide and methanol. Antigenic recovery for osteocalcin was performed by heating it in a glass vat containing 10 mMph 6.0 citric acid in the microwave (700 w) for 2 min; after heating, the slides were placed in the vat with citric acid heated for 40 min. Then, the primary antibodies were incubated, which occurred according to the dilution at 1:100 overnight and at a temperature of 4 °C. Subsequently, the incubation with the secondary antibody (Universal LSAB TM Kit/HRP, Rb/Mo/Goat—DAKO) was carried out for 30 min at room temperature, the biotinylated anti-goat secondary antibody produced in donkeys (Jackson Immunoresearch Laboratories) was used, and the amplifiers selected were avidin and biotin (Vector Laboratories, Burlingame, CA, USA). The chromogen used was the diaminobenzidine solution (Dako, Glostrup, Denmark). Subsequently, the cuts were stained with Mayer's hematoxylin, dehydrated in ethanol, bleached in xylol, and mounted with Permount. Negative controls were performed by replacing the primary antibody with bovine serum albumin (BSA), and positive controls were performed as suggested by the manufacturers of the primary antibodies.

For the analysis, a standardized photo of each slide was performed, with a ×40 magnification of the region of the interface between the bone defect and the old bone, to observe the expression of the osteocalcin (OC) and RankL. Evaluations were performed on the edges of the defects in both periods by means of visual evaluation and were made by the same evaluator in blind conditions. For qualitative analysis, immunostaining for cells was considered [59–63]: (a) (−) = negative; (b) (+) = positive; (c) (++) = superpositive; (d) (+++) = hyperpositive. After the qualitative analysis, semi-quantitative analysis was carried out by converting the scores into percentages: 0% to negative, 20% to positive, 60% to superpositive, and 90% to hyperpositive. A higher percentage reflected an increase in the number of cells stained positively by diaminobenzidine in the area.

2.5.6. Surgical Procedure at Tibiae

After 60 days of the ovariectomy and sham procedure, 20 rats were again anesthetized using the same technique and submitted to the surgical procedure for making the critical defect. With a carbide bur, a critical size defect of 3.0 mm in diameter was performed under copious irrigation with 0.9% sodium chloride. Critical bone defects in the left tibiae of OVX rats were filled with BG or BGT. On the right side, clot stabilization (control) was expected. The Sham group was filled with clot. The flap was repositioned and sutured with #4 silk thread (Ethicon/Johnson & Johnson). Two weeks after surgery, the animals were euthanized, and the tibiae were submitted to a three-point flexion test to verify the influence of the biomaterial on the mechanical property of the newly formed bone tissue.

2.5.7. Biomechanical Properties

After euthanasia, the tibiae were kept in Ringer's solution at −20 °C, until the moment in which the analyzes were performed in the Research Laboratory in Dental Materials and Prosthesis of the ICT/Unesp. To carry out the test, each specimen was placed centrally, along its length, on a support containing two supports (15 mm from each other), with its anterior face facing downwards. The load was applied transversely along the long axis of the tibia on its posterior face at a midpoint between the two supports. The load application support and the supports used had a cylindrical shape with a diameter of 3 mm. The test was conducted on a universal testing machine (Emic®—model DL 200 MF, Testing Equipment and Systems Ltd., São José dos Pinhais, Brazil), which provided a force of 50 kg/F with a constant application speed of 5.08 mm/min until specimen failure.

2.6. Statistical Analysis

All statistically analyzed tests were submitted to the Kolmogorov–Smirnov normality test, and homogeneity of the results was observed ($p > 0.05$). To analyze the Zeta potential in the characterization of the material, the data were analyzed by one-way ANOVA ($p < 0.05$). Data from in vitro tests and histomorphometry and immunohistochemical analysis were analyzed by two-factor ANOVA ($p < 0.05$), considering the period and biomaterial analyzed as variables, and the data are presented as mean value ± standard deviation. The biomechanical evaluation was carried out within only 2 weeks, so it was submitted to one-way ANOVA ($p < 0.05$). When necessary, they were submitted to Tukey's post hoc test. All tests were performed using the GraphPad Prism 6.0 software (GraphPad Software, San Diego, CA, USA) and for all statistical tests, a significance level of 5% and a power of 80% were adopted.

3. Results

3.1. Sample Characterization

Based on Figure 1, it was possible to observe that the surface of the 45S5 bioglass without functionalization presented particles of geometric shapes and right angles compatible with the morphological characteristics of glassy materials (Figure 1(A1,B1)). The agglomeration of the glass particles could also be observed due to less surface energy. However, the functionalized bioglass (BG) and functionalized bioglass groups were associated with

10% teriparatide (BGT). Additionally, Figure 1(A2,A3,B2,B3) show particles with rounded angles and more solubilized, suggesting that the surface was attacked chemically by the process of functionalization.

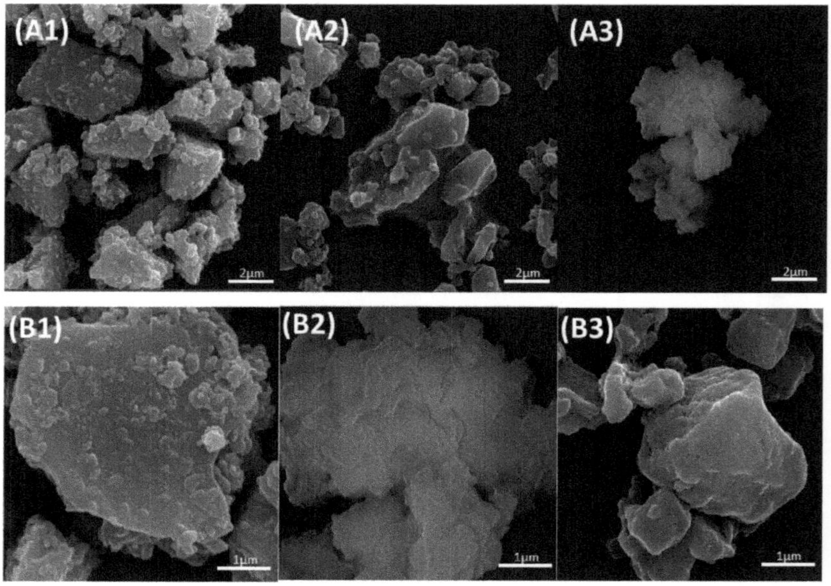

Figure 1. Micrographs of the surfaces of materials. Images obtained by SEM-FEG: (**A1**) 45S5 Bioglass group increase of 20,000 times (scale bars = 2 μm); (**A2**) BG group increase of 20,000 times (scale bars = 2 μm); (**A3**) BGT group increase of 20,000 times (scale bars = 2 μm); (**B1**) 45S5 Bioglass group increase of 50,000 times (scale bars = 1 μm); (**B2**) BG group increase of 50,000 times (scale bars = 1 μm); (**B3**) BGT group increase of 50,000 times (scale bars = 1 μm).

Figure 2 shows three FTIR spectra corresponding to the 45S5 bioglass, functionalized bioglass (BG), and functionalized bioglass groups associated with 10% teriparatide (BGT). The sample composition before and after the functionalization of the bioglass in the absence of the drug has chemical bonds that formed networks characteristic of the bioglass. It is then possible to confirm, by means of the infrared spectroscopy technique, the presence of bands that represent the bonding groups of BO (Si-O-Si), phosphate (PO_4), and carbonate (CO_3) bonds and non-oxygen bridge bonds NBO (SiO-Ca^{2+} O-Si and SiO-Na^+ O-Si) in all groups. It was observed that there was a small reduction in the relative intensity of the NBO bands in the functionalized material in the presence of the drug teriparatide (BGT) when compared to the other samples and the presence of bands in the amide group (C=O) consisting of bonds between carbon and hydrogen (CH); nitrogen and hydrogen (NH); and carbon, oxygen, and nitrogen (CON) (1660 cm^{-1}) in this group. The changes in the control of bands around 1000 cm^{-1} and 1500 cm^{-1} occurred by overlapping the medication bands with those of the bioglass. The carbonate group (CO^{3-}) showed no significant change in any of the groups.

The samples, when evaluated by EDS, show the presence of chemical elements that make up the 45S5 bioglass. Mainly, energy peaks characteristic of the elements calcium (Ca), silicon (Si), and oxygen (O) were observed in the bioglass 45S5 and BG groups (Table 1), as well as the presence of Ca and P in all groups. Meanwhile, samples that were functionalized with the drug teriparatide exhibited sodium peaks (Na) as the main element and decreased Si element. Furthermore, the presence of the chemical element gold (Au) was not observed in the particles of the 45S5 bioglass group. Particles in the graph in Table 1 the chemical element gold (Au) were not observed.

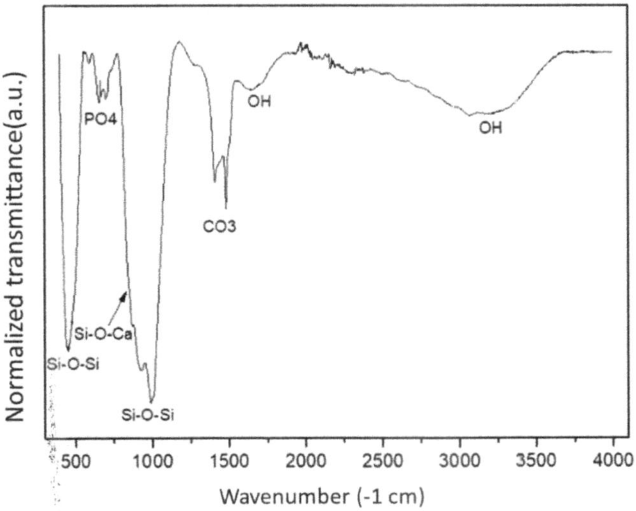

Figure 2. Fourier Transform Infrared Spectrophotometry results.

Table 1. Energy-dispersive analysis (EDS).

EDS Groups	O (Wt%)	Na (Wt%)	Ca (Wt%)	Si (Wt%)	Au (Wt%)	P (Wt%)	C (Wt%)
45S5 bioglass	47.8 ± 0.2	11.7 ± 0.1	12.7 ± 0.1	11.3 ± 0.1	0	1.6 ± 0	0
BG	45.2 ± 0.2	10.5 ± 0.1	13.9 ± 0.1	9.8 ± 0.1	2.5 ± 0.2	1.4 ± 0	16.6 ± 0.2
BGT	49.8 ± 0.2	24.4 ± 0.2	9.7 ± 0.1	6.5 ± 0.1	6.2 ± 0.3	3.4 ± 0.1	0

Mean and standard deviation values of measures' percentage of mass (Wt%).

All samples in Zeta potential test demonstrated the surface reactivity of the particles. The values obtained in the functionalized bioglass particles with the presence of 10% teriparatide (BGT) showed a higher magnitude ($p < 0.05$; One-way ANOVA, Tukey's test) (Figure 3). The 45S5 bioglass showed fewer functional groups, whereas the functionalized groups had an equivalent number of functional groups, since there was no statistical difference in the samples.

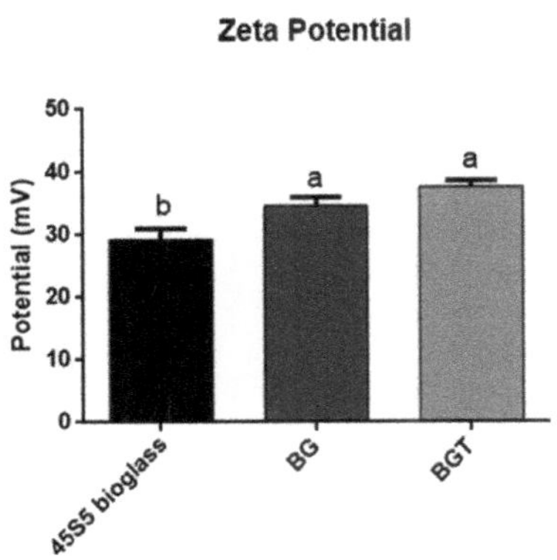

Figure 3. Values of mean and ± standard deviation of the Zeta potential (mV).

3.2. In Vitro Analysis

After 3 and 5 days of culture, the samples were analyzed by scanning electron microscopy (SEM) to show the cellular interaction on the materials (Figure 4). In this analysis, it was evidenced that all the samples allowed cell spreading. Although the materials exhibited irregular macrostructures and reveal a rough surface with many pores and reduced coalescence between the particles, it was possible to observe cellular extensions permeating the particles of the material.

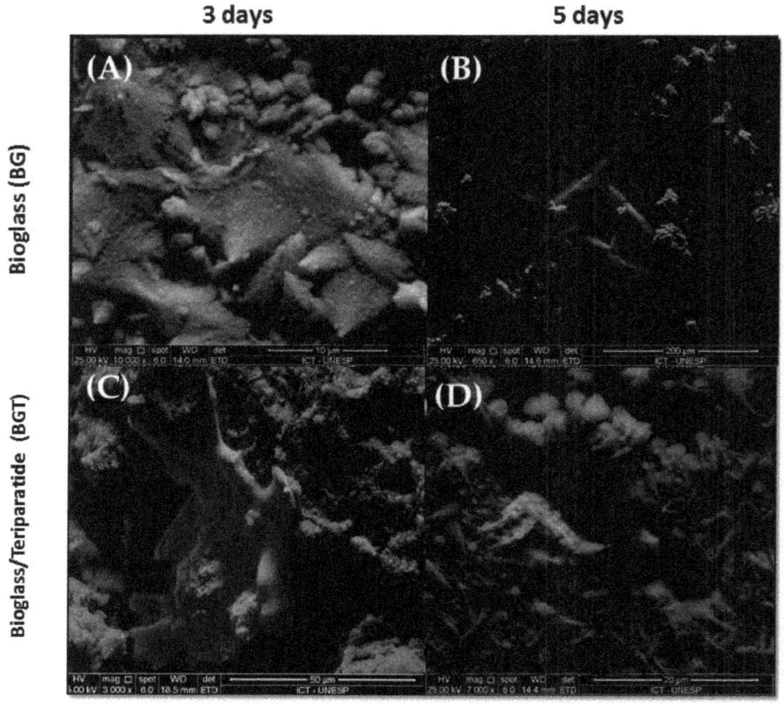

Figure 4. SEM images of cells adherent to the biomaterial: (**A**) BG group in the period of 3 days with an increase of 10,000 times; (scale bars = 10 μm); (**B**) group BG in the period of 5 days with an increase of 650 times (scale bars = 200 μm); (**C**) BGT group in the period of 3 days with an increase of 3000 times (scale bars = 50 μm); (**D**) BGT group within 5 days with an increase of 7000 times (scale bars = 20 μm).

In the period of 3 and 7 days, the groups did not show statistically significant differences ($p > 0.05$). When performing the analysis between the periods, it was verified that the BG and BGT groups in the 7-day period differed statistically from the BG group and the control group in the 3-day period ($p < 0.05$), presenting the cell with the highest viability value. The results are shown in Figure 5A.

Within the evaluated periods, cellular metabolic activity was verified in all groups. In both periods, the mean value of the total protein amount was higher in the bioglass group associated with the drug teriparatide 10% (BGT), but within 3 days, there were no statistically significant differences ($p > 0.05$) between the groups analyzed. In the 7-day period, there was also no statistical difference between the BGT and the BG group ($p > 0.05$), but there was statistical difference with control group ($p < 0.05$). It was possible to observe that all experimental groups (BG and BGT) had a statistically significant increase in the largest period analyzed when compared to the period of 3 days ($p < 0.05$), while this fact is not observed in control group. The results are shown in Figure 5B.

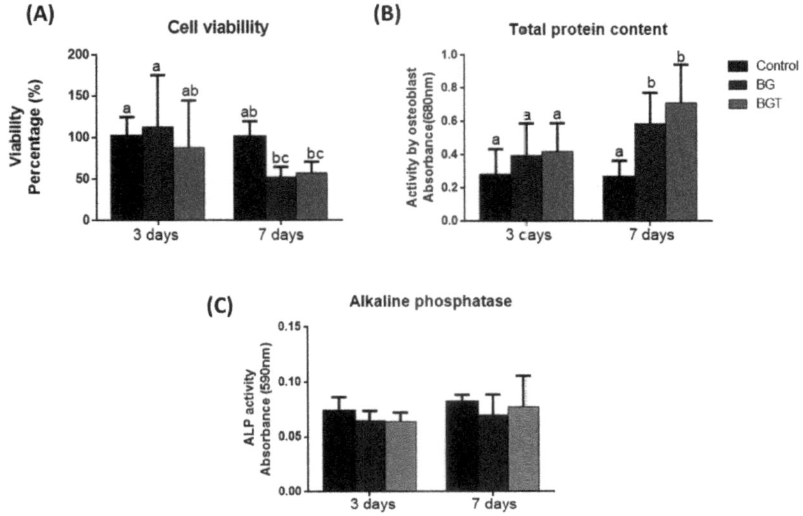

Figure 5. Results obtained through performing in vitro tests. Graphics represent mean values and (±) standard deviation. (**A**) Cell viability (Percentage %) after three days and seven days; (**B**) total protein content (OD—690 nm) after three days and seven days; (**C**) alkaline phosphatase activity (OD—580 nm) after three days and seven days. Letters represent results from Tukey test multiple comparisons; different superscript letters in-dicate significant differences ($p < 0.05$).

In the same lysates from the analysis of total protein content, ALP was measured. In the period of 3 days, the activity of ALP showed no statistical difference from the other groups. It was found that in the 7-day period, the experimental groups showed higher mean values of ALP when compared with the same groups in the period of 3 days, but it also did not differ statistically ($p > 0.05$). Between the periods evaluated, there were no statistically significant differences between any groups ($p > 0.05$). The results are shown in Figure 5C.

After 12 days of cell culture, the formation of mineralization nodules was observed in the experimental groups. It was observed that the quantification of calcium in nodule formation (Figure 6) was greater in the functionalized bioglass group (BG), differing statistically from the control group and BGT group ($p < 0.05$). Figure 6 represents the nodules formed in the wells of the experimental groups.

3.3. Bone Repair Analysis

In the 2-week postoperative period, there was no evidence of inflammation or infection at the surgical site. Microscopically, the pieces evaluated had longitudinal sections of the calvaria of the rats where the biomaterials were positioned. The extremities of the surgical sites contained newformed bone tissue showing bulky osteocytes and connective tissue with the presence of spindle cells. The analyzed period of 2 weeks showed signs of bone tissue remodeling with the proliferation of bone trabeculae from the ends of the bone cortex, close to the region where the material was positioned (Figure 7). This bone showed bulky osteocytes within wide gaps and osteoblasts, arranged in rows around the disorganized trabeculae, and it was also possible to verify the presence of the biomaterial in the areas that contained thicker connective tissue. Specifically, in the experimental groups incorporated with the drug teriparatide (BGTO), the area of the bone defect stands out, with the presence of a thicker connective tissue showing a large amount of blood vessels, while in the other groups, the presence of loose connective tissue and little inflammatory infiltrate (Figure 8).

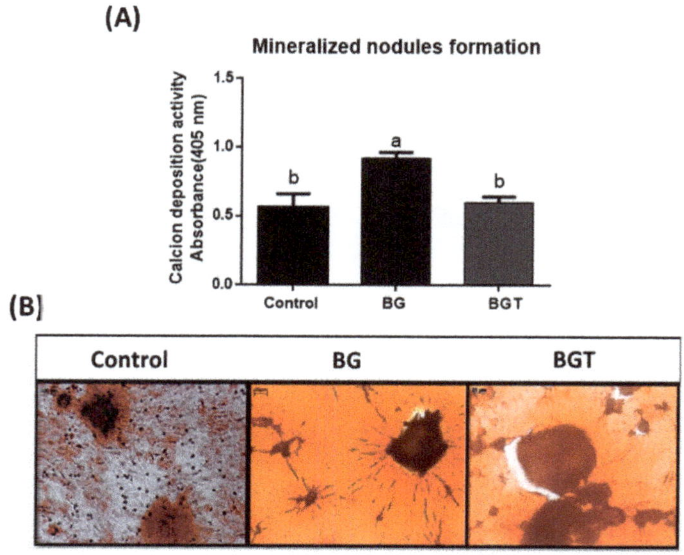

Figure 6. Quantification of calcium in mineralization nodules (OD-405 nm) after 12 days. (**A**) Graphics represent mean values and (±) standard deviation. Letters represent results from Tukey test multiple comparisons; different superscript letters indicate significant differences ($p < 0.05$); (**B**) representation of the formation of mineralization nodules. Cells were stained with red alizarin S to visualize formed calcium deposits. Original magnification ×20 (scale bars = 20 μm).

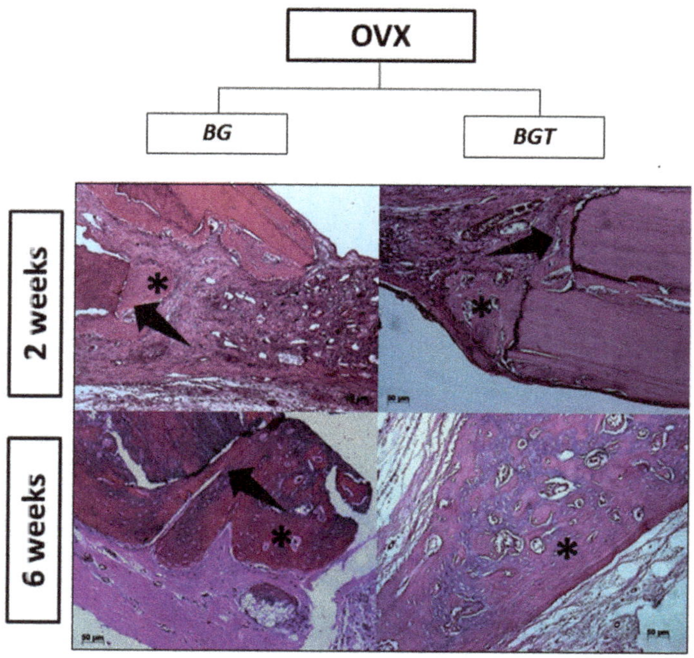

Figure 7. Histological section of the group BGO and BGTO 2 and 6 weeks. Asterisk in the newly formed area. Arrow: black in the delimitation of the critical defect. HE is staining; original magnification ×10 (scale bars = 50 μm).

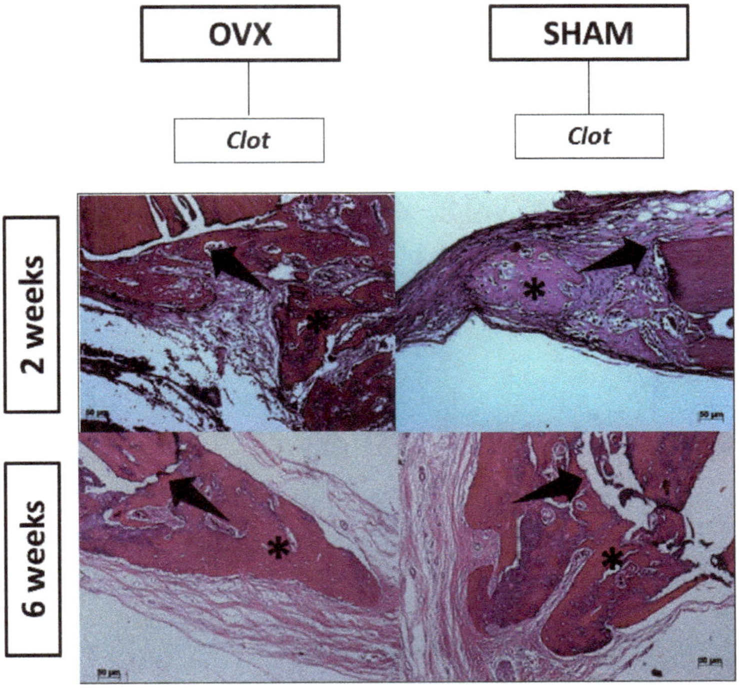

Figure 8. Histological section of the clot group at 2 and 6 weeks. Asterisk in the newly formed area.: Black arrow in the delimitation of the critical defect. HE is staining, original magnification ×10 (scale bars = 50 μm).

In the period of 6 weeks, it was possible to notice that there was still no closure of the defect with bone tissue in any evaluated group; in addition, it was observed that a large part of the biomaterial had already been reabsorbed, with the concentration of bone formed in the stump areas. In these areas, more mature bone tissue was present, with evident Havers canals and lamellae that contained osteocytes within gaps, also arranged in concentric rings. It is possible to observe the presence of compact bone next to the residual biomaterial. The connective tissue is denser and there is a large amount of thicker collagen fibers when compared with groups in the period of 2 weeks. Based on data from descriptive statistics, when considering the influence of the critical-defect filling materials and the periods, the ANOVA-two-factor test of variance was performed, followed by the Tukey multiple comparison test. The results comparing the measured values of the newly formed area between the groups and periods analyzed are shown in Figure 8A. Therefore, when the analysis among the groups from the same period occurred, it was concluded that in both the period of 2 weeks and 6 weeks, there were no statistically significant differences ($p > 0.05$), while when comparing the periods analyzed, statistically significant differences ($p < 0.05$) were observed between all groups in the 2-week period with the BGTO6 group.

The highest maximum force values were observed in the BGTO group, that is, ovariectomized animals with bone defect filled with bioglass functionalized with teriparatide. This group exhibited a statistically significant difference from the other groups ($p < 0.05$). All other groups did not differ from each other ($p > 0.05$) (Figure 9B).

Immunomarking of osteocalcin was concentrated at the edges of the defects in both periods. In all analyzed groups, osteocalcin immnunomarking was observed next to the mineralized bone matrix, being more observed in the later period (6 weeks) of bone neoformation (Figures 10 and 11).

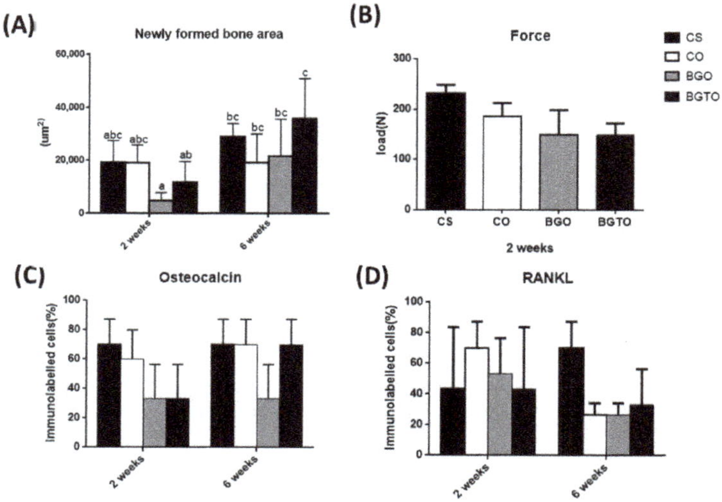

Figure 9. Graphics represent mean values and (±) standard deviation. (**A**) Results obtained by performing bone neoformation in vivo test analyzed at two times: 2 and 6 weeks; Letters represent results from Tukey test multiple comparisons; different superscript letters indicate significant differences ($p < 0.05$). (**B**) force parameter biomechanical test; (**C**) graph showing the distribution of scores (in percentage) referring to the OC immunostaining pattern at 2 weeks and 6 weeks; (**D**) graph showing the distribution of scores (in percentage) referring to the RankL immunostaining pattern at 2 weeks and 6 weeks.

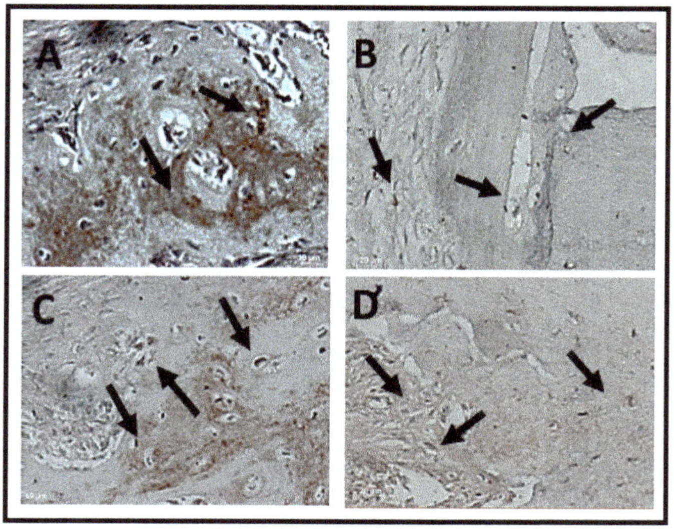

Figure 10. Photomicrographs of the histological sections with positive markings for the polyclonal antibody osteocalcin (OC) over a 2-week period. (**A**) CS, showing evident immunostaiing of cells lining the bone tissue and tissue adjacent to the bone; (**B**) CO, positive immunostaining of OC compatible with the beginning of the bone tissue healing process; (**C**) BGO, immunostaining of OC in adjacent tissue and few lining cells in bone tissue; (**D**) BGTO, osteocytes and lining cells in adjacent tissue are marked by OC, showing few marked regions compatible with bone immaturity in this period. Osteocytes were positive for OC labeling in all groups. Arrows indicate areas and label cells immunomarked by osteocalcin. Original magnification ×40 (scale bars = 20 μm.)

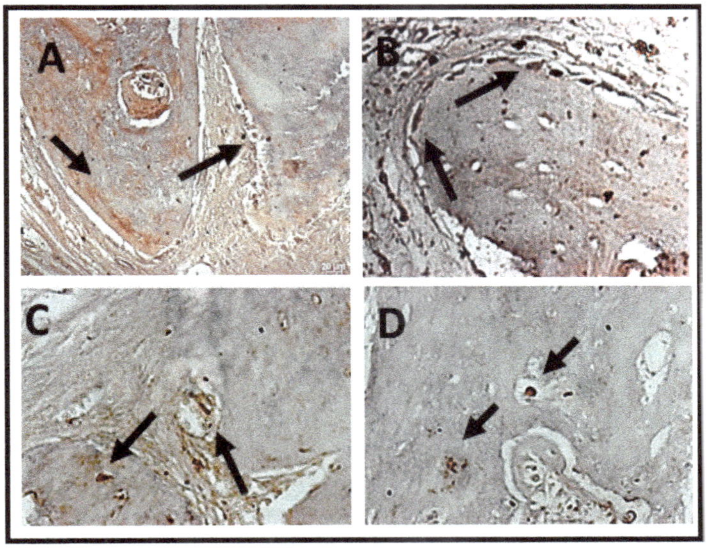

Figure 11. Photomicrographs of the histological sections with positive markings for the polyclonal antibody osteocalcin (OC) over a 6-week period. (**A**) CS, immunostaining osteocalcin with mineralized bone matrix; (**B**) CO, osteoblastic cells around the newly formed bone tissue; (**C**) BGO, immunostaining of OC in cells in bone tissue; (**D**) BGTO, osteocytes within the gaps in the newly formed bone tissue. Osteocytes and osteoblast were positive for OC labeling in all groups. Arrows indicate areas and label cells immunomarked by osteocalcin. Original magnification ×40 (scale bars = 20 μm.)

The average values of cells immunomarked in percentage for osteocalcin demonstrated that in the period of 2 weeks, there were no statistically significant differences ($p > 0.05$), but the Sham and OVX clot groups showed higher mean values than other groups. However, when observing the period of 6 weeks, the BGTO group showed a clear increase in cells immunomarked by osteocalcin compared to the mean values of the period of 2 weeks, but without showing statistical differences ($p > 0.05$). These results can be observed in Figure 11C.

In the qualitative analysis it was possible to observe that osteocalcin (CC) was considered superpositive in the CS groups and positive in the CO group within 2 weeks, with marking on the newly formed bone trabeculae, osteocytes, and in the extracellular matrix. Meanwhile, the experimental groups expressed positive and sometimes negative markings for osteocalcin (OC). The osteocalcin marking pattern indicated that the bone tissue in the period of 2 weeks in the BGO, BGTO groups was immature when compared to the clot groups (Figure 10). In the 6-week period, it was observed that the clot groups showed superpositive marking for osteoblastic cells and osteocytes in the matrix. The same was true of the experimental groups that were superpositive and positive, presenting areas of the extracellular matrix expressing osteocalcin, as well as osteoblasts and osteocytes (Figure 11).

In the qualitative analysis, within 2 weeks, the RankL immunostaining in the nucleus of newly trapped osteocytes in the bone matrix neoformed in osteoblasts in the periphery of the neoformed bone and in fibroblasts present in the connective tissue (Table 2). It was possible to observe that the immunostaining of cells for RankL was considered superpositive in the CO e BGTO group for a 2-week period, while the CS and BGO groups showed less evident markings. It was observed that CO showed a reduction in immunomarked cells in the connective tissue (Figure 12). Within 6 weeks, the presence of more mature bone tissue can be observed, with smaller gaps of osteocytes presenting immunostaining in osteoblasts present in the osteoblastic rhyme of the newly formed bone, as well as in some connective tissue cells for all groups, presenting for all experimental groups and the CO-positive immunostaining group on the RankL scale (Figure 12). On the other hand, the CS group within 6 weeks showed hyperpositive immunostaining for RankL, but more evident markings were observed in the connective tissue and little marking in osteocytes and osteoblasts was present in the matrix and in the periphery of the newly formed bone, respectively (Figure 13). The result of the semi-quantitative analysis of the RankL biomarker showed that the CO group had a higher mean value of immunomarked cells as a percentage in the period of 2 weeks, however, without showing statistically significant differences ($p > 0.05$) with the other groups. In the period of 6 weeks, again, there was no statistically significant difference between the groups ($p > 0.05$), but the CS group had the highest mean value of immunomarked cells, while the other groups showed a decrease in the number of cells marked for RankL in this period (Figure 13).

Table 2. Scores of immunohistochemical Analysis established for experimental groups (OC and RankL) at 2 and 6 weeks, showing: (+) = positive; (c) (++) = superpositive.

	OC		**RankL**	
	2 Weeks	**6 Weeks**	**2 Weeks**	**6 Weeks**
CS	(++)	(++)	(+)	(++)
CO	(++)	(++)	(++)	(+)
BGO	(+)	(+)	(+)	(+)
BGTO	(+)	(++)	(++)	(+)

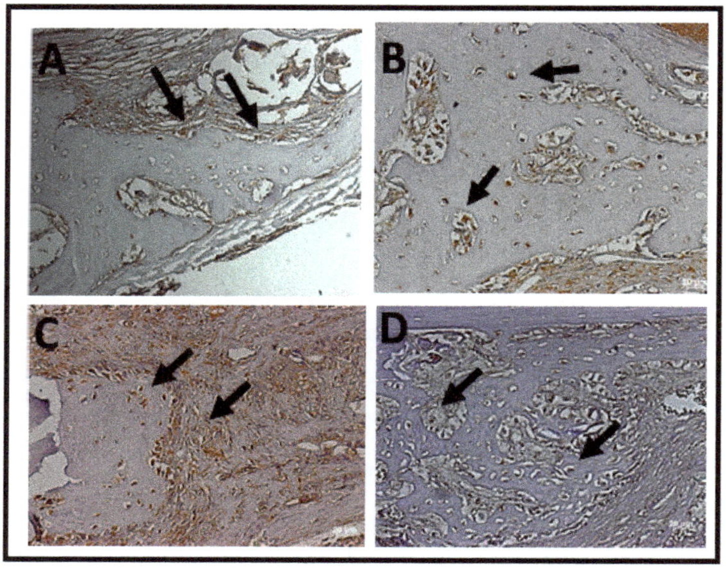

Figure 12. Photomicrographs of the histological sections with positive markings for the polyclonal antibody RankL over a 2-week period. (**A**) CS, immunostaining in the nucleus of osteoblasts in the periphery of the neoformed bone and in fibroblasts present in the connective tissue; (**B**) CO, immunostaining in the nucleus of newly formed osteocytes in the neoformed bone matrix; (**C**) BGO, evident markings in connective tissue and in osteocytes and osteoblasts present in the matrix and in the periphery of the bone neoformed; (**D**) BGTO, markings on osteocytes in the newly formed matrix and less evident in the adjacent connective tissue. Arrows indicate areas and label cells immunomarked by RankL. Original magnification ×40 (scale bars = 20 μm.)

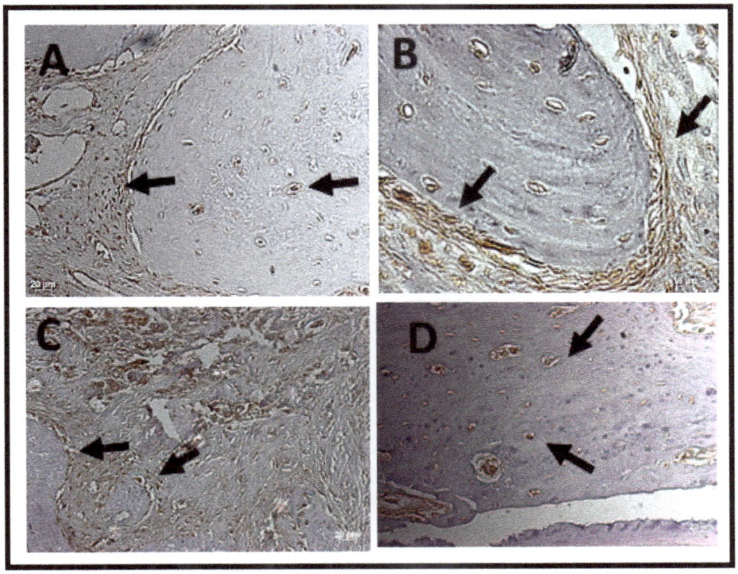

Figure 13. Photomicrographs of the histological sections with positive markings for the polyclonal antibody RankL over a 6-week period. (**A**) CS, more evident markings on connective tissue and little marking in osteocytes and osteoblasts present in the matrix and in the periphery of the newly formed bone; (**B**) CO, positive marking on connective tissue and on cells from the periphery of the neoformed bone; (**C**) BGO, connective tissue with evident RankL immunostained areas; (**D**) BGTO, RankL immunostaining that is less evident in connective tissue and few osteocytes with evident marking. Arrows indicate areas and label cells immunomarked by RankL. Original magnification ×40 (scale bars = 20 μm).

4. Discussion

One of the most prevalent postmenopausal diseases in the world is osteoporosis, a public health problem that sometimes occurs due to estrogen deficiency [64]. However, although it is known that osteoporosis causes an increase in bone loss, the specific mechanism of action of estrogen in bone tissue remains unclear [45]. In cases of individuals who have chronic diseases associated with the need for bone repair, if necessary, it can be difficult [65,66]. Therefore, there is a growing demand for biomaterials that have adequate properties in relation to bone tissue and that can positively influence tissue regeneration, improving biomechanical properties and accelerating the osteogenesis process [31,32,67,68].

One of the biggest challenges for the treatment of injuries resulting from osteoporosis is the low availability of drugs with local action. In this context, recent advances have encouraged the association of biomaterials and drugs as a local drug-release strategy [31]. According to Kyllonen et al. [69] for the release and transport of the active drug, particles of different sizes can be used; therefore, the drug is incorporated, absorbed, or associated with these particles and thus is protected from the degradation from the environment until its release. In a recent study by Mosqueira et al. [32], the authors concluded that the efficiency of the loading of therapeutic agents in strontium-modified bioglass spheres, as well as the drug release rates, were mainly influenced by the pore size of the particles, thus improving the osteogenic potential of the mesenchymal cells of the bone marrow obtained from rats with osteoporosis. In this study, bioglass particles were reduced to nanometer size for surface increase, as the number of surface atoms increases dramatically as the particle size decreases [70], which contributes to drug incorporation and release. In a study by Ajita et al. [71] the authors demonstrated that nanostructured bioglass particles (37.6 ± 0.81 nm) had applications in bone treatments, as well as materials that form the link between bone tissue, showing greater cell proliferation of mesenchymal cells from mice in bioglass dissolution products with smaller particles.

Bioglasses are interesting biomaterials for use in bone repair, as they can stimulate osteogenesis and angiogenesis [29,72,73], improving the adhesion, proliferation, and differentiation of mesenchymal stem cells from stem cells into osteoprogenitor cells and increasing the mineralization rate [74]. Furthermore, the main advantage of bioglass is its ability to bond to bone tissue due to the formation of an interfacial layer of calcium phosphate [22,75–78] As the understanding of the bioglass structure is directly associated with the interaction mechanisms of the material with the fabric, the structural investigation of the particles was carried out through several characterization tests of the samples. SEM-FEG results showed changes in the physical characteristics of the particles, in which the functionalized groups had rounded edges and reduced sizes in relation to the bioglass sample before the sonication process, changes that were characteristic of the functionalization process [56]. The functionalization time of the sample is of paramount importance, as the increase in the sonication period of the bioglass promotes a decrease in the size of the particles and smoothing of the edges of the sample. The sonochemical technique was recently used by the same research group in the evaluation of the local release of PTH 1–34 functionalized to Biogran® in peri-implant bone regeneration in osteoporotic conditions. The results obtained were promising for the osteoporotic group associated with the drug compared to the other groups, with the peri-implant repair for osteoporotic bone functionalized with PTH 1-34 similar to that of healthy bone [76].

While the FT-IR analyses indicated great structural similarity between the 45S5 bioglass and BG groups, similar to the deconvolutions of the FTIR spectra in the study by Lee et al. [77], who showed peaks associated with the structure of the bioglass, the BGT group was different, showing bands from the standard amide group (1660 cm^{-1}) of the teriparatide drug molecule. This same band was observed in the study by Bahari Javan et al. [79]. A decrease in the band of the group not bound to oxygen (NBO) can be observed; these bands are necessary for the dissolution of the bioglass, which will result in the formation of a layer of apatite hydroxycarbonate (HCA) on the surface of the material when in contact with fluids body or aqueous media. It has been suggested that the decrease in the number of NBO bonds is due to the hydrolysis of the bioglass, a process that leads to the formation of a silica network in the topographic area of the material, as described by Fiume et al. [80]. The condensation stage, at this moment of the sonochemical process, incorporates the drug teriparatide on the surface of the biomaterial, which is confirmed by the presence of bands of amide groups in this group. These observed changes may also be indicators of the incorporation of drugs in the silanol groups (Si-OH) of the bioglass [31].

When analyzing the EDS results, it was possible to verify that all groups presented the presence of Ca and P in these samples (Table 1), suggesting the presence of the formation of calcium phosphate phases on the surface of the material, which did not suffer from alterations in the presence of the drug teriparatide [81]. The electrical charge and protein adsorption of the material are also closely associated with the understanding of the biological interaction mechanisms of biomaterials with the tissue, so the analysis of the zeta potential was performed with the objective of obtaining the surface electrical energy properties of the biomaterial, where the zeta potential is influenced according to the medium in which the particles interact. The greater the value of the zeta potential is in magnitude, the greater the stability of the particles in the medium [82]. In this study, the 45S5 bioglass and the functionalized groups exhibited surfaces with different surface charge properties; it was possible through this test to observe that the functionalized groups presented greater amounts of functional groups when compared to the 45S5 bioglass, and this characteristic provides a favorable surface to cell adhesion that can cause greater protein adsorption [82,83].

Furthermore, for the development of a new biomaterial to be used in tissue regeneration, it must be ensured that it does not present potential toxic effects when implanted in the body and stimulates cell activity and differentiation. In the present study, the bioglass was functionalized by the ultrasonic sonochemical technique associated or not with the drug teriparatide 10% and used to verify the behavior of osteoblasts when in contact with samples of these biomaterials through cytotoxicity tests (MTT), total protein content, alkaline phosphatase activity, cell adhesion, and the formation of mineralization nodules, verifying the influence of these materials on osteoblastic cells.

Cytotoxicity tests indicate the effects of samples on the viability of cultured cells and is a predictor of potential toxic or non-toxic effects of biomaterials [84]. The results of the MTT test showed that all samples were biocompatible with values above 70% for viable cells in the two analyzed periods. In the present study, in the longest period of cell culture, the experimental groups exhibited more attenuated means of cell viability, as observed by Westhauser et al. [85], in which they evaluated the 45S5 bioglass associated with scaffolds in mesenchymal cells (hMSC) and observed a decrease in cell viability over longer periods. When the experimental groups were compared to the control group in relation to the total protein content, it was possible to observe that the BG and BGT groups had higher mean values in both periods; when statistically evaluating the groups and periods, it was possible to verify that the BG and BGT in the period of 7 days showed statistical differences with the control group of the same period and all groups in the period of 3 days ($p < 0.05$). This result is plausible, since the synthesis of several proteins that make up the extracellular matrix is influenced by the formation of a superficial layer of hydroxyapatite and silica present in the bioglass, which is responsible for cellular bioactivity [86]. This is because carbonated hydroxyapatite (HCA) is a bioglass dissolution product that positively interferes with the production of extracellular proteins, while BGT, in addition to presenting the bioactivity characteristics of bioglass, also presents the action of the drug teriparatide (PTH 1-34), for which in previous studies it was demonstrated that its local release acts positively on the influence of osteoblastic genes [39,87]. The expression of alkaline phosphatase (ALP) showed no statistically significant intragroup differences ($p > 0.05$) regardless of the period analyzed. ALP is important as it is one of the most reliable markers for osteogenic differentiation [88]. Tsigkou et al. [89] demonstrated that embryonic stem cells of the osteoblastic lineage in vitro start production and maturation after their cell differentiation, expressing several important proteins for the formation of the extracellular matrix, such as extracellular proteins, among them alkaline phosphatase (ALP), sialoproteins, collagenase I, osteopontin, and osteocalcin. The bioactivity of the drug teriparatide on the ALP activity of hMSCs indicated an increase in osteogenic differentiation of hMSCs and stimulation of osteoblastic cell differentiation, both in the continuous release of teriparatide and the pulsatile stimulation of the liberation of this drug [90].

Confirmation of the differentiation of mesenchymal cells (hMSCs) into osteoblastic cells in samples from the experimental groups and the control group was performed by the formation of mineralization nodules, as also observed by Rodrigues et al. [91]. This test is used to assess whether there was acceleration in the process of production and calcification of the cell matrix and characterizes the mesenchymal stem cells as osteoblasts, due to the production of a mineralized matrix [88,92]. In the present study, the BG group presented higher mean values, statistically differing from the other groups ($p < 0.05$). The result of the BGT group may suggest that due to the action of the drug teriparatide, a parathyroid hormone involving specific cellular receptors in osteoblasts and signaling pathways [93], the reproduction of studies only in vitro did not favor the observation of this action of the drug.

Therefore, for a better investigation of the ability to repair the critical defect in calvaria filled with 45S5 bioglass functionalized with 10% teriparatide and considering that the biomaterial can act positively on bone repair in conditions of osteoporotic bone injury, an in vivo analysis was performed. The ovariectomy procedure was performed to suppress the hormone estrogen in these animals. This methodology is already established and results in increased bone loss [43,58,94,95]. This result is due to the production of inflammatory cytokines, such as tumor necrosis factor alpha (TNF-α) from T cells and the production of osteoclastic cytokines, which induce the uncoupling of bone formation in the face of resorption; that is, bone formation does not follow the same pattern, i.e., the abundant rhythm of its reabsorption [96,97]. In in vivo studies, the success of ovariectomy can be confirmed by changes in the animal's estrous cycle [98], using exfoliative cytology as a method of detecting changes in cell morphology during this cycle in the ovariectomized animal [57,99]. In this study, the estrous cycle was evaluated, and it was possible to observe in the vaginal smears that all ovariectomized rats remained in the diestrus phase, confirming the hormonal alteration. According to Luvizuto et al. [58], the permanence of the animal in this phase characterizes the presence of a hormonal state compatible with hypoestrogenism.

In the present study, in the descriptive histological analysis in the period of 2 weeks, the closure of the critical defect was not observed in any group; it was also possible to observe the presence of few inflammatory cells, and in the animals that received the BGT group, there was a large amount of granulation tissue, suggesting that the bioglass dissolution products and the influence of teriparatide on the critical defect were able to increase cell recruitment [17]. For confirmation of cell recruitment, the evaluation was performed at 6 weeks. However, for a material to become promising to be indicated for tissue regeneration, it is expected that after this recruitment, there will be a gradual degradation of the biomaterial with replacement by the original tissue [100]. In the period of 2 weeks, the presence of newly formed bone tissue also occurred mainly in the regions of the stump of the defect, observing many blood vessels and young fibroblasts in the experimental groups. These results agree with the previous study by Day et al. [101], in which the authors reported that 45S5 Bioglass® promoted increased secretion of vascular endothelial growth factor (VEGF) in vitro and increased vascularization in vivo, suggesting that scaffolds containing Bioglass® can stimulate neovascularization, generating beneficial effects for large tissue-engineering constructs. In the period of 6 weeks, in histological sections, it was possible to observe that a large part of the biomaterial had already been reabsorbed. However, the critical defect was not completely regenerated, and the neoformed areas remained at the extremities of the defects in all groups. This result is similar to that reported in the study by Auersvald et al. [102], in which critical defects were made in rat calvaria using collagen sponges soaked in 20 μm of teriparatide as a filling biomaterial to evaluate the local performance of the drug after 15 and 60 days. In both periods, the authors reported that the newly formed bone was restricted to the limits of the defect without closure.

In the intragroup histomorphometric analysis over the period of 2 weeks, the groups did not differ statistically ($p > 0.05$). Studies have shown that the local use of teriparatide intermittently and subcutaneously resulted in an increase in the amount of newly formed bone tissue [38,56]. Ozer et al. [42] also addressed the local use of teriparatide with xenografts in rabbits and found that in periods of 4 and 8 weeks, the group with the presence of the drug showed higher mean values of new bone formation. However, these values were greater in the period of 8 weeks. Furthermore, in 4 weeks, the area and that of the defect containing the material with the drug consisted of regions of interconnected bone trabeculae, mainly at the edges of the defect, with loosely collagenized connective tissue, as observed in our study. In the present study, during the 2-week euthanasia period, we suggest that due to the high molecular weight of the teriparatide drug, there was a delay in its total release, which may have impaired the pharmacokinetics [69,103]. Interferences and limitations in the use of bioglass loaded with teriparatide were also reported in the study by Frigério et al. [104] because the incorporation limited the osteoconductive effect of Biogran®, and despite the beneficial effects of the drug, the results were less expressive in these groups.

In the intragroup histomorphometric analysis, the neoformed area was greater in the BGTO group in the 6-week period but did not differ statistically from the other groups in the same period ($p > 0.05$). When the intergroup comparison was performed in the 2-week and 6-week periods, it was observed that the BGTO group in the 6-week period statistically differed from the BGO and BGTO groups in the 2-week period ($p < 0.05$), showing an significant increase in bone formation in the area of the defect, thus confirming that the osteoporotic bones of the calvaria are highly metabolically responsive to teriparatide treatment [105]. In the study by Gomes-Ferreira et al. [80], the functionalization with PTH 1–34 in implants did not show a significant effect in healthy animals; however, it played a favorable role when osteoporosis was present. It is possible to suggest that, when evaluating tissue neoformation over longer periods, the total release of the drug and a longer duration of its action in the repaired site can be observed. This slow, intermittent release is interesting, as previous studies have shown that intermittent doses are effective for the anabolic function of the drug [56,106,107].

To investigate the participation of proteins in the bone repair process, immunohistochemistry with osteocalcin (OC) and Rank-L were performed. The immunoexpression of osteocalcin (OC) was more evident in the late period (6 weeks) of bone neoformation, since OC is a non-collagenous protein that participates in the maturation of the mineral part of the bone. This fact has established the study of the maturity of osteoblastic cells, which is a marker of late stages of osteoblastic differentiation [89,108,109]. This result agrees with the study by de Oliveira et al. [39], who presented similar results with more evident immunostaining in late periods of bone tissue repair. The action of the drug on the cellular response in the OVX groups was favorable, since in the period of 6 weeks there was a similarity in the average values of the percentage of immunostained cells for OC in the BGTO group when compared to the CS group, which did not have the interference of osteoporosis.

Considering that the OVX group has a deficiency in bone repair, there was an increase in the marking of cells related to CO and BGO in a period of 6 weeks, showing that treatment with the drug was able to increase the secretion of this non-collagen protein by osteoblasts, improving osteogenic activity. Although favorable results were observed for the BGTO group, the CS group presented higher or similar mean results when compared to the OVX group, a result that may be comparable to that observed in the study by Luvizuto et al. [58], in which the evaluation of bone repair in the alveoli of ovariectomized rats showed mean values lower than those observed in the Sham group, even in animals treated with raloxifene and estrogen. Another important aspect was the correlation with the results of histomorphometry and histological analysis, which suggested greater bone neoformation in the experimental groups containing the drug in ovariectomized animals, which is confirmed by the increase in CO marking in this group at 6 weeks in comparison. The results of this study also indicated that the cells present in the calvarial bone defects of all groups were positive for RANKL, with the CO group having higher mean values than the CS group. This was expected since osteoporotic bones undergo greater bone resorption when compared to normal bones [40,110,111]. In the 6-week period, it was observed that all groups showed a decrease in the mean values of the percentage of cells marked by RANKL compared to the 2-week period, except for the CS group; this may be related to the fact that the expression of bone markers can be altered by estrogen deficiency and present interference from the evaluated biomaterial [112].

The drug's ability to stimulate bone formation and increase bone mass and mineral density [45,56,113] corroborate the results of the three-way flexion test points, which demonstrated that maximum force (load/N) in the group of ovariectomized rats with bone defects in tibias filled with bioglass associated with 10% teriparatide (BGTO group) exhibited significantly higher values, statistically differing from the other analyzed groups ($p < 0.05$), which promoted an increase in the biomechanical properties of newly formed bone in rats with endocrine abnormalities when compared to normal animals. Studies with the systemic use of PTH (1–34) for fractures in long bones confirm the increase in mechanical resistance and the amount of newly formed bone [114–117]. A recent study by Leiblein et al. [118] verified that the maximum force was increased in the biomechanical tests performed in animals treated with PTH (1–34) when compared to animals that received doses of simvastatin, alendronate, and strontium ranelate.

In summary, in our study, the results of the in vitro tests reiterate that the experimental groups were favorable to the differentiation of undifferentiated cells from ovaritomized rats into osteogenic cells. However, we suggest that in the in vivo study, it can be inferred that the calvaria bone has a slower healing process when compared to spongy bone, which is because the calvaria has a smaller blood supply and a smaller amount of medullary bone [119]. We suggest, therefore, that longer evaluation times are necessary to verify the formation of a more mature bone tissue and that, despite this, the biomaterial presented itself as a promising material for use in the bone repair of ovariectomized animals.

5. Conclusions

In view of the results obtained and within the experimental conditions of this research, it was concluded that bioglass functionalized with the drug teriparatide 10% prove a positive influence on cell behavior in vitro and bone neoformation in vivo. Additionally, this study contributed to the knowledge of the interactions of osteoblasts with the surfaces of functionalized bio-glasses thus contributing to greater safety in the use of bio-glasses associated with drugs in the health area.

Although these initial results are promising, further studies are needed to evaluate the effectiveness of PTH associated with these biomaterials in new regeneration strategies, to provide additional information on the mechanisms necessary to assess the biological performance of synergistic action between bioglass and the release of the drug teriparatide.

Author Contributions: Conceptualization, L.M.R.d.V.; Data curation, L.A.S.S., J.C.R.d.A. and T.M.B.C.; Formal analysis, J.C.R.d.A. and T.M.B.C.; Funding acquisition, J.C.R.d.A. and L.A.S.S.; Investigation, J.C.R.d.A. and L.A.S.S.; Methodology, L.M.R.d.V., P.N.L.F., R.O. and J.C.R.d.A.; Project administration, L.M.R.d.V.; Resources, J.C.R.d.A. and L.M.R.d.V.; Software, J.C.R.d.A. and V.A.d.B.L.; Supervision, L.M.R.d.V., P.N.L.F. and R.O.; Visualization, P.N.L.F., L.M.R.d.V. and R.O.; Writing—original draft, J.C.R.d.A. and L.M.R.d.V.; Writing—review and editing, J.C.R.d.A., L.M.R.d.V. and R.O. All authors have read and agreed to the published version of the manuscript.

Funding: This work was supported by São Paulo Research Foundation (grant #2019/04185-1 to JCRA. # 2019/07636-4 to LASS. # 2020/12874-9 to TMBC. # 2013/07296 to Centro de Desenvolvimento de Materiais Funcionais (CDMF)) and Coordination for the Improvement of Higher Education Personnel (CAPES).

Institutional Review Board Statement: This study was submitted and approved by the Ethics Committee of Research in Animals of the Institute of Science and Technology of Sao Jose dos Campos/UNESP (01/2019-CEUA/ICT-UNESP).

Informed Consent Statement: Not applicable.

Data Availability Statement: All raw data from characterization are available from the corresponding author upon request.

Acknowledgments: The author acknowledge the São Paulo Research Foundation (FAPESP) and the Coordination for the Improvement of Higher Education Personnel (CAPES).

Conflicts of Interest: The authors declare no conflict of interest.

References

1. Sözen, T.; Özışık, L.; Başaran, N.Ç. An overview and management of osteoporosis. *Eur. J. Rheumatol.* **2017**, *4*, 46–56. [CrossRef] [PubMed]
2. Broken Bones, Broken Lives: A Roadmap to Solve the Fragility Fracture Crisis in Europe. Available online: https://www.osteoporosis.foundation/educational-hub/files/broken-bones-broken-lives-roadmap-solve-fragility-fracture-crisis-europe (accessed on 26 October 2022).
3. Li, L.; Wang, Z. Ovarian Aging and Osteoporosis. *Adv. Exp. Med. Biol.* **2018**, *1086*, 199–215. [CrossRef] [PubMed]
4. Kubota, T.; Hasuike, A.; Tsukune, N.; Ozawa, Y.; Yamamoto, T.; Min, S.; Masako, N.; Shuichi, S. Influence of estrogen deficiency on guided bone augmentation: Investigation of rat calvarial model and osteoblast-like MC3T3-E1 cells. *Eur. J. Oral Sci.* **2018**, *126*, 206–213. [CrossRef] [PubMed]
5. Russow, G.; Jahn, D.; Appelt, J.; Märdian, S.; Tsitsilonis, S.; Keller, J. Anabolic Therapies in Osteoporosis and Bone Regeneration. *Int. J. Mol. Sci.* **2018**, *20*, 83. [CrossRef]
6. Van Houdt, C.I.A.; Gabbai-Armelin, P.R.; Perez, P.M.L.; Ulrich, D.J.O.; Jansen, J.A.; Renno, A.C.M.; Beucken, J. Alendronate release from calcium phosphate cement for bone regeneration in osteoporotic conditions. *Sci. Rep.* **2018**, *8*, 15398. [CrossRef]

7. Mori, H.; Manabe, M.; Kurachi, Y.; Nagumo, M. Osteointegration of dental implants in rabbit bone with low mineral density. *J. Oral Maxillofac. Surg.* **1997**, *55*, 351–361. [CrossRef]
8. Yang, C.; Wu, H.; Chen, S.; Kang, G. Three-dimensional bioglass-collagen-phosphatidylserine scaffolds designed with functionally graded structure and mechanical features. *Biomed. Tech.* **2018**, *63*, 255–259. [CrossRef]
9. Banwart, J.C.; Asher, M.A.; Hassanein, R.S. Iliac crest bone graft harvest donor site morbidity: A statistical evaluation. *Spine* **1995**, *20*, 1055. [CrossRef]
10. Mazzonetto, R.; Netto, H.D.; Nascimento, F.F. *Enxertos Ósseos em Implantodontia*, 1st ed.; Editora Napoleão: São Paulo, Brazil, 2012.
11. Iaquinta, M.R.; Mazzoni, E.; Manfrini, M.; D'Agostino, A.; Trevisiol, L.; Nocini, R.; Trombelli, L.; Barbanti-Brodano, G.; Martini, F.; Tognon, M. Innovative Biomaterials for Bone Regrowth. *Int. J. Mol. Sci.* **2019**, *20*, 618. [CrossRef]
12. de Vasconcellos, L.M.R.; Santana-Melo, G.F.; Silva, E.; Pereira, V.F.; Araújo, J.C.R.; Silva, A.D.R.; Furtado, A.S.A.; Elias, C.M.V.; Viana, B.C.; Marciano, F.R.; et al. Electrospun Poly(butylene-adipate-co-terephthalate)/Nano-hyDroxyapatite/Graphene Nanoribbon Scaffolds Improved the In Vivo Osteogenesis of the Neoformed Bone. *J. Funct. Biomater.* **2021**, *12*, 11. [CrossRef]
13. Silva, A.D.S.; Rodrigues, B.V.M.; Oliveira, F.C.; Carvalho, J.O.; de Vasconcellos, L.M.R.; de Araújo, J.C.R.; Marciano, F.R.; Lobo, A.O. Characterization and in vitro and in vivo assessment of poly(butylene adipate-co-terephthalate)/nano-hydroxyapatite composites as scaffolds for bone tissue engineering. *J. Polym. Res.* **2019**, *26*, 53. [CrossRef]
14. Duracan, C.; Brown, P.W. Biodegradable hydroxyapatite-polymer composites. *Adv. Eng. Mater.* **2001**, *3*, 227–231. [CrossRef]
15. Johari, B.; Kadivar, M.; Lak, S.; Gholipourmalekabadi, M.; Urbanska, A.M.; Mozafari, M.; Ahmadzadehzarajabad, M.; Azarnezhad, A.; Afshari, S.; Zargan, J.; et al. Osteoblast-seeded bioglass/gelatin nanocomposite: A promising bone substitute in critical-size calvarial defect repair in rat. *Int. J. Artif. Organs* **2016**, *39*, 524–533. [CrossRef]
16. Khoshakhlagh, P.; Rabiee, S.M.; Kiaee, G.; Heidari, P.; Miri, A.K.; Moradi, R.; Moztarzadeh, F.; Ravarian, R. Development and characterization of a bioglass/chitosan composite as an injectable bone substitute. *Carbohydr. Polym.* **2017**, *157*, 1261–1271. [CrossRef]
17. Rizwan, M.; Hamdi, M.; Basirun, W.J. Bioglass® 45S5-based composites for bone tissue engineering and functional applications. *J. Biomed. Mater. Res. Part A* **2017**, *105*, 3197–3223. [CrossRef]
18. Zhang, L.; Ke, X.; Lin, L.; Xiao, J.; Yang, X.; Wang, J.; Yang, G.; Xu, S.; Gou, Z.; Shi, Z. Systematic evaluation of the osteogenic capacity of low-melting bioactive glass-reinforced 45S5 Bioglass porous scaffolds in rabbit femoral defects. *Biomed. Mater.* **2017**, *12*, 035010. [CrossRef]
19. Fernandes, H.R.; Gaddam, A.; Rebelo, A.; Brazete, D.; Stan, G.E.; Ferreira, J.M.F. Bioactive Glasses and Glass-Ceramics for Healthcare Applications in Bone Regeneration and Tissue Engineering. *Materials* **2018**, *11*, 2530. [CrossRef]
20. Lee, S.; Matsugaki, A.; Kasuga, T.; Nakano, T. Development of bifunctional oriented bioactive glass/poly(lactic acid) composite scaffolds to control osteoblast alignment and proliferation. *J. Biomed. Mater. Res. Part A* **2019**, *107*, 1031–1041. [CrossRef]
21. Pazarceviren, A.E.; Evis, Z.; Keskin, D.; Tezcaner, A. Resorbable PCEC/gelatin-bismuth doped bioglass-graphene oxide bilayer membranes for guided bone regeneration. *Biomed. Mater.* **2019**, *14*, 035018. [CrossRef]
22. Hench, L.L.; Paschall, H.A. Direct Chemical Bond of Bioactive Glass-Ceramic Materials to Bone and Muscle. *J. Biomed. Mater. Res.* **1973**, *7*, 25–42. [CrossRef]
23. Cerruti, M.; Greenspan, D.; Powers, K. Effect of pH and ionic strength on the reactivity of Bioglass 45S5. *Biomaterials* **2005**, *26*, 1665–1674. [CrossRef] [PubMed]
24. Slosarczyk, A.; Paszkiewicza, Z.; Paluszkiewicz, C. FTIR and XRD evaluation of carbonated hydroxyapatite powders synthesized by wet methods. *J. Mol. Struct.* **2005**, *744–747*, 657–661. [CrossRef]
25. Anderson, A.; Dallmier, A.; Chudzik, S.; Duran, L.; Guire, P.; Hergenrother, R.; Lodhi, M.; Novak, A.; Ofstead, R.; Wormuth, K.; et al. Technologies for the surface modification of biomaterials. In *Biomaterials in Orthopedics*; Yaszemski, M.J., Trantolo, D.J., Lewandrowski, K.U., Hasirci, V., Altobelli, D.E., Wise, D.L., Eds.; Marcel Dekker Inc.: New York, NY, USA, 2004; p. 123.
26. Hench, L.L. The story of Bioglass. *J. Mater. Sci. Mater. Med.* **2006**, *17*, 967–978. [CrossRef] [PubMed]
27. Zamet, J.; Darbar, U.; Griffiths, G.; Bulman, J.; Brägger, U.; Bürgin, W.; Newman, H. Particulate Bioglass® as a grafting material in the treatment of periodontal intrabony defects. *J. Clin. Periodontol.* **1997**, *24*, 410–418. [CrossRef]
28. Tadjoedin, E.S.; De Lange, G.L.; Lyaruu, D.; Kuiper, L.; Burger, E.H. High concentrations of bioactive glass material (Biogran®) vs. Autogenous bone for sinus floor elevation. *Clin. Oral Implants Res.* **2002**, *13*, 428–436. [CrossRef]
29. Baino, F.; Novajra, G.; Miguez-Pacheco, V.; Boccaccini, A.R.; Vitale-Brovarone, C. Bioactive glasses: Special applications outside the skeletal system. *J. Non-Cryst. Solids* **2016**, *432*, 15–30. [CrossRef]
30. Domingues, Z.; Cortés, M.; Gomes, T.; Diniz, H.; Freitas, C.; Gomes, J.; Faria, A.; Sinisterra, R. Bioactive glass as a drug delivery system of tetracycline and tetracycline associated with β-cyclodextrin. *Biomaterials* **2004**, *25*, 327–333. [CrossRef]
31. Lisboa-Filho, P.N.; Gomes-Ferreira, P.H.S.; Batista, F.R.S.; Momesso, G.A.C.; Faverani, L.P.; Okamoto, R. Bone repair with raloxifene and bioglass nanoceramic composite in animal experiment. *Connect. Tissue Res.* **2018**, *59* (Suppl. 1), 97–101. [CrossRef]
32. Mosqueira, L.; Barrioni, B.R.; Martins, T.; Ocarino, N.M.; Serakides, R.; Pereira, M.M. In vitro effects of the co-release of icariin and strontium from bioactive glass submicron spheres on the reduced osteogenic potential of rat osteoporotic bone marrow mesenchymal stem cells. *Biomed. Mater.* **2020**, *15*, 055023. [CrossRef]
33. Rivadeneira, J.; Di Virgilio, A.L.; Audisio, M.C.; Boccaccini, A.R.; Gorustovich, A.A. Evaluation of the antibacterial effects of vancomycin hydrochloride released from agar-gelatin-bioactive glass composites. *Biomed. Mater.* **2015**, *10*, 015011. [CrossRef]

34. Ning, Z.; Tan, B.; Chen, B.; Lau, D.A.S.; Wong, T.M.; Sun, T.; Peng, S.; Li, Z.; Lu, W.W. Precisely Controlled Delivery of Abaloparatide through Injectable Hydrogel to Promote Bone Regeneration. *Macromol. Biosci.* **2019**, *19*, e1900020. [CrossRef]
35. Wang, C.; Yu, S.; Fretwurst, T.; Larsson, L.; Sugai, J.; Oh, J.; Lehner, K.; Jin, Q.; Giannobile, W. Maresin 1 Promotes Wound Healing and Socket Bone Regeneration for Alveolar Ridge Preservation. *J. Dent. Res.* **2020**, *99*, 930–937. [CrossRef]
36. Ersan, N.; van Ruijven, L.J.; Bronckers, A.L.; Olgaç, V.; Ilgüy, D.; Everts, V. Teriparatide and the treatment of bisphosphonate-related osteonecrosis of the jaw: A rat model. *Dentomaxillofac. Radiol.* **2014**, *43*, 20130144. [CrossRef]
37. Mazziotti, G.; Bilezikian, J.; Canalis, E.; Cocchi, D.; Giustina, A. New understanding and treatments for osteoporosis. *Endocrine* **2012**, *41*, 58–69. [CrossRef]
38. Kubota, T.; Hasuike, A.; Naito, M.; Tsunori, K.; Min, S.; Sato, S. Enhancement of Bone Augmentation in Osteoporotic Conditions by the Intermittent Parathyroid Hormone: An Animal Study in the Calvarium of Ovariectomized Rat. *Int. J. Oral Maxillofac. Implants* **2018**, *33*, 1003–1010. [CrossRef]
39. de Oliveira, D.; de Oliveira Puttini, I.; Silva Gomes-Ferreira, P.H.; Palin, L.P.; Matsumoto, M.A.; Okamoto, R. Effect of intermittent teriparatide (PTH 1-34) on the alveolar healing process in orchiectomized rats. *Clin. Oral Investig.* **2019**, *23*, 2313–2322. [CrossRef]
40. Puttini, I.D.O.; Poli, P.P.; Maiorana, C.; de Vasconcelos, I.R.; Schmidt, L.E.; Colombo, L.T.; Hadad, H.; dos Santos, G.M.; de Carvalho, P.S.P.; Souza, F. Evaluation of Osteoconduction of Biphasic Calcium Phosphate Ceramic in the Calvaria of Rats: Microscopic and Histometric Analysis. *J. Funct. Biomater.* **2019**, *10*, 7. [CrossRef]
41. Jiang, X.; Xu, C.; Shi, H.; Cheng, Q. PTH1-34 improves bone healing by promoting angiogenesis and facilitating MSCs migration and differentiation in a stabilized fracture mouse model. *PLoS ONE* **2019**, *14*, e0226163. [CrossRef]
42. Özer, T.; Başlarlı, Ö.; Aktaş, A.; Barış, E.; Çelik, H.H.; Ocak, M. Locally administrated single-dose teriparatide affects critical-size rabbit calvarial defects: A histological, histomorphometric and micro-CT study. *Acta Orthop. Traumatol. Turc.* **2019**, *53*, 478–484. [CrossRef]
43. Tao, Z.-S.; Zhou, W.-S.; Wu, X.-J.; Wang, L.; Yang, M.; Xie, J.-B.; Xu, Z.-J.; Ding, G.-Z. Single-dose local administration of parathyroid hormone (1-34, PTH) with β-tricalcium phosphate/collagen (β-TCP/COL) enhances bone defect healing in ovariectomized rats. *J. Bone Miner. Metab.* **2019**, *37*, 28–35. [CrossRef]
44. Bakker, A.D.; Zandieh-Doulabi, B.; Klein-Nulend, J. Strontium ranelate affects signaling from mechanically-stimulated osteocytes towards osteoclasts and osteoblasts. *Bone* **2013**, *53*, 112–119. [CrossRef] [PubMed]
45. Wang, J.; Li, J.; Yang, L.; Zhou, Y.; Wang, Y. Dose-dependence of PTH-related peptide-1 on the osteogenic induction of MC3T3-E1 cells in vitro. *Medicine* **2017**, *96*, e6637. [CrossRef] [PubMed]
46. Bang, J.H.; Suslick, K.S. Applications of ultrasound to the synthesis of nanostructured materials. *Adv. Mater.* **2010**, *22*, 1039–1059. [CrossRef] [PubMed]
47. Arruda, L.B.; Orlandi, M.O.; Lisboa-Filho, P.N. Morphological modifications and surface amorphization in ZnO sonochemically treated nanoparticles. *Ultrason. Sonochem.* **2013**, *20*, 799–804. [CrossRef]
48. Gonzalo-Juan, I.; Xie, F.; Becker, M.; Tulyaganov, D.U.; Ionescu, E.; Lauterbach, S.; De Angelis Rigotti, F.; Fischer, A.; Riedel, R. Synthesis of Silver Modified Bioactive Glassy Materials with Antibacterial Properties via Facile and Low-Temperature Route. *Materials* **2020**, *13*, 5115. [CrossRef]
49. Spirandeli, B.R.; Campos, T.M.B.; Ribas, R.G.; Thim, G.P.; Trichês, E.S. Evaluation of colloidal and polymeric routes in sol-gel synthesis of a bioactive glass-ceramic derived from 45S5 bioglass. *Ceram. Int.* **2020**, *46*, 20264–20271. [CrossRef]
50. Zhang, L.; Chan, C. Isolation and enrichment of rat mesenchymalstem cells (MSCs) and separation of single-colony derived MSCs. *J. Vis. Exp.* **2010**, *22*, 1852. [CrossRef]
51. do Prado, R.F.; Esteves, G.C.; Santos, E.; Bueno, D.; Cairo, C.; Vasconcellos, L.G.; Sagnori, R.; Tessarin, F.; Oliveira, F.E.; Oliveira, L.D.; et al. In vitro and in vivo biological performance of porous Ti alloys prepared by powder metallurgy. *PLoS ONE* **2018**, *13*, e0196169. [CrossRef]
52. International Organization for Standardization. ISO 10993-5. Biological Evaluation of Medical Devices. Part 5: Tests for Cytotoxicity: In Vitro Methods. Available online: https://www.iso.org/standard/36406.html (accessed on 26 October 2022).
53. Lowry, O.H.; Rosebrough, N.J.; Farr, A.L.; Randall, R.J. Protein measurement with the Folin phenol reagent. *J. Biol. Chem.* **1951**, *193*, 265–275. [CrossRef]
54. Gregory, C.A.; Gunn, W.G.; Peister, A.; Prockop, D.J. An Alizarin red-based assay of mineralization by adherent cells in culture: Comparison with cetylpyridinium chloride extraction. *Anal. Biochem.* **2004**, *329*, 77–84. [CrossRef]
55. Percie du Sert, N.; Hurst, V.; Ahluwalia, A.; Alam, S.; Avey, M.T.; Baker, M.; Browne, W.J.; Clark, A.; Cuthill, I.C.; Dirnagl, U.; et al. The ARRIVE guidelines 2.0: Updated guidelines for reporting animal research. *Br. J. Pharmacol.* **2020**, *177*, 3617–3624. [CrossRef]
56. Gomes-Ferreira, P.H.S.; de Oliveira, D.; Frigério, P.B.; de Souza Batista, F.R.; Grandfield, K.; Okamoto, R. Teriparatide improves microarchitectural characteristics of peri-implant bone in orchiectomized rats. *Osteoporos. Int.* **2020**, *31*, 1807–1815. [CrossRef]
57. Develos Godoy, D.J.; Banlunara, W.; Jaroenporn, S.; Sangvanich, P.; Thunyakitpisal, P. Collagen and mPCL-TCP scaffolds induced differential bone regeneration in ovary-intact and ovariectomized rats. *Biomed. Mater. Eng.* **2018**, *29*, 389–399. [CrossRef]
58. Luvizuto, E.R.; Queiroz, T.P.; Dias, S.M.D.; Okamoto, T.; Dornelles, R.C.M.; Garcia, I.R.; Okamoto, R. Histomorphometric analysis and immunolocalization of RANKL and OPG during the alveolar healing process in female ovariectomized rats treated with oestrogen or raloxifene. *Arch. Oral Biol.* **2010**, *55*, 52–59. [CrossRef]

59. da Cruz Vegian, M.R.; Costa, B.C.A.; de Fátima Santana-Melo, G.; Godoi, F.H.C.; Kaminagakura, E.; Tango, R.N.; do Prado, R.F.; de Oliveira, L.D.; Federico, C.A.; de Oliveira Marco Avelino, S.; et al. Systemic and local effects of radiotherapy: An experimental study on implants placed in rats. *Clin. Oral Investig.* **2020**, *24*, 785–797. [CrossRef]
60. dos Santos, P.L.; Queiroz, T.P.; Margonar, R.; Gomes de Souza Carvalho, A.C.; Okamoto, R.; de Souza Faloni, A.P.; Garcia Júnior, I.R. Guided implant surgery: What is the influence of this new technique on bone cell viability? *J. Oral Maxillofac. Surg.* **2013**, *71*, 505–512. [CrossRef]
61. dos Santos, P.L.; de Molon, R.S.; Queiroz, T.P.; Okamoto, R.; Faloni, A.P.D.S.; Gulinelli, J.L.; Luvizuto, E.R.; Junior, I.R.G. Evaluation of bone substitutes for treatment of peri-implant bone defects: Biomechanical, histological, and immunohistochemical analyses in the rabbit tibia. *J. Periodontal Implant Sci.* **2016**, *46*, 176–196. [CrossRef]
62. Esteves, J.; Marcantonio, E., Jr.; Faloni, A.; Rocha, F.R.G.; Marcantonio, R.A.; Wilk, K.; Intini, G. Dynamics of bone healing after osteotomy with piezosurgery or conventional drilling—Histomorphometrical, immunohistochemical, and molecular analysis. *J. Transl. Med.* **2013**, *11*, 221. [CrossRef]
63. Queiroz, T.P.; Souza, F.A.; Okamoto, R.; Margonar, R.; Pereira-Filho, V.A.; Garcia, I.R., Jr.; Vieira, E.H. Evaluation of immediate bonecell viability and of drill wear after implant osteotomies: Immunohistochemistry and scanning electron microscopy analysis. *J. Oral Maxillofac. Surg.* **2008**, *66*, 1233–1240. [CrossRef]
64. Raisz, L.G. Pathogenesis of osteoporosis: Concepts, conflicts, and prospects. *J. Clin. Investig.* **2005**, *115*, 3318–3325. [CrossRef]
65. Di Munno, O.; Ferro, F. The effect of biologic agents on bone homeostasis in chronic inflammatory rheumatic diseases. *Clin. Exp. Rheumatol.* **2019**, *7*, 502–507.
66. Zerbini, C.A.F.; Clark, P.; Mendez-Sanchez, L.; Pereira, R.M.R.; Messina, O.D.; Uña, C.R.; Adachi, J.D.; Lems, W.F.; Cooper, C.; Lane, N.E. IOF Chronic Inflammation and Bone Structure (CIBS) Working Group. Biologic therapies and bone loss in rheumatoid arthritis. *Osteoporos. Int.* **2017**, *28*, 429–446. [CrossRef] [PubMed]
67. Wang, X.; Zeng, D.; Weng, W.; Huang, Q.; Zhang, X.; Wen, J.; Wu, J.; Jiang, X. Alendronate delivery on amino modified mesoporous bioactive glass scaffolds to enhance bone regeneration in osteoporosis rats. *Artif. Cells Nanomed. Biotechnol.* **2018**, *46* (Suppl. 2), 171–181. [CrossRef] [PubMed]
68. Lin, K.; Xia, L.; Li, H.; Jiang, X.; Pan, H.; Xu, Y.; Lu, W.W.; Zhang, Z.; Chang, J. Enhanced osteoporotic bone regeneration by strontium-substituted calcium silicate bioactive ceramics. *Biomaterials* **2013**, *34*, 10028–10042. [CrossRef]
69. Kyllönen, L.; D'Este, M.; Alini, M.; Eglin, D. Local drug delivery for enhancing fracture healing in osteoporotic bone. *Acta Biomater.* **2015**, *11*, 412–434. [CrossRef]
70. Sun, J.; Fan, W.; Wu, D.; Sun, Y. Structure Control of SiO2 Sol-Gels via Addition of PEG. *Stud. Surf. Sci. Catal.* **1998**, *118*, 617–624. [CrossRef]
71. Ajita, J.; Saravanan, S.; Selvamurugan, N. Effect of size of bioactive glass nanoparticles on mesenchymal stem cell proliferation for dental and orthopedic applications. *Mater. Sci. Eng. C Mater. Biol. Appl.* **2015**, *53*, 142–149 [CrossRef]
72. Hoppe, A.; Güldal, N.S.; Boccaccini, A.R. A review of the biological response to ionic dissolution products from bioactive glasses and glass-ceramics. *Biomaterials* **2011**, *32*, 2757–2774. [CrossRef]
73. Gorustovich, A.A.; Roether, J.A.; Boccaccini, A.R. Effect of bioactive glasses on angiogenesis: A review of in vitro and in vivo evidences. *Tissue Eng. Part B Rev.* **2010**, *16*, 199–207. [CrossRef]
74. Xynos, I.D.; Hukkanen, M.V.; Batten, J.J.; Buttery, L.D.; Hench, L.L.; Polak, J.M. Bioglass 45S5 stimulates osteoblast turnover and enhances bone formation in vitro: Implications and applications for bone tissue engineering. *Calcif. Tissue Int.* **2000**, *67*, 321–329. [CrossRef]
75. Andersson, O.H.; Kangasniemi, I. Calcium phosphate formation at the surface of bioactive glass in vivo. *J. Non-Cryst. Solids* **1990**, *119*, 290–296. [CrossRef]
76. Gomes-Ferreira, P.H.S.; Lisboa-Filho, P.N.; da Silva, A.C.; Bim-Júnior, O.; Batista, F.R.D.S.; Ervolino-Silva, A.C.; Garcia-Junior, I.R.; Okamoto, R. Sonochemical time standardization for bioactive materials used in periimplantar defects filling. *Ultrason. Sonochem.* **2019**, *56*, 437–446. [CrossRef]
77. Lee, E.M.R.; Borges, R.; Marchi, J.; Eduardo, C.D.P.; Marques, M.M. Bioactive glass and high-intensity lasers as a promising treatment for dentin hypersensitivity: An in vitro study. *J. Biomed. Mater. Res. Part B Appl. Biomater.* **2019**, *108*, 939–947. [CrossRef]
78. Andersson, O.H.; Kangasniemi, I.O. Calcium phosphate formation at the surface of bioactive glass in vitro. *J. Biomed. Mater. Res.* **1991**, *25*, 1019–1030. [CrossRef]
79. Bahari Javan, N.; Rezaie Shirmard, L.; Jafary Omid, N.; Akbari Javar, H.; Rafiee Tehrani, M.; Abedin Dorkoosh, F. Preparation, statistical optimisation and in vitro characterisation of poly (3-hydroxybutyrate-co-3-hydroxyvalerate)/poly (lactic-co-glycolic acid) blend nanoparticles for prolonged delivery of teriparatide. *J. Microencapsul.* **2016**, *33*, 460–474. [CrossRef]
80. Fiume, E.; Barberi, J.; Verné, E.; Baino, F. Bioactive glasses: From parent 45S5 composition to scaffold-assisted tissue-healing therapies. *J. Funct. Biomater.* **2018**, *9*, 24. [CrossRef]
81. Gomes-Ferreira, P.H.S.; Micheletti, C.; Frigério, P.B.; de Souza Batista, F.R.; Monteiro, N.G.; Bim-Júnior, O.; Lisboa-Filho, P.N.; Grandfield, K.; Okamoto, R. PTH 1-34-functionalized bioactive glass improves peri-implant bone repair in orchiectomized rats: Microscale and ultrastructural evaluation. *Biomater. Adv.* **2022**, *134*, 112688. [CrossRef]
82. Cai, K.; Frant, M.; Bossert, J.; Hildebrand, G.; Liefeith, K.; Jandt, K.D. Surface functionalized titanium thin films: Zeta-potential, protein adsorption and cell proliferation. *Colloids Surf. B Biointerfaces* **2006**, *50*, 1–8. [CrossRef]

83. Lin, S.; Van den Bergh, W.; Baker, S.; Jones, J.R. Protein interactions with nanoporous sol-gel derived bioactive glasses. *Acta Biomater.* **2011**, *7*, 3606–3615. [CrossRef]
84. Stoddart, M.J. Cell viability assays: Introduction. *Methods Mol. Biol.* **2011**, *740*, 1–6. [CrossRef]
85. Westhauser, F.; Karadjian, M.; Essers, C.; Senger, A.-S.; Hagmann, S.; Schmidmaier, G.; Moghaddam, A. Osteogenic differentiation of mesenchymal stem cells is enhanced in a 45S5-supplemented β-TCP composite scaffold: An in-vitro comparison of Vitoss and Vitoss BA. *PLoS ONE* **2019**, *14*, e0212799. [CrossRef] [PubMed]
86. Furusawa, T.; Mizunuma, K. Osteoconductive properties and efficacy of resorbable bioactive glass as a bone-grafting material. *Implant Dent.* **1997**, *6*, 93–101. [CrossRef] [PubMed]
87. Vanderschueren, D.; Vandenput, L.; Boonen, S.; Lindberg, M.K.; Bouillon, R.; Ohlsson, C. Androgens and bone. *Endocr. Rev.* **2004**, *25*, 389–425. [CrossRef] [PubMed]
88. Reible, B.; Schmidmaier, G.; Prokscha, M.; Moghaddam, A.; Westhauser, F. Continuous stimulation with differentiation factors is necessary to enhance osteogenic differentiation of human mesenchymal stem cells in-vitro. *Growth Factors* **2017**, *35*, 179–188. [CrossRef]
89. Tsigkou, O.; Jones, J.R.; Polak, J.M.; Stevens, M.M. Differentiation of fetal osteoblasts and formation of mineralized bone nodules by 45S5 Bioglass conditioned medium in the absence of osteogenic supplements. *Biomaterials* **2009**, *30*, 3542–3550. [CrossRef]
90. Che, L.; Wang, Y.; Sha, D.; Li, G.; Wei, Z.; Liu, C.; Yuan, Y.; Song, D. A biomimetic and bioactive scaffold with intelligently pulsatile teriparatide delivery for local and systemic osteoporosis regeneration. *Bioact. Mater.* **2022**, *19*, 75–87. [CrossRef]
91. Rodrigues, C.; Naasani, L.I.S.; Zanatelli, C.; Paim, T.C.; Azevedo, J.G.; de Lima, J.C.; Fernandes, M.D.C.; Buchner, S.; Wink, M.R. Bioglass 45S5: Structural characterization of short range order and analysis of biocompatibility with adipose-derived mesenchymal stromal cells in vitro and in vivo. *Mater. Sci. Eng. C Mater. Biol. Appl.* **2019**, *103*, 109781. [CrossRef]
92. Reible, B.; Schmidmaier, G.; Moghaddam, A.; Westhauser, F. Insulin-like growth factor-1 as a possible alternative to bone morphogenetic protein-7 to induce osteogenic differentiation of human mesenchymal stem cells in vitro. *Int. J. Mol. Sci.* **2018**, *19*, 1674. [CrossRef]
93. Blick, S.K.; Dhillon, S.; Keam, S.J. Teriparatide: A review of its use in osteoporosis. *Drugs* **2008**, *68*, 2709–2737. [CrossRef]
94. Roggia, C.; Gao, Y.; Cenci, S.; Weitzmann, M.N.; Toraldo, G.; Isaia, G.; Pacifici, R. Up-regulation of TNF-producing T cells in the bone marrow: A key mechanism by which estrogen deficiency induces bone loss in vivo. *Proc. Natl. Acad. Sci. USA* **2001**, *98*, 13960–13965. [CrossRef]
95. Weitzmann, M.N.; Pacifici, R. Estrogen deficiency and bone loss: An inflammatory tale. *J. Clin. Investig.* **2006**, *116*, 1186–1194. [CrossRef]
96. Weitzmann, M.N.; Roggia, C.; Toraldo, G.; Weitzmann, L.; Pacifici, R. Increased production of IL-7 uncouples bone formation from bone resorption during estrogen deficiency. *J. Clin. Investig.* **2002**, *110*, 1643–1650. [CrossRef]
97. Weitzmann, M.N.; Pacifici, R. Role of the immune system in postmenopausal bone loss. *Curr. Osteoporos. Rep.* **2005**, *3*, 92–97. [CrossRef]
98. Hao, F.; Gu, Y.; Tan, X.; Deng, Y.; Wu, Z.-T.; Xu, M.-J.; Wang, W.-Z. Estrogen replacement reduces oxidative stress in the rostral ventrolateral medulla of ovariectomized rats. *Oxid. Med. Cell. Longev.* **2016**, *2016*, 2158971. [CrossRef]
99. Ramalho-Ferreira, G.; Faverani, L.P.; Momesso, G.A.C.; Luvizuto, E.R.; de Oliveira Puttini, I.; Okamoto, R. Effect of antiresorptive drugs in the alveolar bone healing: A histometric and immunohistochemical study in ovariectomized rats. *Clin. Oral Investig.* **2017**, *21*, 1485–1494. [CrossRef]
100. Huebsch, N.; Mooney, D.J. Inspiration and application in the evolution of biomaterials. *Nature* **2009**, *462*, 426–432. [CrossRef]
101. Day, R.M.; Boccaccini, A.R.; Shurey, S.; Roether, J.A.; Forbes, A.; Hench, L.L.; Gabe, S.M. Assessment of polyglycolic acid mesh and bioactive glass for soft-tissue engineering scaffolds. *Biomaterials* **2004**, *25*, 5857–5866. [CrossRef]
102. Auersvald, C.M.; Santos, F.R.; Nakano, M.M.; Leoni, G.B.; Neto, M.D.D.S.; Scariot, R.; Giovanini, A.F.; Deliberador, T.M. The local administration of parathyroid hormone encourages the healing of bone defects in the rat calvaria: Micro-computed tomography, histological and histomorphometric evaluation. *Arch. Oral Biol.* **2017**, *79*, 14–19. [CrossRef]
103. Malik, D.K.; Baboota, S.; Ahuja, S.; Hasan, S.; Ali, J. Recent advances in protein and peptide drug delivery systems. *Curr. Drug Deliv.* **2007**, *4*, 141–151. [CrossRef]
104. Frigério, P.B.; Gomes-Ferreira, P.H.S.; de Souza Batista, F.R.; Moura, J.; Garcia Júnior, I.R.; Botticelli, D.; Lisboa-Filho, P.N.; Okamoto , R. Effect of Topical PTH 1-34 Functionalized to Biogran® in the Process of Alveolar Repair in Rats Submitted to Orchiectomy. *Materials* **2021**, *15*, 207. [CrossRef]
105. Moore, A.E.; Blake, G.M.; Taylor, K.A.; Rana, A.E.; Wong, M.; Chen, P.; Fogelman, I. Assessment of regional changes in skeletal metabolism following 3 and 18 months of teriparatide treatment. *J. Bone Miner. Res.* **2010**, *25*, 960–967. [CrossRef]
106. Ohata, T.; Maruno, H.; Ichimura, S. Changes over time in callus formation caused by intermittently administering PTH in rabbit distraction osteogenesis models. *J. Orthop. Surg. Res.* **2015**, *10*, 88. [CrossRef] [PubMed]
107. Aleksyniene, R.; Thomsen, J.S.; Eckardt, H.; Bundgaard, K.G.; Lind, M.; Hvid, I. Threedimensional microstructural properties of regenerated mineralizing tissue after PTH (1–34) treatment in a rabbit tibial lengthening model. *J. Musculoskelet. Neuronal Interact* **2009**, *9*, 268–277. [PubMed]
108. Boskey, A.L.; Gadaleta, S.; Gundberg, C.; Doty, S.B.; Ducy, P.; Karsenty, G. Fourier transform infrared microspectroscopic analysis of bones of osteocalcin-deficient mice provides insight into the function of osteocalcin. *Bone* **1998**, *23*, 187–196. [CrossRef]

109. Hauschka, P.V.; Frenkel, J.; DeMuth, R.; Gundberg, C.M. Presence of osteocalcin and related higher molecular weight 4-carboxyglutamic acid-containing proteins in developing bone. *J. Biol. Chem.* **1983**, *258*, 176–182. [CrossRef] [PubMed]
110. Aldini., N.N.; Fini, M.; Giavaresi, G.; Giardino, R.; Greggi, T.; Parisini, P. Pedicular fixation in the osteoporotic spine: A pilot in vivo study on long-term ovariectomized sheep. *J Orthop Res.* **2002**, *20*, 1217–1224. [CrossRef]
111. Weng, S.J.; Yan, D.Y.; Tang, J.H.; Shen, Z.J.; Wu, Z.Y.; Xie, Z.J.; Yang, J.-Y.; Bai, B.-L.; Chen, L.; Boodhun, V.; et al. Combined treatment with cinnamaldehyde and β-TCP had an additive effect on bone formation and angiogenesis in critical size calvarial defect in ovariectomized rats. *Biomed. Pharmacother.* **2019**, *109*, 573–581. [CrossRef]
112. Tera, T.M.; Prado, R.F.; De Marco, A.C.; Santamaria, M.P.; Jardini, M.A. The RANK/ RANKL/ OPG interaction in the repair of autogenous bone grafts in female rats with estrogen deficiency. *Braz. Oral Res.* **2014**, 28. [CrossRef]
113. Yolcu, Y.; Alvi, M.A.; Wanderman, N.; Carlson, B.; Sebastian, A.; Bydon, M.; Freedman, B. Effect of teriparatide use on bone mineral density and spinal fusion: A narrative review of animal models. *Int. J. Neurosci.* **2019**, *129*, 814–820. [CrossRef]
114. Seebach, C.; Skripitz, R.; Andreassen, T.T.; Aspenberg, P. Intermittent parathyroid hormone (1-34) enhances mechanical strength and density of new bone after distraction osteogenesis in rats. *J. Orthop. Res.* **2004**, *22*, 472–478. [CrossRef]
115. Andreassen, T.T.; Fledelius, C.; Ejersted, C.; Oxlund, H. Increases in callus formation and mechanical strength of healing fractures in old rats treated with parathyroid hormone. *Acta Orthop. Scand.* **2001**, *72*, 304–307. [CrossRef]
116. Ejersted, C.; Andreassen, T.; Oxlund, H.; Jørgensen, P.; Bak, B.; Häggblad, J.; Tørring, O.; Nilsson, M. Human parathyroid hormone (1–34) and (1–84) increase the mechanical strength and thickness of cortical bone in rats. *J. Bone Miner. Res.* **1993**, *9*, 1097–1101. [CrossRef]
117. Skripitz, R.; Andreassen, T.T.; Aspenberg, P. Parathyroid hormone (1–34) increases the density of rat cancellous bone in a bone chamber. *J. Bone Jt. Surgery.* **2000**, *82-B*, 138–141. [CrossRef]
118. Leiblein, M.; Henrich, D.; Fervers, F.; Kontradowitz, K.; Marzi, I.; Seebach, C. Do antiosteoporotic drugs improve bone regeneration in vivo? *Eur. J. Trauma Emerg. Surg.* **2020**, *46*, 287–299. [CrossRef]
119. Kadiroğlu, E.T.; Akbalık, M.E.; Karaöz, E.; Kanay, B.E.; Dağ, A.; Ketani, M.A.; Eroğlu, E.G.; Uysal, E.; Tuncer, M.C. Calvarial bone defects in ovariectomised rats treated with mesenchymal stem cells and demineralised freeze-dried bone allografts. *Folia Morphol.* **2020**, *79*, 720–735. [CrossRef]

Journal of *Functional Biomaterials*

MDPI

Article

Mesenchymal Stem Cells Combined with a P(VDF-TrFE)/BaTiO$_3$ Scaffold and Photobiomodulation Therapy Enhance Bone Repair in Rat Calvarial Defects

Leticia Faustino Adolpho [1,†], Larissa Mayra Silva Ribeiro [2,†], Gileade Pereira Freitas [1,3], Helena Bacha Lopes [1], Maria Paula Oliveira Gomes [1], Emanuela Prado Ferraz [1], Rossano Gimenes [2], Marcio Mateus Beloti [1] and Adalberto Luiz Rosa [1,*]

1 Bone Research Lab, School of Dentistry of Ribeirão Preto, University of São Paulo, Ribeirão Preto 14040-904, SP, Brazil; leticia.adolpho@usp.br (L.F.A.); gileade@ufg.br (G.P.F.); helena.lopes@usp.br (H.B.L.); maria.paula.gomes@usp.br (M.P.O.G.); emanuelaferraz@usp.br (E.P.F.); mmbeloti@usp.br (M.M.B.)

2 Institute of Physics and Chemistry, University of Itajubá, Itajubá 37500-903, MG, Brazil; larissamayra@unifei.edu.br (L.M.S.R.); rossano@unifei.edu.br (R.G.)

3 School of Dentistry, Federal University of Goiás, Goiânia 74605-020, GO, Brazil

* Correspondence: adalrosa@forp.usp.br; Tel.: +55-16-3315-4106

† These authors contributed equally to this work.

Abstract: Background: Tissue engineering and cell therapy have been the focus of investigations on how to treat challenging bone defects. This study aimed to produce and characterize a P(VDF-TrFE)/BaTiO$_3$ scaffold and evaluate the effect of mesenchymal stem cells (MSCs) combined with this scaffold and photobiomodulation (PBM) on bone repair. Methods and results: P(VDF-TrFE)/BaTiO$_3$ was synthesized using an electrospinning technique and presented physical and chemical properties suitable for bone tissue engineering. This scaffold was implanted in rat calvarial defects (unilateral, 5 mm in diameter) and, 2 weeks post-implantation, MSCs were locally injected into these defects (n = 12/group). Photobiomodulation was then applied immediately, and again 48 and 96 h post-injection. The µCT and histological analyses showed an increment in bone formation, which exhibited a positive correlation with the treatments combined with the scaffold, with MSCs and PBM inducing more bone repair, followed by the scaffold combined with PBM, the scaffold combined with MSCs, and finally the scaffold alone (ANOVA, $p \leq 0.05$). Conclusions: The P(VDF-TrFE)/BaTiO$_3$ scaffold acted synergistically with MSCs and PBM to induce bone repair in rat calvarial defects. These findings emphasize the need to combine a range of techniques to regenerate large bone defects and provide avenues for further investigations on innovative tissue engineering approaches.

Keywords: bone repair; electrospinning; laser therapy; mesenchymal stem cell; PVDF scaffold

Citation: Adolpho, L.F.; Ribeiro, L.M.S.; Freitas, G.P.; Lopes, H.B.; Gomes, M.P.O.; Ferraz, E.P.; Gimenes, R.; Beloti, M.M.; Rosa, A.L. Mesenchymal Stem Cells Combined with a P(VDF-TrFE)/BaTiO$_3$ Scaffold and Photobiomodulation Therapy Enhance Bone Repair in Rat Calvarial Defects. *J. Funct. Biomater.* **2023**, *14*, 306. https://doi.org/10.3390/jfb14060306

Academic Editor: Sorina Dinescu

Received: 4 May 2023
Revised: 24 May 2023
Accepted: 30 May 2023
Published: 1 June 2023

1. Introduction

Bone is a specialized connective tissue that exhibits great capacity to repair and regenerate when damaged, which may be surpassed by the extension of the lesion, demanding further interventions to achieve restoration in terms of content, anatomy and function [1–3]. The concepts of tissue engineering and cell therapy have been extensively employed to investigate and develop new approaches to treat challenging bone defects [4–8]. In this scenario, the combination of biomaterials and cells offers a promising alternative to autogenous bone graft, the current gold standard material [9,10].

Among a plethora of biomaterials, piezoelectric materials are of interest as they can transduce electrical stimuli to physiological systems in response to events such as cell migration and due to the susceptibility of bone cells to this property [11–13]. The piezoelectric composite poly(vinylidene-trifluoroethylene)/barium titanate (P(VDF-TrFE)/BaTiO$_3$) favors osteoblast differentiation of human alveolar bone-derived cells compared with polytetrafluoroethylene (PTFE) [14,15]. Additionally, the membrane of P(VDF-TrFE)/BaTiO$_3$

induces more bone formation than PTFE in calvarial defects of healthy and osteoporotic rats [16,17]. Such beneficial effects on bone repair are due, at least in part, to the bone resorption inhibition triggered by P(VDF-TrFE)/$BaTiO_3$ through the regulation of a microRNA-34a/receptor activator of nuclear factor kappa B ligand (RANKL) circuit [18].

The good bone response to P(VDF-TrFE)/$BaTiO_3$ composite made it a candidate to be combined with cells to enhance bone repair. Mesenchymal stem cells (MSCs) combined with a P(VDF-TrFE)/$BaTiO_3$ membrane produce more bone repair than the membrane alone when implanted in rat calvarial defects under healthy and osteoporotic conditions [19,20]. The increment in bone repair observed when cells and P(VDF-TrFE)/$BaTiO_3$ were combined suggests that fine-tuning the composite modification and cell response stimulation could result in complete regeneration of the calvarial defect.

Considering bone tissue engineering, scaffolds are more suitable than membranes, and an ideal scaffold should exhibit properties and an architecture that mimic the extracellular matrix, creating a favorable environment for tissue growth [21,22]. Scaffolds can be created using an electrospinning technique, which is a simple and low-cost method of producing fibers at the micro and nanoscale that generate structures with increased surface area based on the material volume [23,24]. Thus, this technique may be employed to produce P(VDF-TrFE)/$BaTiO_3$ scaffolds that may favor bone formation. Regarding cell response stimulation, the use of inflammatory mediators, drugs and specific culture conditions may increase the efficiency of MSCs in regenerative medicine procedures [25,26]. In this context, photobiomodulation (PBM) therapy acts on several signaling pathways that regulate cellular events such as proliferation and differentiation, which are involved in bone formation [27,28]. PBM therapy restores the osteogenic capacity of MSCs derived from diabetic rats and enhances bone formation in rat dental alveolus filled with hydroxyapatite [29,30]. Additionally, PBM increases bone repair in critical size defects treated with either MSCs derived from dental pulp combined with hydrogel or MSCs from adipose tissue combined with decellularized bone matrix [31,32].

Although MSCs can be combined with a P(VDF-TrFE)/$BaTiO_3$ scaffold as well as PBM therapy to enhance bone formation, the combination of these three approaches, cells, scaffolds and PBM, is underexplored in the field of regenerative medicine. Thus, this study aimed to synthesize and characterize a P(VDF-TrFE)/$BaTiO_3$ scaffold produced using an electrospinning technique and evaluate the effect of MSCs combined with this scaffold and submitted to PBM therapy on bone repair.

2. Materials and Methods

2.1. Synthesis of the P(VDF-TrFE)/$BaTiO_3$ Scaffold

The P(VDF-TrFE)/$BaTiO_3$ (90/10, % in volume) composite was obtained by dissolving the copolymer P(VDF-TrFE) (Arkema Piezotech, Pierre-Benite Cedex, France) dispersed in N, N-dimethylformamide (DMF, Sigma-Aldrich, Saint Louis, MO, USA) and acetone (Synth, Diadema, SP, Brazil) at a ratio of 7:3 at 50 °C in a water bath for 3 h. After the complete dissolution of the copolymer, $BaTiO_3$ powder (Sigma-Aldrich) was added to the solution. The copolymer/solvent ratio was 20 g/100 mL. The resulting solution was homogenized using an ultrasonic processor VCX 750 (Sonics & Materials Inc., Newtown, CT, USA) for 6 min in a water–ice bath. The solution was electrospun using a setup composed of a NE-300 syringe pump (New Era, Farmingdale, NY, USA), adjusted to a flow rate of 1500 μL/h, 100 mm from the collector. The voltage applied to the needle (inner diameter of 1.20 mm) was 12 kV using a high-power supply ES40P-5W (Gamma High Voltage, Ormond Beach, FL, USA). The temperature was kept at 25 °C with 42% humidity during electrospinning and the deposition time required to obtain scaffold plates with 0.6 mm in thickness was 30 min. The fibers were collected on nonstick paper and maintained in a vacuum oven for 24 h at 30 °C to ensure that any residual solvent evaporated. Discs of P(VDF-TrFE)/$BaTiO_3$ scaffold were cut with a hole punch (5 mm in diameter) in liquid nitrogen bath to preserve the three-dimensional structure on the edges. Prior to implantation, scaffolds were submitted to ethylene oxide sterilization.

2.2. P(VDF-TrFE)/$BaTiO_3$ Scaffold Characterization

The fiber morphology was examined using scanning electron microscopy (SEM, Phenom ProX, Thermo Fisher Scientific, Waltham, MA, USA). The fiber diameter was estimated in the SEM micrographs using ImageJ software (National Institutes of Health, Bethesda, MD, USA). The pore size distribution was evaluated via microtomographic (μCT) analysis using the SkyScan 1172 system (Bruker-Skyscan, Kontich, Belgium) and the three-dimensional reconstructions were generated using the NRecon Cluster software (Micro Photonics Inc., Allentown, PA, USA). The mean centric linear roughness (R_a) was measured in 5 locations of 3 samples employing three-dimensional images generated by the Phenom ProX SEM software. The elemental analysis was performed on two different areas of the scaffold (fibers and particles) using energy-dispersive X-ray spectroscopy (SEM-EDS, Superscan SSX-550, Shimadzu Corp., Kyoto, Japan) at an acceleration voltage of 20 kV, a working distance of 8 mm and an integration time of 50 s. The P(VDF-TrFE)/$BaTiO_3$ wettability was assessed using a sessile Easy Drop Shape Analyzer (Krüss Scientific, Hamburg, Germany) through contact angle measurements and compared with P(VDF-TrFE) and PVDF. The contact angle was measured at 5 positions of 5 samples (n = 5) using 10 μL of deionized water drop, immediately after drop placement, at 18 °C. Additionally, the contact angle on P(VDF-TrFE)/$BaTiO_3$ was measured at 5 positions of 5 samples (n = 5) every 10 min up to 40 min.

2.3 Evaluation of Bone Repair

2.3.1. Animals

This study involved 60 male Sprague Dawley rats weighing 150–200 g according to the rules of the Committee of Ethics in Animal Research of the School of Dentistry of Ribeirão Preto (Protocol # 0031/2021; date of approval: 11/10/2021).

2.3.2. Isolation and Culture of MSCs

MSCs were harvested from bone marrow of the femurs of 12 rats and cultured in non-inducing culture medium until they reached 70% confluence, as previously described [20]. First-passage MSCs were enzymatically detached and directly injected into rat calvarial defects, as described below. The culture medium was changed every 48 h.

2.3.3. Creation and Treatment of Calvarial Defects

Forty-eight rats were anesthetized with ketamine (75 mg/kg, intraperitoneal; Agener União, Embu-Guaçu, SP, Brazil) and xylazine (6 mg/kg, intraperitoneal; Calier, Juatuba, MG, Brazil) and a unilateral 5-mm-diameter defect was created using a trephine drill (Neodent, Curitiba, PR, Brazil). Then, the defects were implanted with the scaffold and the skin was sutured with mononylon 4.0 (Ethicon Ltd.a, São José dos Campos, SP, Brazil). Two weeks post-calvarial defect creation and scaffold implantation, the animals were randomly grouped (n = 12 per group) according to the treatment of bone defects: (1) only the scaffold (Scaffold); (2) the scaffold combined with MSCs (Scaffold + MSCs); (3) the scaffold combined with PBM therapy (Scaffold + PBM); and (4) the scaffold combined with MSCs and PBM therapy (Scaffold + MSCs + PBM). The rats were anesthetized and 5×10^6 MSCs in 50 μL of phosphate-buffered saline (PBS, Gibco-Life Technologies) were locally injected into each defect, except for those treated with the scaffold without cells, which were injected with 50 μL of PBS (Gibco-Life Technologies). The PBM was based on the local applications of gallium–aluminum–arsenide laser (GaAIAs, Photon III, DMC, São Carlos, SP, Brazil) in continuous contact and punctual operation mode in four points of the calvarial defect, one central and three equidistant around the defect. Irradiations were performed immediately after, as well as 48 and 96 h post-injection, according to the following parameters: wavelength: 808 nm, power: 40 mW, power density: 1.42 W/cm^2, energy density: 4 J/cm^2, irradiation time: 3 s, energy per point: 0.12 J, and spot area: 0.028 cm^2 [33,34]. Four weeks post injection, the animals were euthanized. The calvarias were harvested and fixed in 10% buffered formalin to evaluate the newly formed bone.

2.3.4. μCT Analysis

The μCT analysis was carried out by a single blinded operator using the SkyScan 1172 system (Bruker-Skyscan) and the three-dimensional reconstructions were created using NRecon Cluster software (Micro Photonics Inc.) as previously described [19]. Bone volume (BV, mm^3), percentage of bone volume (BV/TV, %), bone surface (BS, mm^2), trabecular number (Tb.N, 1/mm), trabecular separation (Tb.Sp, mm) and bone mineral density (BMD, g/cm^3) were evaluated in the region of interest: the 5 mm diameter of the calvarial defect [35].

2.3.5. Histological Analysis

After μCT analysis, undecalcified calvariae were dehydrated, embedded in resin (LR White Hard Grade, London, UK) and sectioned using the Exakt cutting system (Exakt, Norderstedt, Germany) to produce two sections per sample. The 150-μm-thick sections were mounted on glass slides and polished to a thickness of 70 μm. The sections were stained with Stevenel's blue (Sigma-Aldrich) for 15 min at 55 °C and alizarin red (Sigma-Aldrich) for 2 min at room temperature or toluidine blue for 20 min at room temperature (Merck, Darmstadt, Germany). The images were obtained using a light microscope (Axioskop 40, Carl Zeiss Inc., Oberkochen, Germany) coupled with a digital camera (Axiocam ICc3, Carl Zeiss).

2.4. Statistical Analysis

The wettability data from P(VDF-TrFE)/$BaTiO_3$, P(VDF-TrFE) and PVDF (n = 5) were compared by one-way ANOVA followed by Duncan's new multiple range test and from P(VDF-TrFE)/$BaTiO_3$ over time (n = 5) by repeated measures ANOVA followed by Tukey's post hoc test. The data from morphometric parameters (n =12) were compared by one-way ANOVA followed by Duncan's new multiple range test and by Pearson's correlation coefficient. The data were expressed as mean ± standard deviation ($p \leq 0.05$).

3. Results

3.1. P(VDF-TrFE)/$BaTiO_3$ Scaffold Characterization

The electrospun P(VDF-TrFE)/$BaTiO_3$ fibers were uniform and continuous, without interruptions or beads along them (Figure 1A). It was possible to see particles as aggregates uniformly distributed on the fibers (Figure 1B). The average fiber diameter was 1.10 ± 0.38 μm (Figure 1C). The pore sizes were distributed in four ranges, with approximately 50% of the pores ranging from 17.68 to 29.47 μm (Figure 1D), and the average *Ra* was 0.869 ± 0.03 μm. The P(VDF-TrFE)/$BaTiO_3$ scaffold was composed of C, F, O, Ba and Ti regardless of the analyzed area, the fiber (Figure 1E) or the particle (Figure 1F). The elemental distribution varied according to the area analyzed with a higher percentage of F in the fiber (Figure 1G) and of C, Ba and Ti in the particle (Figure 1H).

The contact angle was lower on P(VDF-TrFE)/$BaTiO_3$ compared with either P(VDF-TrFE) or PVDF, and lower on P(VDF-TrFE) than PVDF (p < 0.001 for all comparisons) (Figure 2A). The contact angle progressively decreased (p < 0.002 for all comparisons) on P(VDF-TrFE)/$BaTiO_3$ over time, but it is still possible to observe the drop even after 40 min (Figure 2B).

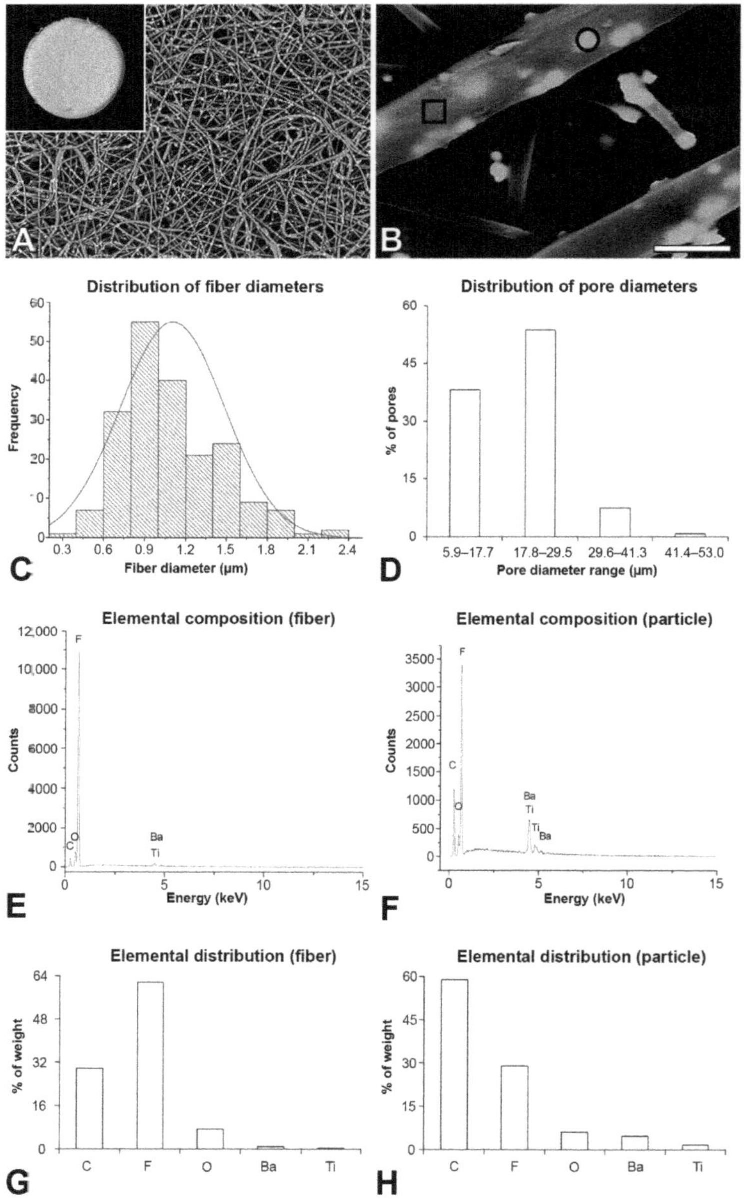

Figure 1. P(VDF-TrFE)/$BaTiO_3$ scaffold characterization. Macroscopic view ((**A**), inset) and scanning electron microscopy of the P(VDF-TrFE)/$BaTiO_3$ scaffold showing the fiber distribution (**A**) and details of the fibers with particles (**B**). Distribution of the fiber diameter (**C**) and pore size (**D**) of the P(VDF-TrFE)/$BaTiO_3$ scaffold. Energy-dispersive X-ray spectroscopy of the P(VDF-TrFE)/$BaTiO_3$ scaffold showing elemental composition of the fiber ((**E**), square in (**B**)) and the particle ((**F**), circle in (**B**)), and elemental distribution of the fiber (**G**) and the particle (**H**). Scale bar: (**A**) = 80 μm; (**A**) (inset) = 3.75 mm; (**B**) = 2 μm.

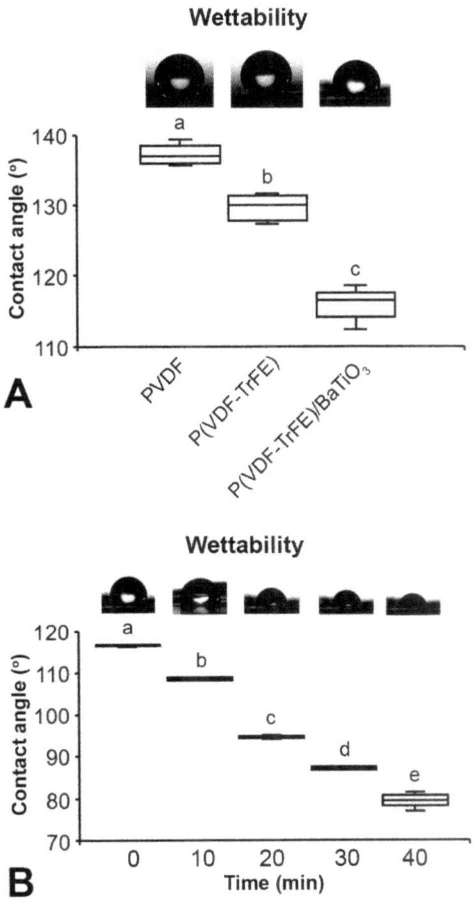

Figure 2. P(VDF-TrFE)/$BaTiO_3$ scaffold characterization. Wettability of the P(VDF-TrFE)/$BaTiO_3$ compared with P(VDF-TrFE) and PVDF measured via the contact angle immediately after placing the deionized water drop (**A**) and time course of the contact angle measured on P(VDF-TrFE)/$BaTiO_3$ (**B**). The data are presented as mean ± standard deviation ($n = 5$). Different letters represent statistically significant differences among P(VDF-TrFE)/$BaTiO_3$, P(VDF-TrFE) and PVDF ((**A**), a–c, $p < 0.001$) and time ((**B**), a–e, $p < 0.002$).

3.2. Evaluation of Bone Repair

The three-dimensional reconstructions indicated more bone repair in defects treated with Scaffold + MSCs + PBM followed by Scaffold + PBM, Scaffold + MSCs and Scaffold (Figure 3A–D), and the morphometric parameters confirmed this finding (Figure 3E–J). The BV was higher in defects treated with Scaffold + MSCs + PBM and Scaffold + PBM compared with Scaffold ($p \leq 0.05$) and showed a positive correlation with treatments (r = 0.482, $p = 0.001$) (Figure 3E). The BV/TV was higher in defects treated with Scaffold + MSCs + PBM and Scaffold + PBM compared with Scaffold ($p \leq 0.05$) and presented a positive correlation with treatments (r = 0.482, $p = 0.001$) (Figure 3F). The BS was higher in defects treated with Scaffold + MSCs + PBM compared with Scaffold + MSCs and Scaffold ($p \leq 0.05$). The BS was higher in defects treated with Scaffold + PBM and Scaffold + MSCs compared with Scaffold ($p \leq 0.05$) and exhibited a positive correlation with treatments (r = 0.584, $p = 0.001$) (Figure 3G). The Tb.N was higher in defects treated with Scaffold + MSCs + PBM and Scaffold + PBM compared with Scaffold ($p \leq 0.05$) and showed a positive correlation with treatments (r = 0.523, p = 0.001) (Figure 3H). The Tb.Sp was lower in defects treated with Scaffold + MSCs + PBM compared with Scaffold + MSCs and Scaffold ($p \leq 0.05$). The Tb.Sp was lower in defects treated with Scaffold + PBM compared to Scaffold ($p \leq 0.05$) and presented a negative correlation with treatments (r = −0.563, p = 0.001) (Figure 3I). The BMD was higher in defects treated with Scaffold + MSCs + PBM compared with Scaffold + MSCs and Scaffold ($p \leq 0.05$). The BMD was higher in defects treated with Scaffold + PBM compared with Scaffold ($p \leq 0.05$) and exhibited a positive correlation with treatments (r = 0.579, p = 0.001) (Figure 3J).

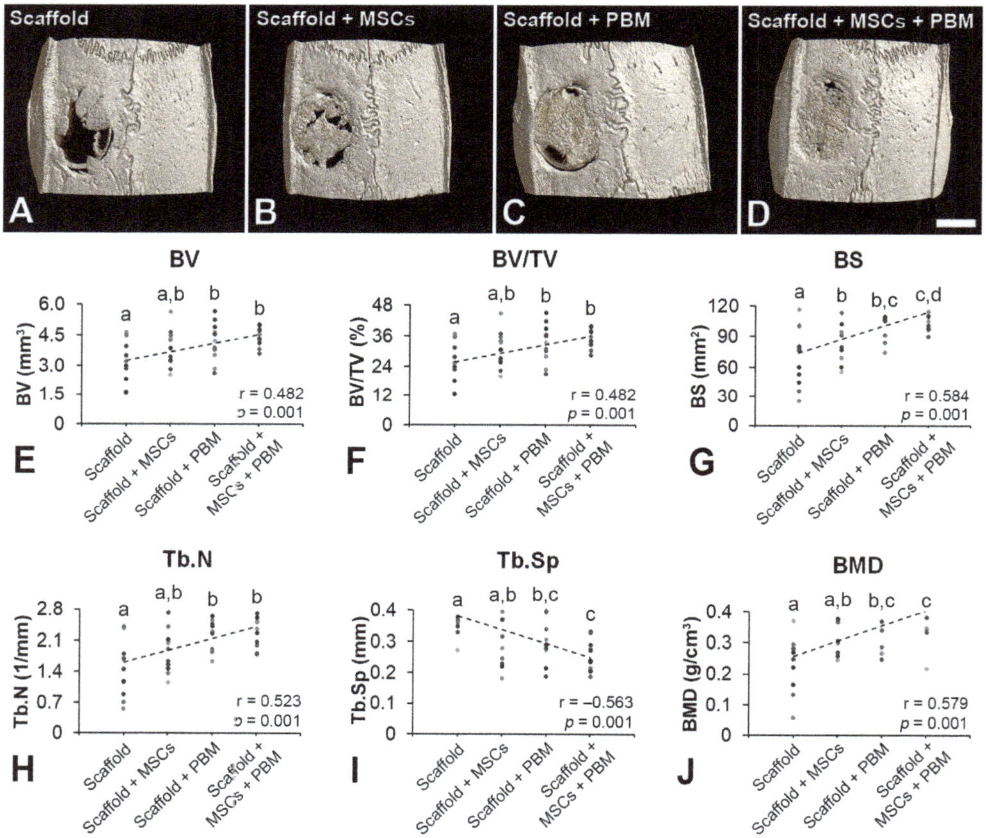

Figure 3. Evaluation of bone repair. Three-dimensional microtomographic reconstructions (**A–D**) of bone formation in rat calvarial defects treated with P(VDF-TrFE)/$BaTiO_3$ scaffold (Scaffold, (**A**)); scaffold combined with bone marrow-derived mesenchymal stem cells (Scaffold + MSCs, (**B**)); scaffold combined with photobiomodulation (Scaffold + PBM, (**C**)); and scaffold combined with MSCs and PBM (Scaffold + MSCs + PBM, (**D**)). Morphometric parameters bone volume (BV, (**E**)), percentage of bone volume (BV/TV, (**F**)), bone surface (BS, (**G**)), trabecular number (Tb.N, (**H**)), trabecular separation (Tb.Sp, (**I**)) and bone mineral density (BMD, (**J**)) evaluated in the region of interest, the 5 mm diameter of the calvarial defect. The data are presented as mean ± standard deviation (n = 12). Different letters (a–d) represent statistically significant differences among the treatments ((**E–J**), $p \leq 0.05$). Scale bar: (**A–D**) = 2.50 mm.

The histological sections stained with Stevenel's blue and alizarin red showed the presence of bone in the edges of the defects and in close contact with the scaffold, irrespective of the treatments (Figure 4A–I,K,M,O). The newly formed bone exhibited characteristics of healthy tissue with areas of immature and lamellar bone, and the presence of osteoblasts, osteocytes and blood vessels, without signs of adverse reactions. Multinucleated giant cells were observed in close contact with the scaffold in histological sections stained with toluidine blue in all of the evaluated groups (Figure 4J,L,N,P).

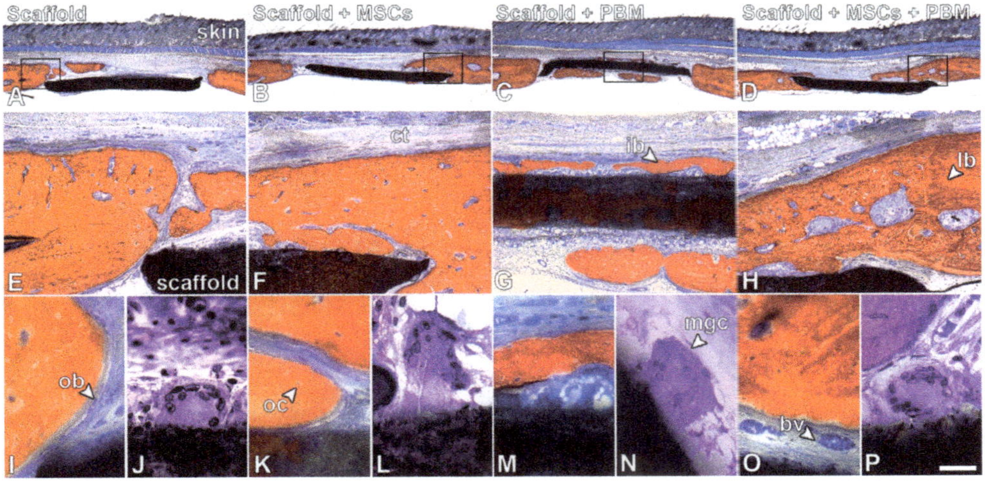

Figure 4. Evaluation of bone repair. Light microscopy of bone formation in rat calvarial defects treated with P(VDF-TrFE)/$BaTiO_3$ scaffold (Scaffold, (**A**,**E**,**I**)), Scaffold + MSCs (**B**,**F**,**K**), Scaffold + PBM (**C**,**G**,**M**) and Scaffold + MSCs + PBM (**D**,**H**,**O**). Multinucleated giant cells were observed in close contact with the scaffold surface in all defects, irrespective of treatment (**J**,**L**,**N**,**P**). Stevenel's blue and alizarin red (**A**–**I**,**K**,**M**,**O**) and toluidine blue (**J**,**L**,**N**,**P**) staining. Squares in (**A**–**D**) are represented in (**E**–**H**). Scale bar: (**A**–**D**) = 1.25 mm; (**E**–**H**) = 200 μm; (**I**,**K**,**M**,**O**) = 50 μm; (**J**,**L**,**N**,**P**) = 20 μm. bv = blood vessel; ct = connective tissue; ib = immature bone; lb = lamellar bone; mgc = multinucleated giant cell; ob = osteoblast; oc = osteocyte.

4. Discussion

Treating large bone defects still represents a significant clinical challenge in orthopedics and oral and maxillofacial surgery, and several strategies focused on tissue engineering and cell therapy have been proposed to manage this issue. As scaffolds, cells and PBM therapy effectively favor bone repair, a synergistic effect is expected when combining them during bone formation. Here, we synthesized a P(VDF-TrFE)/$BaTiO_3$ scaffold using an electrospinning technique with physical and chemical properties that make it suitable to be employed in bone tissue engineering. Then, we demonstrated that the combination of this scaffold with MSCs and PBM therapy is a good strategy to enhance bone formation in a rat calvarial defect model.

Considering that biomaterials should promote a favorable environment for cell adhesion and growth, the properties of scaffolds produced using an electrospinning technique should affect these cell events, with a positive effect on bone formation. As expected, the P(VDF-TrFE)/$BaTiO_3$ scaffold was composed of C, F, O, Ba and Ti regardless of the analyzed area; however, the elemental distribution varied with the fibers exhibiting a higher percentage of F and the particles more C, Ba and Ti. The fiber diameters of the P(VDF-TrFE)/$BaTiO_3$ scaffold synthesized here are in the same range as a biodegradable polyhydroxybutyrate composite that allowed MSC adhesion and spreading [36]. Regarding the pore sizes, the majority ranged from 17 to 29 μm; this did not allow cell infiltration, since MSCs vary from 18 to 30 μm [37]. Such characteristics could explain the bone formation observed only on the P(VDF-TrFE)/$BaTiO_3$ scaffold surface, making the increasing pore size relevant to further investigations. The surface wettability is one of the most important factors determining cell adhesion. The contact angle on PVDF and P(VDF-TrFE)-based materials is related to the hydrophobicity of fluorinated polymers, as alkyl and fluorinated alkyl groups exhibit low interaction energy with water [38]. The contact angle on PVDF was higher than on P(VDF-TrFE) due to the presence of free alkyl groups in the PVDF structure. The addition of $BaTiO_3$ reduced the contact angle thanks to its super-hydrophilic characteristic [39]. The detected contact angle of 79° after 40 min suggests that the P(VDF-TrFE)/$BaTiO_3$ scaffold may favor cell adhesion, since effective adhesion occurs on surfaces with water contact angles ranging from 40 to 80° [40]. Because the presence of nanoparticles affects the thermal and mechanical properties of biomaterials, the addition of $BaTiO_3$ may have modified these scaffold features, potentially influencing the bone response [41]. Thus, further characterizations of the P(VDF-TrFE)/$BaTiO_3$ scaffold should consider the use of molecular dynamics simulation, a low-risk/-cost approach compared with experimental methods, to evaluate its thermal and mechanical properties [41,42]. Together, the features of the P(VDF-TrFE)/$BaTiO_3$ scaffold seem adequate for a biomaterial to be employed in bone tissue engineering approaches.

Several studies have demonstrated the osteogenic potential of P(VDF-TrFE) composites in different in vitro and in vivo models. P(VDF-TrFE)/$BaTiO_3$ membrane favors osteoblast differentiation compared with PTFE, and P(VDF-TrFE)/boron nitride nanotubes promote differentiation of SaOS-2 osteoblast-like cells [14,15,43]. Additionally, P(VDF-TrFE)/$BaTiO_3$ membranes induce more bone formation than PTFE when implanted in calvarial defects of either healthy or osteoporotic rats [16,17,44]. In agreement with these findings, we showed that calvarial defects treated with the P(VDF-TrFE)/$BaTiO_3$ scaffold and PBS injection exhibit significant bone repair compared with defects without the scaffold that were injected with PBS, which are generally filled with connective tissue as previously demonstrated [45,46]. In addition to upregulating osteoblast differentiation, P(VDF-TrFE)/$BaTiO_3$ enhances bone formation by inhibiting bone resorption through the regulation of microRNA-34a/RANKL crosstalk [18].

Although good results were observed in defects treated with the P(VDF-TrFE)/$BaTiO_3$ scaffold, the combination of scaffolds and cells proved to be better in terms of promoting bone repair. Indeed, polymer/hydroxyapatite scaffolds combined with MSCs were more effective than the scaffold alone in the repair of rat calvarial defects [47,48]. Additionally, P(VDF-TrFE)/$BaTiO_3$ membrane combined with osteoblasts differentiated from bone marrow MSCs promotes more bone formation than the membrane alone in calvarial defects of healthy animals; this was also true of MSCs in osteoporotic rats [19,20]. Corroborating these findings, we demonstrated that MSCs combined with the P(VDF-TrFE)/$BaTiO_3$ scaffold resulted in more bone formation compared with the scaffold alone. As the cells stay in the bone defect for approximately 25 days in an experimental model such as this one, the increased bone repair induced by the presence of MSCs could be related to their ability to home and integrate into damaged tissues and provide osteogenic and immunomodulatory effects [19,49,50].

To increase the bone repair induced by the P(VDF-TrFE)/$BaTiO_3$ scaffold combined with MSCs, the PBM was employed as an adjunctive therapy, since its stimulatory effects on host stem cell recruitment, osteoblast differentiation and bone formation have been demonstrated in cell culture and animal models [51–53]. The combination of a collagen membrane with PBM therapy using the same irradiation parameters employed here induced more bone formation than the membrane alone in calvarial defects of osteoporotic rats [34]. In keeping with this, our results showed that PBM therapy increases bone repair in defects implanted with the P(VDF-TrFE)/$BaTiO_3$ scaffold compared with non-irradiated defects. Notably, PBM promoted slightly more bone formation than MSCs when combined with the scaffold, although a non-statistically significant difference was detected. In contrast, PBM did not affect bone repair in calvarial defects of osteoporotic rats implanted with P(VDF-TrFE)/$BaTiO_3$ membranes, which could be partially explained by the differences in irradiation parameters. This supports the relevance of stablishing a precise protocol for PBM therapy application [54].

Different combinations of two out of these three approaches have been the subject of several studies; however, the combination of all three—biomaterials, cells and PBM therapy—is underexplored in the field of bone regeneration. The PBM application in calvarial defects treated with adipose-derived MSCs encapsulated in methacrylated gelatin hydrogels increased bone formation; this also occurred in defects treated with MSCs from dental pulp encapsulated in an injectable BMP-4-loaded hydrogel [32,55]. Here, we observed a progressive increment in bone formation which was confirmed by the correlation between treatments for all evaluated morphometric parameters. This increase was greater than that for defects treated with the P(VDF-TrFE)/$BaTiO_3$ scaffold alone, peaking in defects treated with a combination of the scaffold, MSCs and PBM. The BS and BMD significantly increased, while Tb.Sp decreased when irradiation was applied in defects treated with the scaffold and MSCs. This showcased the synergistic effect of these three elements on bone repair, as the morphometric analysis considered only the bone tissue that formed on top of the scaffold. The positive effect of PBM as an adjunctive therapy to increase bone formation in calvarial defects treated with MSCs combined with P(VDF-TrFE)/$BaTiO_3$ scaffold could be related to the PBM capacity of regulating the osteogenic and immunomodulatory potential of MSCs [29,56,57]. Although no histological differences were observed in the new bone tissue regardless of the treatments, the presence of multinucleated giant cells in contact with the scaffold suggests a foreign body reaction and calls for further investigations on the capacity of these cells to degrade P(VDF-TrFE)/$BaTiO_3$ [58].

We demonstrated the viability of synthesizing a low-cost P(VDF-TrFE)/$BaTiO_3$ scaffold using an electrospinning technique. We also showed that MSCs and PBM therapy acted synergistically when combined with the P(VDF-TrFE)/$BaTiO_3$ scaffold to promote bone repair in rat calvarial defects. Indeed, this combination induced more bone formation than the scaffold alone or in combination with either MSCs or PBM. These findings highlight the need to combine different approaches to achieve complete regeneration of challenging bone defects and find avenues for further investigations into innovative therapies in the field of bone tissue engineering.

Author Contributions: L.F.A.: data curation; formal analysis; investigation, methodology; writing—original draft; and writing—review and editing. L.M.S.R.: data curation; formal analysis; investigation; methodology; and writing—review and editing. G.P.F.: investigation; methodology; and writing—review and editing. H.B.L.: investigation; methodology; and writing—review and editing. M.P.O.G.: investigation; methodology; and writing—review and editing. E.P.F.: conceptualization; funding acquisition; data curation; formal analysis; supervision; and writing—review and editing. R.G.: conceptualization; funding acquisition; data curation; formal analysis; supervision; writing—original draft; and writing—review and editing. M.M.B.: conceptualization; data curation; formal analysis; supervision; writing—original draft; and writing—review and editing. A.L.R.: conceptualization; funding acquisition; data curation; formal analysis; project administration; supervision; writing—original draft; and writing—review and editing. All authors have read and agreed to the published version of the manuscript.

Funding: This work was supported by the University of São Paulo (Grant # 2021.1.10424.1.9), State of São Paulo Research Foundation (FAPESP, Brazil, # 2017/12622-7), National Council for Scientific and Technological Development (CNPq, grant # 303115/2019-8) and Minas Gerais Research Foundation (FAPEMIG, grant # APQ-02154-22).

Institutional Review Board Statement: This study was approved by the Committee of Ethics in Animal Research of the School of Dentistry of Ribeirão Preto (Protocol # 0031/2021; date of approval: 11/10/2021).

Informed Consent Statement: Not applicable.

Data Availability Statement: The datasets used and/or analyzed during the current study are available from the corresponding author on reasonable request.

Acknowledgments: The authors would like to thank Adriana L. G. Almeida, Roger R. Fernandes and Sebastião C. Bianco for their technical assistance during the execution of the study.

Conflicts of Interest: All authors declare no conflict of interest.

References

1. Schuind, F.; Burny, F.; Quintin, J.; Potaznik, A.; Pasteels, J.L. Single stage reconstruction of a large tibial defect using a free vascularised osteomyocutaneous ulnar transfer. *Int. Orthop.* **1989**, *13*, 239–245. [CrossRef] [PubMed]
2. Marcacci, M.; Kon, E.; Zaffagnini, S.; Giardino, R.; Rocca, M.; Corsi, A.; Benvenuti, A.; Bianco, P.; Quarto, R.; Martin, I.; et al. Reconstruction of extensive long-bone defects in sheep using porous hydroxyapatite sponges. *Calcif. Tissue Int.* **1999**, *64*, 83–90. [CrossRef]
3. Tarr, J.T.; Hagan, M.; Zhang, B.; Tanna, N.; Andrews, B.T.; Lee, J.C.; Bradley, J.P. Syndrome of the trephined: Quantitative functional improvement after large cranial vault reconstruction. *Plast. Reconstr. Surg.* **2020**, *145*, 1486–1494. [CrossRef] [PubMed]
4. Langer, R.; Vacanti, J.P. Tissue engineering. *Science* **1993**, *260*, 920–926. [CrossRef]
5. Nevo, Z.; Robinson, D.; Horowitz, S.; Hasharoni, A.; Yayon, A. The manipulated mesenchymal stem cells in regenerated skeletal tissues. *Cell Transplant.* **1998**, *7*, 63–70. [CrossRef]
6. Radice, M.; Brun, P.; Cortivo, R.; Scapinelli, R.; Battaliard, C.; Abatangelo, G. Hyaluronan-based biopolymers as delivery vehicles for bone-marrow-derived mesenchymal progenitors. *J. Biomed. Mater. Res.* **2000**, *50*, 101–109. [CrossRef]
7. Langer, R.; Vacanti, J. Advances in tissue engineering. *J. Pediatr. Surg.* **2016**, *51*, 8–12. [CrossRef] [PubMed]
8. Zhang, X.; Wang, G.; Wang, W.; Ran, C.; Piao, F.; Ma, Z.; Zhang, Z.; Zheng, G.; Cao, F.; Xie, H.; et al. Bone marrow mesenchymal stem cells paracrine TGF-β1 to mediate the biological activity of osteoblasts in bone repair. *Cytokine* **2023**, *164*, 156139. [CrossRef]
9. Yamada, Y.; Ueda, M.; Naiki, T.; Takahashi, M.; Hata, K.; Nagasaka, T. Autogenous injectable bone for regeneration with mesenchymal stem cells and platelet-rich plasma: Tissue-engineered bone regeneration. *Tissue Eng.* **2004**, *10*, 955–964. [CrossRef]
10. Suzuki, S.; Venkataiah, V.S.; Yahata, Y.; Kitagawa, A.; Inagaki, M.; Njuguna, M.M.; Nozawa, R.; Kakiuchi, Y.; Nakano, M.; Handa, K.; et al. Correction of large jawbone defect in the mouse using immature osteoblast-like cells and a 3D polylactic acid scaffold. *PNAS Nexus* **2022**, *1*, pgac151. [CrossRef]
11. Liu, W.; Li, X.; Jiao, Y.; Wu, C.; Guo, S.; Xiao, X.; Wei, X.; Wu, J.; Gao, P.; Wang, N.; et al. Biological effects of a three-dimensionally printed Ti6Al4V scaffold coated with piezoelectric $BaTiO_3$ nanoparticles on bone formation. *ACS Appl. Mater. Interfaces* **2020**, *12*, 51885–51903. [CrossRef] [PubMed]
12. Lim, J.; Liu, Y.C.; Chu, Y.C.; Lin, Y.X.; Hwang, W.H.; Wang, J.L. Piezoelectric effect stimulates the rearrangement of chondrogenic cells and alters ciliary orientation via atypical PKCζ. *Biochem. Biophys. Rep.* **2022**, *30*, 101265. [CrossRef] [PubMed]
13. Bhaskar, N.; Kachappilly, M.C.; Bhushan, V.; Pandya, H.J.; Basu, B. Electrical field stimulated modulation of cell fate of pre-osteoblasts on PVDF/BT/MWCNT based electroactive biomaterials. *J. Biomed. Mater. Res. A* **2023**, *111*, 340–353. [CrossRef]
14. Beloti, M.M.; de Oliveira, P.T.; Gimenes, R.; Zaghete, M.A.; Bertolini, M.J.; Rosa, A.L. In vitro biocompatibility of a novel membrane of the composite poly(vinylidene-trifluoroethylene)/barium titanate. *J. Biomed. Mater. Res. A* **2006**, *79*, 282–288. [CrossRef] [PubMed]
15. Teixeira, L.N.; Crippa, G.E.; Gimenes, R.; Zaghete, M.A.; de Oliveira, P.T.; Rosa, A.L.; Beloti, M.M. Response of human alveolar bone-derived cells to a novel poly(vinylidene fluoride-trifluoroethylene)/barium titanate membrane. *J. Mater. Sci. Mater. Med.* **2011**, *22*, 151–158. [CrossRef] [PubMed]
16. Lopes, H.B.; Santos Tde, S.; de Oliveira, F.S.; Freitas, G.P.; de Almeida, A.L.; Gimenes, R.; Rosa, A.L.; Beloti, M.M. Poly(vinylidene-trifluoroethylene)/barium titanate composite for in vivo support of bone formation. *J. Biomater. Appl.* **2014**, *29*, 104–112. [CrossRef]
17. Scalize, P.H.; Bombonato-Prado, K.F.; de Sousa, L.G.; Rosa, A.L.; Beloti, M.M.; Semprini, M.; Gimenes, R.; de Almeida, A.L.; de Oliveira, F.S.; Hallak Regalo, S.C.; et al. Poly(vinylidene fluoride-trifluorethylene)/barium titanate membrane promotes de novo bone formation and may modulate gene expression in osteoporotic rat model. *J. Mater. Sci. Mater. Med.* **2016**, *27*, 180. [CrossRef]
18. Lopes, H.B.; Ferraz, E.P.; Almeida, A.L.; Florio, P.; Gimenes, R.; Rosa, A.L.; Beloti, M.M. Participation of MicroRNA-34a and RANKL on bone repair induced by poly(vinylidene-trifluoroethylene)/barium titanate membrane. *J. Biomater. Sci. Polym. Ed.* **2016**, *27*, 1369–1379. [CrossRef]
19. Freitas, G.P.; Lopes, H.B.; Almeida, A.L.G.; Abuna, R.P.F.; Gimenes, R.; Souza, L.E.B.; Covas, D.T.; Beloti, M.M.; Rosa, A.L. Potential of osteoblastic cells derived from bone marrow and adipose tissue associated with a polymer/ceramic composite to repair bone tissue. *Calcif. Tissue Int.* **2017**, *101*, 312–320. [CrossRef]
20. Almeida, A.L.G.; Freitas, G.P.; Lopes, H.B.; Gimenes, R.; Siessere, S.; Sousa, L.G.; Beloti, M.M.; Rosa, A.L. Effect of stem cells combined with a polymer/ceramic membrane on osteoporotic bone repair. *Braz. Oral. Res.* **2019**, *33*, e079. [CrossRef]
21. Hu, Y.; Zhang, C.; Zhang, S.; Xiong, Z.; Xu, J. Development of a porous poly(L-lactic acid)/hydroxyapatite/collagen scaffold as a BMP delivery system and its use in healing canine segmental bone defect. *J. Biomed. Mater. Res. A* **2003**, *67*, 591–598. [CrossRef] [PubMed]
22. Ekambaram, R.; Dharmalingam, S. Design and development of biomimetic electrospun sulphonated polyether ether ketone nanofibrous scaffold for bone tissue regeneration applications: In vitro and in vivo study. *J. Biomater. Sci. Polym. Ed.* **2022**, *33*, 947–975. [CrossRef] [PubMed]
23. Valerini, D.; Tammaro, L.; Vitali, R.; Guillot, G.; Rinaldi, A. Sputter-deposited Ag nanoparticles on electrospun PLC scaffolds: Morphology, wettability and antibacterial activity. *Coatings* **2021**, *11*, 345. [CrossRef]

24. Wilk, S.; Benko, A. Advances in fabricating the electrospun biopolymer-based biomaterials. *J. Funct. Biomater.* **2021**, *12*, 26. [CrossRef]
25. Baldari, S.; Di Rocco, G.; Piccoli, M.; Pozzobon, M.; Muraca, M.; Toietta, G. Challenges and strategies for improving the regenerative effects of mesenchymal stromal cell-based therapies. *Int. J. Mol. Sci.* **2017**, *18*, 2087. [CrossRef]
26. Lee, B.C.; Kang, K.S. Functional enhancement strategies for immunomodulation of mesenchymal stem cells and their therapeutic application. *Stem Cell Res. Ther.* **2020**, *11*, 397. [CrossRef]
27. Hosseinpour, S.; Fekrazad, R.; Arany, P.R.; Ye, Q. Molecular impacts of photobiomodulation on bone regeneration: A systematic review. *Prog. Biophys. Mol. Biol.* **2019**, *149*, 147–159. [CrossRef]
28. Fallahnezhad, S.; Jajarmi, V.; Shahnavaz, S.; Amini, A.; Ghoreishi, S.K.; Kazemi, M.; Chien, S.; Bayat, M. Improvement in viability and mineralization of osteoporotic bone marrow mesenchymal stem cell through combined application of photobiomodulation therapy and oxytocin. *Lasers Med. Sci.* **2020**, *35*, 557–566. [CrossRef]
29. Bueno, N.P.; Copete, I.N.; Lopes, H.B.; Arany, P.R.; Marques, M.M.; Ferraz, E.P. Recovering the osteoblastic differentiation potential of mesenchymal stem cells derived from diabetic rats by photobiomodulation therapy. *J. Biophotonics* **2021**, *14*, e202000393. [CrossRef]
30. Dalapria, V.; Marcos, R.L.; Bussadori, S.K.; Anselmo, G.; Benetti, C.; da Silva Santana, A.C.A.; Marinho, N.S.F.; Pinto, R.S.; de Sales, R.S.; de França, L.S.; et al. LED photobiomodulation therapy combined with biomaterial as a scaffold promotes better bone quality in the dental alveolus in an experimental extraction model. *Lasers Med. Sci.* **2022**, *37*, 1583–1592. [CrossRef]
31. Khosravipour, A.; Amini, A.; Farahani, R.M.; Mostafavinia, A.; Asgari, M.; Rezaei, F.; Abrahamse, H.; Chien, S.; Bayat, M. Evaluation of the effects of preconditioned human stem cells plus a scaffold and photobiomodulation administration on stereological parameters and gene expression levels in a critical size bone defect in rats. *Lasers Med. Sci.* **2022**, *37*, 2457–2470. [CrossRef]
32. Diniz, I.M.A.; Carreira, A.C.O.; Sipert, C.R.; Uehara, C.M.; Moreira, M.S.N.; Freire, L.; Pelissari, C.; Kossugue, P.M.; de Araújo, D.R.; Sogayar, M.C.; et al. Photobiomodulation of mesenchymal stem cells encapsulated in an injectable rhBMP4-loaded hydrogel directs hard tissue bioengineering. *J. Cell. Physiol.* **2018**, *233*, 4907–4918. [CrossRef] [PubMed]
33. Martins, M.A.; Martins, M.D.; Lascala, C.A.; Curi, M.M.; Migliorati, C.A.; Tenis, C.A.; Marques, M.M. Association of laser phototherapy with PRP improves healing of bisphosphonate-related osteonecrosis of the jaws in cancer patients: A preliminary study. *Oral. Oncol.* **2012**, *48*, 79–84. [CrossRef] [PubMed]
34. Alves, F.A.M.; Marques, M.M.; Cavalcanti, S.C.S.X.B.; Pedroni, A.C.F.; Ferraz, E.P.; Miniello, T.G.; Moreira, M.S.; Jerônimo, T.; Deboni, M.C.Z.; Lascala, C.A. Photobiomodulation as adjunctive therapy for guided bone regeneration. A microCT study in osteoporotic rat model. *J. Photochem. Photobiol. B.* **2020**, *213*, 112053. [CrossRef] [PubMed]
35. Bouxsein, M.L.; Boyd, S.K.; Christiansen, B.A.; Guldberg, R.E.; Jepsen, K.J.; Müller, R. Guidelines for assessment of bone microstructure in rodents using micro-computed tomography. *J. Bone Miner. Res.* **2010**, *25*, 1468–1486. [CrossRef]
36. Sheng, R.; Mu, J.; Chernozem, R.V.; Mukhortova, Y.R.; Surmeneva, M.A.; Pariy, I.O.; Ludwig, T.; Mathur, S.; Xu, C.; Surmenev, R.A.; et al. Fabrication and characterization of piezoelectric polymer composites and cytocompatibility with mesenchymal stem cells. *ACS Appl. Mater. Interfaces* **2023**, *15*, 3731–3743. [CrossRef] [PubMed]
37. Ge, J.; Guo, L.; Wang, S.; Zhang, Y.; Cai, T.; Zhao, R.C.; Wu, Y. The size of mesenchymal stem cells is a significant cause of vascular obstructions and stroke. *Stem Cell Rev. Rep.* **2014**, *10*, 295–303. [CrossRef]
38. Crick, C.R.; Parkin, I.P. Preparation and characterisation of super-hydrophobic surfaces. *Chemistry* **2010**, *16*, 3568–3588. [CrossRef]
39. Li, R.; Wei, W.; Hai, J.; Gao, L.; Gao, Z.; Fan, Y. Preparation and electric-field response of novel tetragonal barium titanate. *J. Alloys Compd.* **2013**, *574*, 212–216. [CrossRef]
40. Arima, Y.; Iwata, H. Effect of wettability and surface functional groups on protein adsorption and cell adhesion using well-defined mixed self-assembled monolayers. *Biomaterials* **2007**, *28*, 3074–3082. [CrossRef]
41. Shojaei, S.; Shahgholi, M.; Karimipour, A. The effects of atomic percentage and size of Zinc nanoparticles, and atomic porosity on thermal and mechanical properties of reinforced calcium phosphate cement by molecular dynamics simulation. *J. Mech. Behav. Biomed. Mater.* **2023**, *141*, 105785. [CrossRef] [PubMed]
42. Shahshahani, S.; Shahgholi, M.; Karimipour, A. The thermal performance and mechanical stability of methacrylic acid porous hydrogels in an aqueous medium at different initial temperatures and hydrogel volume fraction using the molecular dynamics simulation. *J. Mol. Liq.* **2023**, *382*, 122001. [CrossRef]
43. Genchi, G.G.; Sinibaldi, E.; Ceseracciu, L.; Labardi, M.; Marino, A.; Marras, S.; De Simoni, G.; Mattoli, V.; Ciofani, G. Ultrasound-activated piezoelectric P(VDF-TrFE)/boron nitride nanotube composite films promote differentiation of human SaOS-2 osteoblast-like cells. *Nanomedicine* **2018**, *14*, 2421–2432. [CrossRef] [PubMed]
44. Bai, Y.; Dai, X.; Yin, Y.; Wang, J.; Sun, X.; Liang, W.; Li, Y.; Deng, X.; Zhang, X. Biomimetic piezoelectric nanocomposite membranes synergistically enhance osteogenesis of deproteinized bovine bone grafts. *Int. J. Nanomed.* **2019**, *14*, 3015–3026. [CrossRef] [PubMed]
45. Souza, A.T.P.; Freitas, G.P.; Lopes, H.B.; Ferraz, E.P.; Oliveira, F.S.; Beloti, M.M.; Rosa, A.L. Effect of cell therapy with allogeneic osteoblasts on bone repair of rat calvaria defects. *Cytotherapy* **2018**, *20*, 1267–1277. [CrossRef]
46. Freitas, G.P.; Lopes, H.B.; Souza, A.T.P.; Gomes, M.P.O.; Quiles, G.K.; Gordon, J.; Tye, C.; Stein, J.L.; Stein, G.S.; Lian, J.B.; et al. Mesenchymal stem cells overexpressing BMP-9 by CRISPR-Cas9 present high in vitro osteogenic potential and enhance in vivo bone formation. *Gene Ther.* **2021**, *28*, 748–759. [CrossRef]

47. Andalib, N.; Kehtari, M.; Seyedjafari, E.; Motamed, N.; Matin, M.M. In vivo bone regeneration using a bioactive nanocomposite scaffold and human mesenchymal stem cells. *Cell Tissue Bank.* **2021**, *22*, 467–477. [CrossRef]
48. Naudot, M.; Garcia Garcia, A.; Jankovsky, N.; Barre, A.; Zabijak, L.; Azdad, S.Z.; Collet, L.; Bedoui, F.; Hébraud, A.; Schlatter, G.; et al. The combination of a poly-caprolactone/nano-hydroxyapatite honeycomb scaffold and mesenchymal stem cells promotes bone regeneration in rat calvarial defects. *J. Tissue Eng. Regen. Med.* **2020**, *14*, 1570–1580. [CrossRef]
49. Yagi, H.; Soto-Gutierrez, A.; Parekkadan, B.; Kitagawa, Y.; Tompkins, R.G.; Kobayashi, N.; Yarmush, M.L. Mesenchymal stem cells: Mechanisms of immunomodulation and homing. *Cell Transplant.* **2010**, *19*, 667–679. [CrossRef]
50. Cassiede, P.; Dennis, J.E.; Ma, F.; Caplan, A.I. Osteochondrogenic potential of marrow mesenchymal progenitor cells exposed to TGF-beta 1 or PDGF-BB as assayed in vivo and in vitro. *J. Bone Miner. Res.* **1996**, *11*, 1264–1273. [CrossRef]
51. Tuby, H.; Maltz, L.; Oron, U. Induction of autologous mesenchymal stem cells in the bone marrow by low-level laser therapy has profound beneficial effects on the infarcted rat heart. *Lasers Surg. Med.* **2011**, *43*, 401–409. [CrossRef] [PubMed]
52. Amid, R.; Kadkhodazadeh, M.; Ahsaie, M.G.; Hakakzadeh, A. Effect of low-level laser therapy on proliferation and differentiation of the cells contributing in bone regeneration. *J. Lasers Med. Sci.* **2014**, *5*, 163–170. [PubMed]
53. De Marco, A.C.; Torquato, L.C.; Gonçalves, P.R.; Ribeiro, T.C.; Nunes, C.M.; Bernardo, D.V.; Gomes, M.F.; Jardini, M.A.N.; Santamaria, M.P. The Effect of photobiomodulation therapy in different doses on bone repair of critical size defects in rats: A histomorphometric study. *J. Lasers Med. Sci.* **2021**, *12*, e53. [CrossRef] [PubMed]
54. Rufato, F.C.T.; de Sousa, L.G.; Scalize, P.H.; Gimenes, R.; Regalo, I.H.; Rosa, A.L.; Beloti, M.M.; de Oliveira, F.S.; Bombonato-Prado, K.F.; Regalo, S.C.H.; et al. Texturized P(VDF-TrFE)/BT membrane enhances bone neoformation in calvaria defects regardless of the association with photobiomodulation therapy in ovariectomized rats. *Clin. Oral. Investig.* **2022**, *26*, 1053–1065. [CrossRef]
55. Calis, M.; Irmak, G.; Demirtaş, T.T.; Kara, M.; Üstün, G.G.; Gümüşderelioğlu, M.; Türkkanı, A.; Çakar, A.N.; Özgür, F. Photobiomodulation combined with adipose-derived stem cells encapsulated in methacrylated gelatin hydrogels enhances in vivo bone regeneration. *Lasers Med. Sci.* **2022**, *37*, 595–606. [CrossRef]
56. Stancker, T.G.; Vieira, S.S.; Serra, A.J.; do Nascimento Lima, R.; Dos Santos Feliciano, R.; Silva, J.A., Jr.; Dos Santos, S.A.; Dos Santos Vieira, M.A.; Simões, M.C.B.; Leal-Junior, E.C.; et al. Can photobiomodulation associated with implantation of mesenchymal adipose-derived stem cells attenuate the expression of MMPs and decrease degradation of type II collagen in an experimental model of osteoarthritis? *Lasers Med. Sci.* **2018**, *33*, 1073–1084. [CrossRef] [PubMed]
57. Park, I.S.; Kim, D.K.; Kim, J.H.; Bae, J.S.; Kim, E.H.; Yoo, S.H.; Chung, Y.J.; Lyu, L.; Mo, J.H. Increased anti-allergic effects of secretome of low-level light treated tonsil-derived mesenchymal stem cells in allergic rhinitis mouse model. *Am. J. Rhinol. Allergy* **2022**, *36*, 261–268. [CrossRef]
58. Tanneberger, A.M.; Al-Maawi, S.; Herrera-Vizcaíno, C.; Orlowska, A.; Kubesch, A.; Sader, R.; Kirkpatrick, C.J.; Ghanaati, S. Multinucleated giant cells within the in vivo implantation bed of a collagen-based biomaterial determine its degradation pattern. *Clin. Oral. Investig.* **2021**, *25*, 859–873. [CrossRef]

Journal of
*Functional
Biomaterials*

MDPI

Review

Global Trends and Future Research Directions for Temporomandibular Disorders and Stem Cells

Zuleni Alexandre da Silva, Wallacy Watson Pereira Melo, Hadassa Helez Neves Ferreira, Rafael Rodrigues Lima and Renata Duarte Souza-Rodrigues *

Laboratory of Functional and Structural Biology, Institute of Biological Sciences, Federal University of Pará, Belém 66075-110, Brazil
* Correspondence: renataduarte@ufpa.br

Abstract: Temporomandibular disorder (TMD) is an umbrella term used to describe various conditions that affect temporomandibular joints, masticatory muscles, and associated structures. Although the most conservative and least invasive treatment is preferable, more invasive therapies should be employed to refractory patients. Tissue engineering has been presented as a promising therapy. Our study aimed to investigate trends and point out future research directions on TMD and stem cells. A comprehensive search was carried out in the Web of Science Core Collection (WoS-CC) in October 2022. The bibliometric parameters were analyzed through descriptive statistics and graphical mapping. Thus, 125 papers, published between 1992 and 2022 in 65 journals, were selected. The period with the highest number of publications and citations was between 2012 and 2022. China has produced the most publications on the subject. The most frequently used keywords were "cartilage", "temporomandibular joint", "mesenchymal stem cells", and "osteoarthritis". Moreover, the primary type of study was in vivo. It was noticed that using stem cells to improve temporomandibular joint repair and regeneration is a significant subject of investigation. Nonetheless, a greater understanding of the biological interaction and the benefits of using these cells in patients with TMD is required.

Keywords: temporomandibular joint; temporomandibular disorders; stem cells; tissue engineering

Citation: da Silva, Z.A.; Melo, W.W.P.; Ferreira, H.H.N.; Lima, R.R.; Souza-Rodrigues, R.D. Global Trends and Future Research Directions for Temporomandibular Disorders and Stem Cells. *J. Funct. Biomater.* **2023**, *14*, 103. https://doi.org/10.3390/jfb14020103

Academic Editor: Marco Tatullo

Received: 30 December 2022
Revised: 24 January 2023
Accepted: 30 January 2023
Published: 13 February 2023

1. Introduction

The collective term temporomandibular disorder (TMD) encompasses a set of heterogeneous musculoskeletal and neuromuscular conditions that involves the masticatory muscles, the temporomandibular joint (TMJ), and/or the structures associated with them [1–3]. TMD is considered the main cause of non-dental pain in the orofacial region [1,2,4].

The prevalence of TMD varies between studies, depending on the studied population and the chosen evaluation method. However, a recent study [5] pointed out that TMD occurs in 31% of adults and the elderly. Another study showed that, although the prevalence of TMJ is higher in women, this difference between females and males is only marginally more significant [6].

The etiology of TMD is still controversial. However, studies show that its origin is multifactorial, including biopsychosocial factors [7,8]. Risk factors that exert influence for a few years before the manifestation of TMD signs and symptoms (predisponents) must be taken into consideration, as well as those that act for the condition to develop (initiators) and those that cause the continuation of TMD (perpetuators), making it difficult for the outcome of the treatment [9,10]. TMD signs and symptoms are varied and can include TMJ pain, masticatory muscle pain and/or fatigue, limited mouth opening, headache, joint noises (clicking, popping, or crepitus), TMJ locking, otalgia, tinnitus, ear fullness, and vertigo [9,10].

Due to the multifactorial etiology of TMD, several treatment strategies have been adopted and the choice of these varies according to the degree of severity of the disorder [11].

Preferably, more conservative methods have been chosen as the first line of treatment [12]. Among these, we can mention non-invasive ones such as patient education, pharmacotherapy, physiotherapy, interocclusal splints, prosthetic rehabilitation, and/or minimally invasive such as arthrocentesis, hyaluronic acid injections, intra-articular injections of corticosteroids, platelet-rich plasma (PRP), oxygen–ozone therapy, and arthroscopy [11–14]. In some cases considered as severe, when patients do not respond to the conservative treatments, more invasive and intra-articular surgical interventions are required [13,15].

From this perspective, tissue engineering with stem cells emerges as an interesting and promising regenerative therapy strategy. Stem cells can be derived from two main sources, which are embryonic stem cells (ESCs) and adult stem cells, including mesenchymal stem cells (MSCs), the most widely used in TMJ [16]. The use of these cells occurs at the process of repairing damaged tissues or those with impaired function, in order to replace and repair their normal physiology, in addition to suppressing and modulating the inflammatory process [11,15,17]. This would provide a decrease in painful symptoms, prevent progressive degeneration of cartilage and subchondral bone, therefore favoring the reestablishment of TMJ function [18].

Bibliometric studies are carried out to identify the existing knowledge on a given topic quantitatively and qualitatively, to clarify research trends, as well as highlight themes that are already obsolete and without relevance for the academic and scientific environment [19]. The quantitative data of these studies presents the intensity of research on the subject, such as the authors, journals, and countries that have published the most about it. Also, the number of citations and the impact they cause related to the quantity and quality of the productions and the construction of the knowledge are to be mapped [20]. In addition, bibliometric analysis can provide a current overview of a given subject and point out suggestions for new topics to be addressed and researched to fill the gaps in existing studies [19,21].

Bibliometric TMD mapping studies are already found in the scientific literature [12,22–24]. It was observed that they worked with the theme in general, sometimes selecting the most cited articles in each period. Bibliometric studies on stem cells were also found; however, these studies were focused on the mechanisms of action, characteristics, differentiation, and cell signaling [25,26].

To the best of our knowledge, no previously published bibliometric study has investigated the specific relationship of TMDs with the treatments currently used, nor with the use of stem cells for that purpose. Thus, the present study aimed to investigate research trends and indicate future directions for studies on TMD and stem cells.

2. Materials and Methods

A comprehensive search was performed on the main collection of the Web of Science database on 23 October 2022. Scopus and Google Scholar were used to compare the number of paper citations. Two independent researchers conducted the selection of articles, and cases of disagreement were resolved via consensus. Articles dealing only with TMD and stem cells were included, regardless of the TMD classification or the involved cell type. There was no restriction regarding the year of publication of the articles or the language. Nevertheless, as exclusion criteria, we considered letters to the editor, conference articles, and studies without abstracts.

The search strategy used was elaborated with mesh terms related to the stem cells and temporomandibular joint (Supplementary file S1). Titles, year of publication, author's name, number of citations, the density of citations, country, affiliations, abstract, journal's name, study design, and author's keywords were extracted from papers and imported into the program Microsoft Excel®and VOS viewer (CWTS, Leiden University, Leiden, The Netherlands; https://www.vosviewer.com/ (accessed on 23 October 2022)). With the VOS viewer, three images were generated: the first one was about the network of authors who produced the most on the subject, in which each node represented an author. The size and color of the nodes indicated the frequency of occurrence and to which group each author belonged. The links between each node showed the collaboration network between the authors. Thus, the thickness and proximity between them indicated more frequent collaborations. The second image is related to the articles' country of origin. Each rectangle showed a country, their size pointed to the publications' frequency, each color represented the period of greatest publication and the lines the collaboration networks formed between them. Again, the thickness of these and the proximity between the rectangles indicated more collaboration between countries. The third image is about the most cited keywords, in which the more intense the color and the larger the font size of the letters, the greater the occurrence density is. In contrast, the letters' weaker colors and smaller font sizes indicate a lower occurrence.

After selection, the articles were read in full to search for information such as the type of stem cell used, tissue of origin of these cells, conditions caused in the TMJ, and proposed treatment method and biomarkers used. The main findings of in vitro studies were categorized according to cell origin, whether human or animal, the purpose of treatment, and the differentiation of stem cells. The biomarkers were organized according to their overexpression or inhibition, highlighting the possible influence of this result on TMD.

3. Results

The search carried out in WoS-CC resulted in 243 articles, of which 125 were selected according to the eligibility criteria (Figure 1). The articles were organized according to the year of publication, from the oldest to the most recent (Supplementary file S2). The 125 articles selected in our study received a total of 1321 citations in WoS-CC, 1366 in Scopus, and 2103 in Google Scholar. It is worth mentioning that among the one hundred twenty-five articles, seven were not found in Scopus [27–33], and one was not found in Google Scholar [34].

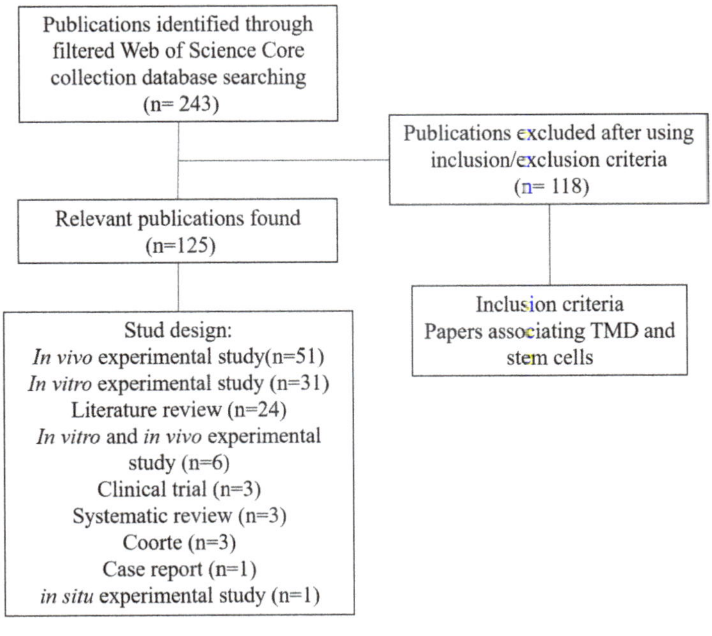

Figure 1. Flowchart of the article selection process on TMD and stem cells.

The selected articles were published between 1992 and 2022. The oldest article, entitled "The relationship of undifferentiated mesenchymal cells to TMJ articular tissue thickness" by Bibb et al., 1992, [35] was published in the "Journal of Dental Research" and received 22 citations in WoS-CC (1.66%), 23 citations in Scopus (1.68%) and 43 citations in Google

Scholar (2.04%). It aimed at evaluating the relationship between joint tissue thickness and the existence of undifferentiated cells in the TMJs of young adults. This study showed the beginning of investigations on the presence of stem cells in the joint region, which was an essential step for experiments currently being carried out that aim to characterize, differentiate, and treat TMJ disorders with undifferentiated or already-differentiated stem cells.

The most recent article, published in October 2022, was entitled "Stem Cells in Temporomandibular Joint Engineering: State of Art and Future Perspectives" and was published in the "Journal of Craniofacial Surgery" and received one citation in WoS-CC (0.075%), one in Scopus (0.073%), and three citations in Google Scholar (0.142%) [11]. It is a narrative review that brings several regenerative approaches to TMJs, including stem cell therapy. The authors report the existence and ability to extract mesenchymal stem cells from various tissues of the human body, including evidence of the ability of stem cells from fibrocartilage to form bone tissue in animal studies.

The period in which the highest number of citations occurred was between 2012 and 2022 (n = 1.051; 79.56%). This was also the period when the highest number of articles were published (n = 109; 87.20%) (Figure 2). The most cited paper (67 citations in WoS-CC, 5,07%) was by Wu et al. in 2014, entitled "The pilot study of fibrin with temporomandibular joint derived synovial stem cells in repairing TMJ disc perforation" [36], and published in the journal "BioMed research international" which deals with a pilot study for the use of scaffold fibrin/chitosan hybrids and mesenchymal stem cells, to evaluate the repair of the TMJ joint disc. Twelve articles received only one citation [11,18,28,37–45]. Ten articles were not cited in Wos-CC [27,29,46–52].

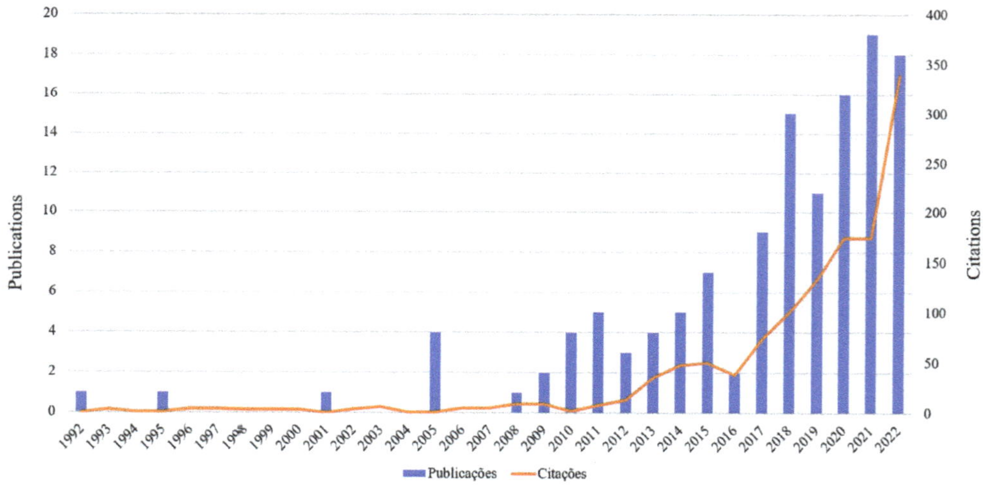

Figure 2. Distribution of articles published annually on TMD and stem cells, as well as their citations between 1992 and 2022.

A total of 65 journals were published on TMD and stem cells between the years 1992 to 2022. The ones that obtained the highest number of articles were the "Journal of Dental Research" (n = 11; 8.8%), "Biomed Research International" (n = 5; 4%), and the "International Journal of Molecular Science (n = 5; 4%). Among the studies published by the Journal of Dental Research, five articles were in vivo studies that aimed to characterize and stimulate stem cell differentiation [31,50,53–55]. Two were in vitro studies [56,57] and one in vivo/in vitro [58]; they also studied the cell differentiation and gene expression outcomes. The other three articles were review papers and aimed to elucidate the search for new therapeutic strategies for treating TMD and repair/regeneration of TMJ tissues [17,46,59].

The authors with the most significant number of articles on the subject can be seen in Figure 3, which was elaborated considering a minimum of two publications per author. The authors with the highest number of published articles are Zhang, M with eight (6.4%) and Zhang, J with seven (5.6%). It was observed that these same authors shared co-authorship in six articles, demonstrating great interaction between them. In addition, it was identified that such authors corresponded with others from different research groups, such as YU, S (n = six; 4.8%) and JIAO, K (n = five; 4%).

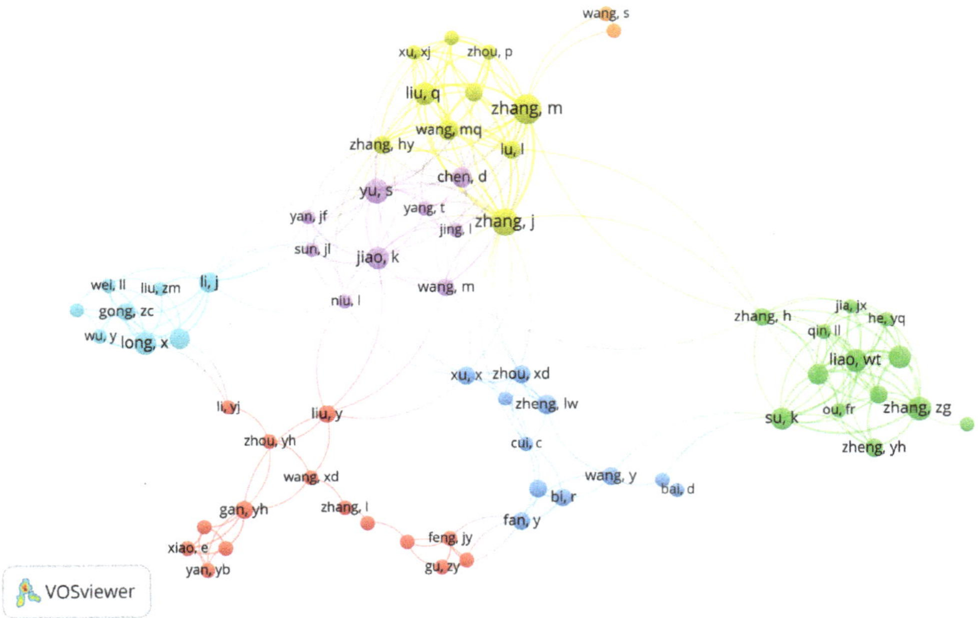

Figure 3. Network visualization of the authors who published the most on TMD and stem cells.

Regarding the countries of origin of the articles, 34 were identified in total. The country with the highest number of published articles was China, with 61. The first publication took place in 2005 and the country has led in the number of publications since 2010. In turn, the USA ranked second, with 32 articles published from 1992 until 2022. Interestingly, the USA led in the number of publications only in the years 1992 (n = 1), 2001(n = 1), and 2009 (n = 2); years in which China did not publish any articles about TMD and stem cells.

Specifically, when analyzing Figure 4, we can see that from 2017 to 2019, China had a peak in publications, which led to a number of articles published. The USA also increased the number of articles published between 2015 and 2017 but remained in second place. Even so, Figure 4 shows that the US has the most connections with other countries, the ten most important being: China (n = sixty-one), Germany (N = five), Canada (n = three), Brazil (n = three), Finland (n = one), Israel (n = one), Saudi Arabia (n = one), and Japan and Bosnia (n = one).

184 keywords were found on the 125 selected articles (Figure 5). In that image, the warmest colors represent the highest occurrence of keywords in the articles. Thus, it was identified that the most used were "cartilage" (n = 41; 4.42%), "temporomandibular joint" (n = 36; 3.88%), "mesenchymal stem cells" (n = 33; 3.56 %), and "osteoarthritis" (n = 26; 2.80%). These words may be related to the most searched points of interest in the analyzed theme.

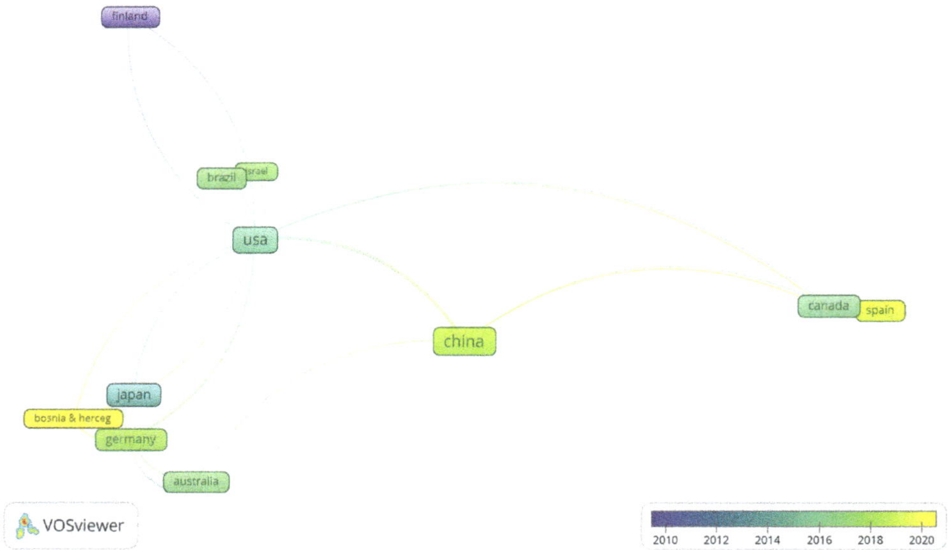

Figure 4. Countries with the highest number of articles published between 2010 and 2020.

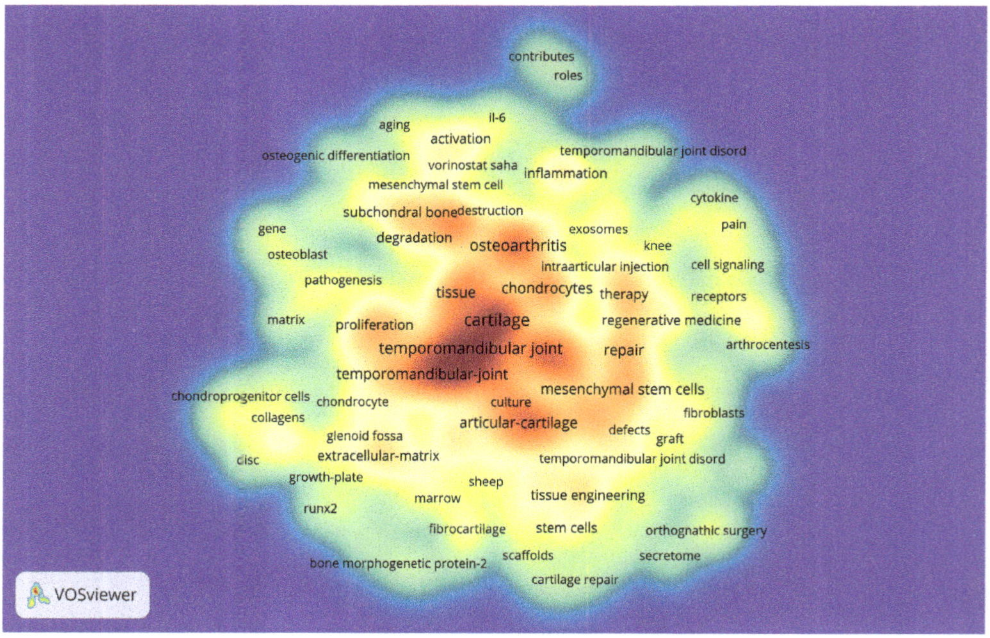

Figure 5. Keyword density map. Words most frequently found in the 125 selected articles. The colors indicate the citation density of the authors, ranging from blue (lowest density) to red (highest density).

In vivo studies (n = 51; 40.8%; 574 citations in WoS-CC), followed by an in vitro study (n = 31; 24.8%, 274 citations in WoS-CC) and the literature review (n = 24; 19.2%, 234 citations in WoS-CC), were the three most frequent types of studies. Also, some papers performed both types of study (in vivo and in vitro) together (n = 6; 4.8%, 56 citations in WoS-CC). Clinical trial (n = three, twenty-three citations in WoS-CC), systematic review (n = three, twenty-five citations in WoS-CC), cohort (n = three, seventeen citations in WoS-CC), case report (n = two, four citations in WoS-CC) and in situ (n = two, sixteen citations in WoS-CC) were the least frequent types of studies, with 2.4%, 2.4%, 2.4%, 1.6%, and 1.6%, respectively.

Analyzing only the in vitro studies (n = 37; 30%), we observed that the mesenchymal stem cells collected for culture and cell differentiation were obtained both from humans and animals (Table 1). The sources of stem cells of animal origin were bone marrow stroma (n = 14; 37.8%), fibrocartilage (n = 1; 2.7%), synovial fluid (n = 1; 2.7%), and myelomonocytes (n = 1; 2.7%). As for human sources, the regions of origin were synovial fluid (n = 9; 24.3%), bone marrow (n = 6; 16.2%), adipose tissue (n = 3; 8.1%), mandibular condylar chondrocytes (n = 1; 2.7%), Wharton's jelly (n = 1; 2.7%), and periodontal ligament (n = 1; 2.7%). After collection, these cells were differentiated into chondrocytes (n = 16; 43.2%), osteoblasts (n = 8; 21.6%), adipocytes (n = 7; 18.9%), neurons (n = 3; 8.1%), chondroblasts (n = 3; 8.1%), fibroblasts (n = 2; 5.4%), fibrochondrocytes (n = 2; 5.4%), osteoclasts (n = 1; 2.7%), macrophages (n = 1; 2.7%), and synoviocytes (n = 1; 2.7%). These studies aimed to treat osteoarthritis (n = 9; 24.3%), favor osteochondral neoformation/remodeling (n = 4; 10.8%), prevent fibrous ankylosis (n = 2; 5.4%), prevent subchondral bone resorption (n = 1; 2.7%), and treat anterior disc displacement (n = 1; 2.7%).

Table 1. Main findings of in vitro studies regarding the origin of stem cells, differentiated cell types, and treatment purposes.

Characteristics of the In Vitro Studies Analyzed	Absolute Frequency	Relative Frequency (%)
Source of origin of mesenchymal stem cells		
Animal origin		
Bone marrow stroma	14	37.8
Fibrocartilage	1	2.7
Synovial fluid	1	2.7
Myelomonocytes	1	2.7
Human origin		
Synovial fluid	9	24.3
Bone marrow	6	16.2
Adipose tissue	3	8.1
Mandibular condylar chondrocytes	1	2.7
Wharton jam	1	2.7
Periodontal ligament	1	2.7
Cell types differentiated from stem cells		
Chondrocytes	16	43.2
Osteoblasts	8	21.6
Adipocytes	7	18.9
Neural	3	8.1
Chondroblasts	3	8.1
Fibroblasts	2	5.4
Fibrochondrocytes	2	5.4
Osteoclasts	1	2.7
Macrophages	1	2.7
Purpose of TMD treatment		
Osteoarthritis	8	21.6
Osteochondral Neoformation/remodeling	4	10.8
Fibrous ankylosis	2	5.4
Bone ankylosis	2	5.4
Subchondral bone resorption	1	2.7
Joint inflammation	1	2.7
Anterior disc displacement	1	2.7

Analysis of in vivo studies (n = 57; 46.3%) showed that mesenchymal stem cells were collected from animals and humans (Table 2). Sources of animal origin were femoral bone (n = 6; 10.5%), tibia (n = 4; 7%), ilium (n = 3; 5.2%), TMJ subchondral bones (n = 1; 1.7%), synovial fluid (n = 1; 1.7%), condyle (n = 1; 1.7%) and glenoid fossa (n = 1; 1.7%). In turn, the sources of human origin were umbilical cord (n = 2; 3.5%), dental pulp (n = 1; 1.7%), condylar cartilage (n = 1; 1.7%) and adipose tissue (n = 1; 1.7%). Such studies aimed at treating osteoarthritis (n = 16; 28%), condylar cartilage defect (n = 5; 8.7%), joint ankylosis (n = 2; 3.5%), unilateral excision of the condyle (n = 1; 1.7%), subchondral bone deterioration (n = 1; 1.7%), hemimandible excision (n = 1; 1.7%), osteochondral defects (n = 1; 1.7%), excision condylar head (n = 1; 1.7%), hemifacial microsomia (n = 1; 1.7%), muscular or dental injuries such as anterior crossbite (n = 5; 8.7%), tension-responsive muscle hypertrophy mechanical (n = 2; 3.5%), lateral pterygoid hyperfunction (n = 1; 1.7%) and masseter myofascial pain (n = 1; 1.7%). In addition, age-related joint changes were also analyzed, such as postnatal condylar growth (n = 1; 1.7%), postnatal craniomandibular joint disc growth (n = 1; 1.7%), and condylar aging. (n = 1; 1.7%).

Table 2. Main findings of in vivo studies regarding the source of stem cells and TMJ disorders to be treated.

In Vivo Studies Characteristics	Absolute Frequency	Relative Frequency (%)
Source of origin of mesenchymal stem cells		
Animal origin	6	10.5
Femoral bone	4	7
Tibia	3	5.2
Ilium	1	1.7
TMJ subchondral bones	1	1.7
Synovial fluid	1	1.7
Condyle	1	1.7
Glenoid fossa		
Human origin	2	3.5
The umbilical cord	1	1.7
Dental pulp	1	1.7
Condylar cartilage	1	1.7
TMJ bone changes		
Osteoarthritis	16	28
Condylar cartilage defect	5	8.7
Joint ankylosis	2	3.5
Unilateral excision of condyle	1	1.7
Subchondral bone deteriorationn	1	1.7
Hemimandible excision	1	1.7
Osteochondral defects	1	1.7
Condylar head excision	1	1.7
Hemifacial microsomia	1	1.7
Malocclusions		
Anterior crossbite	5	8.7
Muscles disorders		
Muscular hypertrofy	2	3.5
Lateral pterygoid hyperfunction	1	1.7
Masseter myofascial pain	1	1.7
Age-related joint disorders		
Condylar postnatal growth	1	1.7
Postnatal growth of Craniomandibular articular disk	1	1.7
Condylar aging	1	1.7

Among the results obtained in our study, expressions and inhibitions of biomarkers and their possible influence on different TMJ disorders were also analyzed (Table 3).

Table 3. The main biomarkers evaluated in selected studies, their expression or inhibition, and their possible influence on TMJ disorders.

Biomarkers	Expression/Inhibition	Relationship
Ror2	Expression	Induction of osteoclast formation
Tn-C	Expression	Chondrocyte formation
Sox9	Expression	Chondrocyte formation/cartilaginous regeneration
Proteoglycan 4 (Prg4)-null	Expression	Ectopic formation of mineralized tissues and osteophytes in the articular disc, mandibular condyle and glenoid fossa
TRPS1	Expression	Participates in ATM development
Notch1	Inhibition	Temporary delay in the progress of cartilage degradation
TNF-α	Expression	Inflammatory factor that delays improvement in osteoarthritis
IFN-γ	Expression	Inflammatory factor that delays improvement in osteoarthritis
Adrb2	Expression	Induces subchondral bone loss in osteoarthritis
HIF-1alfa	Expression	May induce stem cells to promote chondrogenic repair of condylar cartilage and inhibit bone sclerosis
GDF11	Expression	Inhibits chondrocyte adipogenesis
ki67	Expression	Cartilaginous regeneration
FGF 18	Expression	Cartilaginous regeneration
MicroRNA-29b	Expression	Increased subchondral bone loss and osteoclast hyperfunction
Norepinefrina	Expression	Degenerative changes of the condylar subchondral bone
Osteopontina	Expression	Induction in the differentiation of chondrogenic and osteogenic cells
Colágeno tipo I	Expression	Cartilaginous regeneration
Colágeno tipo II	Expression	Induction in the differentiation of chondrogenic and osteogenic cells

Among the clinical studies (n = 3; 2.4%), two were carried out in adults and one in children, with mesenchymal stem cells collected from the donor himself from regions of the spinal cord (n = 1), from adipose tissues (n = 1), and neurogenic synovial membranes (n = 1). The main objectives of these studies were to treat condylar hyperplasia (n = 1), internal TMJ dysfunction (n = 1), joint ankylosis, and first arch dysplasia syndrome (n = 1).

Figure 6 represents the collection scheme, according to findings from clinical studies, in vitro and in vivo studies, and donor tissues, differentiation, and applicability of mesenchymal stem cells.

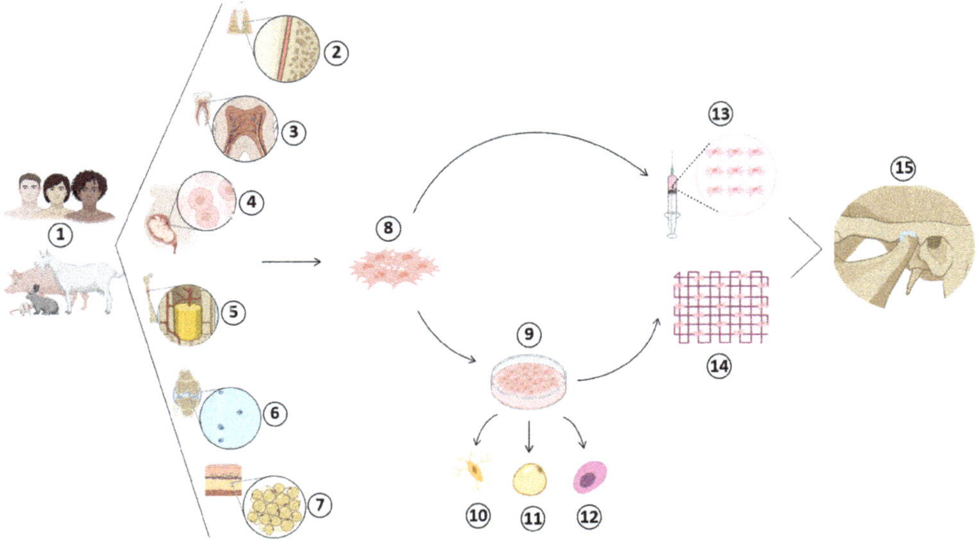

Figure 6. Stem cells scheme collection: source ①, donor area ②–⑦, differentiation ⑧, mode of application ①-①, and receptor region of mesenchymal stem cells ①. ①: Human or animal sources. ②: Periodontal ligament. ③: Dental pulp. ④: Umbilical cord. ⑤: Bone marrow. ⑥: Synovial fluid. ⑦: Adipose tissue. ⑧: Mesenchymal stem cell. ⑨: Cell culture and differentiation. ⑩: Osteocytes. ⑪: Adipocytes. ⑫: Chondrocytes. ⑬: Injection. ⑭ Scaffold. ⑮: Temporomandibular joint.

4. Discussion

In recent years, studies indicated that using MSC is a promising strategy for treating TMD [47,60]. It is because they are multipotent cells that can be extracted from different sources such as bone marrow, the umbilical cord, muscle, adipose tissue, dermis, peripheral blood, liver, dental pulp, synovium, and synovial fluid of TMJ [16,61], in addition to being able to differentiate into other cell types, such as adipocytes, osteocytes, and chondrocytes [21]. Depending on the damage caused to the joint, stem cells can act in tissue repair and regeneration, suppressing the inflammatory process and modulating the immune system [17]. For example, in the initial stage of joint osteoarthritis, these cells have a protective, homeostatic, and regenerative function, while in the more advanced stages, they act by delaying tissue degeneration [62].

In our study, qualitative and quantitative analyses of the 125 selected articles on "TMD" and "stem cells" were carried out. In general, we observed that the clinical and experimental studies evaluated the action of mesenchymal stem cells collected from different sources and their possible reparative/regenerating potential against various TMJ disorders. In vivo and in vitro experimental studies were the main types found. The main donor sites were sources for bone marrow stroma, femoral bone, tibia, and synovial fluid. Chondrocytes and osteoblasts were the main cell types differed from stem cells for in vitro analysis. The central TMJ joint disorder investigated regarding the role of stem cells was osteoarthritis and, among muscle disorders, muscle hypertrophy. Joint modifications resulting from age, such as postnatal condylar growth and condylar aging, were also analyzed within the studies, as well as the evaluation of biomarkers, which, when expressed or inhibited, were associated—or not—with TMJ disorders.

The main database used in our analysis was WoS-CC. This was the main database chosen since it was designed to satisfy users in citation analysis and even provides graphic images of the analyzed results [63]. Scopus and Google Scholar databases were used to compare the number of citations. As already reported by other bibliometric studies [64,65], the database with the highest number of citations was Google Scholar. A possible explanation for this result may be related to the fact that Google Scholar presents citations of open-access books, thesis, dissertations, and online journals [65], journals that are not covered or not yet indexed by other databases, e-print archives, universities, or governmental and non-governmental organization web sites [66].

Although we selected all articles on the subject available at WoS-CC, regardless of whether they have citations, identifying studies with the highest number of citations is essential, as these data suggest that it gives it greater relevance in the field of knowledge [67]. Moreover, the most cited articles can be considered classic articles, exerting more significant scientific influence on the field and generating discussions about future research projections in the area [68]. However, it is worth mentioning that the time of publication can strongly influence the number of citations, since recently published studies do not have enough time to reach a large number of citations [69].

Analyzing articles published in WoS-CC on TMD and stem cells over the years brought some interesting findings. For example, even though the first article was published in 1992, only from 2008 onwards did the number of publications show regularity, with 2021 being the year with the highest number. In that year, in vivo studies were the majority (n = 8) and highlighted the ability to repair bone defects and TMJ tissue regeneration when treated with stem cells [39,41,42,53,70–73]. Such treatments were performed by injecting stem cells into the TMJ or using scaffolds enriched with them, thereby promoting guided tissue regeneration. A bibliometric study on artificial extracellular matrices showed a great trend in research on stem cells in tissue engineering and pointed out that in vitro studies are gradually being replaced by in vivo studies with a more clinical focus, justifying the increase in this type of research [74].

In general, when measured over the years, the type of study that most prevailed was once again in vivo (n = 51), followed by in vitro (n = 37, 30%). Such studies aimed at using mesenchymal stem cells to treat various joint disorders of the TMJ, mainly osteoarthritis, a disease characterized by chronic inflammation of the synovial tissue, as well as progressive cartilage degradation and remodeling of the subchondral bone [75]. Osteoarthritis is classified as articular TMD by the Diagnostic Criteria for Temporomandibular Disorders (DC/DTM) [76], an essential diagnostic tool used by clinical dentists and researchers. The prevalence of these studies contrasts with the epidemiological reality regarding TMD, since TMJ osteoarthritis is a highly prevalent disease [77].

The most published authors were Zhang M (n = 8; 6.4%) and Zhang J (n = 7; 5.6%). The two are part of the same team and share authorship in six articles whose central themes refer to the consequences and treatment of joint osteoarthritis. These were in vivo studies carried out with mice and published between 2014 and 2022. Interestingly, this was the period in which there were more publications on the subject, and the articles in which the authors shared authorship obtained a more significant number of quotes. As our study is about a specific theme, TMD and stem cells, when we compare our results with other bibliometric studies on each of these themes individually [23–26], the results are different, and several other authors appear. This is probably because the themes are very broad and cover classic studies, which have great impact, and were cited many times, serving as a basis and guideline to determine clinical conducts or even research protocols.

In the present study, the country that most published on the subject was China, which accounted for 61 published articles, reinforcing the result of other bibliometric studies on stem cells, in which China also ranked first with 1247 and 899 articles published, respectively [25,26]. Also, the USA obtained second place in these same studies, with 392 and 278 articles published, respectively. Our results indicate that despite having obtained second place in the number of publications on the subject, the USA had the most significant interaction with other countries. In a bibliometric study on TMD [24], the US leads in the number of articles published, with China in 15th place. Another study, also on TMD [23], again pointed to the USA being first place in the number of published articles and China in fourth place. A comparison between our results and the other cited bibliometric studies seems to indicate that China only leads studies on TMD specifically related to stem cells.

Regarding keywords, they are used to facilitate the search for specific fields since the increase in the occurrence of one of these words can help to understand the search trend on a particular topic [12]. In our study, this analysis was performed using the density map, and it was observed that the most frequent words related to TMD were cartilage (n = 41; 4.42%), temporomandibular joint (n = 36; 3.88%), and osteoarthritis (n = 26; 2.80%). These words are closely related to each other, as osteoarthritis is a type of TMJ disorder characterized by progressive cartilage degeneration, abnormal subchondral bone remodeling, and synovitis [78]. Given the complexity of the disease, the most used conservative treatments are aimed at relieving symptoms [79]. However, the search for different treatments, such as those using stem cells with the aim of promoting tissue regeneration in the case of osteoarthritis, has recently grown [80], which may justify the search index on the field. Another bibliometric study on the performance of publications and research trends in TMD [81] found that the most frequent keywords were "orofacial pain", "chronic pain", "bruxism", and "myofascial pain". These words are related, respectively, to the classification of TMD as an orofacial and chronic pain; bruxism, which for years was confused with TMD, but is known to be different and may or may not be associated; and myofascial pain, which is one of the main types of muscular TMD.

Analyzing stem cells' effects on TMJ tissues is crucial for understanding the formation and regeneration potential of such cell types on joint structures. Thus, it is worth highlighting the growing number of studies to describe the action of specific biomarkers in the natural development of healthy TMJ tissues or the presence of dysfunctions. In our study, the main biomarkers studied among the 125 articles were Ror2, TnC, Sox9, Protoglican4 (prr4) –null, TRPS1, Notch1, TNF-α, IFN-γ, Adrb2, HIF-1 alpha, GDF11, Ki67, FGF 18, Micro RNA-29b, Norepinephrine, Osteopontin, Type I Collagen, and Type II Collagen. Yang et al. [82] analyzing the Ror2 biomarker, found that its overexpression increases osteoclastogenic activity and subchondral bone loss, indicating that its presence is part of the formation process of bone defects in joint osteoarthritis. Other biomarkers investigated were Adrb2, whose increased expression induced subchondral bone loss in osteoarthritis [83], and Micro RNA-29b expression, which increases subchondral bone loss and osteoclast hyperfunction [84]. The analysis of the biomarkers HIF-1alpha, Ki67, FGF 18, collagen type I, and type II demonstrated that their expression causes cartilaginous regeneration through the induction and differentiation of chondrogenic cells. Furthermore, the inhibition of Notch1 promoted the temporary delay of the progress of cartilage degradation [37,72,74,85].

Although it was not selected among the 125 articles, it is worth mentioning an interesting recent article [86] that investigated the hypothesis that the polymorphism in the PAX7 gene would be associated with muscular TMD patients. As the authors explain, satellite cells (SC) are skeletal muscle stem cells, activated in cases of muscle injury. However, for them to be activated, growth factors must be present at the injury site, such as the expression of the transcription factor PAX7 in the differentiation of muscle cells. When they migrate to the site of injury, SCs also begin to express another myogenic regulatory factor, MyoD (myogenic differentiation). Thus, in the process of the differentiation of SC cells into myoblasts, there is an increase in MyoD expression and a reduction in PAX7 expression. Based on this principle, the authors selected two polymorphisms of a single nucleotide in the PAX7 gene (rs766325, rs6659735) and, as a result, they perceived that alterations in PAX7 can influence the muscle pathophysiology, and the homozygous genotype (GG) rs6659735 is apparently associated with the disorder muscle in individuals with TMD.

Another interesting point observed in the selected studies was that collecting stem cells was easy and the various properties found in these cells make them the target of the investigation. The source depends on the ability of stem cells to differentiate, the modulation of the immune and inflammatory response, and the power of paracrine communication [87].

In our study, the primary sources of stem cells were dental pulp, periodontal ligament, synovia, and synovial fluid.

Mesenchymal stem cells derived from dental pulp are easily accessible, as they can be extracted from deciduous teeth naturally lost by children or third molars commonly indicated for extraction [88]. These cells have excellent differentiation potential and can be used to treat various tissues, including bones and cartilage [89]. Moreover, in dental-derived stem cells, NURR1 downregulation favors osteoblastic differentiation, with the further characteristic that these cells seem to possess an expression of stemness genes, such as the transcription factors Kruppel-like factor 4 (Klf-4), octamer-binding transcription factor 4 (Oct-4), homeobox transcription factor Nanog (Nanog), associated with the maintenance of self-renewal and multi-differentiation capacity and reported in MSCs of different origin, confirming the great potential of dental-derived stem cells in tissue regeneration [90,91]. Due to these characteristics, several studies have been developed to investigate these cells [92,93].

On the other hand, stem cells derived from the periodontal ligament are very potent cells that exhibit properties similar to those of mesenchymal stem cells derived from bone marrow, that is, they are clonogenic and have a high proliferative capacity and, specifically, are capable of regenerating periodontal tissues [94]. In addition, they exhibit the capacity for long-term survival, self-renewal, and show an ability to differentiate into specialized cells and regenerate various functional tissues [95,96].

Stem cells derived from the synovium have a high power of proliferation and chondrogenic differentiation. They can be extracted from tissue fragments removed in arthroscopy and can adhere to the cartilage when in a favorable environment [86]. Following, the first study on the characterization of cells derived from TMJ synovial fluid as stem cells was carried out by Koyama et al. [97]. In this, the authors confirmed that human synovial fluid is a good source of cells capable of differentiating into several other lineages, such as osteoblasts, chondrocytes, neurons, and adipocytes. Subsequently, Sun et al. [98] investigated both cells derived from fragments of synovia and mesenchymal stem cells derived from synovial fluid (SFCs) and identified that both SFCs can be induced to differentiate down osteogenic, chondrogenic, adipogenic, and neurogenic lineages in vitro and that probably, the intima is the most likely tissue origin of SFMSCs in the TMJ.

We also observed that some selected studies highlighted the benefits of an injection of stem cells into TMJs affected by some type of experimentally induced dysfunction and, for this purpose, each of these studies used different specific strategies, both to induce the lesion and to promote the injection of the cells. With the objective of evaluating in vitro and in vivo the effects of 20 minutes daily of low-intensity pulsed ultrasound (LIPUS) treatment, El-Bialy et al. [99] carried out a pilot study using bone marrow stromal cells (BMC) isolated from rabbits. These cells were expanded, differentiated into chondrogenic and osteogenic lineages, inserted into biodegradable scaffolds and later implanted into the amputated TMJ articular condylar defect. The results of this cited pilot study suggest that LIPUS can enhance chondrogenic and osteogenic differentiation of BMSCs in vitro as well as to enhance functional integration of these in vivo in rabbits. Zhang et al. [100] used a mouse model of TMJ osteoarthritis to weekly inject green fluorescent protein-labeled exogenous bone marrow stromal cells (GFP-BMSC) for 4, 8, and 12 successive weeks. They concluded that BMSCs were able to repair damaged cartilage by increasing matrix production and scavenging activity and by not increasing cell proliferation. Wang et al. [101] evaluated in vivo and in vitro the effects of an injection of bone marrow mesenchymal stem cell (BMSC)-derived small extracellular vesicles (BMSC-sEVs) on cartilage reconstruction with a TMJOA model in rabbits and confirmed that the BMSC-sEVs may play an important role in cartilage reconstruction in TMJOA via the autotaxin–YAP signaling axis. [49] also analyzed in vitro and in vivo the effects of an intra-articular injection of MSCs by placing these cells on the surface of TGFβ-loaded GelMA microspheres in a model of TMJ arthritis bilaterally induced in rats. They concluded that BMSCs-coated microspheres can effectively promote the repair and reconstruction of cartilage defects within the TMJ arthritis.

It is worth mentioning that among the selected articles, one tested the effectiveness of the intra-articular infiltration of autologous stem cells derived from adipose tissue in the upper compartment of the TMJ in four patients Carboni et al [102]. This study focused on verifying the results of this procedure regarding pain, joint noises, maximum mandibular opening, and mandibular movements. The authors concluded that the results of this preliminary study demonstrated that autologous stem cells derived from adipose tissue were able to resolve the symptoms presented by the patients, as well as being effective in restoring the structural anatomical integrity verified through the functional magnetic resonance imaging (MRI).

Thus, when analyzing all the results of our study, it is observed that the use of stem cells as a treatment strategy for TMD has been increasingly studied both in vitro and in vivo due to a certain ease of obtaining (depending on origin), the capacity for cell proliferation and differentiation and the regenerative potential of joint structures.

5. Conclusions

As far as we are concerned, this is the first bibliometric study that relates stem cells and TMD. With regard to the metrics, we observed that among the selected articles, although the first study was carried out in 1992, a more significant number of publications and citations occurred between the years 2012 and 2022. Furthermore, although 34 countries produce articles on the subject, China was in the lead, followed by the USA. The two authors with the highest number of published articles shared co-authorship in several studies. The main keywords found were "cartilage", "temporomandibular joint", "mesenchymal stem cells", and "osteoarthritis", which were closely related to the topic and indicated research trends in the area. When it comes to evaluating the knowledge produced, the selected articles evaluated cells extracted from different sources regarding their potential for differentiation and therapeutic intent. It was found that the most commonly used cell type were mesenchymal stem cells, which have a high proliferative capacity and a high power of differentiation into different cell types. In the selected studies, the main sources of these cells were the dental pulp, the periodontal ligament, the synovium, and the synovial fluid. In vivo studies were the most numerous, and this result possibly indicates a growing interest in using these cells to treat refractory patients with TMD. Based on our findings, it is suggested that future research in this area should be carried out in order to determine if there is any type and/or subtype of TMD that would benefit more from the treatment with stem cells, as well as if there would be any stage of development of TMD, i.e., more recent or more advanced, in which the use of stem cells would provide greater morphological and functional benefits to patients, consequently bringing benefits to their quality of life.

Supplementary Materials: The following supporting information can be downloaded at: https://www.mdpi.com/article/10.3390/jfb14020103/s1. Supplementary file S1: The Search strategy and Supplementary file S2: Table S1—The 125 papers on the directions of Temporomandibular Disorders and Stem Cells.

Author Contributions: Conceptualization: Z.A.d.S., R.R.L. and R.D.S.-R.; methodology: Z.A.d.S., W.W.P.M., H.H.N.F. and R.D.S.-R.; literature search and data analysis: Z.A.d.S., W.W.P.M., R.D.S.-R. and H.H.N.F.; writing—original draft preparation: Z.A.d.S. and W.W.P.M.; writing—review and edition: R.D.S.-R. and R.R.L.; supervision: R.D.S.-R. All authors have read and agreed to the published version of the manuscript.

Funding: This research was funded by the Coordenação de Aperfeiçoamento de Pessoal de Nível Superior Brazil (CAPES)—Finance Code 001. The APC was funded by Pró-Reitoria de Pesquisa e Pós-graduação from Federal University of Pará (PROPESP-UFPA).

Institutional Review Board Statement: Not applicable.

Informed Consent Statement: Not applicable.

Data Availability Statement: Not applicable.

Acknowledgments: This work was supported by Coordenação de Aperfeiçoamento de Pessoal de Nível Superior Brazil (CAPES) and Pró-Reitoria de Pesquisa e Pós-Graduação da UFPA (PROPESP, UFPA, Brazil).

Conflicts of Interest: The authors declare no conflict of interest.

References

1. De Leeuw, R.; Klasser, G.D. *Orofacial Pain: Guidelines for Assessment, Diagnosis, and Management*; Quintessence: Hanover Park, IL, USA, 2018.
2. Okeson, J.P.; de Leeuw, R. Differential diagnosis of temporomandibular disorders and other orofacial pain disorders. *Dent. Clin. N. Am.* **2011**, *55*, 105–120. [CrossRef] [PubMed]
3. Nagai, H.; Miyamoto, Y. Isolation and characterization of synovial cells from the human temporomandibular joint. *J. Oral Pathol. Med.* **2006**, *35*, 104–110. [CrossRef] [PubMed]
4. Craane, B.; De Laat, A. Physical therapy for the management of patients with temporomandibular disorders and related pain. *Cochrane Database Syst. Rev.* **2018**, *2018*, CD005621. [CrossRef]
5. Valesan, L.F.; Da-Cas, C.D. Prevalence of temporomandibular joint disorders: A systematic review and meta-analysis. *Clin. Oral Investig.* **2021**, *25*, 441–453. [CrossRef]
6. Slade, G.D.; Ohrbach, R. Painful Temporomandibular Disorder: Decade of Discovery from OPPERA Studies. *J. Dent. Res.* **2016**, *95*, 1084–1092. [CrossRef]
7. Reis, P.H.F.; Laxe, L.A.C. Distribution of anxiety and depression among different subtypes of temporomandibular disorder: A systematic review and meta-analysis. *J. Oral Rehabil.* **2022**, *49*, 754–767. [CrossRef]
8. Dos Santos, E.A.; Peinado, B.R.R. Association between temporomandibular disorders and anxiety: A systematic review. *Front. Psychiatry* **2022**, *13*, 990430. [CrossRef] [PubMed]
9. Carrara, S.V.; Cotti, P.C.R.; Barbosa, J.S. Statement of the 1st consensus on temporomandibular disorders and orofacial pain. *Dental Press J. Orthod.* **2010**, *15*, 114–120. [CrossRef]
10. Durham, J.; Newton-John, T.R. Temporomandibular disorders. *BMJ Clin. Res. Ed.* **2015**, *350*, h1154. [CrossRef] [PubMed]
11. Minervini, G.; Del Mondo, D. Stem Cells in Temporomandibular Joint Engineering: State of Art and Future Persectives. *J. Craniofac. Surg.* **2022**, *33*, 2181–2187. [CrossRef] [PubMed]
12. Ferrillo, M.; Gallo, V. The 50 most-cited articles on temporomandibular disorders: A bibliometric analysis. *J. Back Musculoskelet. Rehabil.* **2022**, 1–19, *pre-press*. [CrossRef]
13. Abouelhuda, A.M.; Khalifa, A.K. Non-invasive different modalities of treatment for temporomandibular disorders: Review of literature. *J. Korean Assoc. Oral Maxillofac. Surg.* **2018**, *44*, 43–51. [CrossRef]
14. Minervini, G.; Fiorillo, L. Prosthodontic Treatment in Patients with Temporomandibular Disorders and Orofacial Pain and/or Bruxism: A Review of the Literature. *Prosthesis* **2022**, *4*, 253–262. [CrossRef]
15. Pagotto, L.E.C.; de Santana Santos, T. The efficacy of mesenchymal stem cells in regenerating structures associated with the temporomandibular joint: A systematic review. *Arch. Oral Biol.* **2021**, *125*, 105104. [CrossRef]
16. Zhang, S.; Yap, A.U. Stem Cells for Temporomandibular Joint Repair and Regeneration. *Stem Cell Rev. Rep.* **2015**, *11*, 728–742. [CrossRef]
17. Gong, S.; Emperumal, C.P. Regeneration of temporomandibular joint using in vitro human stem cells: A review. *J. Tissue Eng. Regen. Med.* **2022**, *16*, 591–604. [CrossRef] [PubMed]
18. Chęciński, M.; Chęcińska, K.; Turosz, N.; Kamińska, M.; Nowak, Z.; Sikora, M.; Chlubek, D. Autologous Stem Cells Transplants in the Treatment of Temporomandibular Joints Disorders: A Systematic Review and Meta-Analysis of Clinical Trials. *Cells* **2022**, *11*, 2709. [CrossRef] [PubMed]
19. Wang, L.; Jiang, J.; Li, Y.; Huang, J.; Wang, R.; Liang, Y.; He, C.; Liu, S. Global trends and hotspots in research on osteoporosis rehabilitation: A bibliometric study and visualization analysis. *Front. Public Health* **2022**, *10*, 1022035. [CrossRef] [PubMed]
20. Leclerc, A.; Chastang, J.F. The bibliographic impact of epidemiological studies: What can be learnt from citations? *Occup. Environ. Med.* **2010**, *67*, 213–216. [CrossRef]
21. Liu, J.; Gao, J.; Niu, Q.; Wu, F.; Wu, Z.; Zhang, L. Bibliometric and visualization analysis of mesenchymal stem cells and rheumatoid arthritis (from 2012 to 2021). *Front. Immunol.* **2022**, *13*, 1001598. [CrossRef]
22. Machado, N.A.; Lima, F.F. Current panorama of temporomandibular disorders' field in Brazil. *J. Appl. Oral Sci.* **2014**, *22*, 146–151. [CrossRef] [PubMed]
23. Bai, B.; Bai, X. Mapping research trends of temporomandibular disorders from 2010 to 2019: A bibliometric analysis. *J. Oral Rehabil.* **2021**, *48*, 517–530. [CrossRef] [PubMed]
24. Al-Sharaee, Y.; Al-Moraissi, E.A. Top 100 Cited Publications in the Field of Temporomandibular Disorders: A Bibliometric Analysis. *Front. Oral Health* **2022**, *3*, 864519. [CrossRef]
25. Wang, R.; Wang, Y. Bibliometric and visual analysis of mesenchymal stem cells in the treatment of osteoporosis based on CiteSpace software. *Medicine* **2022**, *101*, e31859. [CrossRef]
26. Zhang, X.; Lu, Y. An Overview of Current Research on Mesenchymal Stem Cell-Derived Extracellular Vesicles: A Bibliometric Analysis From 2009 to 2021. *Front. Bioeng. Biotechnol.* **2022**, *10*, 1109. [CrossRef]

27. Sun, Y.P.; Lu, Y.Y. Effects of interleukin 1β on long noncoding RNA and mRNA expression profiles of human synovial fluid derived mesenchymal stem cells. *Sci. Rep.* **2022**, *12*, 8432. [CrossRef] [PubMed]
28. Liao, W.T.; Sun, J.D. Histone deacetylase inhibitors attenuated interleukin-1β-induced chondrogenesis inhibition in synovium-derived mesenchymal stem cells of the temporomandibular joint. *Bone Jt. Res.* **2022**, *11*, 40–48. [CrossRef]
29. Yang, J.; Xu, Y. MicroRNA-26b regulates BMSC osteogenic differentiation of TMJ subchondral bone through β-catenin in osteoarthritis. *Bone* **2022**, *162*, 116448. [CrossRef] [PubMed]
30. Jia, J.; Sun, J. Knockdown of long non coding RNA AK094629 attenuates the interleukin 1β induced expression of interleukin 6 in synovium derived mesenchymal stem cells from the temporomandibular joint. *Mol. Med Rep.* **2020**, *22*, 1195–1204. [CrossRef]
31. Sun, J.; Liao, W. Suberoylanilide Hydroxamic Acid Attenuates Interleukin-1β-Induced Interleukin-6 Upregulation by Inhibiting the Microtubule Affinity-Regulating Kinase 4/Nuclear Factor-κB Pathway in Synovium-Derived Mesenchymal Stem Cells from the Temporomandibular Joint. *Inflammation* **2020**, *43*, 1246–1258. [CrossRef]
32. Kim, Y.H.; Bang, J.I. Protective effects of extracorporeal shockwave on rat chondrocytes and temporomandibular joint osteoarthritis; preclinical evaluation with in vivo99mTc-HDP SPECT and ex vivo micro-CT. *Osteoarthr. Cartl.* **2019**, *27*, 1692–1701. [CrossRef]
33. Xu, T.; Wu, M. RhoA/Rho kinase signaling regulates transforming growth factor-β1-induced chondrogenesis and actin organization of synovium-derived mesenchymal stem cells through interaction with the Smad pathway. *Int. J. Mol. Med.* **2012**, *30*, 1119–1125. [CrossRef] [PubMed]
34. Takahashi, M.; Fujikawa, K. An in situ hybridization study of MMP-2, -9, -13, -14, TIMP-1, and -2 mRNA in fetal mouse mandibular condylar cartilage as compared with limb bud cartilage. *Gene Expr. Patterns* **2019**, *32*, 1–11. [CrossRef] [PubMed]
35. Bibb, C.A.; Pullinger, A.G.; Baldioceda, F. The relationship of undifferentiated mesenchymal cells to TMJ articular tissue thickness. *J. Dent. Res.* **1992**, *71*, 1816–1821. [CrossRef]
36. Wu, Y.; Gong, Z. The pilot study of fibrin with temporomandibular joint derived synovial stem cells in repairing TMJ disc perforation. *BioMed Res. Int.* **2014**, *2014*, 454021. [CrossRef]
37. Wang, H.; Shi, Y. GDF11 inhibits abnormal adipogenesis of condylar chondrocytes in temporomandibular joint osteoarthritis. *Bone Jt. Res.* **2022**, *11*, 453–464. [CrossRef]
38. Li, Y.; Zhou, Y. Synovial macrophages in cartilage destruction and regeneration-lessons learnt from osteoarthritis and synovial chondromatosis. *Biomed. Mat.* **2021**, *17*, 012001. [CrossRef]
39. Katagiri, W.; Endo, S. Conditioned medium from mesenchymal stem cells improves condylar resorption induced by mandibular distraction osteogenesis in a rat model. *Heliyon* **2021**, *7*, e06530. [CrossRef]
40. Zhang, P.P.; Liang, S.X. Differences in the biological properties of mesenchymal stromal cells from traumatic temporomandibular joint fibrous and bony ankylosis: A comparative study. *Anim. Cells Syst.* **2021**, *25*, 296–311. [CrossRef] [PubMed]
41. Zhang, Y.; Xu, X. Elder Mice Exhibit More Severe Degeneration and Milder Regeneration in Temporomandibular Joints Subjected to Bilateral Anterior Crossbite. *Front. Physiol.* **2021**, *12*, 2039. [CrossRef] [PubMed]
42. Cha, S.; Lee, S.M. Enhanced Circadian Clock in MSCs-Based Cytotherapy Ameliorates Age-Related Temporomandibular Joint Condyle Degeneration. *Int. J. Mol. Sci.* **2021**, *22*, 10632. [CrossRef]
43. Helgeland, E.; Pedersen, T.O. Angiostatin-functionalized collagen scaffolds suppress angiogenesis but do not induce chondrogenesis by mesenchymal stromal cells in vivo. *J. Oral Sci.* **2020**, *62*, 371–376. [CrossRef]
44. Yoshitake, H.; Kayamori, K. Biomarker expression related to chondromatosis in the temporomandibular joint. *Cranio J. Craniomandib. Pract.* **2021**, *39*, 362–366. [CrossRef]
45. Yang, M.C.; Wang, D.H. Role of Link N in Modulating Inflammatory Conditions. *J. Oral Facial Pain Headache* **2019**, *33*, 114–122. [CrossRef]
46. Karic, V.; Chandran, R. Photobiomodulation and Stem Cell Therapy for Temporomandibular Joint Disc Disorders. *Photobiomodulation Photomed. Laser Surg.* **2020**, *38*, 398–408. [CrossRef]
47. AbuBakr, N.; Fares, A.E. Mesenchymal stem cells-derived microvesicles versus platelet-rich plasma in the treatment of monoiodoacetate-induced temporomandibular joint osteoarthritis in Albino rats. *Heliyon* **2022**, *8*, e10857. [CrossRef]
48. Zhang, J.; Pi, C. PTHrP promotes subchondral bone formation in TMJ-OA. *Int. J. Oral Sci.* **2022**, *14*, 37. [CrossRef]
49. Yang, Y.; Huang, C. Superwettable and injectable GelMA-MSC microspheres promote cartilage repair in temporomandibular joints. *Front. Bioeng. Biotechnol.* **2022**, *10*, 1026911. [CrossRef] [PubMed]
50. Thamm, J.R.; Jounaidi, Y. Temporomandibular Joint Fibrocartilage Contains CD105 Positive Mouse Mesenchymal Stem/Progenitor Cells with Increased Chondrogenic Potential. *J. Maxillofac. Oral Surg.* **2022**, 1–12. [CrossRef]
51. Putnová, B.; Hurník, P. Effect of human adipose-derived regenerative cells on temporomandibular joint healing in immunodeficient rabbits. *Acta Vet. Brno* **2019**, *88*, 49–56. [CrossRef]
52. Lavi, A.; Pelled, G. Isolation and characterization of mesenchymal stromal progenitors from the temporomandibular joint disc. *J. Tissue Eng. Regen. Med.* **2017**, *11*, 1553–1561. [CrossRef]
53. Yu, X.; Hu, Y. A bilayered scaffold with segregated hydrophilicity-hydrophobicity enables reconstruction of goat hierarchical temporomandibular joint condyle cartilage. *Acta Biomater.* **2021**, *121*, 288–302. [CrossRef]
54. Ruscitto, A.; Scarpa, V. Notch Regulates Fibrocartilage Stem Cell Fate and Is Upregulated in Inflammatory TMJ Arthritis. *J. Dent. Res.* **2020**, *99*, 1174–1181. [CrossRef]

55. Zhou, Y.; Chen, M. Profiling of Stem/Progenitor Cell Regulatory Genes of the Synovial Joint by Genome-Wide RNA-Seq Analysis. *BioMed. Res. Int.* **2018**, *2018*, 9327487. [CrossRef]
56. Lin, Y.; Umebayashi, M. Combination of polyetherketoneketone scaffold and human mesenchymal stem cells from temporomandibular joint synovial fluid enhances bone regeneration. *Sci. Rep.* **2019**, *9*, 472. [CrossRef] [PubMed]
57. He, L.H.; Xiao, E. Osteoclast Deficiency Contributes to Temporomandibular Joint Ankylosed Bone Mass Formation. *J. Dent. Res.* **2015**, *94*, 1392–1400. [CrossRef]
58. Yang, H.; Zhang, M. Inhibition of Ihh Reverses Temporomandibular Joint Osteoarthritis via a PTH1R Signaling Dependent Mechanism. *Int. J. Mol. Sci.* **2019**, *20*, 3797. [CrossRef] [PubMed]
59. Dashnyam, K.; Lee, J.H. Intra-articular biomaterials-assisted delivery to treat temporomandibular joint disorders. *J. Tissue Eng.* **2018**, *9*, 1–12. [CrossRef]
60. Ciocca, L.; Donati, D. Mesenchymal stem cells and platelet gel improve bone deposition within CAD-CAM custom-made ceramic HA scaffolds for condyle substitution. *BioMed Res. Int.* **2013**, *2013*, 549762. [CrossRef]
61. Li, N.; Hua, J. Interações entre células-tronco mesenquimais e o sistema imunológico. *Cell. Mol. Life Sci.* **2017**, *74*, 2345–2360. [CrossRef] [PubMed]
62. Jiang, Y.; Tuan, R.S. Origin and function of cartilage stem/progenitor cells in osteoarthritis. *Nat. Rev. Rheumatol.* **2015**, *11*, 206–212. [CrossRef] [PubMed]
63. Falagas, M.E.; Pitsouni, E.I. Comparison of PubMed, Scopus, Web of Science, and Google Scholar: Strengths and weaknesses. *FASEB J. Off. Publ. Fed. Am. Soc. Exp. Biol.* **2008**, *22*, 338–342. [CrossRef] [PubMed]
64. Ahmad, P.; Vincent Abbott, P. Uma análise bibliométrica dos 50 artigos mais citados publicados na revista Dental Traumatology. *Dent. Traumatol.* **2020**, *36*, 89–99. [CrossRef] [PubMed]
65. Melo, W.W.P.; Aragão, W.A.B. Effects of Photobiomodulation on Oral Mucositis: Visualization and Analysis of Knowledge. *Life* **2020**, *12*, 1940. [CrossRef]
66. Bakkalbasi, N.; Bauer, K. Three options for citation tracking: Google Scholar, Scopus and Web of Science. *Biomed. Digit. Libr.* **2006**, *3*, 7. [CrossRef]
67. Li, T.H.; Li, Z.M. Global Analyses and Latest Research Hot Spots of Adipose-Derived Stem Cells in Fat Grafting: A Bibliometric and Visualized Review. *Aesth. Plast. Surg.* **2022**, 1–13. [CrossRef]
68. Fardi, A.; Kodonas, K.; Economides, N. Top-cited articles in endodontic journals. *J. Endod.* **2011**, *37*, 1183–1190. [CrossRef]
69. Schwarz, G.M.; Hajdu, S. The top fifty most influential articles on hip fractures. *Int. Orthop.* **2022**, *46*, 2437–2453. [CrossRef]
70. Sumarta, N.P.M.; Kamadjaja, D.B. Human umbilical cord mesenchymal stem cells over platelet rich fibrin scaffold for mandibular cartilage defects regenerative medicine. *Pesqui. Bras. Odontopediatria Clínica Integr.* **2021**, *21*, e0034. [CrossRef]
71. Gomez, M.; Wittig, O. Mesenchymal Stromal Cell Transplantation Induces Regeneration of Large and Full-Thickness Cartilage Defect of the Temporomandibular Joint. *Cartilage* **2021**, *13*, 18145–18215. [CrossRef]
72. Cheng, M.S.; Yi, X. Overexpression of HIF-1alpha in Bone Marrow Mesenchymal Stem Cells Promote the Repair of Mandibular Condylar Osteochondral Defect in a Rabbit Model. *J. Oral Maxillofac. Surg. Off. J. Am. Assoc. Oral Maxillofac. Surg.* **2021**, *79*, 345.e1–345.e15. [CrossRef]
73. Köhnke, R.; Ahlers, M.O. Temporomandibular Joint Osteoarthritis: Regenerative Treatment by a Stem Cell Containing Advanced Therapy Medicinal Product (ATMP)—An In Vivo Animal Trial. *Int. J. Mol. Sci.* **2021**, *22*, 443. [CrossRef] [PubMed]
74. Simmons, P.; McElroy, T. A Bibliometric Review of Artificial Extracellular Matrices Based on Tissue Engineering Technology Literature: 1990 through 2019. *Materials* **2020**, *13*, 2891. [CrossRef]
75. Vîrlan, M.J.R.; Costea, D.E. Degenerative bony changes in the temporal component of the temporomandibular joint—Review of the literature. *Rom. J. Morphol. Embryol.* **2022**, *63*, 61–69. [CrossRef] [PubMed]
76. Schiffman, E.; Ohrbach, R. Orofacial Pain Special Interest Group, International Association for the Study of Pain. Diagnostic Criteria for Temporomandibular Disorders (DC/TMD) for Clinical and Research Applications: Recommendations of the International RDC/TMD Consortium Network* and Orofacial Pain Special Interest Group. *J. Oral Facial Pain Headache* **2014**, *28*, 6–27. [PubMed]
77. Alzahrani, A.; Yadav, S. Incidental findings of temporomandibular joint osteoarthritis and its variability based on age and sex. *Imaging Sci. Dent.* **2020**, *50*, 245–253. [CrossRef]
78. Lee, Y.H.; Park, H.K. Emerging Potential of Exosomes in Regenerative Medicine for Temporomandibular Joint Osteoarthritis. *Int. J. Mol. Sci.* **2020**, *21*, 1541. [CrossRef]
79. Yuan, W.; Wu, Y. A new frontier in temporomandibular joint osteoarthritis treatment: Exosome-based therapeutic strategy. *Front. Bioeng. Biotechnol.* **2022**, *10*, 1074536. [CrossRef]
80. Xia, D.; Wu, J. Mapping Thematic Trends and Analysing Hotspots Concerning the Use of Stem Cells for Cartilage Regeneration: A Bibliometric Analysis From 2010 to 2020. *Front. Pharmacol.* **2022**, *12*, 3563. [CrossRef]
81. Al-Moraissi, E.A.; Christidis, N. Publication performance and trends in temporomandibular disorders research: A bibliometric analysis. *J. Stomatol. Oral Maxillofac. Surg.* 2022, 101273, *in press*. [CrossRef]
82. Yang, T.; Zhang, J. Wnt5a/Ror2 mediates temporomandibular joint subchondral bone remodeling. *J. Dent. Res.* **2015**, *94*, 803–812. [CrossRef]
83. Sun, J.L.; Yan, J.F. Conditional deletion of Adrb2 in mesenchymal stem cells attenuates osteoarthritis-like defects in temporomandibular joint. *Bone* **2020**, *133*, 115229. [CrossRef]

84. Sun, J.L.; Yan, J.F. MicroRNA-29b Promotes Subchondral Bone Loss in TMJ Osteoarthritis. *J. Dent. Res.* **2020**, *99*, 1469–1477. [CrossRef] [PubMed]
85. Luo, X.; Jiang, Y. Inhibition of notch signaling pathway temporally postpones the cartilage degradation progress of temporomandibular joint arthritis in mice. *J. Craniomaxillofac. Surg.* **2018**, *46*, 1132–1138. [CrossRef]
86. Quinelato, V.; Bonato, L.L. PAX7 gene polymorphism in muscular temporomandibular disorders as potentially related to muscle stem cells. *BMC Musculoskelet. Disord.* **2021**, *22*, 959. [CrossRef] [PubMed]
87. Fernandes, T.L.; Kimura, H.A. Human Synovial Mesenchymal Stem Cells Good Manufacturing Practices for Articular Cartilage Regeneration. *Tissue Eng. Part C Methods* **2018**, *24*, 709–716. [CrossRef]
88. Tian, J.; Chen, W. Small extracellular vesicles derived from hypoxic preconditioned dental pulp stem cells ameliorate inflammatory osteolysis by modulating macrophage polarization and osteoclastogenesis. *Bioact. Mater.* **2022**, *22*, 326–342. [CrossRef]
89. Nuti, N.; Corallo, C. Multipotent Differentiation of Human Dental Pulp Stem Cells: A Literature Review. *Stem Cell Ver. Rep.* **2016**, *12*, 511–523. [CrossRef]
90. Di Benedetto, A.; Posa, F. NURR1 Downregulation Favors Osteoblastic Differentiation of MSCs. *Stem Cells Int.* **2017**, *2017*, 7617048. [CrossRef]
91. Ballini, A.; Di Benedetto, A. Stemness genes expression in naïve vs. osteodifferentiated human dental-derived stem cells. *Eur. Rev. Med. Pharmacol. Sci.* **2019**, *23*, 2916–2923. [PubMed]
92. Shi, X.; Mao, J.; Liu, Y. Pulp stem cells derived from human permanent and deciduous teeth: Biological characteristics and therapeutic applications. *Stem Cells Transl. Med.* **2020**, *9*, 445–464. [CrossRef]
93. Cui, S.J.; Zhang, T. DPSCs Attenuate Experimental Progressive TMJ Arthritis by Inhibiting the STAT1 Pathway. *J. Dent. Res.* **2020**, *99*, 446–455. [CrossRef] [PubMed]
94. Seo, B.M.; Miura, M.; Gronthos, S.; Bartold, P.M.; Batouli, S.; Brahim, J.; Young, M.; Robey, P.G.; Wang, C.Y.; Shi, S. Investigation of multipotent postnatal stem cells from human periodontal ligament. *Lancet* **2004**, *364*, 149–155. [CrossRef] [PubMed]
95. Menicanin, D.; Mrozik, K.M. Periodontal-ligament-derived stem cells exhibit the capacity for long-term survival, self-renewal, and regeneration of multiple tissue types in vivo. *Stem Cells Dev.* **2014**, *23*, 1001–1011. [CrossRef]
96. Zhang, J.; Guo, F. Periodontal ligament mesenchymal stromal cells increase proliferation and glycosaminoglycans formation of temporomandibular joint derived fibrochondrocytes. *BioMed Res. Int.* **2014**, *2014*, 410167. [CrossRef] [PubMed]
97. Koyama, N.; Okubo, Y. Pluripotency of mesenchymal cells derived from synovial fluid in patients with temporomandibular joint disorder. *Life Sci.* **2011**, *89*, 741–747. [CrossRef]
98. Sun, Y.P.; Zheng, Y.H. Synovium fragment-derived cells exhibit characteristics similar to those of dissociated multipotent cells in synovial fluid of the temporomandibular joint. *PLoS ONE* **2014**, *9*, e101896. [CrossRef]
99. El-Bialy, T.; Uludag, H. In vivo ultrasound-assisted tissue-engineered mandibular condyle: A pilot study in rabbits. *Tissue Eng. Part C Methods* **2010**, *16*, 1315–1323. [CrossRef]
100. Zhang, M.; Yang, H. Matrix replenishing by BMSCs is beneficial for osteoarthritic temporomandibular joint cartilage. *Osteoarthr. Cartl.* **2017**, *25*, 1551–1562. [CrossRef]
101. Wang, Y.; Zhao, M. BMSC-Derived Small Extracellular Vesicles Induce Cartilage Reconstruction of Temporomandibular Joint Osteoarthritis via Autotaxin-YAP Signaling Axis. *Front. Cell Dev. Biol.* **2021**, *9*, 656153. [CrossRef]
102. Carboni, A.; Amodeo, G. Temporomandibular Disorders Clinical and Anatomical Outcomes After Fat-Derived Stem Cells Injection. *J. Craniofac. Surg.* **2019**, *30*, 793–797. [CrossRef] [PubMed]

Journal of
Functional Biomaterials

MDPI

Article

Strontium Carbonate and Strontium-Substituted Calcium Carbonate Nanoparticles Form Protective Deposits on Dentin Surface and Enhance Human Dental Pulp Stem Cells Mineralization

Tatiane Cristina Dotta [1], Larwsk Hayann [2], Leonardo de Padua Andrade Almeida [1], Lucas Fabrício B. Nogueira [2], Mayara M. Arnez [1], Raisa Castelo [1], Ana Flávia B. Cassiano [3], Gisele Faria [3], Milena Martelli-Tosi [2,4], Massimo Bottini [5], Pietro Ciancaglini [2], Alma B. C. E. B. Catirse [1] and Ana Paula Ramos [2,*]

1 Department of Dental Materials and Prosthodontics, Ribeirão Preto School of Dentistry, University of São Paulo, Ribeirão Preto 14040-904, Brazil
2 Department of Chemistry, Ribeirão Preto Faculty of Philosophy, Sciences and Letters at Ribeirão Preto, University of São Paulo, Ribeirão Preto 14040-901, Brazil
3 Department of Restorative Dentistry, School of Dentistry at Araraquara, Sao Paulo State University (UNESP), Araraquara 14801-385, Brazil
4 Department of Food Engineering, Faculty of Animal Science and Food Engineering, University of São Paulo, Pirassununga 13645-900, Brazil
5 Department of Experimental Medicine, University of Rome Tor Vergata, 00133 Rome, Italy
* Correspondence: anapr@ffclrp.usp.br

Citation: Dotta, T.C.; Hayann, L.; de Padua Andrade Almeida, L.; Nogueira, L.F.B.; Arnez, M.M.; Castelo, R.; Cassiano, A.F.B.; Faria, G.; Martelli-Tosi, M.; Bottini, M.; et al. Strontium Carbonate and Strontium-Substituted Calcium Carbonate Nanoparticles Form Protective Deposits on Dentin Surface and Enhance Human Dental Pulp Stem Cells Mineralization. *J. Funct. Biomater.* **2022**, *13*, 250. https://doi.org/10.3390/jfb13040250

Academic Editor: Patricia Pereira

Received: 25 October 2022
Accepted: 14 November 2022
Published: 17 November 2022

Abstract: Strontium acetate is applied for dental hypersensitivity treatment; however, the use of strontium carbonates for this purpose has not been described. The use of Sr-carbonate nanoparticles takes advantage of both the benefits of strontium on dentin mineralization and the abrasive properties of carbonates. Here in, we aimed to synthesize strontium carbonate and strontium-substituted calcium carbonate nanoparticles and test them as potential compounds in active dentifrices for treating dental hypersensitivity. For this, $SrCO_3$, $Sr_{0.5}Ca_{0.5}CO_3$, and $CaCO_3$ nanoparticles were precipitated using Na_2CO_3, $SrCl_2$, and/or $CaCl_2$ as precursors. Their morphology and crystallinity were evaluated by electron microscopy (SEM) and X-ray diffraction, respectively. The nanoparticles were added to a poly (vinyl alcohol) gel and used to brush dentin surfaces isolated from human third molars. Dentin chemical composition before and after brushing was investigated by infrared spectroscopy (FTIR) and X-ray dispersive energy spectroscopy. Dentin tubule morphology, obliteration, and resistance of the coatings to acid attack were investigated by SEM and EDS. The cytotoxicity and ability of the particles to trigger the mineralization of hDPSCs in vitro were studied. Dentin brushed with the nanoparticles was coated by a mineral layer that was also able to penetrate the tubules, while $CaCO_3$ remained as individual particles on the surface. FTIR bands related to carbonate groups were intensified after brushing with either $SrCO_3$ or $Sr_{0.5}Ca_{0.5}CO_3$. The shift of the phosphate-related FTIR band to a lower wavenumber indicated that strontium replaced calcium on the dentin structure after treatment. The coating promoted by $SrCO_3$ or $Sr_{0.5}Ca_{0.5}CO_3$ resisted the acid attack, while calcium and phosphorus were removed from the top of the dentin surface. The nanoparticles were not toxic to hDPSCs and elicited mineralization of the cells, as revealed by increased mineral nodule formation and enhanced expression of COL1, ALP, and RUNX2. Adding $Sr_{0.5}Ca_{0.5}CO_3$ as an active ingredient in dentifrices formulations may be commercially advantageous since this compound combines the well-known abrasive properties of calcium carbonate with the mineralization ability of strontium, while the final cost remains between the cost of $CaCO_3$ and $SrCO_3$. The novel $Sr_{0.5}Ca_{0.5}CO_3$ nanoparticles might emerge as an alternative for the treatment of dental hypersensitivity.

Keywords: biomaterials; dental hypersensivity; dentin; mineralization; strontium-containing nanoparticles

1. Introduction

Loss of dental structure related to diet, incorrect tooth brushing, aging, and occlusal disorders, among other factors, can cause dental hypersensitivity. This disorder, which is characterized by intense pain after osmotic, thermal, or chemical stimulus, affects up to 30% of the global population [1–8]. The hydrodynamic theory proposed by Brännström in the 1960s explains that hypersensitivity is associated with dentin fluid reaching nerve fibers located either inside or at the interface of dentin tubules, thereby triggering pain [3–6,9–11]. Several treatments have been proposed for hypersensitivity. Among them, the use of dentifrices containing strontium acetate, arginine, potassium nitrate, or tin monofluorophosphate fluoride as a desensitizing agent is the most common because they are relatively inexpensive and can be applied at home [6,8,10,12–15]. Desensitizing agents can also be used to prevent tooth erosion caused by ingestion of foods and drinks with acid pH, including citrus fruits, wine, and fruit juice, or even by stomach pH deregulation [16–21].

Strontium, especially from water and food, is naturally incorporated into mineralized tissues such as bone and teeth because Sr^{2+} can substitute Ca^{2+} in the apatite structure [22]. In this sense, Sr^{2+} is also associated with calcified tissue remineralization [2,22–26]. Strontium chloride was the first strontium-based desensitizing agent to be applied in hypersensitivity treatment [27,28]. However, there was no clinical evidence of the positive effects of this compound as compared to the fluorides employed for the same purpose. Two theories explain the positive effects of strontium during hypersensitivity treatment: (a) incorporation of Sr^{2+} into vacant Ca^{2+} sites due to demineralization and reprecipitation of strontium-apatite mineral [2,14,29–31]; and (b) precipitation of organic compounds of the dental matrix and odontoblastic activation, to form a blocking layer that prevents liquids from circulating inside dentin tubules [31]. Nowadays, dentifrices contain strontium as the acetate salt [27,32], but other strontium-containing compounds have also been applied for bone tissue remineralization. Strontium ranelate is one example, but its use has been suspended almost worldwide due to some side effects, particularly those related to cardiovascular diseases [24]. In turn, strontium carbonate positively affects bone regeneration, so it has been associated with a reduced risk of fracture in healthy and osteoporotic animal models and with accelerated osteointegration of dental implants [33]. Despite these advantages, strontium carbonate has not yet been included in the formulation of dentifrices yet [33–36]. Using strontium carbonate and strontium/calcium carbonate in dentifrices would allow the well-known abrasive effect of calcium carbonate and the remineralizing and desensitizing properties of strontium to be combined. Moreover, compared to strontium acetate, the lower solubility of strontium carbonate would prevent Sr^{2+} removal from the dentin surface. For instance, Seong et al. have found augmented pain alleviation in patients treated with potassium nitrate-mixed hydroxyapatite compared to the pure potassium nitrate toothpaste, which is also a highly soluble salt [37].

More recently, the action of Sr^{2+} on dental mineralization at a cell level has also been studied [38–40]. Martin-del-Campo et al. prepared 3D scaffolds containing strontium folate embedded with human dental pulp stem cells (hDPSCs) as mineralizing cell models [39]. Cell differentiation to a mineralizing phenotype was stimulated by the presence of Sr^{2+}. The mechanisms of action of Sr^{2+} released from bioactive glasses on the proliferation and differentiation of hDPSCs have been explained through the stimulus of the expression of key odontogenic markers, including dentine sialophosphoprotein (DSPP) and dentine matrix protein 1 (DMP-1) [38,40].

Here, we aimed to synthesize strontium carbonate and strontium-substituted calcium carbonate nanoparticles and testing them as potential compounds in active dentifrices for treating dental hypersensitivity. The use of nanoparticles allowed strontium deposition on the dentin surface and penetration into dentin tubules, as well as tubule obliteration. Moreover, the particles induced osteogenic differentiation of hDPSCs. Strontium carbonate may promote dentin surface abrasion and remineralization simultaneously. The new mineral coatings deposited at the dentin surfaces after the treatment were resistant to acid attack.

2. Materials and Methods

2.1. Materials

Dentin discs were prepared from third molar teeth from the tooth bank of the Dental School (University of São Paulo, Ribeirão Preto, Brazil) after approval of the Ethics in Research Committee of the same institute (06245218.2.0000.5419). To synthesize $SrCO_3$ and $Sr_{0.5}Ca_{0.5}CO_3$ nanoparticles, calcium chloride, strontium chloride, and sodium carbonate were acquired from Labsynth and used without further purification. Poly (vinyl) alcohol (MW 9000–10,000, 80% hydrolyzed) was purchased from Sigma-Aldrich. All the aqueous solutions were prepared with ultrapure deionized water from a Milli-Q filtration system (resistivity 18.2 M·Ω·cm).

2.2. Methods

2.2.1. Preparation of Dentin Discs

To obtain 1 mm thick dentin discs, the third molars were sectioned at the cementoenamel junction; an acrylic template (Dencor®, Methyl Methacrylate, *Artigos Odontológicos Clássico* Ltda, São Paulo, Brazil) was used as sample holder. The best dentin discs were selected with the aid of a magnifying glass; discs containing tertiary dentin, caries, and cracks were eliminated [41]. The dentin discs were cleaned in deionized water in an ultrasound bath (ALTSonic Clean- ALT Equipamentos, Ribeirão Preto, Brazil) for 30 s. Then, the discs were kept in 6 wt.% aqueous citric acid solution for 2 min to remove the dental slurry [41]. Next, the discs were immersed in deionized water and cleaned in an ultrasound bath for 10 min. After cleaning, all the discs were conditioned in 0.5 mL of artificial saliva [42] and maintained at 37 ± 1 °C.

2.2.2. Synthesis of $SrCO_3$, $Sr_{0.5}Ca_{0.5}CO_3$, and $CaCO_3$ Nanoparticles

$SrCO_3$ nanoparticles were prepared by mixing 100 mL of 2.66 mol·L^{-1} aqueous $SrCl_2$ solution and 100 mL of 1.06 mol·L^{-1} Na_2CO_3. The formation of a white precipitate evidenced that a poorly soluble product assigned to $SrCO_3$ was produced. The reaction is depicted below:

$$SrCl_2(aq) + Na_2CO_3\,(aq) \rightarrow SrCO_3(s) + 2NaCl(aq) \quad (1)$$

$Sr_{0.5}Ca_{0.5}CO_3$ nanoparticles were obtained by mixing 100 mL of 1.33 mol·L^{-1} aqueous $SrCl_2$ solution and 100 mL of 0.73 mol·L^{-1} $CaCl_2$. Then, 100 mL of 1.06 mol·L^{-1} aqueous Na_2CO_3 solution was added to this mixture. The formation of a white precipitate evidenced that a poorly soluble product assigned to $SrCO_3$ was produced. The reaction is depicted below:

$$\frac{1}{2}\,SrCl_2(aq) + \frac{1}{2}\,CaCl_2(aq) + Na_2CO_3\,(aq) \rightarrow Sr_{0.5}Ca_{0.5}CO_3(s) + 2NaCl(aq) \quad (2)$$

$CaCO_3$ nanoparticles were prepared by mixing 100 mL of 1.47 mol·L^{-1} aqueous $CaCl_2$ solution and 100 mL of 1.06 mol·L^{-1} Na_2CO_3.

$$CaCl_2(aq) + Na_2CO_3\,(aq) \rightarrow CaCO_3(s) + 2NaCl(aq) \quad (3)$$

All the reactions were carried out under vigorous stirring at 20,000 rpm for 10 min on an Ultra Turrax® device (IKA® T18 basic, IKA® Works Brazil Ltda, Campinas, Brazil). The solids were filtered, washed three times with deionized water for purification, and dried under controlled conditions at 50 °C for 40 h before being characterized and applied in further experiments.

2.2.3. Characterization of Nanoparticles

The crystalline structure of the $SrCO_3$, $Sr_{0.5}Ca_{0.5}CO_3$, and $CaCO_3$ nanoparticles was investigated by X-ray diffraction (XRD) in a D2 PHASER, Bruker-AXS powder diffractometer, with CuKα as source (30 kV, 10 mA). The *Crystallography Open Database* (COD) was used to index the peaks. The size and the charge of the nanoparticles dispersed in an aqueous medium were investigated by dynamic light scattering (DLS) in a ZetaSizer Nano device (Malvern, UK) equipped with a 580 nm laser at a scattering angle of 172°. The morphology of the gold-coated nanoparticles was studied by scanning electron microscopy (SEM) on a microscope (Shimadzu SS-550, Kyoto, Japan) operating at 20 kV.

2.2 4. Preparation of Gels Containing Nanoparticles for Application on Dentin Discs

First, an aqueous gel containing 10 wt.% PVA was prepared by dissolving the polymer in ultrapure deionized water under mechanical stirring at 90 °C. Then, 30 wt.% $SrCO_3$, $Sr_{0.5}Ca_{0.5}CO_3$, or $CaCO_3$ nanoparticles were added to this gel under stirring at 25 °C until a homogeneous gel was obtained. The rheological behavior was analyzed with a Modular Compact Rheometer MCR 52 (Antom-Paar GmbH, Graz, Austria) using the cone and plate (CP50-1, 50 mm, angle 1°) geometry. Steady-state flow measurements were carried out with shear rates ranging from 1 to 100 s^{-1} at 25 °C ± 3. The samples presented pseudoplastic behavior, and the apparent viscosity ranged from 2.2 up to 2.9 Pa.s (at 100 s^{-1}) without statistical differences.

2.2 5. Application of Gel Containing Nanoparticles to Dentin Surface by Brushing

A gel containing $SrCO_3$, $Sr_{0.5}Ca_{0.5}CO_3$, or $CaCO_3$ nanoparticles (0.3 g) was deposited on the surface of the dentin discs. Then, the discs were coupled to a mechanical brushing simulator (Biopdi, São Carlos, Brazil) and positioned perpendicularly to the toothbrush bristle under a force corresponding to 200 g. Each brushing cycle involved 40 movements at the machine at a rate of 4.5 movements/s^{-1}, in a total of 10 cycles/s^{-1}. After brushing, the discs were washed with deionized water for 10 s and immersed in artificial saliva at 37 °C for 4.5 h until the next brushing [16]. This procedure was repeated twice a day for 14 days. The discs were divided Into four groups, according to the nanoparticle that was tested as a desensitizing agent, namely $SrCO_3$, $Sr_{0.5}Ca_{0.5}CO_3$, $CaCO_3$, and $Sr(CH_3COO)_2$ (Sensodyne® Rapid Relief).

2.2 6. Resistance of Coating to Acid Attack

The resistance of the coatings deposited on the surface of the dentin discs was tested under an acid attack promoted by immersing the discs in 3 mL of Coca-Cola® for 1 h under slow stirring (50%) in a stirring table (Cintec, Ribeirão Preto, Brazil); the procedure described by Olley et al. was followed [16]. After this procedure, the discs were washed with deionized water for 10 s and stored in artificial saliva at 37 °C.

2.2.7. Chemical Characterization of Dentin Discs

Changes in the chemical composition of the dentin surfaces before and after brushing with a desensitizing agent were followed by Fourier-transform infrared spectroscopy (FTIR); an attenuated total reflectance (ATR) accessory (IRPrestige-21, Shimadzu) was used. The presence of the chemical elements Ca, Sr, and P on the surfaces was semi-quantitatively analyzed by X-ray scattering spectroscopy (EDS) (500 Digital Processing; IXRF Systems) coupled to a scanning electron microscope (Zeiss EVO 50—Zeiss, Oberkochen, Germany). The samples were coated with a thin conductive graphite layer before the EDS analyses (Bal-Tec SCD 050 Sputter Coater).

2.2.8. Obliteration of Dentin Tubules

The morphology of gold-coated (Bal-Tec SCD 050 Sputter Coater) dentin discs was analyzed by SEM. The images were also used for the calculus of the dentin tubules obliteration after brushing with a desensitizing agent [43]. To this end, two observers analyzed the images and defined each individual tubule as partially obliterated (P), completely obliterated (O), and non-obliterated (A). The sum of P + A + O should be the total number of tubules (T) observed in the image. Therefore, the number (n) of obliterated tubules was calculated by Equation (4):

$$n = \frac{(P) + (O)}{T} \times 100 \quad (4)$$

The average value of n was calculated by using the value obtained by each observer.

2.2.9. Cell Culture Conditions

Human dental pulp stem cells (hDPSCs) were used in the experiments. The pulp tissue was collected from the third molars of young patients, extracted for orthodontic reasons. Cells were isolated using the enzymatic dissociation method and analyzed by flow-cytometric immunophenotyping to confirm the mesenchymal identity (data not shown). The hDPSCs were cultured in α-minimum essential medium (α-MEM, Sigma-Aldrich, St. Louis, MO, USA), supplemented with 10% fetal bovine serum (FBS), penicillin (100 IU/mL), and streptomycin (100 μg/mL), under 5% CO_2, in a 95% humidified atmosphere at 37 °C. The α-MEM osteogenic medium—OM (with 50 μg/mL L-ascorbic acid and 10 mM β-glycerophosphate) was used to conduct all the cell-culture-related assays. Cells from the 4th passage were used in the experiments.

2.2.10. Cell Viability Measurement

To test the cytotoxicity of the nanoparticles ($CaCO_3$, $Sr_{0.5}Ca_{0.5}CO_3$, and $SrCO_3$), cells were trypsinized, resuspended in α-MEM, seeded on 96-well plate at the density 1.3×10^4 cells per well and incubated in the air at 37 °C and 5% CO_2. Cells were allowed to adhere on the bottom surface of the plate for 24 h, followed by the treatment with the nanoparticles (0.1, 1, and 10 μg/mL). To assess cell viability, we applied MTT assay after 24 and 48 h of culture. Cell viability was expressed as the percentage of the average of three experiments compared to the untreated control for each day of the culture.

2.2.11. Tissue Non-Specific Alkaline Phosphatase (TNAP) Activity

To determine TNAP activity, firstly, we exposed the hDPSCs (2×10^4 cells/well on 24-well plate) to the nanoparticles ($CaCO_3$, $Sr_{0.5}Ca_{0.5}CO_3$, and $SrCO_3$ at 1 μg/mL) for 7 days and 14 days. The medium was changed every 48 h supplemented with or without (control) the nanoparticles. Then, we harvested the plasma membrane fraction in triplicate from cells after 7 and 14 days of culture, according to (71). TNAP activity was accomplished by the degradation of p-nitrophenylphosphate (pNPP), and its subproduct was analyzed by reading its absorbance at 410 nm. The activity was expressed as U mg^{-1} of total protein content so that one unit of enzyme is defined as the amount of enzyme capable of hydrolyzing 1.0 nmol of substrate per mg of protein at 37 °C.

2.2.12. Mineralized Nodule Formation

Alizarin Red staining (ARS) was used to assess the calcium deposits. The hDPSCs were plated (2×10^4 cells/well) in 24-well plates. The medium with and without the nanoparticles (1 μg/mL) was renewed every two days. After 14 days, the cells were fixed with 70% ethanol at 4 °C for 1 h, rinsed with deionized water, and stained with 2% ARS (pH 4.2, Sigma-Aldrich). The nodules were dissolved with 10% cetylpyridinium chloride (Sigma-Aldrich) under shaking for 15 min. Aliquots of the resuspension were collected, and the optical density was measured at 562 nm wavelength in a spectrophotometer. The results obtained from the absorbance of the control were considered as 100%.

2.2.13. Messenger RNA (mRNA) Expression by Real-Time Polymerase Chain Reaction (RT-qPCR)

The effect of the nanoparticles on the expression of osteogenic biomarkers collagen type 1 alpha 1 (COL1A1), alkaline phosphatase (ALP), runt-related transcription factor 2 (RUNX2), and glyceraldehyde 3-phosphate dehydrogenase (GAPDH) was evaluated by RT-qPCR. hDPSCs were incubated with 1 μg/mL of the nanoparticles and cultivated for 7 and 14 days. For RNA extraction, the PureLink™ RNA Mini Kit (Invitrogen, Life Technologies, Grand Island, NY, USA) was used, following the manufacturer's instructions. Complementary DNA (cDNA) was synthesized from 1 μg of total RNA using the High-Capacity cDNA Reverse Transcription Kit (Applied Biosystems, Life Technologies, Grand Island, NY, USA). Gene expression (n = 3) was analyzed in StepOne equipment (Applied Biosystems) using TaqMan PCR Master Mix and probes (Applied Biosystems) for the target genes. Transcripts were normalized to GAPDH, and data were shown as relative mRNA expression using the cycle threshold ($2^{-\Delta\Delta Ct}$) method [44].

3. Results

3.1. Characterization of $SrCO_3$, $Sr_{0.5}Ca_{0.5}CO_3$, and $CaCO_3$ Nanoparticles

3.1.1. Investigation of the Crystalline Structure of Nanoparticles by X-Ray Diffraction

Figure 1 depicts the X-ray diffraction pattern of the $SrCO_3$, $Sr_{0.5}Ca_{0.5}CO_3$, and $CaCO_3$ nanoparticles. The reaction between Ca^{2+} and CO_3^{2-} in an aqueous solution resulted in the precipitation of pure calcium carbonate ($CaCO_3$) as calcite [COD-1010962] (Figure 1—black line). In turn, the reaction between Sr^{2+} and CO_3^{2-} resulted in the precipitation of pure strontium carbonate as strontianite [COD-5000093] (Figure 1—green line). Precipitation of a mixed calcium-strontium carbonate mineral [COD 9015600] from the mixture of Ca^{2+} and Sr^{2+} with CO_3^{2-} was evident in the diffraction pattern depicted in Figure 1—red line. The peaks corresponding to strontianite (Figure 1—green line) shifted to higher 2θ values, the corroborating substitution of Sr^{2+} with a lighter ion, i.e., Ca^{2+}, in the structure of the mixed carbonate (Figure 1—red line). Phase segregation was not evident.

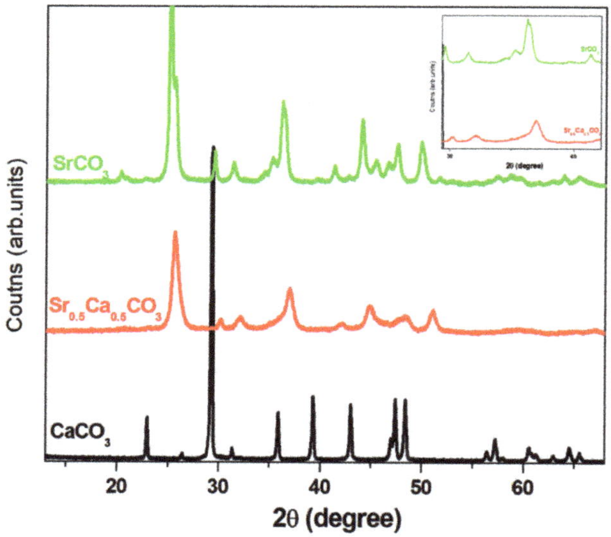

Figure 1. Structural characterization of the particles. X-ray diffraction patterns of $SrCO_3$ (green line), $Sr_{0.5}Ca_{0.5}CO_3$ (red line), and $CaCO_3$ (black line). The structures correspond to $CaCO_3$ calcite (COD-1010962), $SrCO_3$ strontianite (COD-5000093), and a mixed $Sr_{0.5}Ca_{0.5}CO_3$ mineral (COD 9015600). The displacement of the peaks to higher 2θ in the case of $Sr_{0.5}Ca_{0.5}CO_3$ as compared to $SrCO_3$ evidences the presence of a lighter element in the structure (Ca) (see image inserted in the figure).

3.1.2. Nanoparticle Size and Charge

The synthesis of particles presenting a diameter smaller than the diameter of the tubules is mandatory to allow the particles to penetrate the dentin structure. Moreover, either lower negative or positive charges also might favor the deposition on the negatively charged dentin surface [45]. Table 1 lists the mean diameter of the carbonate nanoparticles.

$CaCO_3$ presented size distribution by number centered at 590 nm. The strontium-containing carbonate nanoparticles were the smallest among the ones we studied; their size distribution was centered at 66.7 and 387.0 nm for $Sr_{0.5}Ca_{0.5}CO_3$ and $SrCO_3$, respectively. The small diameter of $Sr_{0.5}Ca_{0.5}CO_3$ can be assigned to their slightly higher zeta-potential (ζ) (Table 1), which may have prevented aggregation.

Table 1. Hydrodynamic diameter, polydyspersity index (PDI), and zeta-potential (ζ) determined by DLS. The number between parenthesis in the diameter column corresponds to the full width at the half maximum of the size distribution.

Sample	Diameter (nm)	PDI	ζ (mV)
$SrCO_3$	387.0 (166)	0.56	-2.1 ± 3.2
$Sr_{0.5}Ca_{0.5}CO_3$	66.7 (16.7)	0.58	-9.1 ± 4.5
$CaCO_3$	590.8 (156.5)	0.90	-2.5 ± 4.8

3.1.3. Nanoparticle Morphology

We investigated the nanoparticles' morphology by SEM (Figure 2). $SrCO_3$ (Figure 2A) and $Sr_{0.5}Ca_{0.5}CO_3$ (Figure 2B) nanoparticles exhibited a spherical morphology, in agreement with the DLS data (Table 1). We verified the typical orthorhombic shape for the calcite-$CaCO_3$ nanoparticles (Figure 2C). We assigned the presence of aggregates (red arrows) to the drying process as well as to the low ζ value.

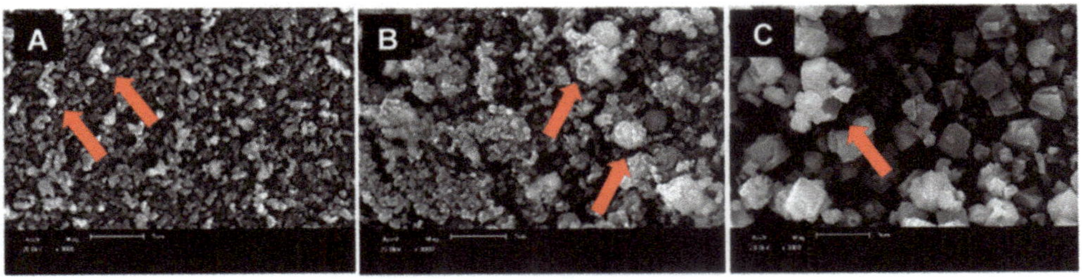

Figure 2. SEM images of the $SrCO_3$ (**A**), $Sr_{0.5}Ca_{0.5}CO_3$ (**B**), and $CaCO_3$ (**C**) nanoparticles. The red arrows indicate the presence of micrometric particles due to aggregated nanosized particles. The scale bar corresponds to 5 µm. The images were acquired using accelerating voltage of 20 kV and 3000× magnification.

3.1.4. Effects of the Nanoparticles on the Viability of hDPSCs

We tested the potential cytotoxic effect of the nanoparticles using cell assays in vitro. For this, we carried MTT assay at 24 and 48 h. The data depicted in Figure 3a,b showed that the nanoparticles did not significantly affect the viability of hDPSCs in concentrations up to 10 µg/mL, except for the $SrCO_3$.

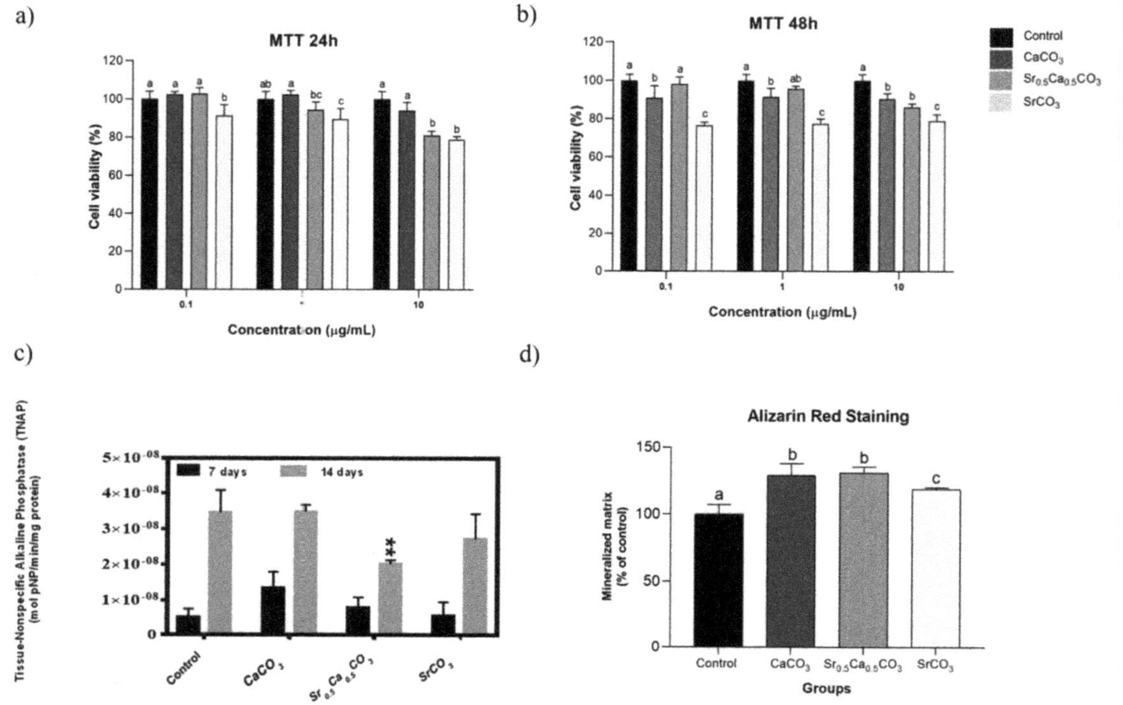

Figure 3. Effect of the nanoparticles on human dental stem cells (hDPSCs) in vitro. Cell viability of hDPSCs after 24 h (**a**) and 48 h (**b**) of culturing the presence of the nanoparticles measured by MTT assay. The results represent mean ± SD of five experiments. Statistical analysis was carried out by two-way ANOVA, and Tukey's post-test were performed to compare the results. Bars with different letters represent significant differences between groups for each concentration of particles tested. Tissue-Non-specific Alkaline Phosphatase (TNAP) activity after 7 and 14 days of culturing in the presence and absence (control) of 1 μg/mL of particles (**c**). For the determination of TNAP activity, multiple statistical comparisons were performed by one-way ANOVA in comparison with the control, ** = 0.0014. The results represent mean ± SD of three experiments. Evaluation of the formation of the mineralized nodules by alizarin red staining after 14 days of culturing in the presence and absence (control) of 1 μg/mL of particles (**d**). The results represent mean ± SD of five experiments. Statistical analysis was carried out by two-way ANOVA, and Tukey's post-test was performed to compare the results. Bars with different letters represent significant differences between groups. All the controls correspond to hDPSC cultured on the polystyrene plate in absence of the nanoparticles.

3.1.5. TNAP Activity and Mineralized Nodules Formation

We studied the effect of the nanoparticles upon mineralization by measuring the enzymatic activity of tissue non-specific alkaline phosphatase (TNAP), the main enzyme responsible for inorganic phosphate generation in mineralization-competent cells, and quantifying the formation of mineral nodules by alizarin red staining [46,47]. The results presented in the Figure 3C revealed that TNAP enzymatic activity was not significantly affected by the presence of the particles at the 7th day of culture. However, after 14 days, a slight reduction was observed for the samples treated with $Sr_{0.5}Ca_{0.5}CO_3$ nanoparticles. Quantification of mineral nodules revealed the ability of all the nanoparticles to stimulate mineralization by hDPSCs (Figure 3D). The formation of mineralized nodules was increased up to 30% in the presence of $CaCO_3$ and $Sr_{0,5}Ca_{0,5}CO_3$ nanoparticles and 18% in the presence of $SrCO_3$ nanoparticles, compared to the control in the absence of the particles.

3.1.6. Expression of RUNX2, COL1, and ALP by hDPSCs

The expression of the main osteogenic markers RUNX2, COL1, and ALP by hDPSCs was evaluated by RT-PCR analysis. The data presented in Figure 4 revealed increased expression of COL1 (Figure 4A) and ALP (Figure 4B) genes after 7 days of culturing in the presence of all the nanoparticles, compared to the control. This result corroborates the enhanced formation of mineral nodules by the cells cultivated in the presence of the nanoparticles (Figure 3D). After 14 days of culturing, the expression of COL1 and ALP was reduced, following the fate of the cells. It is worth noting that $SrCO_3$ nanoparticles, which had the highest amount of Sr, sustained the increased expression of ALP and increased the expression of the transcription factor RUNX2 after 14 days of culture, following the positive effects of strontium on mineralization [48].

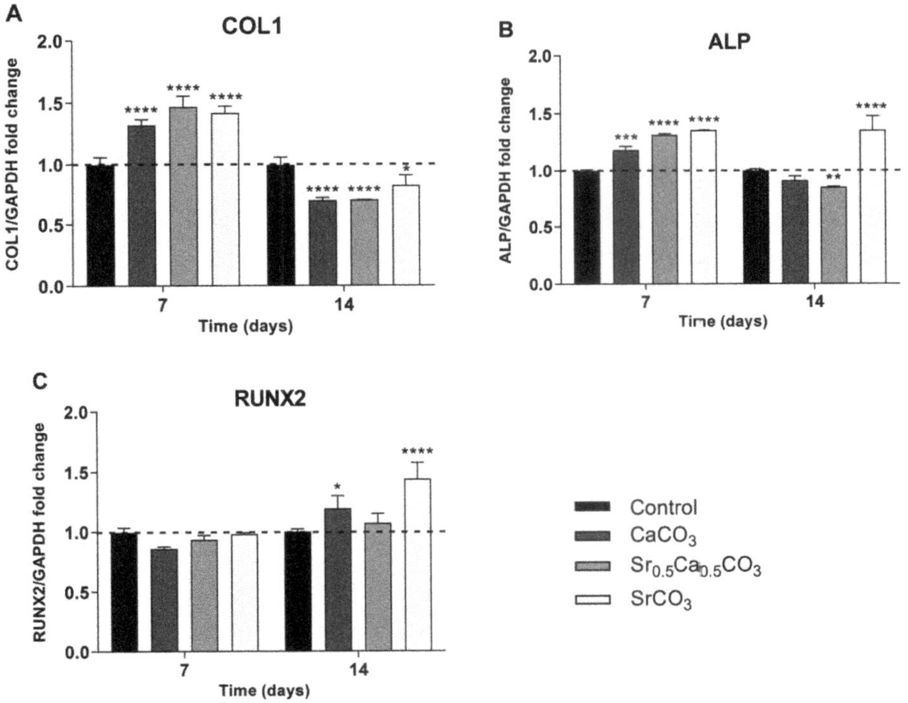

Figure 4. Expression of the osteogenic markers COL1 (**A**), ALP (**B**), and RUNX2 (**C**) by hDPSCs cultivated in the presence of 1 μg/mL Sr-carbonates for 7 and 14 days. The control corresponds to the cells cultivated in polystyrene plate in absence of the particles. Statistical analyses were carried out by two-way ANOVA, and Tukey's post-test was performed in relation to the control; **** $p < 0.0001$, *** $p < 0.001$, ** $p < 0.01$, and * $p < 0.05$.

3.2. Chemical Analysis of Dentin Surface before and after Brushing with a Desensitizing Agent

Figure 5A–C illustrate the ATR-FTIR spectra of the dentin surface before (black line) and after brushing with a desensitizing agent for 1 (red lines), 7 (green lines), or 14 days (blue lines). For comparison, we also present results for dentin discs brushed with Sensodyne® Rapid Relief under the same experimental conditions (Figure 5D).

All the discs displayed bands attributed to asymmetric stretching of the PO_4^{3-} group at 1070 cm^{-1}. Furthermore, we detected the low intense band at 1250 cm^{-1}, which is related to the amide-I group of collagens and is important for the characterization of the dentin surface. Other low-intensity bands in the range of 1750 and 1400 cm^{-1}, ascribed to the presence of carboxylate and amine groups in the organic matrix, emerged in the spectra.

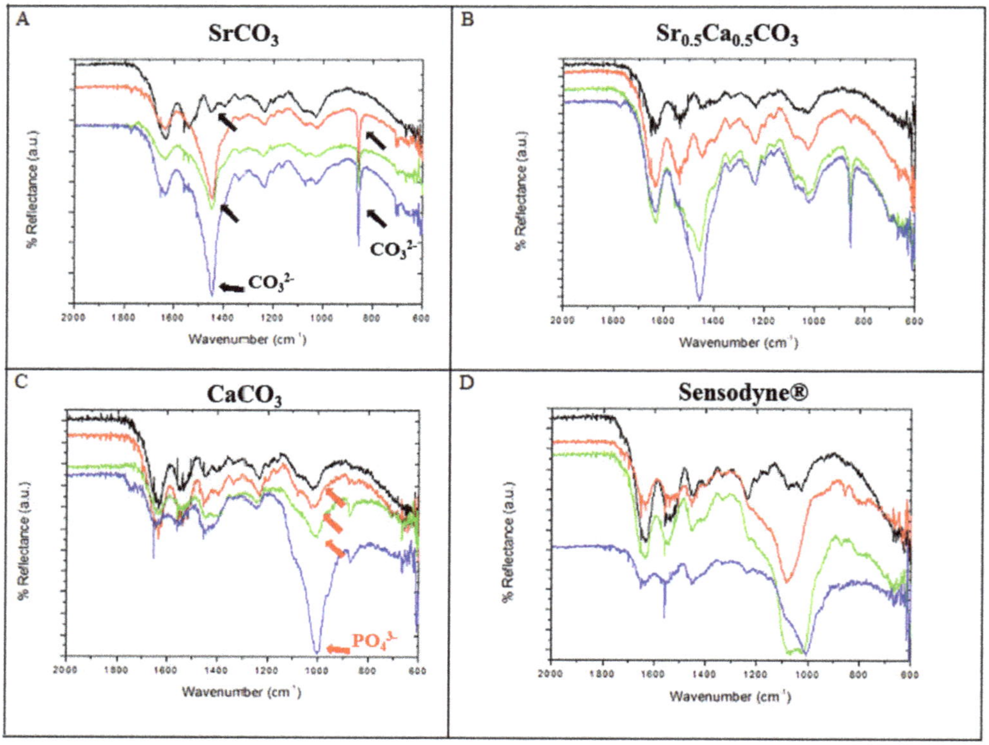

Figure 5. Chemical analysis of the dentin discs before and after brushing. ATR-FTIR of the dentin surface before (black lines) and after brushing for 1 (red lines), 7 (green lines), and 14 (blue lines) days with $SrCO_3$ nanoparticles (**A**), $Sr_{0.5}Ca_{0.5}CO_3$ nanoparticles (**B**), $CaCO_3$ nanoparticles (**C**), and Sensodyne® Rapid Relief (**D**). The black and red arrows indicate changes in the bands assigned to the presence of CO_3^{2-} and PO_4^{3-} groups, respectively, after brushing.

After brushing with $SrCO_3$, the bands at 1480 and 850 cm^{-1} intensified, as indicated by the black arrows in Figure 5A. This was associated with asymmetric stretching of the CO_3^{2-} groups, thus indicating the deposition of carbonate particles on the dentin surface. The band related to the PO_4^{3-} group did not change significantly. We observed a similar behavior after brushing with $Sr_{0.5}Ca_{0.5}CO_3$ nanoparticles (Figure 5B). Interestingly, brushing with $CaCO_3$ nanoparticles did not alter the intensity of the bands at 1480 and 850 cm^{-1} significantly (Figure 5C), but the band at 1070 cm^{-1}, assigned to PO_4^{3-} asymmetric stretching, increased, as indicated by the red arrows in Figure 5C. The ATR-FTIR results obtained after brushing with Sensodyne® Rapid Relief, well-known for its desensitizing properties, revealed the effect of this product on the structure of the phosphate groups, as demonstrated by the changes in the intensity and shape of the band at 1070 cm^{-1}.

3.3. Morphological Characterization and Quantitative Analysis of Dentin Tubule Obliteration after Brushing with a Desensitizing Agent

We analyzed the morphology of the dentin discs before (Figure 6A_1–A_4) and after brushing with $SrCO_3$ nanoparticles (Figure 6B), $Sr_{0.5}Ca_{0.5}CO_3$ nanoparticles (Figure 6D), $CaCO_3$ nanoparticles (Figure 6F), or Sensodyne® Rapid Relief (Figure 6H).

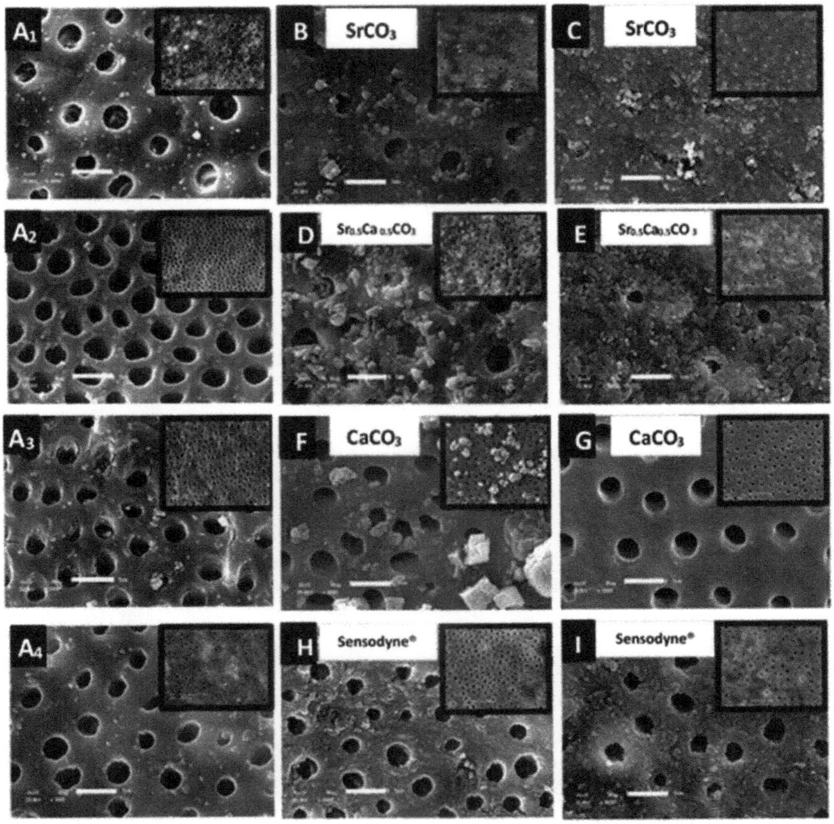

Figure 6. SEM imagens of the dentin specimens before (**A_1–A_4**) and after brushing with $SrCO_3$ nanoparticles (**B**,**C**), $Sr_{0.5}Ca_{0.5}CO_3$ nanoparticles (**D**,**E**), $CaCO_3$ nanoparticles (**F**,**G**), and Sensodyne® Rapid Relief (**H**,**I**). Images C, E, G, and I were acquired after acid challenge. The inserts correspond to images with low amplification. The scale bar corresponds to 5 μm.

For the discs brushed with $SrCO_3$ nanoparticles (Figure 6B) or $Sr_{0.5}Ca_{0.5}CO_3$ nanoparticles (Figure 6), the nanoparticles were deposited at the top as well as inside the dentin tubules, which produced a coating and led to 67.7% and 93.3% tubule obliteration, respectively (Table 2). The acid attack did not remove the coatings (Figure 6C,E), and obliteration was higher after this procedure.

After brushing with $CaCO_3$ nanoparticles, micrometric particles were present at the top surface. The larger size of these particles did not allow them to penetrate the dentin tubules. Moreover, the particles were removed, and a completely clean surface emerged after acid attack (Figure 6G).

Brushing with Sensodyne® Rapid Relief (Figure 6H) promoted the deposition of a coating in the intertubular spaces of the dentin surface, but obliteration was low (19.5%) (Table 2). Although coating morphology changed after acid attack, the tubules remained open (Figure 6I).

We also acquired SEM images of transversal cuts of the dentin discs before (Figure 7A_1–A_4) and after (Figure 7B,D,F,G) brushing. Tubule morphology clearly changed after brushing with either the particles or Sensodyne® Rapid Relief. These changes were related to the deposition of the desensitizing nanoparticles inside the tubules and their adhesion to the tubule walls. The acid attack did not remove the nanoparticles from the tubules, particularly for the discs brushed with $Sr_{0.5}Ca_{0.5}CO_3$ nanoparticles (Figure 7E).

Table 2. Percentage of obliterated dentin tubules (*n*) on the surface of the specimens after brushing with gel containing nanoparticles, followed (+) or not (−) by acid attack (AA).

Sample	AA	*n* (%)
Control	−	0% ± 0.0
$SrCO_3$nanoparticles	−	67.7% ± 0.01
	+	65.7% ± 0.08
$Sr_{0.5}Ca_{0.5}CO_3$ nanoparticles	−	93.3% ± 0.09
	+	100% ± 0.0
$CaCO_3$nanoparticles	−	24.7% ± 0.10
	+	0% ± 0.0
Sendodyne®	−	19.5% ± 0.08
	+	31.9% ± 0.15

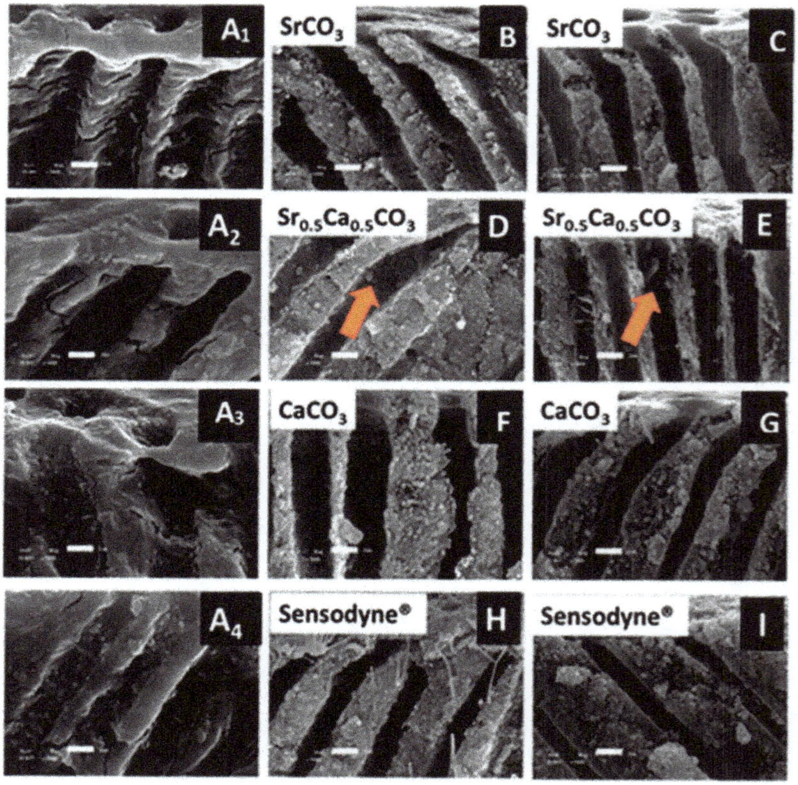

Figure 7. Transverse view SEM imagens of the dentin specimens before (**A_1–A_4**) after brushing with $SrCO_3$ nanoparticles (**B,C**), $Sr_{0.5}Ca_{0.5}CO_3$ nanoparticles (**D,E**), $CaCO_3$ nanoparticles (**F,G**) and Sensodyne® Rapid Relief (**H,I**). Images C, E, G, and I were acquired after acid challenge. The scale bar corresponds to 2 μm. The orange arrows indicate the presence of particles inside the dentin tubules.

3.4. Quantification and Mapping of Ca, Sr, and P in Dentin Discs

By using EDS, we quantified and mapped the chemical elements present on the dentin disc surface and inside the tubules before and after brushing with the nanoparticles. We compared the results to data obtained after brushing with Sensodyne® Rapid Relief. Table 3 summarizes the (Ca + Sr)/P molar ratio calculated for the dentin discs. The control group, consisting of non-brushed dentin discs, presented a (Ca + Sr)/P molar ratio close to 1.81, in agreement with the values reported for biological apatite. The discs brushed with $SrCO_3$ nanoparticles or $Sr_{0.5}Ca_{0.5}CO_3$ nanoparticles had a (Ca + Sr)/P ratio close to the control. However, the amount of strontium was higher in the brushed than in the non-brushed teeth, as expected. After the acid attack, the relative amount of calcium reduced significantly in the samples brushed with strountium-containing nanoparticles, which reflected in an increased amount of strontium since the calculations are relative. Therefore, strontium remained on the dentin surface even after contact with acidic fluid. Discs brushed with $CaCO_3$ nanoparticles presented a lower relative amount of calcium and phosphorus, and the acid challenge reduced the amount of these elements on the surface of the discs even further because of the abrasive process.

Table 3. Quantification by EDS of the chemical elements present on the surface of dentin specimens before (control) and after brushing with gel containing nanoparticles, followed (+) or not (−) by acid attack (AA).

Sample	AA	Ca	Sr	P	(Ca + Sr)/P
Control	−	63.47	1.00	35.52	1.81 ± 0.17
$SrCO_3$nanoparticles	−	59.20	6.00	34.90	1.87 ± 0.17
	+	24.90	26.92	48.18	1.07 ± 0.17
$Sr_{0.5}Ca_{0.5}CO_3$ nanoparticles	−	59.67	4.54	35.78	1.79 ± 0.17
	+	43.07	47.03	9.90	9.10 ± 0.17
$CaCO_3$nanoparticles	−	59.50	0.90	39.61	1.52 ± 0.17
	+	53.30	4.36	42.33	1.36 ± 0.17
Sendodyne®	−	40.58	16.67	42.75	1.34 ± 0.17
	+	50.49	5.56	43.95	1.27 ± 0.17

For the samples treated with Sensodyne® Rapid Relief, the strontium content increased after brushing compared to the control as an effect for strontium acetate deposition. However, the strontium content decreased after the acid attack, which was reflected in increased calcium relative amount. Nevertheless, the decreased strontium content after the acid attack evidenced weak attachment of this element to the dentin surface when strontium acetate is used as an active ingredient.

We performed EDS mapping of Ca, Sr, and P on the surface of dentin discs brushed with strontium-containing carbonate nanoparticles before (Figure 8) and after (Figure 9) acid attack to illustrate the quantitative results presented in Table 3. It is important to note that the maps obtained from the top view analysis are a sum of the contributions of the elemental content at the surface and the depth by which the electron beam can penetrate. The images evidenced strontium deposition on the surface (Figure 8—dark blue) and inside the tubules (Figure 9—dark blue) of the dentin discs brushed with $SrCO_3$ nanoparticles or $Sr_{0.5}Ca_{0.5}CO_3$ nanoparticles. With respect to brushing with Sensodyne® Rapid Relief, the presence of open tubules and low obliteration was clear and corroborated the SEM analysis (Figure 6). More interestingly, observation of the transversal cut of the dentin demonstrated calcium and phosphorus removal (Figure 9—green and red) from the top surface after the acid attack for all the dentin samples. However, strontium remained on the surface (Figure 9—dark blue), revealing the specific binding of this element to the discs and the resistance of the coatings to acidic conditions.

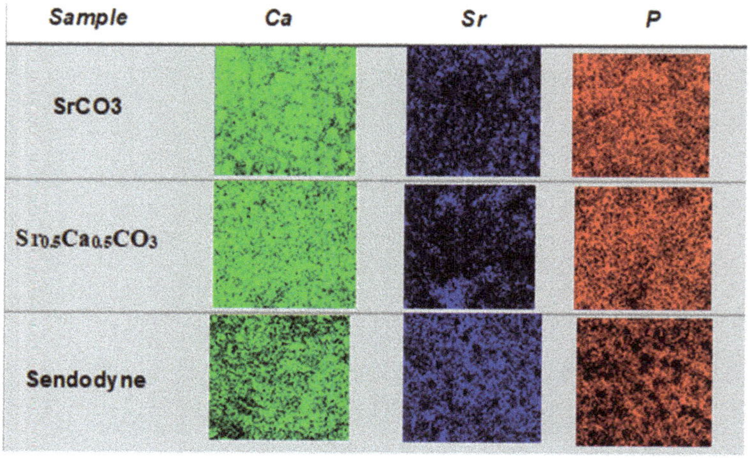

Figure 8. Calcium (Ca), Strontium (Sr), and Phosphorous (P) deposition on the surfaces. EDS maps of Ca (green), Sr (dark blue), and P (red) present on the dentin surface (top view) after brushing with strontium-containing carbonate nanoparticles.

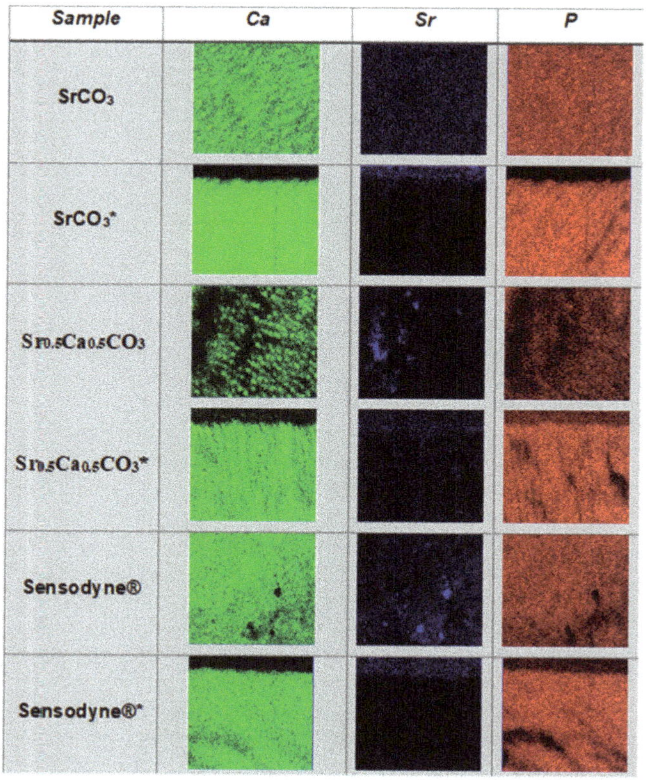

Figure 9. Resistance of the coatings to the acidic treatment. EDS maps of Ca (green), Sr (dark blue), and P (red) present at the dentin top and side view after brushing with Sr-containing nanoparticles and after acid attack. * Indicates the images obtained from the transversal cut of the samples.

4. Discussion

Different active ingredients and several methodologies have been proposed to treat dental hypersensitivity [5,7,49]. Most treatments involve blocking the exposed dentin tubules and desensitizing the pulp nerve. However, the efficacy and durability of these treatments are still under discussion [6,50], motivating the search for new biocompatible materials with permanent and fast effects associated with specific binding to the dentin surface, biocompatibility, and low irritability [6,7,50,51]. Here, we propose using strontium-containing carbonate nanoparticles that might exhibit all these properties.

We structurally characterized the $SrCO_3$, $Sr_{0.5}Ca_{0.5}CO_3$, and $CaCO_3$ nanoparticles by X-ray diffraction, as depicted in Figure 1. We confirmed that pure $CaCO_3$ (calcite) and $SrCO_3$ (strontianite) minerals were formed by comparing the diffraction pattern of the nanoparticles with patterns extracted from the *Crystallography Open Database* (COD). We also indexed the mixed mineral $Sr_{0.5}Ca_{0.5}CO_3$ based on the COD. Phase segregation was not evident. We assigned the shift of the diffraction peaks to higher 2θ values upon the addition of $SrCO_3$ to substitution of Sr^{2+} for the lighter Ca^{2+} in the crystal lattice [52–54].

Particle size determined by DLS revealed diameters in the nanometric scale, which should favor penetration of the nanoparticles into the micro-sized dentin tubules [15,55]. The mean hydrodynamic diameter of $SrCO_3$ nanoparticles was close to 387 nm (Table 1), resembling the diameter of the particles in Sensodyne® Rapid Relief [15]. Moreover, the mean diameter of $CaCO_3$ nanoparticles was 590 nm (Figure 2C). Small particles were present in the $Sr_{0.5}Ca_{0.5}CO_3$ samples, which had hydrodynamic diameters centered at 66.7 nm (Table 1), thus guaranteeing their penetration into the dentin tubules. The polydispersity index (PDI) of the samples ranged from 0.6 to 0.9, revealing broad size distribution (Table 1). Although the PDI values were higher than those reported for homogeneous nanoparticle dispersions [15,55] (e.g., the PDI reported for the aqueous dispersion of the particles in Sensodyne® Rapid Relief is 0.28 [15]), the whole size distribution was still at the nanometric scale, allowing the particles to be applied for dentin tubule obliteration.

Another important point regarding surface/particle interaction is charge. Since the dentin is negatively charged [56], positively charged or even neutral particles may be attracted to the surface by electrostatic or van der Waals forces, thus resulting in enhanced interaction. The zeta-potential (ζ) is influenced by the chemical composition of the particles and the dispersing medium dielectric constant, pH, and ionic strength [15,55]. The ζ values of $SrCO_3$, $CaCO_3$, and $Sr_{0.5}Ca_{0.5}CO_3$ nanoparticles were −2.08, −2.46, and −9.10 mV, respectively (Table 1). The ζ value increased upon substitution of Ca^{2+} for the less mobile Sr^{2+} in the mixed-carbonate particles. The low negative charge of all the particles should prevent repulsion and allow them to interact with the highly negative charge of the dentin surface [56]. Regarding the application of the nanoparticles as remineralizing and desensitizing agents with abrasive properties, their low ζ values allowed them to approach the dentin surface and deposit inside the dentin tubules [15,55]. The ζ value of Sensodyne® Rapid Relief is −16.5 mV. Despite this higher ζ, this toothpaste still has positive effects when it comes to treating hypersensitivity [15].

Both $SrCO_3$ nanoparticles and $Sr_{0.5}Ca_{0.5}CO_3$ nanoparticles consisted of spherical nanoparticles (Figure 2A,B). In contrast, for $CaCO_3$ nanoparticles, SEM revealed the formation of typical calcite orthorhombic micrometric structures (Figure 2C). Aggregation in the samples was a consequence of their low ζ [53]. The SEM images supported the hydrodynamic sizes found by DLS (Table 1). The larger size and orthorhombic morphology of $CaCO_3$ nanoparticles partially explained their small effect on tubule obliteration and their ability to form continuous deposits on the dentin surface (Figure 6F).

Going further, we investigated the cytotoxicity of the nanoparticles on hDPSCs cultures in vitro. Concentrations close to 10 μg/mL resulted in 20% in the cell viability compared to the control for the $Sr_{0,5}Ca_{0,5}CO_3$ e $SrCO_3$ nanoparticles (Figure 3a,b). This might indicate the interaction of the nanoparticles with the cell membrane, thus promoting a slight decrease in cell viability rather than a significant drop [57]. The ability of the nanoparticles to induce dentin remineralization has been investigated by their effect on TNAP activity [58]. The time-course events associated with TNAP activity have been previously reported [46,47]. Even though the TNAP messenger RNA transcription happens as early as the 7th day of cell culture, its highest activity occurs on the the14th day. Based on this, we sought to test both 7 days and 14 days. After 7 days of culture, the TNAP activity of odontoblast treated in the presence of $CaCO_3$ nanoparticles, $SrCO_3$ nanoparticles, and $Sr_{0.5}Ca_{0.5}CO_3$ nanoparticles was like the control group (Figure 3C); however, on the day 14th, $SrCO_3$ nanoparticles promoted increased TNAP activity compared to $Sr_{0,5}Ca_{0,5}CO_3$ nanoparticles, although lower than the control group (Figure 3C).

Several studies report the influence of different strontium concentration in TNAP activity, which corroborate our hypothesis. Bonnelye et al. demonstrated that TNAP gene expression and activity significantly increased when higher doses of strontium ranelate were provided [59]. Our group has also found increased TNAP activity on osteoblasts treated with a flavonoid-strontium complex [48]. Moreover, alizarin red staining (Figure 3D) showed that the presence of the particles significantly enhanced the mineral nodules deposition in comparison to the control group on the 14th day of culturing. The stimulus upon mineralization was more significant for $CaCO_3$ nanoparticles and $Sr_{0.5}Ca_{0.5}CO_3$ nanoparticles, which might be related to the Ca^{2+} ions already present in the particles' structure. The formation of mineralized nodules indicates the differentiation of hDPSCs in an osteogenic phenotype [40]. Moreover, the deposition of minerals is an indicator of matrix vesicle secretion by the cells, which usually occurs when the cells switch from an immature to a mature phenotype [60]. These results were confirmed by the analysis of the expression of the main osteogenic markers COL1, ALP, and RUNX2. All the particles stimulated the expression of COL1 and ALP and sustained the expression of the transcriptional factor RUNX2-related genes after 7 days of culturing, compared to the control. These genes are earlier markers of differentiation of hDPSCs to an osteoblastic phenotype [61–63], which supports the enhanced mineral nodule formation after treatment with the particles (Figure 3D). $SrCO_3$ nanoparticles stimulated the expression of the genes related to these three osteogenic markers even after 14 days of culturing (Figure 4).

After conducting biological assays in vitro, we mixed the $SrCO_3$, $Sr_{0.5}Ca_{0.5}CO_3$, or $CaCO_3$ nanoparticles with PVA to obtain gel-like samples, which we further used to brush the dentin disc surface. The ATR-FTIR spectra of the dentin samples before and after brushing helped to investigate changes in the chemical composition of the dentin surface after contact with the nanoparticles. Figure 5A illustrates the spectra of the dentin discs brushed with $SrCO_3$ nanoparticles. Bands assigned to biological apatite emerged in the 1100–1000 cm^{-1} range and were related to asymmetric stretching of the phosphate group [53,64,65]. The typical amide III band of collagen also appeared at 1250 cm^{-1}, attesting to the presence of this structural protein on the dentin surface [66,67]. Bands ascribed to carboxylate and amine groups were observed between 1750 and 1500 cm^{-1} [68]. The natural substitution of phosphate with carbonate groups is a fingerprint of biological apatite [53,64,65,69–71]. However, the intensity of the bands at 1480 and 850 cm^{-1} increased after brushing with $SrCO_3$ (Figure 5A—red line and black arrows), while there was no evidence of changes in the band at 1070 cm^{-1} attributed to phosphate. The intensity of carbonate-related bands increased as a function of the number of brushing cycles (Figure 5A—green and blue lines). This result supported the accumulation of nanoparticles due to sequential brushing, reinforcing the ability of $SrCO_3$ nanoparticles to bind to the dentin surface without modifying the apatite mineral structure.

The samples brushed with $Sr_{0.5}Ca_{0.5}CO_3$ nanoparticles behaved similarly: the bands related to carbonate groups intensified upon sequential brushing cycles for 7 (Figure 5B—green line) and 14 days (Figure 5B—blue line). We then compared the effects of $SrCO_3$ nanoparticles and $Sr_{0.5}Ca_{0.5}CO_3$ nanoparticles: after the first brushing cycle, the application of $SrCO_3$ nanoparticles inverted the relative intensity between the bands at 1070 cm^{-1}, assigned to phosphate, and 850 cm^{-1}, assigned to carbonate, whereas brushing with $Sr_{0.5}Ca_{0.5}CO_3$ nanoparticles elicited this inversion only after 7 days. We attributed this result to the higher negative ζ value of $Sr_{0.5}Ca_{0.5}CO_3$ nanoparticles, which translated into lower ability to bind to the dentin surface as compared to $SrCO_3$ nanoparticles. The fact that the phosphate band slightly shifted to lower wavenumbers after brushing with $Sr_{0.5}Ca_{0.5}CO_3$ was noteworthy and indicated that Ca^{2+} was replaced with the heavier Sr^{2+} in the apatite lattice [53,65,69]. The intensity of the band at 1480 cm^{-1} (black arrow) did not change significantly after brushing with $CaCO_3$ nanoparticles, which indicated a lower deposition of the particles on the dentin surface. Given that $SrCO_3$ and $CaCO_3$ nanoparticles have similar ζ values, this result attested to the affinity of Sr^{2+} for the dentin surface and the enhanced binding ability of the strontium-containing carbonate nanoparticles. Moreover, the larger size and orthorhombic morphology of $CaCO_3$ nanoparticles may have prevented their contact with the dentin surface and penetration into the dentin tubules, as also suggested by SEM analysis (Figure 6F). The higher intensity of the phosphate band after brushing with $CaCO_3$ nanoparticles for 14 days could result from higher exposure of the apatite mineral, assigned to the removal of the smear layer due to the well-known abrasive effect of this compound [2,31]. Similar to the effect of the strontium-containing carbonate nanoparticles prepared herein, brushing with Sensodyne® Rapid Relief (strontium acetate as an active ingredient) intensified the band attributed to phosphate and shifted it to a lower wavenumber. Nevertheless, the bands related to carbonate remained unchanged.

We calculated tubule obliteration by Equation (1) by counting the total number of tubules and classifying them into the open, partially obliterated, and obliterated in a 10 mm^2 area of an SEM image. The results presented in Table 2 evidenced the low obliteration ability (24.7%) of $CaCO_3$. This finding supported the absence of changes in the bands associated with carbonate in the FTIR spectrum obtained for dentin after brushing with $CaCO_3$ nanoparticles. Nevertheless, the spherical strontium-containing nanoparticles promoted higher obliteration (67.7%) for the samples brushed with $SrCO_3$ nanoparticles, which remained stable after acidic treatment (Table 2). The samples brushed with $Sr_{0.5}Ca_{0.5}CO_3$ nanoparticles had the highest obliteration values, which were even higher after the acidic challenge when they reached 100% (Table 2). This could be due to the higher solubility of $CaCO_3$ nanoparticles in an acidic medium, which led to the re-deposition of smaller particles that were able to penetrate into the dentin tubules as a result of a well-known dissolution/reprecipitation process [72,73]. Surprisingly, although brushing with Sensodyne® Rapid Relief promoted intertubular deposition of the particles, this toothpaste presented the lowest obliteration ability, which was close to 19.5% before and 31.9% after acidic treatment. The side-view SEM images evidenced changes in the morphology of the internal tubule walls and the penetration of particles (Figure 7). In particular, the red arrow in Figure 6 indicates the presence of nanoparticles inside the tubules of the samples brushed with $Sr_{0.5}Ca_{0.5}CO_3$ nanoparticles, evidencing the high affinity of these nanoparticles for the dentin structure.

We carried out elemental analysis of the dentin surface before and after brushing by EDS coupled with SEM. The relative amount of Ca, Sr, and P depicted in Table 3 revealed a (Ca + Sr)/P molar ratio close to 1.8 for clean dentin samples before brushing. The stoichiometric ratio expected for hydroxyapatite is 1.67. The higher values could be correlated to cationic and anionic substitutions in the biological apatite structure. The FTIR data supported this finding: bands related to B-type substitution, in which phosphate groups are replaced with carbonate, were present in the spectra [24,74] Furthermore, the presence of a small amount of strontium before brushing was related to the natural substitution of Ca^{2+} for Sr^{2+} in mineralized tissue. Strontium uptake at concentrations close to 2.4 mg/day from water and diet has been reported [2,23–25,74]. Brushing with $CaCO_3$ nanoparticles reduced the (Ca + Sr)/P molar ratio due to exposure of phosphate because of abrasion, as also evidenced by FTIR analysis. After the acid attack, the relative amount of Ca decreased, indicating partial dentin demineralization [2,23–25]. The (Ca + Sr)/P molar ratio after brushing with $SrCO_3$ nanoparticles was like the ratio in clean dentin (Table 3). However, the relative strontium content increased, indicating that strontium deposited on the dentin surface. We identified homogeneous strontium deposits in the EDS images in Figure 8. After the acid attack, the relative amount of calcium decreased, which translated into increased strontium and phosphorus content (Figure 9 and Table 3). Calcium removal and maintenance of strontium deposited on the dentin surface were evident in the EDS images (Figure 9). The lack of a green color, related to calcium, and the presence of a blue color, related to strontium, in the side view of Figure 9 after the acid attack supported the higher affinity of strontium for the dentin surface, which could result in remineralization and obliteration [69,75]. We verified a similar behavior after we brushed dentin discs with the mixed $Sr_{0.5}Ca_{0.5}CO_3$ nanoparticles: calcium and strontium were homogeneously deposited (Table 3, Figure 8), and calcium was partially removed after the acid attack (Figure 9). Calcium removal was less accentuated as compared to the samples brushed with $SrCO_3$ nanoparticles, which indicated that the coating formed by deposition of $Sr_{0.5}Ca_{0.5}CO_3$ nanoparticles on dentin was more resistant to acid attack (Figure 9). The presence of green spots in the side view images indicated that higher amounts of calcium remained deposited on the dentin surface after acid attack as compared to the samples brushed with $SrCO_3$ nanoparticles. The reduction in the phosphorus content (Table 3 and Figure 9) could be assigned to the formation of carbonated apatite as supported by the FTIR data [76]. Brushing with Sensodyne® Rapid Relief resulted in a higher amount of strontium in dentin as compared to the samples brushed with the tested nanoparticles (Table 3). Nevertheless, intertubular despite intratubular deposition was evident from the presence of the black holes observed in the EDS image attributed to non-obliterated tubules (Figure 8). After the acid attack, a reduced amount of strontium remained on the dentin samples brushed with Sensodyne® Rapid Relief as compared to the samples brushed with $SrCO_3$ nanoparticles or $Sr_{0.5}Ca_{0.5}CO_3$ nanoparticles (Table 3), revealing the lower resistance of the coating formed by the commercial toothpaste as compared to the coating formed by the new active ingredients containing strontium proposed herein.

5. Conclusions

Dentin samples brushed with the strontium-containing nanoparticles were coated with a mineral layer that was also able to penetrate dentin tubules, while $CaCO_3$ nanoparticles remained as individual particles on the dentin surface. Moreover, after brushing with $SrCO_3$ nanoparticles or $Sr_{0.5}Ca_{0.5}CO_3$ nanoparticles, the bands related to carbonate groups intensified. The slight shift of the phosphate-related FTIR band to the lower wavenumber indicated that strontium replaced calcium on the dentin structure after treatment. The coating promoted by $SrCO_3$ nanoparticles or $Sr_{0.5}Ca_{0.5}CO_3$ nanoparticles resisted an acidic environment, while calcium and phosphorus were removed from the top of the dentin surface. The nanoparticles were not toxic to hDPSCs, sustained TNAP activity, and promoted mineralization. The performance of the strontium-containing nanoparticles in terms of tubule obliteration and resistance to acid attack was even better as compared to Sensodyne® Rapid Relief. Adding $Sr_{0.5}Ca_{0.5}CO_3$ nanoparticles as an active ingredient in dentifrice formulations may be commercially advantageous because this compound combines the well-known abrasive properties of calcium carbonate with the mineralization ability of strontium, while the final cost remains between the costs of $CaCO_3$ nanoparticles and $SrCO_3$ nanoparticles.

Author Contributions: Conceptualization, A.P.R. and A.B.C.E.B.C.; Methodology, T.C.D., G.F., L.F.B.N., A.F.B.C., R.C. and M.M.-T.; Validation, T.C.D., L.H. and L.d.P.A.A.; Formal Analysis, T.C.D., L.H. and A.P.R.; Investigation, T.C.D.; L.H., A.F.B.C. and M.M.A.; Resources, T.C.D. and M.M.-T.; Data Curation, A.P.R.; Writing—Original Draft Preparation, T.C.D., L.H. and A.P.R.; Writing—Review and Editing, M.B., P.C. and A.B.C.E.B.C.; Visualization, A.P.R. and M.B.; Supervision, A.P.R., P.C., M.B., G.F. and A.B.C.E.B.C.; Project Administration, A.P.R.; Funding Acquisition, A.P.R. and P.C. All authors have read and agreed to the published version of the manuscript.

Funding: This research was funded by São Paulo Research Foundation (FAPESP-Brazil), grants number [2019/25054-2], [2019/08568-2] and National Council for Scientific and Technological Development (CNPq-Brazil), grant number [408440/2021-0]. T.C. Dotta acknowledges CAPES for the studentship (88887.480310/2020-00).

Institutional Review Board Statement: The study was conducted in accordance with the Declaration of Helsinki, and approved by the Ethics Committee of the "Tooth Bank of the Dentistry School" (University of São Paulo, Ribeirão Preto, Brazil) (protocol code 06245218.2.0000.5419 on 17 February 2019).

Data Availability Statement: The data presented in this study are available on request from the corresponding author.

Acknowledgments: The authors thank Cynthia M.C.P. Manso for linguistic advice.

Conflicts of Interest: No conflict to disclosure.

References

1. Amarasena, N.; Spencer, J.; Ou, Y.; Brennan, D. Dentine hypersensitivity—Australian dentists' perspective. *Aust. Dent. J.* **2010**, *55*, 181–187. [CrossRef] [PubMed]
2. Saeki, K.; Marshall, G.W.; Gansky, S.A.; Parkinson, C.R.; Marshall, S.J. Strontium effects on root dentin tubule occlusion and nanomechanical properties. *Dent. Mater.* **2016**, *32*, 240–251. [CrossRef] [PubMed]
3. Chen, C.L.; Parolia, A.; Pau, A.; Porto, I.C.C.D.M. Comparative evaluation of the effectiveness of desensitizing agents in dentine tubule occlusion using scanning electron microscopy. *Aust. Dent. J.* **2015**, *60*, 65–72. [CrossRef] [PubMed]
4. Hall, C.; Mason, S.; Cooke, J. Exploratory randomised controlled clinical study to evaluate the comparative efficacy of two occluding toothpastes—A 5% calcium sodium phosphosilicate toothpaste and an 8% arginine/calcium carbonate toothpaste—for the longer-term relief of dentine hypersensitivity. *J. Dent.* **2017**, *60*, 36–43. [CrossRef]
5. Alencar, C.D.M.; de Paula, B.L.F.; Ortiz, M.I.G.; Magno, M.B.; Silva, C.M.; Maia, L.C. Clinical efficacy of nano-hydroxyapatite in dentin hypersensitivity: A systematic review and meta-analysis. *J. Dent.* **2019**, *82*, 11–21. [CrossRef]
6. Hu, M.-L.; Zheng, G.; Zhang, Y.-D.; Yan, X.; Li, X.-C.; Lin, H. Effect of desensitizing toothpastes on dentine hypersensitivity: A systematic review and meta-analysis. *J. Dent.* **2018**, *75*, 12–21. [CrossRef]
7. Zeola, L.F.; Soares, P.V.; Cunha-Cruz, J. Prevalence of dentin hypersensitivity: Systematic review and meta-analysis. *J. Dent.* **2019**, *81*, 1–6. [CrossRef]
8. Lynch, E.; Brauer, D.S.; Karpukhina, N.; Gillam, D.G.; Hill, R.G. Multi-component bioactive glasses of varying fluoride content for treating dentin hypersensitivity. *Dent. Mater.* **2012**, *28*, 168–178. [CrossRef]
9. Hughes, N.; Mason, S.; Jeffery, P.; Welton, H.; Tobin, M.; O'Shea, C.; Browne, M. A comparative clinical study investigating the efficacy of a test dentifrice containing 8% strontium acetate and 1040 ppm sodium fluoride versus a marketed control dentifrice containing 8% arginine, calcium carbonate, and 1450 ppm sodium monofluorophospha. *J. Clin. Dent.* **2010**, *21*, 49–55.
10. Pinto, S.C.S.; Silveira, C.M.M.; Pochapski, M.T.; Pilatt, G.L.; Santos, F.A. Effect of desensitizing toothpastes on dentin. *Braz. Oral Res.* **2012**, *26*, 410–417. [CrossRef]
11. Bossù, M.; Saccucci, M.; Salucci, A.; DI Giorgio, G.; Bruni, E.; Uccelletti, D.; Sarto, M.S.; Familiari, G.; Relucenti, M.; Polimeni, A. Enamel remineralization and repair results of Biomimetic Hydroxyapatite toothpaste on deciduous teeth: An effective option to fluoride toothpaste. *J. Nanobiotechnology* **2019**, *17*, 17. [CrossRef]
12. Palazon, M.T.; Scaramucci, T.; Aranha, A.C.C.; Prates, R.A.; Lachowski, K.M.; Hanashiro, F.S.; Youssef, M.N. Immediate and Short-Term Effects of In-Office Desensitizing Treatments for Dentinal Tubule Occlusion. *Photomed. Laser Surg.* **2013**, *31*, 274–282. [CrossRef]
13. Mantzourani, M.; Sharma, D. Dentine sensitivity: Past, present and future. *J. Dent.* **2013**, *41*, S3–S17. [CrossRef]
14. Davies, M.; Paice, E.M.; Jones, S.B.; Leary, S.; Curtis, A.R.; West, N.X. Efficacy of desensitizing dentifrices to occlude dentinal tubules. *Eur. J. Oral Sci.* **2011**, *119*, 497–503. [CrossRef]
15. Mockdeci, H.; Polonini, H.; Martins, I.; Granato, A.-P.; Raposo, N.; Chaves, M.D.G.A.M. Evaluation of ex vivo effectiveness of commercial desensitizing dentifrices. *J. Clin. Exp. Dent.* **2017**, *9*, e503–e510. [CrossRef]
16. Olley, R.C.; Pilecki, P.; Hughes, N.; Jeffery, P.; Austin, R.S.; Moazzez, R.; Bartlett, D. An in situ study investigating dentine tubule occlusion of dentifrices following acid challenge. *J. Dent.* **2012**, *40*, 585–593. [CrossRef]
17. Rajguru, S.A.; Padhye, A.M.; Gupta, H.S. Effects of two desensitizing dentifrices on dentinal tubule occlusion with citric acid challenge: Confocal laser scanning microscopy study. *Indian J. Dent. Res.* **2017**, *28*, 450. [CrossRef]
18. Pinto, S.C.S.; Batitucci, R.G.; Pinheiro, M.C.; Zandim, D.L.; Spin-Neto, R.; Sampaio, J.E.C. Effect of an Acid Diet Allied to Sonic Toothbrushing on Root Dentin Permeability: An In Vitro Study. *Braz. Dent. J.* **2010**, *21*, 390–395. [CrossRef] [PubMed]
19. Wang, Z.; Sa, Y.; Sauro, S.; Chen, H.; Xing, W.; Ma, X.; Jiang, T.; Wang, Y. Effect of desensitising toothpastes on dentinal tubule occlusion: A dentine permeability measurement and SEM in vitro study. *J. Dent.* **2010**, *38*, 400–410. [CrossRef] [PubMed]
20. Gandolfi, M.G.; Silvia, F.; Pashley, D.H.; Gasparotto, G.; Carlo, P. Calcium silicate coating derived from Portland cement as treatment for hypersensitive dentine. *J. Dent.* **2008**, *36*, 565–578. [CrossRef] [PubMed]
21. Sauro, S.; Mannocci, F.; Tay, F.R.; Pashley, D.H.; Cook, R.; Carpenter, G.H.; Watson, T.F. Deproteinization Effects of NaOCl on Acid-etched Dentin in Clinically-relevant vs Prolonged Periods of Application. A Confocal and Environmental Scanning Electron Microscopy Study. *Oper. Dent.* **2009**, *34*, 166–173. [CrossRef]

22. Parkinson, C.; Willson, R. A comparative in vitro study investigating the occlusion and mineralization properties of commercial dentifrices in a four-day dentin model. *J. Clin. Dent.* **2011**, *22*, 68–73.
23. Al-Khafaji, T.J.; Wong, F.; Fleming, P.S.; Karpukhina, N.; Hill, R. Novel fluoride and strontium-containing bioactive glasses for dental varnishes-design and bioactivity in Tris buffer solution. *J. Non-Cryst. Solids* **2019**, *503–504*, 120–130. [CrossRef]
24. Denry, I.; Goudouri, O.-M.; Fredericks, D.C.; Akkouch, A.; Acevedo, M.R.; Holloway, J.A. Strontium-releasing fluorapatite glass-ceramic scaffolds: Structural characterization and in vivo performance. *Acta Biomater.* **2018**, *75*, 463–471. [CrossRef]
25. Moghanian, A.; Firoozi, S.; Tahriri, M.; Sedghi, A. A comparative study on the in vitro formation of hydroxyapatite, cytotoxicity and antibacterial activity of 58S bioactive glass substituted by Li and Sr. *Mater. Sci. Eng. C* **2018**, *91*, 349–360. [CrossRef] [PubMed]
26. Fujioka-Kobayashi, M.; Tsuru, K.; Nagai, H.; Fujisawa, K.; Kudoh, T.; Ohe, G.; Ishikawa, K.; Miyamoto, Y. Fabrication and evaluation of carbonate apatite-coated calcium carbonate bone substitutes for bone tissue engineering. *J. Tissue Eng. Regen. Med.* **2018**, *12*, 2077–2087. [CrossRef] [PubMed]
27. Gjorgievska, E.S.; Nicholson, J.W.; Slipper, I.J.; Stevanovic, M.M. Remineralization of Demineralized Enamel by Toothpastes: A Scanning Electron Microscopy, Energy Dispersive X-Ray Analysis, and Three-Dimensional Stereo-Micrographic Study. *Microsc. Microanal.* **2013**, *19*, 587–595. [CrossRef] [PubMed]
28. Markowitz, K. The original desensitizers: Strontium and potassium salts. *J. Clin. Dent.* **2009**, *20*, 145–151.
29. Orsini, G.; Procaccini, M.; Manzoli, L.; Sparabombe, S.; Tiriduzzi, P.; Bambini, F.; Putignano, A. A 3-Day Randomized Clinical Trial to Investigate the Desensitizing Properties of Three Dentifrices. *J. Periodontol.* **2013**, *84*, e65–e73. [CrossRef]
30. West, N.; Newcombe, R.G.; Hughes, N.; Mason, S.; Maggio, B.; Sufi, F.; Claydon, N. A 3-day randomised clinical study investigating the efficacy of two toothpastes, designed to occlude dentine tubules, for the treatment of dentine hypersensitivity. *J. Dent.* **2013**, *41*, 187–194. [CrossRef]
31. Medvecky, L.; Stulajterova, R.; Giretova, M.; Mincik, J.; Vojtko, M.; Balko, J.; Briancin, J. Effect of tetracalcium phosphate/monetite toothpaste on dentin remineralization and tubule occlusion in vitro. *Dent. Mater.* **2018**, *34*, 442–451. [CrossRef]
32. Earl, J.S.; Ward, M.; Langford, R. Investigation of dentinal tubule occlusion using FIB-SEM milling and EDX. *J. Clin. Dent.* **2010**, *21*, 37–41. [CrossRef] [PubMed]
33. Liu, Y.; Zhou, Y.; Jiang, T.; Liang, Y.-D.; Zhang, Z.; Wang, Y.-N. Evaluation of the osseointegration of dental implants coated with calcium carbonate: An animal study. *Int. J. Oral Sci.* **2017**, *9*, 133–138. [CrossRef] [PubMed]
34. Rushdi, M.; Abdel-Fattah, A.; Soliman, Y. Radiation-induced defects in strontium carbonate rod for EPR dosimetry applications. *Radiat. Phys. Chem.* **2017**, *131*, 1–6. [CrossRef]
35. Schofield, E.J.; Sarangi, R.; Mehta, A.; Jones, A.M.; Smith, A.; Mosselmans, J.F.W.; Chadwick, A.V. Strontium carbonate nanoparticles for the surface treatment of problematic sulfur and iron in waterlogged archaeological wood. *J. Cult. Herit.* **2015**, *18*, 306–312. [CrossRef]
36. Liu, M.; Zhang, F.; Wang, P.; Meng, J.; Zuo, B. Preparation and characterization of silk fibroin/strontium carbonate film through rapid formation. *Mater. Lett.* **2017**, *189*, 46–49. [CrossRef]
37. Seong, J.; Newcombe, R.G.; Foskett, H.L.; Davies, M.; West, N.X. A randomised controlled trial to compare the efficacy of an aluminium lactate/potassium nitrate/hydroxylapatite toothpaste with a control toothpaste for the prevention of dentine hypersensitivity. *J. Dent.* **2021**, *108*, 103619. [CrossRef]
38. Huang, M.; Hill, R.G.; Rawlinson, S.C.F. Strontium (Sr) elicits odontogenic differentiation of human dental pulp stem cells (hDPSCs): A therapeutic role for Sr in dentine repair? *Acta Biomater.* **2016**, *38*, 201–211. [CrossRef]
39. Martin-Del-Campo, M.; Rosales-Ibañez, R.; Alvarado, K.; Sampedro, J.G.; Garcia-Sepulveda, C.A.; Deb, S.; Román, J.S.; Rojo, L. Strontium folate loaded biohybrid scaffolds seeded with dental pulp stem cells induce in vivo bone regeneration in critical sized defects. *Biomater. Sci.* **2016**, *4*, 1596–1604. [CrossRef]
40. Di Tinco, R.; Sergi, R.; Bertani, G.; Pisciotta, A.; Bellucci, D.; Carnevale, G.; Cannillo, V.; Bertoni, L. Effects of a Novel Bioactive Glass Composition on Biological Properties of Human Dental Pulp Stem Cells. *Materials* **2020**, *13*, 4049. [CrossRef]
41. Gillam, D.G.; Tang, J.Y.; Mordan, N.J.; Newman, H.N. The effects of a novel Bioglass® dentifrice on dentine sensitivity: A scanning electron microscopy investigation. *J. Oral Rehabil.* **2002**, *29*, 305–313. [CrossRef] [PubMed]
42. Amaechi, B.T.; Higham, S.M.; Edgar, W.M. Factors influencing the development of dental erosion in vitro: Enamel type, temperature and exposure time. *J. Oral Rehabil.* **1999**, *26*, 624–630. [CrossRef] [PubMed]
43. Toledano-Osorio, M.; Osorio, E.; Aguilera, F.S.; Medina-Castillo, A.L.; Toledano, M.; Osorio, R. Improved reactive nanoparticles to treat dentin hypersensitivity. *Acta Biomater.* **2018**, *72*, 371–380. [CrossRef]
44. Leguizamón, N.D.P.; Rodrigues, E.M.; de Campos, M.L.; Nogueira, A.V.B.; Viola, K.S.; Schneider, V.K.; Neo-Justino, D.M.; Tanomaru-Filho, M.; Zambuzzi, W.F.; Henrique-Silva, F.; et al. In vivo and in vitro anti-inflammatory and pro-osteogenic effects of citrus cystatin CsinCPI-2. *Cytokine* **2019**, *123*, 154760. [CrossRef]
45. Nagavi-Alhoseiny, A.A.; Torshabi, M.; Rasoulianboroujeni, M.; Tayebi, L.; Tabatabaei, F.S. Effect of sodium chloride on gene expression of Streptococcus mutans and zeta potential of demineralized dentin. *J. Oral Biol. Craniofacial Res.* **2019**, *9*, 1–4. [CrossRef]
46. Millán, J.L. The Role of Phosphatases in the Initiation of Skeletal Mineralization. *Calcif. Tissue Res.* **2013**, *93*, 299–306. [CrossRef] [PubMed]
47. Millán, J.L. Alkaline Phosphatases: Structure, substrate specificity and functional relatedness to other members of a large superfamily of enzymes. *Purinergic Signal.* **2006**, *2*, 335–341. [CrossRef]

48. Cruz, M.A.E.; Tovani, C.B.; Favarin, B.Z.; Soares, M.P.R.; Fukada, S.Y.; Ciancaglini, P.; Ramos, A.P. Synthesis of Sr–morin complex and its in vitro response: Decrease in osteoclast differentiation while sustaining osteoblast mineralization ability. *J. Mater. Chem. B* **2019**, *7*, 823–829. [CrossRef]
49. Cummins, D. Dentin hypersensitivity: From diagnosis to a breakthrough therapy for everyday sensitivity relief. *J. Clin. Dent.* **2009**, *20*, 1–9. [PubMed]
50. Cummins, D. Recent advances in dentin hypersensitivity: Clinically proven treatments for instant and lasting sensitivity relief. *Am. J. Dent.* **2010**, *23*, 3–13.
51. Davari, A.; Ataei, E.; Assarzadeh, H. Dentin Hypersensitivity: Etiology, Diagnosis and Treatment; A Literature Review. *J. Dent.* **2013**, *14*, 136–145.
52. Tovani, C.B.; Gloter, A.; Azaïs, T.; Selmane, M.; Ramos, A.P.; Nassif, N. Formation of stable strontium-rich amorphous calcium phosphate: Possible effects on bone mineral. *Acta Biomater.* **2019**, *92*, 315–324. [CrossRef] [PubMed]
53. Cruz, M.; Zanatta, M.; da Veiga, M.; Ciancaglini, P.; Ramos, A. Lipid-mediated growth of SrCO3/CaCO3 hybrid films as bioactive coatings for Ti surfaces. *Mater. Sci. Eng. C* **2019**, *99*, 762–769. [CrossRef] [PubMed]
54. Tovani, C.B.; Oliveira, T.M.; Soares, M.P.R.; Nassif, N.; Fukada, S.Y.; Ciancaglini, P.; Gloter, A.; Ramos, A.P. Strontium Calcium Phosphate Nanotubes as Bioinspired Building Blocks for Bone Regeneration. *ACS Appl. Mater. Interfaces* **2020**, *12*, 43422–43434. [CrossRef]
55. Planchon, M.; Ferrari, R.; Guyot, F.; Gélabert, A.; Menguy, N.; Chanéac, C.; Thill, A.; Benedetti, M.F.; Spalla, O. Interaction between Escherichia coli and TiO_2 nanoparticles in natural and artificial waters. *Colloids Surf. B Biointerfaces* **2013**, *102*, 158–164. [CrossRef]
56. Weerkamp, A.H.; Uyen, H.M.; Busscher, H.J. Effect of Zeta Potential and Surface Energy on Bacterial Adhesion to Uncoated and Saliva-coated Human Enamel and Dentin. *J. Dent. Res.* **1988**, *67*, 1483–1487. [CrossRef]
57. Fadeel, B.; Garcia-Bennett, A.E. Better safe than sorry: Understanding the toxicological properties of inorganic nanoparticles manufactured for biomedical applications. *Adv. Drug Deliv. Rev.* **2010**, *62*, 362–374. [CrossRef]
58. Nogueira, L.F.B.; Cruz, M.A.E.; Trovani, C.B.; Lopes, H.B.; Beloti, M.M.; Ciancaglini, P.; Bottini, M.; Ramos, A.P. Curcumin-loaded carrageenan nanoparticles: Fabrication, characterization, and assessment of the effects on osteoblasts mineralization. *Colloids Surf. B Biointerfaces* **2022**, *217*, 112622. [CrossRef]
59. Bonnelye, E.; Chabadel, A.; Saltel, F.; Jurdic, P. Dual effect of strontium ranelate: Stimulation of osteoblast differentiation and inhibition of osteoclast formation and resorption in vitro. *Bone* **2008**, *42*, 129–138. [CrossRef] [PubMed]
60. Bottini, M.; Mebarek, S.; Anderson, K.L.; Strzelecka-Kiliszek, A.; Bozycki, L.; Simão, A.M.S.; Bolean, M.; Ciancaglini, P.; Pikula, J.B.; Pikula, S.; et al. Matrix vesicles from chondrocytes and osteoblasts: Their biogenesis, properties, functions and biomimetic models. Biochim. et Biophys. *Acta (BBA)-Gen. Subj.* **2018**, *1862*, 532–546. [CrossRef] [PubMed]
61. Ono, N.; Balani, D.H.; Kronenberg, H.M. Stem and progenitor cells in skeletal development. *Curr. Top. Dev. Biol.* **2019**, *133*, 1–24. [CrossRef] [PubMed]
62. Meka, S.R.K.; Jain, S.; Chatterjee, K. Strontium eluting nanofibers augment stem cell osteogenesis for bone tissue regeneration. *Colloids Surf. B: Biointerfaces* **2016**, *146*, 649–656. [CrossRef] [PubMed]
63. Ilmer, M.; Karow, M.; Geissler, C.; Jochum, M.; Neth, P. Human Osteoblast–Derived Factors Induce Early Osteogenic Markers in Human Mesenchymal Stem Cells. *Tissue Eng. Part A* **2009**, *15*, 2397–2409. [CrossRef]
64. De Souza, I.D.; Cruz, M.A.; de Faria, A.N.; Zancanela, D.C.; Simão, A.M.; Ciancaglini, P.; Ramos, A.P. Formation of carbonated hydroxyapatite films on metallic surfaces using dihexadecyl phosphate–LB film as template. *Colloids Surf. B: Biointerfaces* **2014**, *118*, 31–40. [CrossRef] [PubMed]
65. López-Sánchez, M.; Dominguez-Vidal, A.; Ayora-Cañada, M.; Molina-Díaz, A. Assessment of dentifrice adulteration with diethylene glycol by means of ATR-FTIR spectroscopy and chemometrics. *Anal. Chim. Acta* **2008**, *620*, 113–119. [CrossRef] [PubMed]
66. Chavez, M.B.; Kramer, K.; Chu, E.Y.; Thumbigere-Math, V.; Foster, B.L. Insights into dental mineralization from three heritable mineralization disorders. *J. Struct. Biol.* **2020**, *212*, 107597. [CrossRef] [PubMed]
67. Ruiz, G.C.; Cruz, M.A.; Faria, A.N.; Zancanela, D.C.; Ciancaglini, P.; Ramos, A.P. Biomimetic collagen/phospholipid coatings improve formation of hydroxyapatite nanoparticles on titanium. *Mater. Sci. Eng. C* **2017**, *77*, 102–110. [CrossRef]
68. Tovani, C.B.; Oliveira, T.M.; Gloter, A.; Ramos, A.P. Sr2+-Substituted CaCO3 Nanorods: Impact on the Structure and Bioactivity. *Cryst. Growth Des.* **2018**, *18*, 2932–2940. [CrossRef]
69. Frangopol, P.T.; Mocanu, A.; Almasan, V.; Garbo, C.; Balint, R.; Borodi, G.; Bratu, I.; Horovitz, O.; Tomoaia-Cotisel, M. Synthesis and structural characterization of strontium substituted hydrozyapatites. *Rev. Roum. Chim.* **2016**, *61*, 337–344.
70. Rehman, I.; Bonfield, W. Characterization of hydroxyapatite and carbonated apatite by photo acoustic FTIR spectroscopy. *J. Mater. Sci. Mater. Med.* **1997**, *8*, 1–4. [CrossRef] [PubMed]
71. Landi, E.; Celotti, G.; Logroscino, G.; Tampieri, A. Carbonated hydroxyapatite as bone substitute. *J. Eur. Ceram. Soc.* **2003**, *23*, 2931–2937. [CrossRef]
72. Bertazzo, S.; Zambuzzi, W.F.; Campos, D.D.P.; Ogeda, T.L.; Ferreira, C.V.; Bertran, C.A. Hydroxyapatite surface solubility and effect on cell adhesion. *Colloids Surf. B Biointerfaces* **2010**, *78*, 177–184. [CrossRef] [PubMed]
73. Mozafari, M.; Banijamali, S.; Baino, F.; Kargozar, S.; Hill, R.G. Calcium carbonate: Adored and ignored in bioactivity assessment. *Acta Biomater.* **2019**, *91*, 35–47. [CrossRef] [PubMed]

74. Wang, Y.-L.; Chang, H.-H.; Chiang, Y.-C.; Lin, C.-H.; Lin, C.-P. Strontium ion can significantly decrease enamel demineralization and prevent the enamel surface hardness loss in acidic environment. *J. Formos. Med. Assoc.* **2018**, *118*, 39–49. [CrossRef]
75. Zhang, W.; Shen, Y.; Pan, H.; Lin, K.; Liu, X.; Darvell, B.W.; Lu, W.W.; Chang, J.; Deng, L.; Wang, D.; et al. Effects of strontium in modified biomaterials. *Acta Biomater.* **2011**, *7*, 800–808. [CrossRef]
76. Wingender, B.; Azuma, M.; Krywka, C.; Zaslansky, P.; Boyle, J.; Deymier, A. Carbonate substitution significantly affects the structure and mechanics of carbonated apatites. *Acta Biomater.* **2021**, *122*, 377–386. [CrossRef]

Journal of
Functional Biomaterials

MDPI

Article

Healing Patterns of Non-Collagenated Bovine and Collagenated Porcine Xenografts Used for Sinus Floor Elevation: A Histological Study in Rabbits

Yuhei Miyauchi [1], Takayuki Izutani [1], Yuki Teranishi [1], Takahisa Iida [1], Yasushi Nakajima [1,2], Samuel Porfirio Xavier [2,3,*] and Shunsuke Baba [1]

1 Department of Oral Implantology, Osaka Dental University, 8-1 Kuzuhahanazonocho, Hirakata, Osaka 573-1121, Japan
2 ARDEC Academy, Viale Giovanni Pascoli 67, 47923 Rimini, Italy
3 Department of Oral and Maxillofacial Surgery and Periodontology, Faculty of Dentistry of Ribeirão Preto, University of São Paulo, Ribeirão Preto 14040-904, São Paulo, Brazil
* Correspondence: spx@forp.usp.br

Abstract: Objective: To compare healing of collagenated and non-collagenated xenografts used for maxillary sinus floor elevation. Materials and Methods: Two different xenografts were used: deproteinized bovine bone (DBBM group) and collagenated corticocancellous porcine bone (collagenated group). Healing was studied after 2, 4, and 8 weeks. The loss of dimensions of the elevated area and the percentages of new bone, xenograft remnants, osteoclastic zones, vessels, inflammatory infiltrates, and soft tissues were analyzed. Three regions were evaluated: close to the bone walls (bone wall region), subjacent the sinus mucosa (submucosa region), and the center of the elevated area (middle region). The primary variables were the percentage of new bone and xenograft remnants. Results: Between 2 and 8 weeks, the elevated areas showed a reduction of 16.3% and 52.2% in the DBBM and collagenated groups, respectively ($p < 0.01$ between the two areas after 8 weeks). After 8 weeks, the highest content of new bone was observed in the bone wall region, which was higher in the collagenated group than in the DBBM group (41.6% and 28.6%, respectively; $p < 0.01$). A similar quantity of new bone was found between the two groups in other regions. A higher percentage of vessels in all regions evaluated ($p < 0.01$) and soft tissue in the sub-mucosa region ($p < 0.05$) was found in the collagenated group than in the DBBM group. Conclusions: The present study showed that both xenografts allowed new bone formation. In comparison with the non-collagenated xenograft, the collagenated xenograft underwent higher resorption, resulting in greater shrinkage of the elevated space after sinus lifting and a higher content of new bone in the regions close to the bone walls. Clinical relevance: In this study, the region adjacent to the bone wall showed the highest new bone content. This region resembles the base of the sinus, closest to the sinus floor and walls, and is the most important region from a clinical point of view because it is where the implant will be installed. Residues of the biomaterial remained after 8 weeks of healing. Other reports have shown that these biomaterial residues may interfere with the integration of implants.

Keywords: bone substitutes; sinus floor augmentation; bone formation; bone regeneration; bone resorption; animal model

Citation: Miyauchi, Y.; Izutani, T.; Teranishi, Y.; Iida, T.; Nakajima, Y.; Xavier, S.P.; Baba, S. Healing Patterns of Non-Collagenated Bovine and Collagenated Porcine Xenografts Used for Sinus Floor Elevation: A Histological Study in Rabbits. *J. Funct. Biomater.* **2022**, *13*, 276. https://doi.org/10.3390/jfb13040276

Academic Editor: Lidy Fratila-Apachitei

Received: 22 November 2022
Accepted: 1 December 2022
Published: 5 December 2022

1. Introduction

The presence of sufficient bone volume is a necessary prerequisite for implant installation and to provide a long-term successful outcome [1]. When no sufficient bone volume is available in the posterior region of the maxilla, sinus floor elevation has been shown to be a reliable surgical approach that allows oral rehabilitation using implants [2,3]. Lateral access is one of the most common approaches used for sinus lifting [2].

A series of anatomical characteristics should be considered before performing a surgical procedure. Among them, we should consider the height of the residual bone crest, the angle and position of the palatal-nasal recess in relation to the floor of the sinus, the angle of the floor sinus formed by the lateral and medial sinus walls, the width of the lateral wall, the position of the posterior superior alveolar artery, the presence of intra-sinus pathologies and septa, and the patency of the antrum [4,5].

The space gained after sinus floor elevation tends to be lost over time [6–8]. Volumetric changes after sinus lifting evaluated using computed tomography (CTs) scans or cone beam CTs (CBCTs) were reported in a systematic review [9]. To maintain the elevated space, biomaterials [10–13], implants without filler material [3,13–21], or resorbable or non-resorbable devices [22–25], were placed within the elevated space.

Histometric outcomes of various grafting materials used for sinus floor elevation have been discussed in various systematic reviews [10–12]. A systematic review with network meta-analyses [10] found differences in bone formation and graft resorption among the various types of xenografts included in the assessments. A systematic review with a meta-analysis [12] concluded that autogenous bone was the best choice when a high content of newly formed bone was needed. Furthermore, other biomaterials might be considered reliable bone substitutes in cases of concern regarding the donor site. The outcomes from the review showed significantly greater new bone formation with bovine bone than with hydroxyapatite alone, and better results with a biphasic graft composed of β-tricalcium phosphate and hydroxyapatite than with bovine bone. In a Bayesian network meta-analysis [11], however, it was concluded that autogenous bone showed reduced new bone formation than several other biomaterials used as grafts.

Different healing features and resorption rates of different biomaterials have also been reported in rabbits experiments [26–29]. In one of these experiments [26], sinus augmentation was performed either with deproteinized bovine bone mineral (DBBM) or with a collagen sponge. Healing was evaluated after different periods and showed stable dimensions of the elevated space between 7 and 40 days, while at the collagen site, >50% of the dimensions were lost. The total amount of new bone was higher at DBBM sites than at collagen sponge sites.

In a similar study in rabbits [27], sinus augmentation was performed using either DBBM or autogenous bone, and simultaneous implants were installed. Healing was evaluated after 7 and 40 days. After 40 days of healing, a higher percentage of new bone was found at autogenous sites than at DBBM sites. However, two thirds of the area of the elevated space was lost at the autogenous sites, while the dimensions were maintained at the DBBM sites.

In another similar study [28], sinus floor augmentation was performed bilaterally using collagenated corticocancellous porcine bones. At the test site, a collagen membrane was placed adjacent to the sinus mucosa prior to grafting. Similar amounts of new bone were found in both the sinuses. However, approximately 50% of the elevated space was lost during healing.

These studies showed similar percentages of new bone [26,28] but a higher loss in the dimension of the elevated space in the collagenated group than in the non-collagenated group. A comparison between these two different xenografts on different parameters, such as bone formation, graft resorption, vessel formation, presence of resorptive zones, and inflammatory infiltrates, might still provide useful information. Moreover, detailed data on these parameters in different regions of the elevated regions, such as those close to the bone walls or sinus mucosa, might allow the disclosure of differences in healing between the two biomaterials. Hence, the aim of the present study was to compare the healing of collagenated and non-collagenated xenografts used for maxillary sinus floor elevation.

The hypothesis of this study was that collagenated and non-collagenated xenografts used for sinus floor augmentation might result in different percentages of newly formed bone.

2. Materials and Methods

2.1. Ethical Statement

The protocols for both experiments were approved by the Ethical Committee of the Faculty of Dentistry in Ribeirão Preto of the University of São Paulo (USP, São Paulo, Brazil). The protocol numbers were 2017.1.278.58.9 [30] and 2015.1.834.58.7 [31] for the non-collagenated (DBBM) and collagenated xenografts, respectively. The ARRIVE guidelines and SYRCLE risk of bias tool for animal studies were adopted.

2.2. Power Calculation and Sample Size

A one-tail analysis of the raw data on the total bone formed within the elevated space from the two experiments [30,31] was performed, with a power of 0.84, $\alpha = 0.05$, and an effect size of 1.68. It was assumed that different proportions of new bone could be found between the two groups in the various regions evaluated. A sample size of six animals was considered sufficient to reject the null hypothesis that the population means of the experimental and control groups were equal. G*Power 3.1 was used for the calculations.

2.3. Experimental Design

The histological slides from two experiments, in which two different biomaterials were used as grafts for sinus floor elevation, were further evaluated to analyze other variables and different regions. Other data have been reported elsewhere [30,31]. In both studies, 18 male New Zealand rabbits (3.5–4.0 kg with a mean age of 4–5 months) were used. The maxillary sinuses of rabbits were elevated using xenografts with different features. In one study [30], a non-collagenated deproteinized bovine bone mineral (DBBM) with either 0.250–1.0 mm or 1.0–2.0 mm granules (Bio-Oss®, Geistlich Biomaterials, Wolhusen, LU, Switzerland) was used as filler material, while in the other study [31], a collagenated corticocancellous porcine bone (OsteoBiol Gen-Os, 0.250–1.0 µm, Tecnoss, Giaveno, Italy) was applied. Both studies were randomized.

Randomization was performed electronically by an operator not involved in the surgeries and was maintained in sealed opaque envelopes that were opened after exposure of the nasal bone [31] or after elevation of the sinus mucosa bilaterally [30].

In each study, three groups were formed; each was composed of six animals that were euthanized after 2, 4, and 8 weeks.

2.4. Xenografts

Bio-Oss (Geistlich Biomaterials, Wolhusen, LU, Switzerland) is a bovine bone deproteinized using a strong alkalis and organic solvent treatment at a temperature of 300 °C. The total porosity, as evaluated by the mercury intrusion method, was 63.5%, whereas the intraparticle porosity was 51%. The real density was 3.21 $g{\cdot}cm^{-3}$ and the mineral content was 95% [32].

Osteobiol Gen-Os (Tecnoss, Giaveno, Italy) is a porcine bone treated at a low temperature of up to 130 °C to eliminate the pathogens that allow the preservation of the structure and composition of both collagen and hydroxyapatite. The total porosity, as evaluated by the mercury intrusion method, was 33.1%, whereas the intraparticle porosity was 21%. The real density was 2.43 $g{\cdot}cm^{-3}$ and the mineral content was 64.6% [32].

2.5. Surgical Procedures

Acepromazine (1.0 mg/kg; Acepran®, Vetnil, Louveira, São Paulo, Brazil) was administered subcutaneously to induce anesthesia, followed by intramuscular (IM) injection of a mixture of xilazine (3.0 mg/kg; Dopaser®, Hertape Calier, Juatuba, Minas Gerais, Brazil) and 60 mg/kg ketamine (50.0 mg/kg; União Química Farmacêutica Nacional S/A, Embuguaçú, São Paulo, Brazil). Local anesthesia was injected into the experimental zone.

In the DBBM group, the skin, muscle, and periosteum on the nasal dorsum were incised and the nasal bone was exposed. Access windows were prepared bilaterally with drills, the sinus mucosa was elevated with small curettes, and the DBBM, either small or large granules, was placed inside the elevated spaces. The access windows were covered with a collagen membrane (Bio-Gide® Geistlich Biomaterials, Wolhusen, LU, Switzerland), and the overlying soft tissues were subsequently sutured. Similar surgical procedures were performed for the collagenated group. However, the access windows were prepared using either drills or an air sonic instrument (Sonosurgery, TeKne Dental, Calenzano, Florence, Italy) and micro-saw (SFS 102, Komet-Brasseler-GmbH, Rastatt, Germany). The elevated space was filled with collagenated corticocancellous porcine bone. The access windows were covered using a collagen membrane (Evolution, 0.3 mm, OsteoBiol Gen-Os; Tecnoss, Giaveno, Italy) and the soft tissues were closed with internal resorbable sutures (Polyglactin 910 5-0, Vicryl®, Ethicon, Johnson & Johnson, São José dos Campos, Brazil) and the skin was sutured with nylon (Ethilon 4-0®, Ethicon, Johnson & Johnson, São José dos Campos, Brazil).

Three healing periods were analyzed after 2, 4, and 8 weeks. In both studies, six animals were used for each healing period.

2.6. Maintenance Care

The animals were maintained in an acclimatized room at the animal facilities of the Faculty of Dentistry in Ribeirão Preto, University of São Paulo. The animals were housed in individual cages with ad libitum access to food and water. The monitoring of wounds and biological functions was performed daily by expert personnel. In the post-operative period, ketoprofen twice a day for 2 days (3.0 mg/kg, Ketofen 10%, Merial, Campinas, São Paulo, Brazil) was injected IM while Tramadol 2% twice a day for 2 days (1.0 mg/kg, Cronidor, Agener União Saúde Animal, Apucarana, Paraná, Brazil) was injected subcutaneously.

2.7. Euthanasia

Animals were anesthetized in a manner similar to that of the surgical session. Subsequently, an overdose of sodium thiopental (1.0 g, 2 mL, Thiopentax®, Cristália Produtos Químicos Farmacêuticos, Itapira, São Paulo, Brazil) was used to euthanize the animals.

2.8. Histological Preparation

The biopsies were placed in 10% buffered formalin and scanned for micro-CT analysis. The samples were dehydrated in increasing concentrations of ethanol (50% to 100%) for 6 days and subsequently included in an ascending series of resin solutions (50% to 100%; LR White™ hard grid, London Resin Co Ltd., Berkshire, UK). Thermal polymerization was performed in an oven at 60 °C for 24 h. After polymerization, the biopsies were cut in a coronal plane, crossing the small screw placed between the two sinuses, using high precision cutting equipment (Exakt, Apparatebau, Norderstedt, Germany). Two hemibiopsies were obtained from which two slides of approximately 150 µm were prepared. Using a sequence of sandpapers of decreasing grain size mounted on a grinding machine (Exakt, Apparatebau, Norderstedt, Germany), the two slices were ground to a thickness of approximately 60 µm and polished. Two sections were obtained from each sample and stained with toluidine blue or Stevenel's blue and alizarin red.

2.9. Data Analysis

The surgery in both groups was performed by the same surgeon (E.R.S.; see acknowledgements), and both histological measurements were performed by a well-trained examiner (K.A.A.A.; see acknowledgements). Before starting the measurements, a calibration with another professional (D.B.; see acknowledgements) was performed, and it was carried out until the inter-rater agreement in the identification of the tissues achieved a Cohen's coefficient $k > 0.80$.

To maintain high similarity and homogeneity, only one sinus was used in each experiment in the present study, that is, with the small granules in the DBBM experiment (DBBM group) and with the access window prepared with drills in the collagenated corticocancellous porcine bone experiment (collagenated group).

Histological measurements were performed using NIS-Elements D software (v 4.0, Laboratory Imaging, Nikon Corporation, Tokyo, Japan) on an Eclipse Ci microscope (Nikon Corporation, Tokyo, Japan) equipped with a video camera (Digital Sight DS-2Mv, Nikon Corporation, Tokyo, Japan). A point-counting procedure was used, superposing a lattice with squares of 75 µm in dimension on the image of the histological slide at ×100 magnification.

The following tissues were evaluated in both studies: mineralized new bone, xenografts, inflammatory infiltrate areas, vessels, osteoclastic zones, and soft tissues (marrow spaces, connective tissues, and matrix tissue). The tissues were evaluated in three different regions: close to the sinus bone walls (bone walls region), subjacent sinus mucosa (submucosa region), and in the center of the elevated space (middle region).

Two groups were defined: the DBBM and collagenated groups. The primary variables were the percentage of new bone and xenograft remnants. All the other parameters were secondary variables. Mean values and standard deviations were calculated for each variable using the software Excel 2013 (Microsoft Corporation, Redmond, WA, USA).

Statistical analyses were performed for both primary and secondary variables using the Prism software (Version 9.4.1, IBM Inc., Chicago, IL, USA). The Shapiro-Wilk test for the assessment of normality was applied for all variables. The unpaired t-test or Mann-Whitney test was used for statistical analyses. The significance level was set at $p < 0.05$.

3. Results

All periods are presented as n = 6, and data were available for all specimens. Data from three regions were evaluated (Figure 1A–D): bone walls (Table 1), sub-mucosa (Table 2), and middle (Table 3) regions.

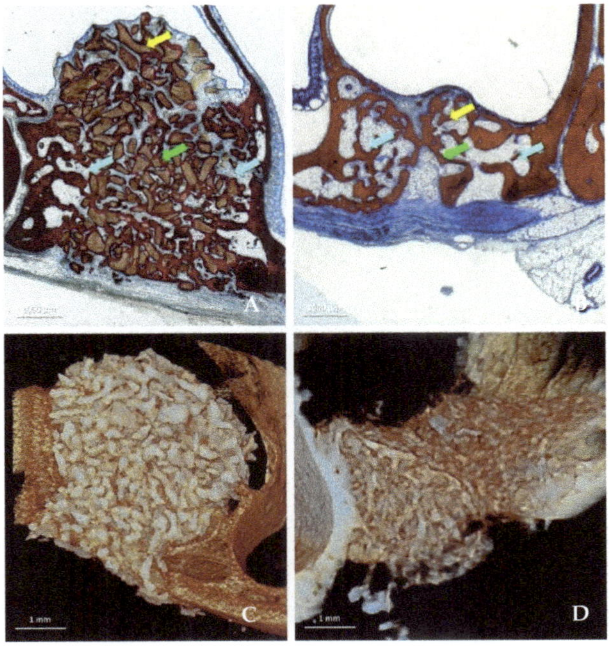

Figure 1. Healing after 8 weeks in the DBBM (**A**,**C**) and collagenated (**B**,**D**) groups illustrated in histological (**A**,**B**) and microCT images (**C**,**D**). The tissues were evaluated in three different regions: close to the sinus bone walls (bone walls region, light blue arrows), subjacent the sinus mucosa (sub-mucosa region, yellow arrow) and in the center of the elevated space (middle region, green arrow). Stevenel's blue and alizarin red stain.

Table 1. Tissues components in the bone walls region in the DBBM and collagenated groups at the various periods of healing. Mean values ± standard deviation in percentages.

	New Bone	Xenograft	Soft Tissues	Osteoclastic Zones	Vessels	Inflammatory Infiltrate
2 Weeks DBBM	11.5 ± 7.6 *	49.8 ± 5.3 *	33.9 ± 7.5 *	0.7 ± 0.9 *	4.2 ± 3.8	0.0 ± 0.0 *
2 Weeks Collagenated	1.6 ± 1.7 *	43.5 ± 4.3 *	48.8 ± 3.5 *	2.7 ± 0.8 *	2.6 ± 1.5	0.7 ± 0.6 *
p value	0.011	0.048	0.001	0.010	0.810	0.002
4 Weeks DBBM	24.7 ± 6.2	43.9 ± 4.1 *	26.8 ± 6.6 *	0.0 ± 0.1 *	4.6 ± 1.3	0.0 ± 0.0 *
4 Weeks Collagenated	20.4 ± 15.2	20.2 ± 12.9 *	44.4 ± 7.3 *	5.3 ± 2.6 *	8.7 ± 5.0	0.9 ± 1.0 *
p value	0.535	0.002	0.001	0.003	0.079	0.007
8 Weeks DBBM	28.6 ± 5.3 *	43.9 ± 4.4 *	26.3 ± 2.5 *	0.0 ± 0.0 *	0.9 ± 0.7 *	0.2 ± 0.5
8 Weeks Collagenated	41.6 ± 6.5 *	5.3 ± 4.3 *	43.1 ± 5.5 *	0.6 ± 0.5 *	9.4 ± 3.1 *	0.1 ± 0.3
p value	0.004	0.004	0.004	0.002	<0.0001	0.902

* $p < 0.05$ between deproteinized bovine bone mineral (DBBM) and collagenated corticocancellous porcine bone (collagenated) groups.

Table 2. Tissues components in the sub-mucosa region in the DBBM and collagenated groups at the various periods of healing. Mean values ± standard deviation in percentages.

	New Bone	Xenograft	Soft Tissues	Osteoclastic Zones	Vessels	Inflammatory Infiltrate
2 Weeks DBBM	5.6 ± 13.6	53.0 ± 10.7	37.5 ± 5.5	0.6 ± 0.8 *	3.4 ± 3.1	0.0 ± 0.0
2 Weeks Collagenated	0.0 ± 0.0	52.6 ± 5.5	42.8 ± 4.6	2.0 ± 1.0 *	2.3 ± 1.4	0.3 ± 0.4
p value	0.317	0.631	0.100	0.024	0.457	0.059
4 Weeks DBBM	5.7 ± 7.8	54.6 ± 9.4 *	36.9 ± 8.9 *	1.3 ± 1.4 *	1.5 ± 1.3	0.0 ± 0.0 *
4 Weeks Collagenated	5.7 ± 11.2	28.2 ± 1.8 *	50.1 ± 10.7 *	9.6 ± 7.5 *	3.9 ± 3.9	2.5 ± 2.2 *
p value	0.494	0.002	0.042	0.024	0.186	0.007
8 Weeks DBBM	28.6 ± 14.0	50.3 ± 4.6 *	20.2 ± 10.4 *	0.3 ± 0.9	0.4 ± 1.1 *	0.0 ± 0.0 *
8 Weeks Collagenated	28.1 ± 16.4	5.6 ± 4.4 *	54.6 ± 21.5 *	2.3 ± 3.4	8.8 ± 5.0 *	0.7 ± 0.6 *
p value	0.950	<0.0001	0.006	0.216	0.003	0.022

* $p < 0.05$ between deproteinized bovine bone mineral (DBBM) and collagenated corticocancellous porcine bone (collagenated) groups.

In the DBBM group, the area of the elevated space decreased by 5.6% between 2 and 4 weeks ($p = 0.599$) and by 16.3% between 2 and 8 weeks ($p = 0.108$). In the collagenated group, the reduction in area was 33.7% and 52.2% between 2 and 4 weeks ($p = 0.001$) and between 2 and 8 weeks ($p < 0.001$), respectively.

After 2 weeks of healing, new bone was observed in both groups, especially close to the bone walls (Figure 2A,B).

Table 3. Tissues components in the middle region in the DBBM and collagenated groups at the various periods of healing. Mean values ± standard deviation in percentages.

	New Bone	Xenograft	Soft Tissues	Osteoclastic Zones	Vessels	Inflammatory Infiltrate
2 Weeks DBBM	1.8 ± 4.5	59.8 ± 8.6 *	34.3 ± 6.9 *	1.6 ± 1.6	2.1 ± 1.8	0.3 ± 0.9
2 Weeks Collagenated	0.1 ± 0.2	50.0 ± 6.3 *	46.0 ± 6.5 *	1.2 ± 0.5	2.4 ± 1.1	0.4 ± 0.4
p value	0.902	0.048	0.032	0.571	0.470	0.210
4 Weeks DBBM	11.9 ± 6.0	51.5 ± 14.9 *	31.3 ± 7.5 *	0.0 ± 0.0 *	5.4 ± 4.1	0.0 ± 0.0 *
4 Weeks Collagenated	10.2 ± 18.2	21.0 ± 19.7 *	55.8 ± 16.2 *	5.2 ± 2.8 *	6.5 ± 5.9	1.4 ± 1.4 *
p value	0.149	0.013	0.007	0.002	0.719	0.022
8 Weeks DBBM	23.2 ± 12.8	48.3 ± 7.0 *	27.8 ± 7.0 *	0.2 ± 0.4	0.6 ± 0.8 *	0.0 ± 0.0 *
8 Weeks Collagenated	25.1 ± 8.0	5.6 ± 3.2 *	56.2 ± 12.8 *	1.3 ± 2.5	10.1 ± 4.9 *	1.6 ± 2.0 *
p value	0.764	<0.0001	0.001	0.216	0.006	0.022

* $p < 0.05$ between deproteinized bovine bone mineral (DBBM) and collagenated corticocancellous porcine bone (collagenated) groups.

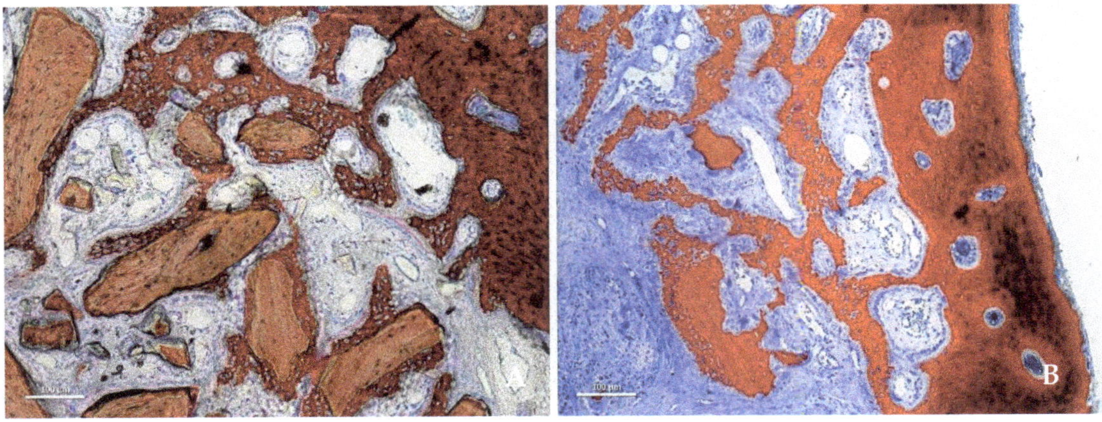

Figure 2. Photomicrographs of ground sections representing the healing after 2 weeks. New bone was formed from the sinus bone walls and surrounded the neighbor granules. (**A**) DBBM group. (**B**) collagenated group. Stevenel's blue and alizarin red stain.

Higher bone formation was observed in the DBBM group than in the collagen group during the healing period ($p = 0.011$). After 8 weeks of healing, new bone increased over time in all regions in both groups, reaching the highest value in the collagenated group. In this period of healing, the only statistically significant difference in bone formation was found in the bone walls region ($p = 0.004$), with a fraction of 28.6 ± 5.3% in the DDBM group and 41.6 ± 6.5% in the collagenated group (Figure 3).

The xenograft content in the DBBM group decreased by ~6% between weeks 2 and 8 in the bone wall region ($p = 0.064$), ~3% in the submucosal region ($p = 0.598$), and ~11.5% in the middle region ($p = 0.029$). The collagenated group presented much higher resorption, reducing the percentages from about 43–53% to 5–6% in the same interval of healing. The differences were statistically significant in all regions after 4 and 8 weeks of healing ($p < 0.01$; Figure 4).

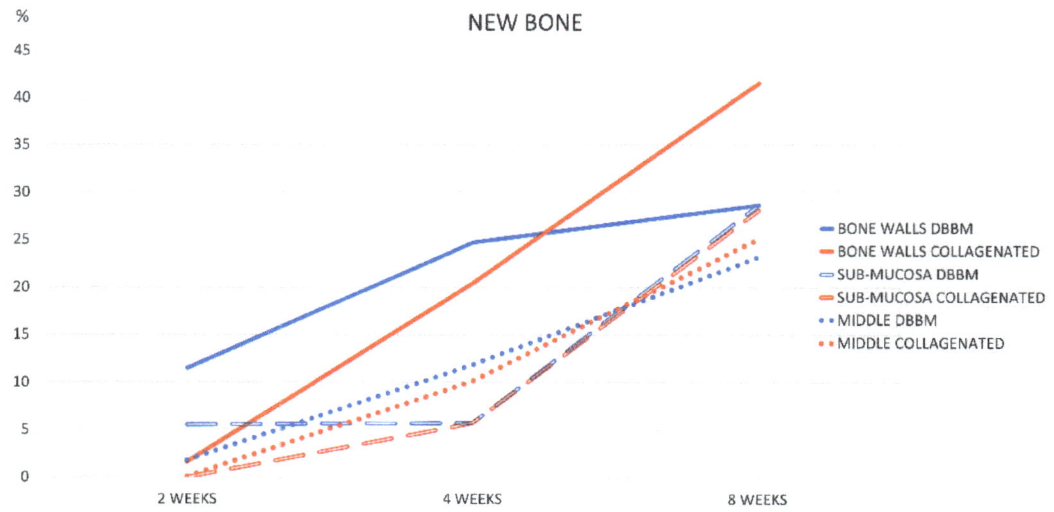

Figure 3. Graph illustrating percentage bone formation in the various regions examined after 2, 4, and 8 weeks of healing.

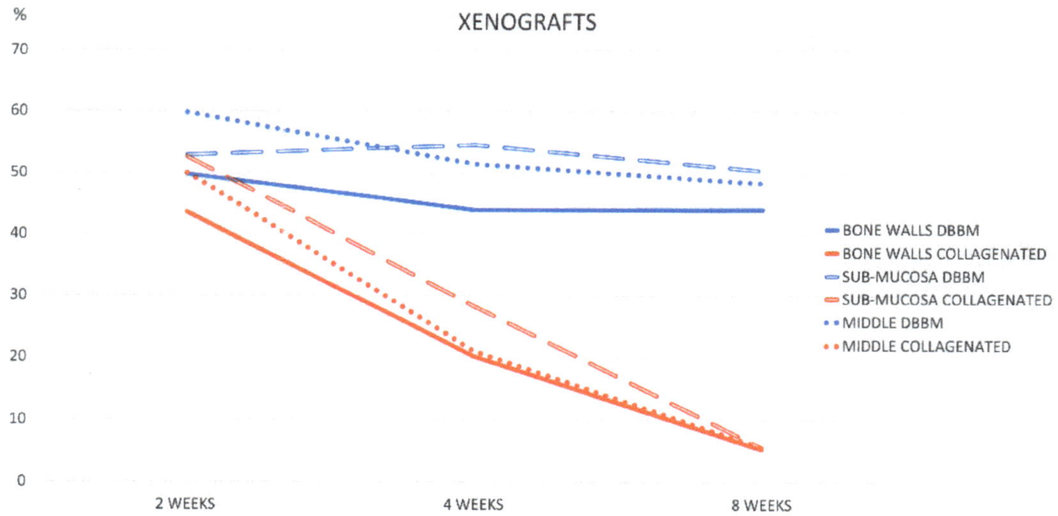

Figure 4. Graph illustrating percentage xenograft resorption in the various regions examined after 2, 4, and 8 weeks of healing.

After 8 weeks of healing, higher soft tissue content was present in the collagenated group than in the DBBM group, and the difference was statistically significant in all regions ($p < 0.001$; Figure 5A,B).

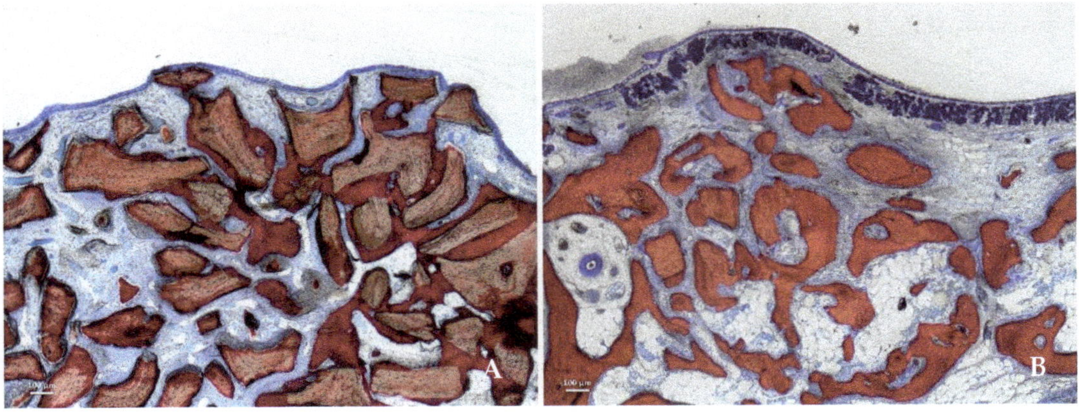

Figure 5. Photomicrographs of ground sections representing the healing after 8 weeks in the sub-mucosa region. Xenograft particles and soft tissues were present in both groups, but in higher percentages in the DBBM group. (**A**) DBBM group. (**B**) Collagenated group. Stevenel's blue and alizarin red stain.

The osteoclastic zones were limited in extension after 2 weeks of healing. A higher concentration of these zones was observed in the bone wall region of the collagenated group (Figure 6).

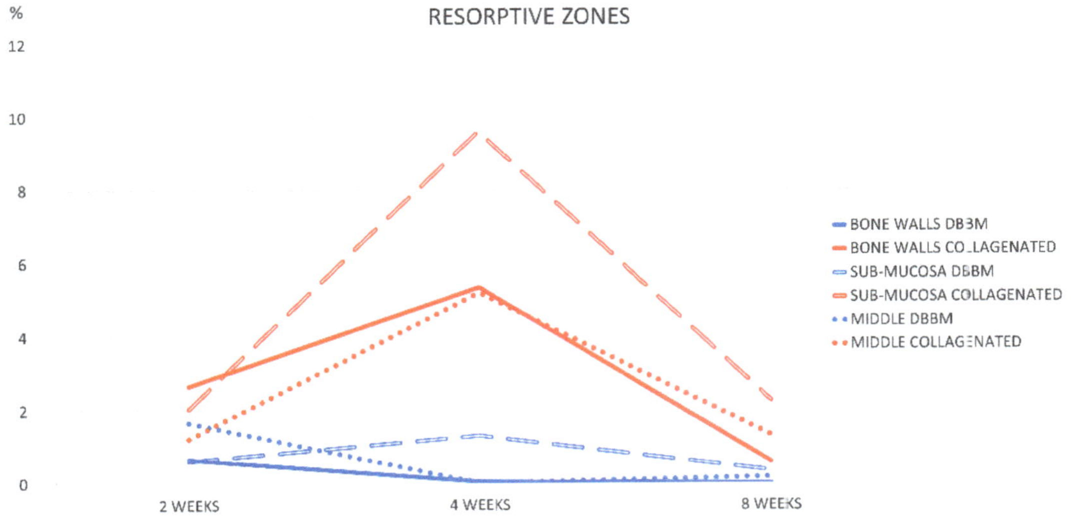

Figure 6. Graph illustrating the percentages of resorptive zones in the various regions examined after 2, 4, and 8 weeks of healing.

After 4 weeks, the osteoclastic zones remained stable in the DBBM group and increased in the collagenated group, reaching 9.6% in the sub-mucosa region (Figure 7A,B).

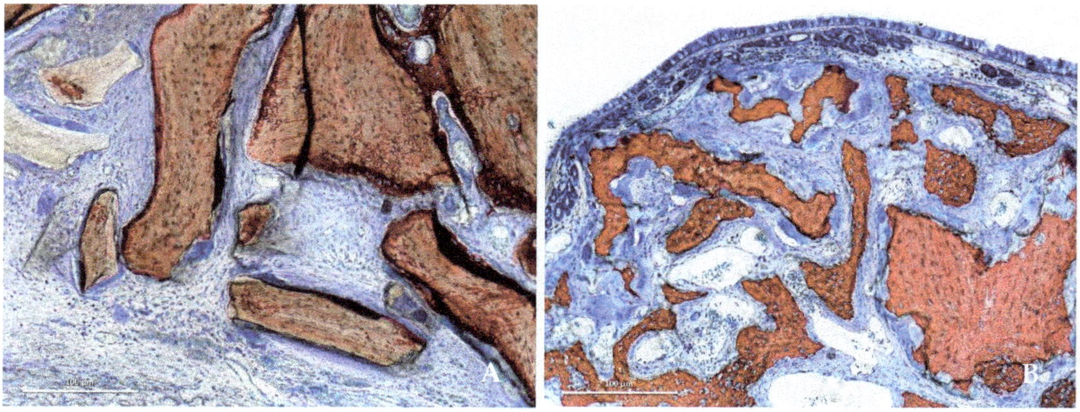

Figure 7. Photomicrographs of ground sections showing resorptive zones after 4 weeks. (**A**) DBBM group. (**B**) Collagenated group. Stevenel's blue and alizarin red stain.

In this period, the differences between the groups were statistically significant in all regions ($p < 0.05$). The osteoclastic zone content decreased after 8 weeks to values similar to or lower than those in the 2-week period.

After 2 weeks of healing, the range of the percentages of vessels was 2.1–4.2% in the DBBM group, and 2.3–2.6% in the collagenated group (Figure 8). In the DBBM group, after 8 weeks of healing, the vessel content was <1% in all regions. Conversely, the proportion of vessels in the collagenated group increased over time, reaching fractions 9–10% after 8 weeks ($p < 0.01$). In this period, the differences between the groups were statistically significant for all the regions evaluated.

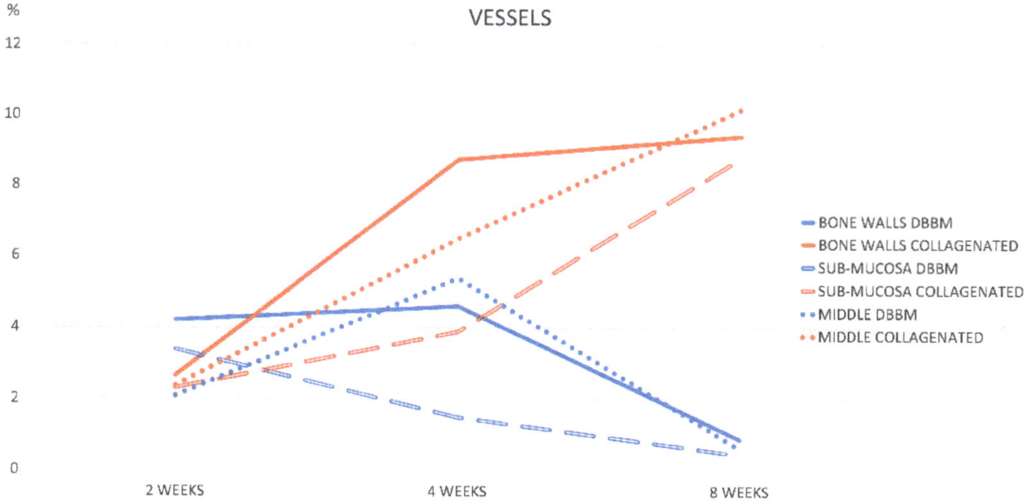

Figure 8. Graph illustrating the percentages of vessels in the various regions examined after 2, 4, and 8 weeks of healing.

After 2 weeks of healing, the inflammatory infiltrate was low (<1%) in all regions. The percentages remained low ($\leq 0.2\%$) in the DBBM group, increased in the collagenated group after 4 weeks, and ranged between 0.7–1.6% after 8 weeks.

4. Discussion

The dimensions of the elevated space decreased over time in both the DBBM and collagenated groups. However, between 2 and 8 weeks of healing, a greater reduction in the dimensions was observed in the collagenated group, with >50% of the total area, whereas in the DBBM group, the reduction was ~16%. In the DBBM group, dimensional variations were similar to those reported for humans [9]. In contrast, in the collagenated group, the shrinkage was higher than that reported in a clinical study in which the dimensional variations evaluated on CBCTs 9 months after sinus lifting ranged between 18% and 34% [33,34]. Nevertheless, it should be considered that in other clinical situations, such as ridge preservation after tooth extraction, no differences between collagenated and DBBM were found regarding the maintenance of soft tissue dimensions [35].

In the present study, after 2 weeks, new bone formation was observed in the sinus bone walls in both groups. This is in agreement with several other experimental studies that showed a similar amount of healing [36–40]. In an experiment on minipigs, sinus augmentation was performed using DBBM or an aqueous paste of synthetic nanoparticulated hydroxyapatite [40]. Healing was evaluated after 6 and 12 weeks in three regions that were progressively more distant from the sinus bone walls. The farther from the bone walls, the lower the bone volume observed, confirming that bone was formed from the sinus bone walls. In the present study, in the first period of healing, higher proportions of new bone were found closer to the bone walls than to the others. The influence of bone walls on bone formation was also shown in a clinical study [41]. Biopsies were collected from 18 patients 6 months after sinus floor elevation. A higher percentage of new bone was observed in the mesial sites, closer to the mesial bone wall, than in the distal sites. Moreover, a strong negative correlation between bone percentage and sinus width was found. This, in turn, means that the greater the distance between the lateral and palatal walls, the lower is bone formation.

In the sub-mucosa region, new bone reached percentages similar to the other regions only in the last period assessed, that is, after 8 weeks of healing. This might be because the sinus mucosa did not participate in bone formation in the earliest period of healing, despite its potential to form bone [42,43]. This finding is in agreement with the data from other experimental studies [8,17,36,37,44]. After 8 weeks, higher percentages of new bone were found in the collagenated group than in the DBBM group in the bone wall region. The percentages of new bone found after 8 weeks of healing in the present study were similar to those reported in human histological studies. In a clinical study [45], 48 sinuses were elevated with either DBBM or synthetic graft. Biopsies were collected 6–8 months after surgery. The percentages of mineralized bone were 19.8% and 21.6%, respectively. In another clinical study [46], bovine cancellous bone obtained at a high temperature (>1200 °C) was used for sinus floor elevation. Mini-implants were installed 6 months after healing and were retrieved after another 3 months. Histological examination of the biopsies revealed 21–23% mineralized bone. In sinuses elevated with a collagenated xenograft similar to that used in the present study, and with mini-implants again installed after 6 months of healing and retrieved after 3 months, percentages of mineralized bone ranging between ~32–39% were found [47–49]. In another clinical study [50], biopsies collected through a crestal approach revealed 40.1% mineralized bone. The percentages of mineralized bone reported in the above-mentioned studies were similar to those found close to the bone walls in the present study. This outcome might be related to the region from which the biopsies were collected in clinical studies, that is, the bone crest. This region contained the residual bone crest, and it was close to the floor and sinus walls, a condition that resembles that in the experiment in rabbits.

This aspect has important clinical relevance, because the region close to the base of the sinus is where the implants will be installed. Therefore, efforts should be made to preserve this region, for example, by making the access windows as small as possible and far away from the sinus floor to increase the source of new bone from the lateral sinus wall [33,34].

The higher percentages of new bone found in the collagenated groups compared to the DBBM group in the bone wall region matches the lower content of xenograft residual in the collagenated group compared to the DBBM group; in fact, a higher resorption of biomaterial was observed in the former than in the latter group. After 8 weeks of healing, >40% of DBBM and <6% of collagenated corticocancellous porcine bone were still present in all regions. This outcome is in agreement with other studies that showed a higher maintenance of DBBM granules over time compared to collagenated xenografts [26,30]. It must be considered that the presence of the biomaterial inside the bone (composite bone) might interfere with the osseointegration of implants installed both in a simultaneous [51] or delayed fashion [46–49]. In a rabbit experiment [51], implants were placed simultaneously with sinus augmentation performed with a non-collagenated xenograft. Granules of the biomaterial were found in contact with the implant surface in percentages ranging between 1.7% and 3.6%. Biopsies of mini-implants installed 6 months after sinus floor elevation were collected from humans after a further 3 months of healing [46–49]. The implant surface was found to be in contact with granules of biomaterials that hampered the growth of newly formed bone on the implant surface. The percentages presented a large range when a collagenated xenograft was used (0.6 to 15.9%) [47–49] but reached 25.1% to 27.2% when a non-collagenated xenograft was used [46].

Higher amounts of soft tissue were found in the collagenated group than in the DBBM group, with a statistically significant difference after 8 weeks in all regions. This tissue was mainly composed of marrow spaces in the bone wall region and, in the middle and submucosa regions, by provisional matrix that was not yet completely involved in the remodeling process (Figure 5A, B).

In the present study, the percentage of osteoclastic zones was higher in the collagenated group than in the DBBM group, which is in accordance with the different levels of resorption of the biomaterial in the two groups. The highest percentages of resorptive zones were observed after 4 weeks of healing in the collagenated group, especially in the sub-mucosa region. Considering the tendency of the sinus to regain its original size [6–8,28,33,34], the higher rate of graft resorption in the collagenated group than in the DBBM group contributed to the greater loss of dimension of the elevated space observed in the former than in the latter.

In the DBBM group, the proportion of vessels decreased between 2 and 8 weeks, while in the collagenated group, the content in vessels continuously increased, which might influence healing. Blood provides nutrients, gas exchange, immune system cells, mesenchymal stem cells, and growth factors [52–54]. It has been shown that various biomaterials, including xenogeneic grafts, were associated pro-angiogenic effect and an increased expression of various grow factors [53].

The higher content of inflammatory infiltrates, although limited in dimension, in the collagenated group than in the DBBM group was related to higher xenograft resorption in the former than in the latter. The present study did not include gene expression analysis. Nevertheless, it should be considered that lymphocytes play an important role in the immune system and determine the specificity of the immune response following infections [55]. Damage to the bone triggers the immune response and the activation of neutrophils, mast cells, monocytes, and macrophages [56]. It has been shown in mice that the absence of lymphocytes B and T compromises bone regeneration in fractures [57]. In an experiment in rabbit femurs [58], gene expression analysis was performed around implants and in sham sites. Upregulation of the immune system and downregulation of bone resorption markers were observed at the implant sites compared with the sham sites.

The present study also demonstrated differences in bone formation and graft resorption between collagenated and non-collagenated xenografts. DBBM presented a very low rate of resorption, and the bone was formed in close contact with the granules of the graft that were interconnected by a bridge of newly formed bone. Bone formation in rabbits was similar to that described in a previous study [26]. In the early stages of healing, the DBBM granules were surrounded by dense tissue, whereas loose tissue filled the spaces among the granules. Over time, the dense tissue was replaced by new bone in contact with the graft surface, whereas the loose tissue was replaced by primitive marrow spaces. The collagenated graft showed mixed characteristics, including bone formation around graft granules, similar to DBBM, and sites with several osteoclastic zones engendering an extensive graft resorption similar to that shown for autogenous bone [59].

The surgical procedures performed in the experiments included in the present study were conducted under sterile conditions, although such conditions could not be completely obtained in the oral environment. Nevertheless, it might help to reduce the possible contamination of instruments, biomaterials, and the operative field [60,61]. Technologies and procedures may be implemented with the aim of better sterilizing the operating field and improving healing [62].

Further studies that include longer healing periods and the evaluation of different biomaterials may be performed. Synthetic biomaterials should be studied further. In fact, the increased demand for bone regeneration, as well as the morbidity of the donor sites for autogenous bone, the possible presence of prions within xenogenous grafts, and patients' religious beliefs have increased interest in studying synthetic (alloplastic) materials. Among them, biphasic β-tricalcium phosphate/hydroxyapatite (β-TCP/HA) and bioactive glasses presented interesting results. It was recently showed that at β-TCP/HA graft new bone was not only surrounding the particles but also spreading inward them forming a structure called by the authors "interpenetrating bone network" [63]. Bioactive glasses have also been widely used in maxillofacial surgery, providing interesting results in bone formation, even better than those obtained with other synthetic biomaterials [64].

In the present study, two experiments were selected for homogeneity in methodology. The experiments were performed in the same laboratory by the same research group. The healing periods were similar, and histological slides were prepared in the same laboratory using the same methods and analyzed by the same expert assessor. To increase homogeneity, only one sinus was used for each experiment, that is, with the small granules in the DBBM experiment (DBBM group) and that with the access window prepared with drills in the collagenated porcine xenograft experiment (collagenated group). In turn, this means that both access windows were prepared using drills.

Despite efforts to increase the homogeneity of the methodology, various limitations must be mentioned. The dimensions of the two access windows were slightly different: 3.5 mm × 6 mm (21 mm^2) in the non-collagenated group and 4 mm × 4 mm (16 mm^2) in the DBBM group. However, access windows of 18 mm^2 and 30 mm^2 did not show differences in bone formation in a similar study in rabbits [65]. The different collagen membranes that covered the osteotomies in the two experiments might be considered another limitation of the study. However, it is unlikely that this difference affected the healing of the internal sides of the elevated spaces. This has been substantiated by both animal [66] and human [46] studies. The model used should also be considered as a limitation. In fact, when interpreting the results of the present study to humans, the different speeds of healing are an important topic to be considered [67]. Other important limitations are the short period of evaluation and the selection of material from previous studies, even though efforts were made to maintain a high homogeneity between the two studies.

5. Conclusions

The present study showed that both xenografts allowed new bone formation. In comparison with a non-collagenated xenograft, a collagenated xenograft underwent higher resorption, resulting in greater shrinkage of the elevated space after sinus lifting and a higher content of new bone in the regions close to the bone walls. However, considering the limitations of the present study, the results should be interpreted with caution.

Author Contributions: Conceptualization, Y.M. and S.B.; methodology, T.I. (Takahisa Iida); validation, T.I. (Takayuki Izutani), Y.T. and T.I. (Takahisa Iida); formal analysis, Y.M. and Y.N.; investigation, S.P.X.; writing—review and editing, Y.M., T.I. (Takahisa Iida), S.P.X. and S.B.; visualization, Y.M., T.I. (Takayuki Izutani), Y.T. and T.I. (Takahisa Iida); supervision, S.P.X. and S.B.; project administration, Y.N. and S.P.X. All authors have read and agreed to the published version of the manuscript.

Funding: Research was funded by ARDEC Academy, Rimini, Italy.

Institutional Review Board Statement: The protocols for both experiments were approved by the Ethical Committee of the Faculty of Dentistry in Ribeirão Preto of the University of São Paulo (USP, São Paulo, Brazil). The protocol numbers were 2017.1.278.58.9 [30] and 2015.1.834.58.7 [31] for the non-collagenated (DBBM) and collagenated xenografts, respectively.

Data Availability Statement: The data are available on reasonable request.

Acknowledgments: We acknowledge the support of Erick Ricardo Silva in carrying out the experiment, and Sebastião Bianco (Faculty of Dentistry of Ribeirão Preto, University of São Paulo, Brazil) for preparation of the histology. We also thank Karol Alí Apaza Alccayhuaman (Competence Center of Oral Biology, Medical University of Vienna, Austria) for the histological measurements. The scientific contribution of Daniele Botticelli (ARDEC Academy, Rimini, Italy) is highly appreciated.

Conflicts of Interest: The authors declare no conflict of interest.

References

1. Buser, D.; Dula, K.; Hess, D.; Hirt, H.P.; Belser, U.C. Localized ridge augmentation with autografts and barrier membranes. *Periodontology 2000* **1999**, *19*, 151–163. [CrossRef]
2. Pjetursson, B.E.; Tan, W.C.; Zwahlen, M.; Lang, N.P. A systematic review of the success of sinus floor elevation and survival of implants inserted in combination with sinus floor elevation. *J. Clin. Periodontol.* **2008**, *35* (Suppl. S8), 216–240. [CrossRef]
3. Lundgren, S.; Cricchio, G.; Hallman, M.; Jungner, M.; Rasmusson, L.; Sennerby, L. Sinus floor elevation procedures to enable implant placement and integration: Techniques, biological aspects and clinical outcomes. *Periodontology 2000* **2017**, *73*, 103–120. [CrossRef]
4. Kawakami, S.; Botticelli, D.; Nakajima, Y.; Sakuma, S.; Baba, S. Anatomical analyses for maxillary sinus floor augmentation with a lateral approach: A cone beam computed tomography study. *Ann. Anat.* **2019**, *226*, 29–34. [CrossRef]
5. Testori, T.; Weinstein, T.; Taschieri, S.; Wallace, S.S. Risk factors in lateral window sinus elevation surgery. *Periodontology 2000* **2019**, *81*, 91–123. [CrossRef]
6. Asai, S.; Shimizu, Y.; Ooya, K. Maxillary sinus augmentation model in rabbits: Effect of occluded nasal ostium on new bone formation. *Clin. Oral Implant. Res.* **2002**, *13*, 405–409. [CrossRef]
7. Xu, H.; Shimizu, Y.; Asai, S.; Ooya, K. Grafting of deproteinized bone particles inhibits bone resorption after maxillary sinus floor elevation. *Clin. Oral Implant. Res.* **2004**, *15*, 126–133. [CrossRef]
8. Scala, A.; Botticelli, D.; Faeda, R.S.; Garcia Rangel, I., Jr.; Américo de Oliveira, J.; Lang, N.P. Lack of influence of the Schneiderian membrane in forming new bone apical to implants simultaneously installed with sinus floor elevation: An experimental study in monkeys. *Clin. Oral Implant. Res.* **2012**, *23*, 175–181. [CrossRef]
9. Shanbhag, S.; Shanbhag, V.; Stavropoulos, A. Volume changes of maxillary sinus augmentations over time: A systematic review. *Int. J. Oral Maxillofac. Implant.* **2014**, *29*, 881–892. [CrossRef]
10. Canellas, J.V.D.S.; Drugos, L.; Ritto, F.G.; Fischer, R.G.; Medeiros, P.J.D. Xenograft materials in maxillary sinus floor elevation surgery: A systematic review with network meta-analyses. *Br. J. Oral Maxillofac. Surg.* **2021**, *59*, 742–751. [CrossRef]
11. Trimmel, B.; Gede, N.; Hegyi, P.; Szakács, Z.; Mezey, G.A.; Varga, E.; Kivovics, M.; Hanák, L.; Rumbus, Z.; Szabó, G. Relative performance of various biomaterials used for maxillary sinus augmentation: A Bayesian network meta-analysis. *Clin. Oral Implant. Res.* **2021**, *32*, 135–153. [CrossRef] [PubMed]
12. Corbella, S.; Taschieri, S.; Weinstein, R.; Del Fabbro, M. Histomorphometric outcomes after lateral sinus floor elevation procedure: A systematic review of the literature and meta-analysis. *Clin. Oral Implant. Res.* **2016**, *27*, 1106–1122. [CrossRef] [PubMed]
13. Lie, S.A.N.; Claessen, R.M.M.A.; Leung, C.A.W.; Merten, H.A.; Kessler, P.A.W.H. Non-grafted versus grafted sinus lift procedures for implantation in the atrophic maxilla: A systematic review and meta-analysis of randomized controlled trials. *Int. J. Oral Maxillofac. Surg.* **2022**, *51*, 122–132. [CrossRef] [PubMed]
14. Ellegaard, B.; Kølsen-Petersen, J.; Baelum, V. Implant therapy involving maxillary sinus lift in periodontally compromised patients. *Clin. Oral Implant. Res.* **1997**, *8*, 305–315. [CrossRef]
15. Lundgren, S.; Andersson, S.; Gualini, F.; Sennerby, L. Bone reformation with sinus membrane elevation: A new surgical technique for maxillary sinus floor augmentation. *Clin. Implant Dent. Relat. Res.* **2004**, *6*, 165–173. [CrossRef]
16. Cricchio, G.; Sennerby, L.; Lundgren, S. Sinus bone formation and implant survival after sinus membrane elevation and implant placement: A 1- to 6-year follow-up study. *Clin. Oral Implant. Res.* **2011**, *22*, 1200–1212. [CrossRef]
17. Scala, A.; Botticelli, D.; Rangel, I.G., Jr.; de Oliveira, J.A.; Okamoto, R.; Lang, N.P. Early healing after elevation of the maxillary sinus floor applying a lateral access: A histological study in monkeys. *Clin. Oral Implant. Res.* **2010**, *21*, 1320–1326. [CrossRef]

18. Jungner, M.; Cricchio, G.; Salata, L.A.; Sennerby, L.; Lundqvist, C.; Hultcrantz, M.; Lundgren, S. On the Early Mechanisms of Bone Formation after Maxillary Sinus Membrane Elevation: An Experimental Histological and Immunohistochemical Study. *Clin. Implant Dent. Relat. Res.* **2015**, *17*, 1092–1102. [CrossRef]
19. Ellegaard, B.; Baelum, V.; Kølsen-Petersen, J. Non-grafted sinus implants in periodontally compromised patients: a time-to-event analysis. *Clin. Oral Implants Res.* **2006**, *17*, 156–164. [CrossRef]
20. Starch-Jensen, T.; Schou, S. Maxillary Sinus Membrane Elevation with Simultaneous Installation of Implants without the Use of a Graft Material: A Systematic Review. *Implant Dent.* **2017**, *26*, 621–633. [CrossRef]
21. Moraschini, V.; Uzeda, M.G.; Sartoretto, S.C.; Calasans-Maia, M.D. Maxillary sinus floor elevation with simultaneous implant placement without grafting materials: A systematic review and meta-analysis. *Int. J. Oral Maxillofac. Surg.* **2017**, *46*, 636–647. [CrossRef] [PubMed]
22. Cricchio, G.; Palma, V.C.; Faria, P.E.; de Oliveira, J.A.; Lundgren, S.; Sennerby, L.; Salata, L.A. Histological findings following the use of a space-making device for bone reformation and implant integration in the maxillary sinus of primates. *Clin. Implant Dent. Relat. Res.* **2009**, *11* (Suppl. S1), e14–e22. [CrossRef] [PubMed]
23. Cricchio, G.; Palma, V.C.; Faria, P.E.; de Olivera, J.A.; Lundgren, S.; Sennerby, L.; Salata, L.A. Histological outcomes on the development of new space-making devices for maxillary sinus floor augmentation. *Clin. Implant Dent. Relat. Res.* **2011**, *13*, 224–230. [CrossRef] [PubMed]
24. Johansson, L.Å.; Isaksson, S.; Adolfsson, E.; Lindh, C.; Sennerby, L. Bone regeneration using a hollow hydroxyapatite space-maintaining device for maxillary sinus floor augmentation—A clinical pilot study. *Clin. Implant Dent. Relat. Res.* **2012**, *14*, 575–584. [CrossRef] [PubMed]
25. Schweikert, M.; Botticelli, D.; de Oliveira, J.A.; Scala, A.; Salata, L.A.; Lang, N.P. Use of a titanium device in lateral sinus floor elevation: An experimental study in monkeys. *Clin. Oral Implant. Res.* **2012**, *23*, 100–105. [CrossRef]
26. Caneva, M.; Lang, N.P.; Garcia Rangel, I.J.; Ferreira, S.; Caneva, M.; De Santis, E.; Botticelli, D. Sinus mucosa elevation using Bio-Oss(®) or Gingistat(®) collagen sponge: An experimental study in rabbits. *Clin. Oral Implant. Res.* **2017**, *28*, e21–e30. [CrossRef]
27. De Santis, E.; Lang, N.P.; Ferreira, S.; Rangel Garcia, I., Jr.; Caneva, M.; Botticelli, D. Healing at implants installed concurrently to maxillary sinus floor elevation with Bio-Oss(®) or autologous bone grafts. A histo-morphometric study in rabbits. *Clin. Oral Implant. Res.* **2017**, *28*, 503–511. [CrossRef]
28. Iida, T.; Carneiro Martins Neto, E.; Botticelli, D.; Apaza Alccayhuaman, K.A.; Lang, N.P.; Xavier, S.P. Influence of a collagen membrane positioned subjacent the sinus mucosa following the elevation of the maxillary sinus. A histomorphometric study in rabbits. *Clin. Oral Implant. Res.* **2017**, *28*, 1567–1576. [CrossRef]
29. Iida, T.; Silva, E.R.; Lang, N.P.; Apaza Alccayhuaman, K.A.; Botticelli, D.; Xavier, S.P. Histological and micro-computed tomography evaluations of newly formed bone after maxillary sinus augmentation using a xenograft with similar density and mineral content of bone: An experimental study in rabbits. *Clin. Exp. Dent. Res.* **2018**, *4*, 284–290. [CrossRef]
30. Iida, T.; Baba, S.; Botticelli, D.; Masuda, K.; Xavier, S.P. Comparison of histomorphometry and microCT after sinus augmentation using xenografts of different particle sizes in rabbits. *Oral. Maxillofac. Surg.* **2020**, *24*, 57–64. [CrossRef]
31. Masuda, K.; Silva, E.R.; Botticelli, D.; Apaza Alccayhuaman, K.A.; Xavier, S.P. Antrostomy Preparation for Maxillary Sinus Floor Augmentation Using Drills or a Sonic Instrument: A Microcomputed Tomography and Histomorphometric Study in Rabbits. *Int. J. Oral Maxillofac. Implant.* **2019**, *34*, 819–827. [CrossRef]
32. Figueiredo, M.; Henriques, J.; Martins, G.; Guerra, F.; Judas, F.; Figueiredo, H. Physicochemical characterization of biomaterials commonly used in dentistry as bone substitutes—Comparison with human bone. *J. Biomed. Mater. Res. B Appl. Biomater.* **2010**, *92*, 409–419. [CrossRef] [PubMed]
33. Kawakami, S.; Lang, N.P.; Ferri, M.; Apaza Alccayhuaman, K.A.; Botticelli, D. Influence of the height of the antrostomy in sinus floor elevation assessed by cone beam computed tomography—A randomized clinical trial. *Int. J. Oral Maxillofac. Implant.* **2019**, *34*, 223–232. [CrossRef] [PubMed]
34. Kawakami, S.; Lang, N.P.; Iida, T.; Ferri, M.; Apaza Alccayhuaman, K.A.; Botticelli, D. Influence of the position of the antrostomy in sinus floor elevation assessed with cone-beam computed tomography: A randomized clinical trial. *J. Investig. Clin. Dent.* **2018**, *9*, e12362. [CrossRef] [PubMed]
35. Sapata, V.M.; Llanos, A.H.; Cesar Neto, J.B.; Jung, R.E.; Thoma, D.S.; Hämmerle, C.H.F.; Pannuti, C.M.; Romito, G.A. Deproteinized bovine bone mineral is non-inferior to deproteinized bovine bone mineral with 10% collagen in maintaining the soft tissue contour post-extraction: A randomized trial. *Clin. Oral Implant. Res.* **2020**, *31*, 294–301. [CrossRef]
36. Scala, A.; Lang, N.P.; Velez, J.U.; Favero, R.; Bengazi, F.; Botticelli, D. Effects of a collagen membrane positioned between augmentation material and the sinus mucosa in the elevation of the maxillary sinus floor. An experimental study in sheep. *Clin. Oral Implant. Res.* **2016**, *27*, 1454–1461. [CrossRef]
37. Favero, V.; Lang, N.P.; Canullo, L.; Urbizo Velez, J.; Bengazi, F.; Botticelli, D. Sinus floor elevation outcomes following perforation of the Schneiderian membrane. An experimental study in sheep. *Clin. Oral Implant. Res.* **2016**, *27*, 233–240. [CrossRef]
38. Omori, Y.; Ricardo Silva, E.; Botticelli, D.; Apaza Alccayhuaman, K.A.; Lang, N.P.; Xavier, S.P. Reposition of the bone plate over the antrostomy in maxillary sinus augmentation: A histomorphometric study in rabbits. *Clin. Oral Implant. Res.* **2018**, *29*, 821–834. [CrossRef]
39. Godoy, E.P.; Alccayhuaman, K.A.A.; Botticelli, D.; Amaroli, A.; Balan, V.F.; Silva, E.R.; Xavier, S.P. Osteoconductivity of Bovine Xenograft Granules of Different Sizes in Sinus Lift: A Histomorphometric Study in Rabbits. *Dent. J.* **2021**, *9*, 61. [CrossRef]

40. Busenlechner, D.; Huber, C.D.; Vasak, C.; Dobsak, A.; Gruber, R.; Watzek, G. Sinus augmentation analysis revised: The gradient of graft consolidation. *Clin. Oral Implant. Res.* **2009**, *20*, 1078–1083. [CrossRef] [PubMed]
41. Stacchi, C.; Rapani, A.; Lombardi, T.; Bernardello, F.; Nicolin, V.; Berton, F. Does new bone formation vary in different sites within the same maxillary sinus after lateral augmentation? A prospective histomorphometric study. *Clin. Oral Implant. Res.* **2022**, *33*, 322–332. [CrossRef] [PubMed]
42. Gruber, R.; Kandler, B.; Fuerst, G.; Fisher, M.B.; Watzek, G. Porcine sinus mucosa holds cells that respond to bone morphogenetic protein (BMP)-6 and BMP-7 with increased osteogenic differentiation in vitro. *Clin. Oral Implant. Res.* **2004**, *15*, 575–580. [CrossRef] [PubMed]
43. Srouji, S.; Ben-David, D.; Lotan, R.; Riminucci, M.; Livne, E.; Bianco, P. The innate osteogenic potential of the maxillary sinus (Schneiderian) membrane: An ectopic tissue transplant model simulating sinus lifting. *Int. J. Oral Maxillofac. Surg.* **2010**, *39*, 793–801. [CrossRef] [PubMed]
44. Costa, M.M.; Botticelli, D.; Moses, O.; Omori, Y.; Fujiwara, S.; Silva, E.R.; Xavier, S.P. Maxillary Sinus Augmentation Using Ceramic Alloplastic Granules or Paste: An Experimental Study in Rabbits. *Dent. J.* **2021**, *9*, 65. [CrossRef] [PubMed]
45. Cordaro, L.; Bosshardt, D.D.; Palattella, P.; Rao, W.; Serino, G.; Chiapasco, M. Maxillary sinus grafting with Bio-Oss or Straumann Bone Ceramic: Histomorphometric results from a randomized controlled multicenter clinical trial. *Clin. Oral Implant. Res.* **2008**, *19*, 796–803. [CrossRef] [PubMed]
46. Kotsu, M.; Apaza Alccayhuaman, K.A.; Ferri, M.; Iezzi, G.; Piattelli, A.; Fortich Mesa, N.; Botticelli, D. Osseointegration at Implants Installed in Composite Bone: A Randomized Clinical Trial on Sinus Floor Elevation. *J. Funct. Biomater.* **2022**, *13*, 22. [CrossRef]
47. Hirota, A.; Iezzi, G.; Piattelli, A.; Ferri, M.; Tanaka, K.; Apaza Alccayhuaman, K.A.; Botticelli, D. Influence of the position of the antrostomy in sinus floor elevation on the healing of mini-implants: A randomized clinical trial. *Oral Maxillofac. Surg.* **2020**, *24*, 299–308. [CrossRef]
48. Imai, H.; Iezzi, G.; Piattelli, A.; Ferri, M.; Apaza Alccayhuaman, K.A.; Botticelli, D. Influence of the Dimensions of the Antrostomy on Osseointegration of Mini-implants Placed in the Grafted Region After Sinus Floor Elevation: A Randomized Clinical Trial. *Int. J. Oral Maxillofac. Implant.* **2020**, *35*, 591–598. [CrossRef]
49. Morimoto, A.; Kobayashi, N.; Ferri, M.; Iezzi, G.; Piattelli, A.; Fortich Mesa, N.; Botticelli, D. Influence on Implant Bone Healing of a Collagen Membrane Placed Subjacent the Sinus Mucosa-A Randomized Clinical Trial on Sinus Floor Elevation. *Dent. J.* **2022**, *10*, 105. [CrossRef]
50. Tanaka, K.; Iezzi, G.; Piattelli, A.; Ferri, M.; Mesa, N.F.; Apaza Alccayhuaman, K.A.; Botticelli, D. Sinus Floor Elevation and Antrostomy Healing: A Histomorphometric Clinical Study in Humans. *Implant Dent.* **2019**, *28*, 537–542. [CrossRef]
51. Masuda, K.; Silva, E.R.; Apaza Alccayhuaman, K.A.; Botticelli, D.; Xavier, S.P. Histologic and Micro-CT Analyses at Implants Placed Immediately After Maxillary Sinus Elevation Using Large or Small Xenograft Granules: An Experimental Study in Rabbits. *Int. J. Oral Maxillofac. Implant.* **2020**, *35*, 739–748. [CrossRef] [PubMed]
52. Sivaraj, K.K.; Adams, R.H. Blood vessel formation and function in bone. *Development* **2016**, *143*, 2706–2715. [CrossRef] [PubMed]
53. Saghiri, M.A.; Asatourian, A.; Garcia-Godoy, F.; Sheibani, N. The role of angiogenesis in implant dentistry part II: The effect of bone-grafting and barrier membrane materials on angiogenesis. *Med. Oral Patol. Oral Cir. Buccal* **2016**, *21*, e526–e537. [CrossRef] [PubMed]
54. Liu, J.; Kerns, D.G. Mechanisms of guided bone regeneration: A review. *Open Dent. J.* **2014**, *8*, 56–65. [CrossRef] [PubMed]
55. Nesti, M.; Carli, E.; Giaquinto, C.; Rampon, O.; Nastasio, S.; Giuca, M.R. Correlation between viral load, plasma levels of CD4–CD8 T lymphocytes and AIDS-related oral diseases: A multicenter study on 30 HIV+ children in the HAART era. *J. Biol. Regul. Homeost. Agents* **2012**, *26*, 527–537.
56. Yang, N.; Liu, Y. The Role of the Immune Microenvironment in Bone Regeneration. *Int. J. Med. Sci.* **2021**, *18*, 3697–3707. [CrossRef]
57. El Khassawna, T.; Serra, A.; Bucher, C.H.; Petersen, A.; Schlundt, C.; Könnecke, I.; Malhan, D.; Wendler, S.; Schell, H.; Volk, H.D.; et al. T Lymphocytes Influence the Mineralization Process of Bone. *Front. Immunol.* **2017**, *8*, 562. [CrossRef]
58. Trindade, R.; Albrektsson, T.; Galli, S.; Prgomet, Z.; Tengvall, P.; Wennerberg, A. Osseointegration and foreign body reaction: Titanium implants activate the immune system and suppress bone resorption during the first 4 weeks after implantation. *Clin. Implant Dent. Relat. Res.* **2018**, *20*, 82–91. [CrossRef]
59. Scala, A.; Lang, N.P.; de Carvalho Cardoso, L.; Pantani, F.; Schweikert, M.; Botticelli, D. Sequential healing of the elevated sinus floor after applying autologous bone grafting: An experimental study in minipigs. *Clin. Oral Implant. Res.* **2015**, *26*, 419–425. [CrossRef] [PubMed]
60. Scharf, D.R.; Tarnow, D.P. Success rates of osseointegration for implants placed under sterile versus clean conditions. *J. Periodontol.* **1993**, *64*, 954–956. [CrossRef] [PubMed]
61. Friberg, B. Sterile operating conditions for the placement of intraoral implants. *J. Oral Maxillofac. Surg.* **1996**, *54*, 1334–1336. [CrossRef]
62. Valenti, C.; Pagano, S.; Bozza, S.; Ciurnella, E.; Lomurno, G.; Capobianco, B.; Coniglio, M.; Cianetti, S.; Marinucci, L. Use of the Er:YAG Laser in Conservative Dentistry: Evaluation of the Microbial Population in Carious Lesions. *Materials* **2021**, *14*, 2387. [CrossRef] [PubMed]
63. Tanaka, K.; Botticelli, D.; Canullo, L.; Baba, S.; Xavier, S.P. New bone ingrowth into β-TCP/HA graft activated with argon plasma: A histomorphometric study on sinus lifting in rabbits. *Int. J. Implant Dent.* **2020**, *6*, 36. [CrossRef] [PubMed]
64. Skallevold, H.E.; Rokaya, D.; Khurshid, Z.; Zafar, M.S. Bioactive Glass Applications in Dentistry. *Int. J. Mol. Sci.* **2019**, *20*, 5960. [CrossRef]

65. Perini, A.; Viña-Almunia, J.; Carda, C.; Martín de Llano, J.J.; Botticelli, D.; Peñarrocha-Diago, M. Influence of the Use of a Collagen Membrane Placed on the Bone Window after Sinus Floor Augmentation-An Experimental Study in Rabbits. *Dent. J.* **2021**, *9*, 131. [CrossRef]
66. Scala, A.; Viña-Almunia, J.; Carda, C.; Martín de Llano, J.J.; Soto-Peñaloza, D.; Peñarrocha-Diago, M.; Peñarrocha-Diago, M.; Botticelli, D. Sequential healing of the elevated sinus floor with different size of antrostomy: A histomorphometric study in rabbits. *Oral Maxillofac. Surg.* **2020**, *24*, 403–410. [CrossRef] [PubMed]
67. Botticelli, D.; Lang, N.P. Dynamics of osseointegration in various human and animal models—A comparative analysis. *Clin. Oral Implant. Res.* **2017**, *28*, 742–748. [CrossRef] [PubMed]

Journal of *Functional Biomaterials*

MDPI

Article

Sinus Mucosal Damage Triggered by Synthetic or Xenogeneic Bone Substitutes: A Histological Analysis in Rabbits

Yuki Omori [1], Daniele Botticelli [2,*], Stefano Migani [2], Vitor Ferreira Balan [3], Eduardo Pires Godoy [4] and Samuel Porfirio Xavier [3]

1 Department of Oral Implantology, Osaka Dental University, 8-1 Kuzuhahanazonocho, Osaka 573-1121, Japan
2 ARDEC Academy, Viale Giovanni Pascoli 67, 47923 Rimini, Italy
3 Department of Oral and Maxillofacial Surgery and Periodontology, Faculty of Dentistry of Ribeirão Preto, University of São Paulo, Av. do Café-Subsetor Oeste-11 (N-11), Ribeirão Preto 14040-904, Brazil
4 Department of Oral Biology, Faculty of Dentistry of Ribeirão Preto, University of São Paulo, Ribeirão Preto 14040-904, Brazil
* Correspondence: daniele.botticelli@gmail.com

Abstract: Background: It has been shown in rabbit models that the sinus mucosa in contact with graft particles might experience a progressive thinning and perforations. The phenomenon depends on the graft used. Hence, the aim of the present study was to compare the damaging effects of a synthetic of a xenogeneic graft. Methods: Forty New Zealand rabbits received a bilateral sinus elevation. Both sinuses of twenty rabbits were grafted with a biphasic 60% hydroxyapatite and 40% β-tricalcium phosphate while the other twenty received a deproteinized bovine bone mineral graft. Thinned sites (<40 μm) and perforations on the mucosa in contact with graft particles were evaluated after 2 and 10 weeks (ten animals each period). The width of the pseudostratified epithelium was also measured as control. Results: After 2 weeks of healing, 61 thinned sites were detected in the Synthetic group and 49 in the Xenogeneic group. After 10 weeks, the number of thinned mucosae increased to 79 sites in the Synthetic group ($p = 0.222$ between periods), and to 114 sites in the Xenogeneic group ($p = 0.030$ between groups; $p = 0.001$ between periods). Perforations were few in the 2-week period, two in two sinuses out of 20 in the Synthetic group, and four in two sinuses out of 20 in the Xenogeneic group ($p = 0.721$). In the 10-week period, the perforations increased to eight in the Synthetic group, distributed in six sinuses out of 20, and to sixteen in the Xenogeneic group, distributed in 11 sinuses out of 20 ($p = 0.082$). The pseudostratified epithelium presented a reduced width at the thinned sites. Conclusions: The contact with synthetic or xenogeneic grafts will induce thinning and possible perforations of the sinus mucosa. This effect will increase over time, and it is stronger at the xenogeneic than the synthetic graft.

Keywords: animal study; sinus floor elevation; bone healing; Schneiderian membrane; histology; sinus mucosa perforation

Citation: Omori, Y.; Botticelli, D.; Migani, S.; Ferreira Balan, V.; Pires Godoy, E.; Xavier, S.P. Sinus Mucosal Damage Triggered by Synthetic or Xenogeneic Bone Substitutes: A Histological Analysis in Rabbits. *J. Funct. Biomater.* **2022**, *13*, 257. https://doi.org/10.3390/jfb13040257

Academic Editor: Xinping Zhang

Received: 2 November 2022
Accepted: 17 November 2022
Published: 19 November 2022
Corrected: 6 May 2024

1. Introduction

The lateral approach for sinus floor elevation has been well documented in literature, resulting in a high success rate [1,2]. Like any surgical technique, this can also present complications, the most frequent being the perforation of the sinus mucosa [3,4]. The perforations might occur during sinus mucosa elevation, grafting procedure, or implant installation [5–7]. Perforations might heal spontaneously [6], and it was not considered a risk factor for dental implant survival in a systematic review with meta-analysis [8]. Nevertheless, sinusitis has been associated with biomaterial extruded into the sinus [8–12], an event that induced some surgeons to remove the extruded graft from the sinus cavity [11,12].

However, it has recently been demonstrated in experiments in rabbits that the sinus mucosa in contact with graft particles might become thinner over time and eventually perforate [13–15]. It has also been shown that the number of thinned and perforated sites depends on the characteristics of the biomaterial in contact with the sinus mucosa [13,15]. In an experiment in rabbits, sinus augmentation was performed bilaterally using either autogenous bone or a deproteinized bovine bone mineral [13]. Implants were also installed simultaneously. After 40 days of healing, due to the higher resorption rate of the autogenous compared to the xenograft grafts, only one thinned site (width <40 μm) and no perforations were found at the sinus mucosa in contact with autogenous grafts, while 96 thinned sites and three perforations were observed at the sinus mucosa in contact with the xenograft particles. A direct contact with the implant apex and threads may also result in perforations of the sinus mucosa [13,16]. Due to the results of these reports, further evaluations should be performed aiming to find the biomaterial that elicits the lower damage to the sinus mucosa. Hence, the aim of the present study was to compare the damaging effects on the sinus mucosa of a synthetic and a xenogeneic graft.

2. Materials and Methods

2.1. Ethical Statements

Two experiments on sinus floor augmentation in rabbits using a synthetic or a xenogeneic graft were evaluated. The experimental protocols were approved by the Ethical Committee of the Faculty of Dentistry of Ribeirão Preto, University of São Paulo (Synthetic study: 2018.1.454.58.2, approved on 19 September 2018; Xenogeneic study: protocol No 2018.1.10.58.7, approved on 21 March 2018). The article was written according to the ARRIVE guidelines. The Brazilian rules for animal care were accurately followed.

2.2. Study Design

Maxillary sinus augmentation was carried out bilaterally in rabbits. A biphasic hydroxyapatite and beta-tricalcium phosphate (HA β-TCP) in the synthetic study, and a deproteinized bovine bone mineral (DBBM) in the xenogeneic study, were used as grafts. Both biomaterials were treated with argon plasma at the test sites while no treatments were provided to the grafts used at the control sites. Sinus mucosa thinning and perforations were assessed. The histomorphometric data describing the healing within the elevated regions were reported elsewhere [17,18].

2.3. Experimental Animals

In each experiment, twenty albino New Zealand rabbits, ~3.5–4 kg of weight and 5–6 months old, were used. Two groups composed of 10 animals were obtained in each experiment and euthanized after 2 or 10 weeks from surgery, respectively.

2.4. Biomaterials

Two grafts have been evaluated: a synthetic group: GUIDOR Calc-i-oss CRYSTAL+ (Sunstar, Etoy, Switzerland) composed of 60% hydroxyapatite and 40% β-tricalcium phosphate irregular-shaped granules with dimensions of 0.450–1.0 mm; and a xenogeneic group: Bio-Oss® (granules 0.250–1.0 mm; Geistlich Biomaterial, Wolhusen, Switzerland) composed of a bovine inorganic, porous hydroxyapatite from cancellous bone deproteinized at a temperature of 300 °C.

2.5. Sample Size

The sample size was determined based on the number of thinning mucosa sites in contact with the biomaterials after 40 days of healing reported by a previously discussed study in rabbits [13]. The determined effect size was 2.983. Applying an $\alpha = 0.05$, a power of 0.9, four animals each group were obtained. A slower resorption of the synthetic biomaterial used in the present study compared to autogenous bone used in that study used for calculation [13] might increase the number of thinning sites, decreasing the difference with the xenogeneic graft. Hence, ten instead of four animals for each group were considered sufficient to reject the null hypothesis that the population means of the two groups are equal.

2.6. Randomization and Allocation Concealment

The randomization plan was applied originally for the argon plasma treatment in both studies. The histological assessor was not informed before measurement about a possible comparison of the data between the two different studies. Mean values or sum of the data were used for the two sinuses of each rabbit so that the animal was the statistical unit used for analyses.

2.7. Clinical Procedures

The anesthetic and surgical procedures were similar in both experiments. Briefly, acepromazine (1.0 mg/kg, Acepran®, Vetnil, Louveira, São Paulo, Brazil) subcutaneously and xylazine (3.0 mg/Kg, Dopaser®, Hertape Calier, Juatuba, Minas Gerais, Brazil) and ketamine hydrochloride (50mg/kg, Ketamin Agener, União Química Farmacêutica Nacional S/A, Embu-Guaçú, São Paulo, Brazil) intramuscular were administrated. The rabbit's muzzle was shaved and disinfected. An incision was performed in the midline of the nasal dorsum by expert surgeons, and the nasal bone was exposed. Osteotomies were prepared using trephines and drills. A small screw was applied in the nasal-incisal suture as a reference for histological process. The sinus mucosa was elevated, and the biomaterial was grafted into the sub-antral spaces. The access windows were subsequently covered using a collagen membrane (Bio-Gide, Geistlich Biomaterial, Wolhusen, LU, Switzerland). For more details, see the previous published articles [17,18].

2.8. Euthanasia

The rabbits were first anesthetized, and euthanasia was performed in a closed transparent acrylic box containing gas carbon dioxide (CO_2) or an overdose of sodium thiopental (1.0 g, 2 mL, Thiopentax®, Cristália Produtos Químicos Farmacêuticos, Itapira, São Paulo, Brazil).

2.9. Housing and Husbandry

The animals were kept in individual cages in a climatized room with access to food and water ad libitum. The biological functions and the wounds were checked daily by specialized operators for the whole period of the experiment. A prophylactic dose of oxytetracycline dehydrate (40 mg/kg, IM, Terramicina LA, Zoetis Indústria e Produtos Veterinários, Campinas, São Paulo, Brazil) and, postoperatively, ketoprofen (3.0 mg/kg, IM., Ketofen 1%, Merial, Monte-Mor, Sao Paulo, Brazil) and tramadol hydrochloride (Tramadol 2%, 1.0 mg/kg, SC., Cronidor, Agener União Saúde Animal, Apucarana, Parana, Brazil) for 2 days were administrated.

2.10. Histological Preparation

After fixation with formalin, the biopsies were dehydrated and then embedded in resin (LR White™ hard grid, London Resin Co., Ltd., Berkshire, UK) and polymerized. Two grounds sections representing the central region of the sinuses were obtained using a cutting and grinding equipment (Exakt®, Apparatebau, Norderstedt, Germany). The sections were stained with either Stevenel's blue and alizarin red or toluidine blue.

2.11 Calibration for Histometric Evaluations

All measurements were carried out by an assessor (S.M.) after a training with an expert (D.B.). The intra-rater reliability in the measurements of the sinus mucosa width and perforation dimensions was $K > 0.90$.

2.12. Histological Analyses

The pristine mucosa was measured at the medial and lateral sinus walls in regions not included in the elevated area. A mean value of the two measurements was used for comparisons with the thinned sites.

The number and the width of the elevated sinus mucosa in close contact with the graft granules was measured, and all measurements <40 μm were recorded. The number and dimensions of the sinus mucosa perforations at the graft granules were also assessed.

2.13. Experimental Outcomes and Statistical Methods

The Wilcoxon test was used for dependent variables while a Mann–Whitney test was applied for independent variables. The software Prism 9.4.1 (GraphPad Software, LLC, San Diego, CA, USA) was used for statistical analyses.

3. Results

3.1. Clinical Outcomes

One sinus of the 2-week xenogeneic group presented a small perforation of ~0.5 mm that was protected with a small piece of collagen membrane. No complications were observed during healing in any group. At the time of histomorphometric analysis presented in a previous paper, one biopsy of the 10-week synthetic groups was not available for analysis due to technical problems. However, for the present analysis, the histological slide was available for analysis allowing n = 10 for all groups and periods.

3.2. Descriptive Histological Evaluation

Several thinned mucosa sites (<40 μm) were identified in both groups (Figure 1a,b).

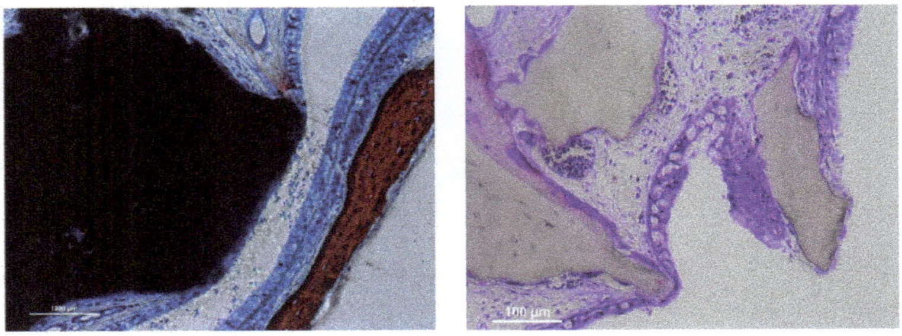

(a) (b)

Figure 1. (**a**) Synthetic site: Stevenel's blue and alizarin red stain. (**b**) Xenogeneic site: toluidine blue stain. Note the progressive decrease in width of both sinus mucosae and pseudostratified epithelia. A loss of cilia is evident in the thinnest sites on both biomaterials. While the process of resorption has a minimal impact on the xenogeneic graft, the synthetic graft has undergone a process already described as an interpenetrating bone network [18] characterized by concurrent bone formation within the biomaterial structure during its resorption.

The number of such sites was slightly higher in the synthetic compared to the xenogeneic groups in the 2-week period. However, that condition was reversed in the 10-week period due to a vast increase in the xenogeneic group of the number of thinned sites. The thinned sites were characterized by a tight contact towards the biomaterial granules resulting in a progressive damage to the sinus mucosa. In the initial stages, only the lamina propria was involved in histological modification, presenting vessels and mucosal glands displaced and deformed as if pressed against the granules surface (Figure 2a,b).

In the more advanced stages, the pseudostratified epithelium decreased in width, presenting a progressive loss of globet cells and cilias. In some cases, only a very thin layer of epithelial cells or connective tissue were limiting the direct continuity with the sinus cavity (Figures 1 and 2). The thinned sites were generally devoid of inflammatory infiltrates.

Few perforations were observed in the 2-week period, while the number was found increased considerably in the 10-week period, especially in the xenogeneic group. A tapered epithelium was generally bordering the exposed surface of the granules in a manner to maintain the integrity of the underlying tissues (Figure 3a–d). Several sites presented few inflammatory cells, while, in other cases, the infiltrates were more evident.

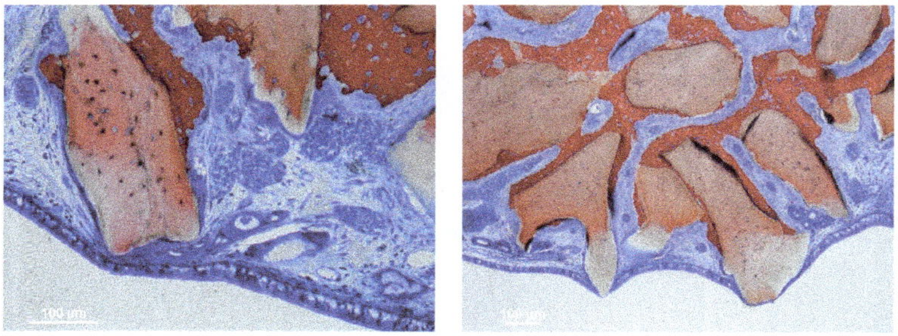

(a) (b)

Figure 2. (**a**,**b**) Xenogeneic sites presenting displacement of mucosal glands and vessels. A progressive thinning of sinus mucosa and pseudostratified epithelium was observed in several sites in close contact with the biomaterial granules. Note the new bone apposition on the biomaterial surface in sites opposite to the sinus mucosa.

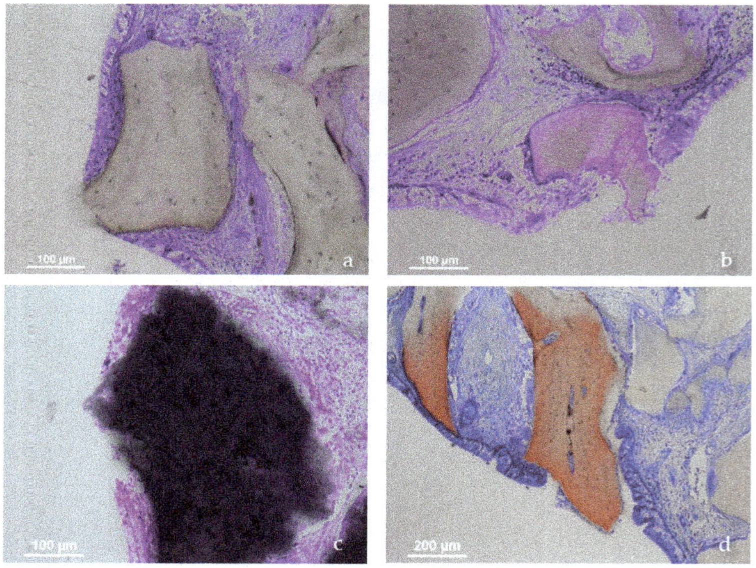

Figure 3. (**a**,**b**,**d**) Xenogeneic sites; (**c**) synthetic site. Note a tapered epithelium bordering the exposed surface of the granules in a manner to maintain the integrity of the underlying tissues.

3.3. Histometric Assessments

The mean width of the pristine sinus mucosa ranged between 73 μm and 96 μm. None of the pristine mucosa evaluated presented a width <40 μm. After 2 weeks of healing, 61 thinned sites were detected in the Synthetic group, presenting a mean width of 21 μm (Figure 4). In the Xenogeneic group, the sites were 49 (p = 0.537 between group) with a mean width of 20 μm (p = 0.561). After 10 weeks, the number of thinned mucosae increased to 79 sites in the Synthetic group (p = 0.222 between periods) maintaining the same mean width, and to 114 sites in the Xenogeneic group (p = 0.030 between groups; p = 0.001 between periods) presenting a mean width of 17 μm (p = 0.070 between groups; p = 0.105

between periods). The number of sites presenting a width <10 μm and <20 μm increased between 2 and 10 weeks (Figure 5).

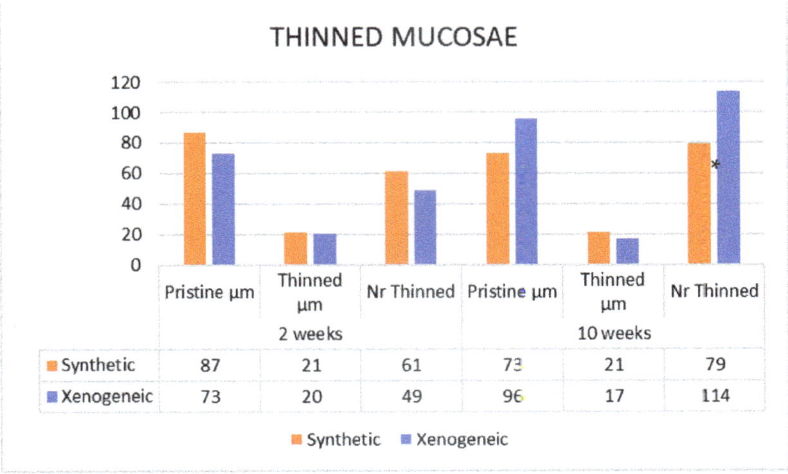

Figure 4. Number of thinned sites in the synthetic and xenogeneic groups after 2 and 10 weeks of healing. * = $p < 0.05$.

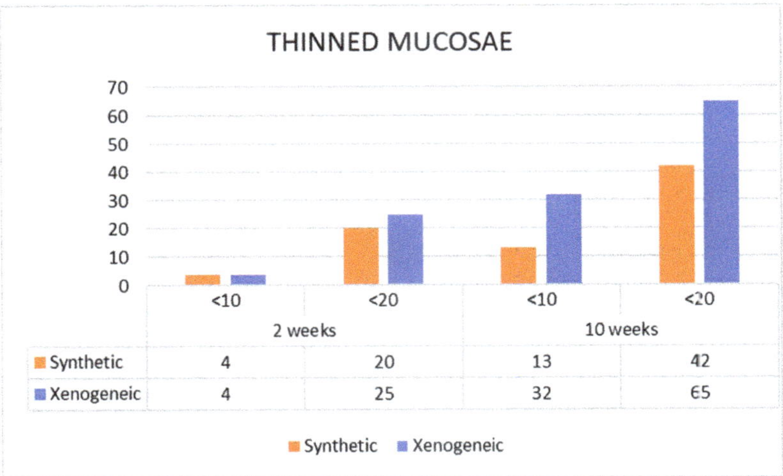

Figure 5. Number of sites presenting sinus mucosa width <10 μm or <20 μm in both synthetic and xenogeneic groups after 2 and 10 weeks of healing.

The pseudo-stratified epithelium at the thinned sites was also thinner compared to the pristine mucosa. The mean values of the pristine epithelium range between 18–20 μm while those at the thinned sites were 12–13 μm (Figure 6). The cilia were not considered in the measurements. The differences between pristine and thinned mucosa were statistically significant for both groups and periods ($p < 0.01$) while no differences were found between groups.

There were few perforations were in the 2-week period (Figure 7), two in two sinuses in the Synthetic group, and four in two sinuses in the Xenogeneic group ($p = 0.721$). In the 10-week period, the perforations increased by 4 in both groups so that eight perforations were detected in the Synthetic group, distributed in 6 sinuses out of 20, and sixteen in the Xenogeneic group distributed in 11 sinuses out of 20 ($p = 0.082$). The difference between periods was statistically significant only for the Xenogeneic group ($p = 0.020$). The mean dimension of the perforations was 171 ± 133 μm for the Synthetic group and 123 ± 47 for the Xenogeneic group.

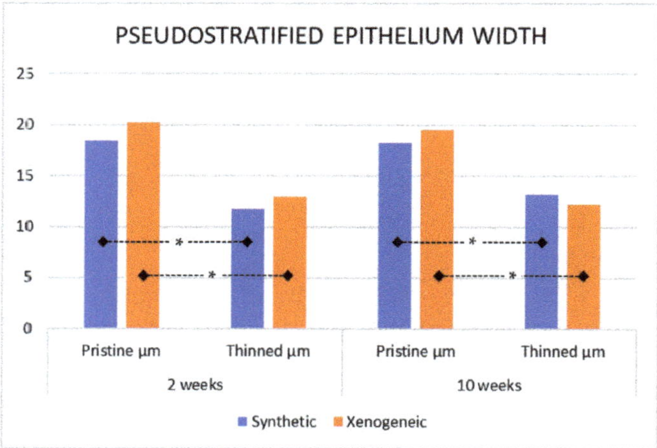

Figure 6. Pseudostratified epithelium width at the pristine not elevated region and at the thinned sites. * = $p < 0.05$.

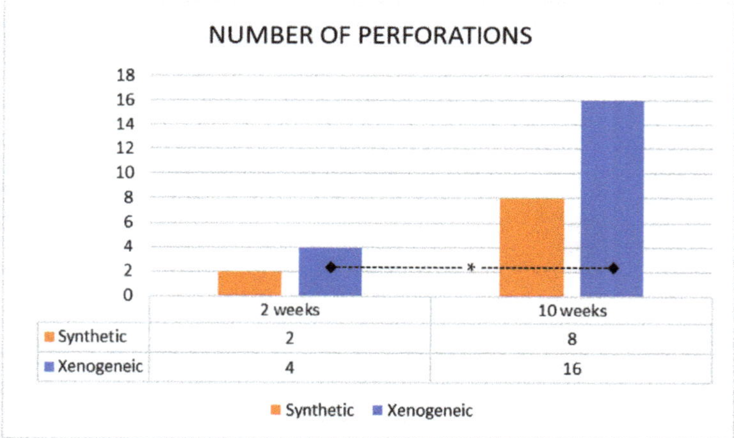

Figure 7. Number of perforations in both the synthetic and xenogeneic groups after 2 and 10 days of healing. * = $p < 0.05$.

4. Discussion

The aim of the present study was to compare the damaging effects on the sinus mucosa after sinus augmentation produced by synthetic or xenogeneic biomaterials used as grafts. It was seen that a close contact with the granules of the biomaterial resulted in thinning of the mucosa (<40 μm of width) and of the pseudostratified epithelium as well as perforations of the sinus mucosa. Both thinning and perforations of the mucosa increased between 2 and 10 weeks of healing.

The results are in agreement with other previous experimental studies in rabbits that showed similar damages to the sinus mucosa produced by biomaterials (13–15) or implants [13,16]. Similarly, in the present study, the thinned mucosa sites and the perforations were mostly located against sharpened edges of the biomaterial that were protruding beyond the dome shape of the elevated space. The results after 2 weeks cannot exclude that the pressure applied to the biomaterial during grafting might have influenced both perforations and thinning of the sinus mucosa. Nevertheless, the increased number of thinning sites and perforations over time related to the tendency of the sinus to regain its dimensions [19–21]. This will result in a forced repositioning of the sinus mucosa onto the elevated space, thoroughly delimiting its outline. It is on the periphery of the elevated space that the sinus mucosa will result in contact with graft. Especially when the granules' projections protrude beyond the outline of the elevated space, the mucosa might be involved in the thinning process. The tissues compressed against the edges of the granules will react over time, starting with a displacement of vessels and mucosa glands, followed by thinning of the pseudostratified epithelium with loss of globet cells and cilias. Eventually, the mucosa will be perforated. Interestingly, at the perforation sites, the sinus mucosa appears to surround the exposed surface with a tapered epithelium completely. This represents a physiological reaction of the mucosa aiming to maintain secured the internal environment, as already described in previously published articles describing the effect on sinus mucosa of biomaterials [13–15] and implants [13,16]. However, with the progression of the exposure, the sinus mucosa will grow around the graft to complete its expulsion and regain continuity behind the granule.

The results from the present study showed that, after 10 weeks, the number of thinning sites and perforations was higher for the Xenogeneic group compared to the Synthetic group. This might be related to the different conformation of the surface of the granules, but also the way that the granules are resorbed or integrated into the newly form tissues within the elevated space. The xenogeneic graft used in the present study remains almost not resorbed during the first period of healing so that new bone will grow onto its surface [22–25], resulting in an osseointegration of the graft. The synthetic material used is instead partly resorbed and partly penetrated by newly formed bone. This event will conduce to the formation of a composite mixture of graft residual and bone that has been called "interpenetrating bone network" [18]. Both factors described above might have offered a certain degree of protection to the synthetic material from thinning and perforation events. Another experiment showed the importance of the biomaterial characteristics on the sinus mucosa damaging. Either autogenous bone or a xenogeneic graft was used for sinus augmentation in rabbits [13]. After 40 days of healing, only one thinning mucosa site and no perforations were detected in the autogenous group. Conversely, in the xenogeneic group, 96 thinned sites and three perforations in two sinuses out of six were observed.

It has to be considered that the corticalization of the new sinus floor subjacent the elevated sinus mucosa was not observed yet after 10 weeks. The formation of this layer of corticalized bone might protect from further perforations. In humans, a total or partial corticalization have been verified in 30% to 75% of cases after 9 months of healing [26–29].

The limitations of the present study are related to the smaller dimensions and thinner mucosa in rabbits compared to humans. The grafts appear huge in relation to the dimensions and the mucosa presents a width of ~80 μm. A histological analysis in human reported widths ranging between 0.45 and 1 mm [30]. A CBCT analysis [31] on eighty-eight sinuses programmed for sinus floor elevation reported a mean width of the sinus mucosa of 2 mm. However, thirty sinuses presented a width < 1 mm, the thinnest being 0.4 mm. It cannot be excluded that, over time, the thinnest sinus mucosa in human might also be damaged by the contact to sharpen projections of not resorbed grafts, resulting eventually in perforations. The sinus mucosa showed repairing processes around the exposed graft that might result in the expulsion of the graft into the sinus cavity with complete restoration of the continuity of the sinus mucosa. The granule might be eliminated through the ostium/infundibulum preventing infective sinusal problems. The extrusion of biomaterial into the sinuses has been reported both during the surgical procedure [6,7] or at a later stage requiring the removal of the graft sometimes [11,32]. Complications might be associated with sinus floor elevation [33]. Short implants have been shown to represent a valid alternative, exhibiting lower marginal bone loss compared to standard-length implants and similar survival rate [34].

5. Conclusions

In conclusion, the contact with xenogeneic or synthetic grafts will induce thinning and possible perforation of the sinus mucosa. This effect will increase over time, and it is stronger at the xenogeneic than the synthetic graft.

Author Contributions: Conceptualization, Y.O. and D.B.; Data curation, S.M. and D.B.; Formal analysis, Y.O. and D.B.; Funding acquisition, D.B.; Investigation, S.M., V.F.B. and E.P.G.; Methodology, D.B. and S.P.X.; Project administration, D.B.; Supervision, S.P.X.; Validation, Y.O., V.F.B. and E.P.G.; Writing—original draft, D.B.; Writing—review and editing, Y.O., D.B. and S.P.X. All authors have read and agreed to the published version of the manuscript.

Funding: This study was funded by ARDEC Academy, viale Giovanni Pascoli 67, 47923 Rimini, Italy.

Institutional Review Board Statement: The experimental protocols were approved by the Ethical Committee of the Faculty of Dentistry of Ribeirão Preto, University of São Paulo (Synthetic study: 2018.1.454.58.2, approved on 19 September 2018; Xenogeneic study: protocol No 2018.1.10.58.7, approved on 21 March 2018).

Informed Consent Statement: Not applicable.

Data Availability Statement: The data are available on reasonable request.

Acknowledgments: We acknowledge the contribution of Sebastião Bianco (Faculty of Dentistry of Ribeirão Preto, SP, Brazil) to histological processing.

Conflicts of Interest: The authors declare no conflict of interest.

References

1. Pjetursson, B.E.; Tan, W.C.; Zwahlen, M.; Lang, N.P. A systematic review of the success of sinus floor elevation and survival of implants inserted in combination with sinus floor elevation. *J. Clin. Periodontol.* **2008**, *35* (Suppl. 8), 216–240. [CrossRef]
2. Del Fabbro, M.; Wallace, S.S.; Testori, T. Long-term implant survival in the grafted maxillary sinus: A systematic review. *Int. J. Periodontics Restor. Dent.* **2013**, *33*, 773–783. [CrossRef]
3. Kim, J.; Jang, H. A review of complications of maxillary sinus augmentation and available treatment methods. *J. Korean Assoc. Oral Maxillofac. Surg.* **2019**, *45*, 220–224. [CrossRef]
4. Stacchi, C.; Andolsek, F.; Berton, F.; Perinetti, G.; Navarra, C.O.; Di Lenarda, R. Intraoperative Complications During Sinus Floor Elevation with Lateral Approach: A Systematic Review. *Int. J. Oral Maxillofac. Implant.* **2017**, *32*, e107–e118. [CrossRef] [PubMed]
5. Reiser, G.M.; Rabinovitz, Z.; Bruno, J.; Damoulis, P.D.; Griffin, T.J. Evaluation of maxillary sinus membrane response following elevation with the crestal osteotome technique in human cadavers. *Int. J. Oral Maxillofac. Implant.* **2001**, *16*, 833–840.
6. Nkenke, E.; Schlegel, A.; Schultze-Mosgau, S.; Neukam, F.W.; Wiltfang, J. The endoscopically controlled osteotome sinus floor elevation: A preliminary prospective study. *Int. J. Oral Maxillofac. Implant.* **2002**, *17*, 557–566.
7. Berengo, M.; Sivolella, S.; Majzoub, Z.; Cordioli, G. Endoscopic evaluation of the bone-added osteotome sinus floor elevation procedure. *Int. J. Oral Maxillofac. Surg.* **2004**, *33*, 189–194. [CrossRef] [PubMed]
8. Díaz-Olivares, L.A.; Cortés-Bretón Brinkmann, J.; Martínez-Rodríguez, N.; Martínez-González, J.M.; López-Quiles, J.; Leco-Berrocal, I.; Meniz-García, C. Management of Schneiderian membrane perforations during maxillary sinus floor augmentation with lateral approach in relation to subsequent implant survival rates: A systematic review and meta-analysis. *Int. J. Implant. Dent.* **2021**, *7*, 91. [CrossRef] [PubMed]
9. Nolan, P.J.; Freeman, K.; Kraut, R.A. Correlation between Schneiderian membrane perforation and sinus lift graft outcome: A retrospective evaluation of 359 augmented sinus. *J. Oral Maxillofac. Surg.* **2014**, *72*, 47–52. [CrossRef] [PubMed]
10. Park, W.B.; Han, J.Y.; Kang, P.; Momen-Heravi, F. The clinical and radiographic outcomes of Schneiderian membrane perforation without repair in sinus elevation surgery. *Clin. Implant. Dent. Relat. Res.* **2019**, *21*, 931–937. [CrossRef]
11. Doud Galli, S.K.; Lebowitz, R.A.; Giacchi, R.J.; Glickman, R.; Jacobs, J.B. Chronic sinusitis complicating sinus lift surgery. *Am. J. Rhinol.* **2001**, *15*, 181–186. [CrossRef] [PubMed]
12. Urban, I.A.; Nagursky, H.; Church, C.; Lozada, J.L. Incidence, diagnosis, and treatment of sinus graft infection after sinus floor elevation: A clinical study. *Int. J. Oral Maxillofac. Implant.* **2012**, *27*, 449–457.
13. Kato, S.; Botticelli, D.; De Santis, E.; Kanayama, M.; Ferreira, S.; Rangel-Garcia, I., Jr. Sinus mucosa thinning and perforation after sinus augmentation. A histological study in rabbits. *Oral. Maxillofac. Surg.* **2021**, *25*, 477–485. [CrossRef] [PubMed]

14. Miki, M.; Botticelli, D.; Silva, E.R.; Xavier, S.P.; Baba, S. Incidence of Sinus Mucosa Perforations During Healing After Sinus Elevation Using Deproteinized Bovine Bone Mineral as Grafting Material: A Histologic Evaluation in a Rabbit Model. *Int. J. Oral Maxillofac. Implant.* **2021**, *36*, 660–668. [CrossRef]
15. Favero, R.; Apaza Alccayhuaman, K.A.; Botticelli, D.; Xavier, S.P.; Ferreira Balan, V.; Macchi, V.; De Caro, R. Sinus Mucosa Thinning and Perforations after Sinus Lifting Performed with Different Xenografts: A Histological Analysis in Rabbits. *Dent. J.* **2021**, *10*, 2. [CrossRef] [PubMed]
16. Omori, Y.; Botticelli, D.; Ferri, M.; Delgado-Ruiz, R.; Ferreira Balan, V.; Xavier, S.P. Argon bioactivation of implants installed simultaneously to maxillary sinus lifting without graft. An experimental study in rabbits. *Dent. J.* **2021**, *9*, 105. [CrossRef] [PubMed]
17. Hirota, A.; Yamada, Y.; Canullo, L.; Xavier, S.P.; Baba, S. Bioactivation of Bovine Bone Matrix Using Argon Plasma: An Experimental Study for Sinus Augmentation in Rabbits. *Int. J. Oral Maxillofac. Implant.* **2020**, *35*, 731–738. [CrossRef] [PubMed]
18. Tanaka, K.; Botticelli, D.; Canullo, L.; Baba, S.; Xavier, S.P. New bone ingrowth into β-TCP/HA graft activated with argon plasma: A histomorphometric study on sinus lifting in rabbits. *Int. J. Implant. Dent.* **2020**, *6*, 36. [CrossRef] [PubMed]
19. Asai, S.; Shimizu, Y.; Ooya, K. Maxillary sinus augmentation model in rabbits: Effect of occluded nasal ostium on new bone formation. *Clin. Oral Implant. Res.* **2002**, *13*, 405–409. [CrossRef]
20. Xu, H.; Shimizu, Y.; Asai, S.; Ooya, K. Grafting of deproteinized bone particles inhibits bone resorption after maxillary sinus floor elevation. *Clin. Oral Implant. Res.* **2004**, *15*, 126–133. [CrossRef] [PubMed]
21. Scala, A.; Botticelli, D.; Faeda, R.S.; Garcia Rangel, I., Jr.; Américo de Oliveira, J.; Lang, N.P. Lack of influence of the Schneiderian membrane in forming new bone apical to implants simultaneously installed with sinus floor elevation: An experimental study in monkeys. *Clin. Oral Implant. Res.* **2012**, *23*, 175–181. [CrossRef] [PubMed]
22. Busenlechner, D.; Huber, C.D.; Vasak, C.; Dobsak, A.; Gruber, R.; Watzek, G. Sinus augmentation analysis revised: The gradient of graft consolidation. *Clin. Oral Implant. Res.* **2009**, *20*, 1078–1083. [CrossRef] [PubMed]
23. Caneva, M.; Lang, N.P.; Garcia Rangel, I.J.; Ferreira, S.; Caneva, M.; De Santis, E.; Botticelli, D. Sinus mucosa elevation using Bio-Oss® or Gingistat® collagen sponge: An experimental study in rabbits. *Clin. Oral Implant. Res.* **2017**, *28*, e21–e30. [CrossRef] [PubMed]
24. Godoy, E.P.; Alccayhuaman, K.A.A.; Botticelli, D.; Amaroli, A.; Balan, V.F.; Silva, E.R.; Xavier, S.P. Osteoconductivity of Bovine Xenograft Granules of Different Sizes in Sinus Lift: A Histomorphometric Study in Rabbits. *Dent. J.* **2021**, *9*, 61. [CrossRef] [PubMed]
25. Ferreira Balan, V.; Botticelli, D.; Peñarrocha-Oltra, D.; Masuda, K.; Pires Godoy, E.; Xavier, S.P. Maxillary Sinus Floor Augmentation with Two Different Inorganic Bovine Bone Grafts: An Experimental Study in Rabbits. *Chin. J. Dent. Res.* **2022**, *25*, 93–105. [CrossRef]
26. Kawakami, S.; Lang, N.P.; Iida, T.; Ferri, M.; Apaza Alccayhuaman, K.A.; Botticelli, D. Influence of the position of the antrostomy in sinus floor elevation assessed with cone-beam computed tomography: A randomized clinical trial. *J. Investig. Clin. Dent.* **2018**, *9*, e12362. [CrossRef] [PubMed]
27. Kawakami, S.; Lang, N.P.; Ferri, M.; Apaza Alccayhuaman, K.A.; Botticelli, D. Influence of the height of the antrostomy in sinus floor elevation assessed by cone beam computed tomography—A randomized clinical trial. *Int. J. Oral Maxillofac. Implant.* **2019**, *34*, 223–232. [CrossRef] [PubMed]
28. Hirota, A.; Lang, N.P.; Ferri, M.; Fortich Mesa, N.; Apaza Alccayhuaman, K.A.; Botticelli, D. Tomographic evaluation of the influence of the placement of a collagen membrane subjacent to the sinus mucosa during maxillary sinus floor augmentation: A randomized clinical trial. *Int. J Implant Dent.* **2019**, *5*, 31. [CrossRef]
29. Imai, H.; Lang, N.P.; Ferri, M.; Hirota, A.; Apaza Alccayhuaman, A.A.; Botticelli, D. Tomographic assessment on the influence on dimensional variations of the use of a collagen membrane to protect the antrostomy after maxillary sinus floor augmentation. A randomized clinical trial. *Int. J. Oral. Maxillofac. Implant.* **2020**, *35*, 350–356. [CrossRef]
30. Aimetti, M.; Massei, G.; Morra, M.; Cardesi, E.; Romano, F. Correlation between gingival phenotype and Schneiderian membrane thickness. *Int. J. Oral Maxillofac. Implant.* **2008**, *23*, 1128–1132.
31. Kawakami, S.; Botticelli, D.; Nakajima, Y.; Sakuma, S.; Baba, S. Anatomical analyses for maxillary sinus floor augmentation with a lateral approach: A cone beam computed tomography study. *Ann. Anat.* **2019**, *226*, 29–34. [CrossRef] [PubMed]
32. Jiam, N.T.; Goldberg, A.N.; Murr, A.H.; Pletcher, S.D. Surgical treatment of chronic rhinosinusitis after sinus lift. *Am. J. Rhinol. Allergy* **2017**, *31*, 271–275. [CrossRef] [PubMed]
33. Hsu, Y.T.; Rosen, P.S.; Choksi, K.; Shih, M.C.; Ninneman, S.; Lee, C.T. Complications of sinus floor elevation procedure and management strategies: A systematic review. *Clin. Implant. Dent. Relat. Res.* 2022; *Epub ahead of print*. [CrossRef]
34. Toledano, M.; Fernández-Romero, E.; Vallecillo, C.; Toledano, R.; Osorio, M.T.; Vallecillo-Rivas, M. Short versus standard implants at sinus augmented sites: A systematic review and meta-analysis. *Clin. Oral Investig.* **2022**, *26*, 6681–6698. [CrossRef] [PubMed]

Journal of *Functional Biomaterials*

MDPI

Article

Hydrothermal Synthesis of Fluorapatite Coatings over Titanium Implants for Enhanced Osseointegration—An In Vivo Study in the Rabbit

Eduardo Santiago [1,2], Victor Martin [1,2], Bruno Colaço [3], Maria Helena Fernandes [1,2], Catarina Santos [4,5,*] and Pedro S. Gomes [1,2,*]

1 BoneLab—Laboratory for Bone Metabolism and Regeneration, Faculty of Dental Medicine, U. Porto Rua Dr. Manuel Pereira da Silva, 4200-393 Porto, Portugal
2 LAQV/REQUIMTE, University of Porto, 4160-007 Porto, Portugal
3 Animal and Veterinary Research Centre (CECAV), Associate Laboratory for Animal and Veterinary Science—AL4AnimalS, University of Trás-os-Montes and Alto Douro (UTAD), 5000-801 Vila Real, Portugal
4 EST Setúbal, CDP2T, Instituto Politécnico de Setúbal, Campus IPS, 2910-761 Setúbal, Portugal
5 Centro Química Estrutural, Instituto Superior Técnico, Universidade de Lisboa, Av. Rovisco Pais, 1049-001 Lisboa, Portugal
* Correspondence: catarina.santos@estsetubal.ips.pt (C.S.); pgomes@fmd.up.pt (P.S.G.)

Abstract: This work aims at the development and characterization of fluorapatite coatings, innovatively prepared by the hydrothermal method, aiming for enhanced osseointegration of titanium implants. Fluoride-containing coatings were prepared and characterized by scanning and transmission electron microscopy, Fourier-transform infrared spectroscopy—attenuated total reflectance, and X-ray photoelectron spectroscopy. The biological response was characterized by microtomographic evaluation and histomorphometric analysis upon orthotopic implantation in a translational rabbit experimental model. Physic-chemical analysis revealed the inclusion of fluoride in the apatite lattice with fluorapatite formation, associated with the presence of citrate species. The in vivo biological assessment of coated implants revealed an enhanced bone formation process—with increased bone-to-implant contact and bone volume. The attained enhancement of the osteogenic process may be attributable to the conjoined modulatory activity of selected fluoride and citrate levels within the produced coatings. In this regard, the production of fluorapatite coatings with citrate, through the hydrothermal method, entails a promising approach for enhanced osseointegration in implant dentistry and orthopedic applications.

Keywords: fluorapatite; hydrothermal synthesis; surface functionalization; titanium; osseointegration; microtomography

Citation: Santiago, E.; Martin, V.; Colaço, B.; Fernandes, M.H.; Santos, C.; Gomes, P.S. Hydrothermal Synthesis of Fluorapatite Coatings over Titanium Implants for Enhanced Osseointegration—An In Vivo Study in the Rabbit. *J. Funct. Biomater.* **2022**, *13*, 241. https://doi.org/10.3390/jfb13040241

Academic Editors: Adalberto Luiz Rosa and Marcio Mateus Beloti

Received: 19 October 2022
Accepted: 10 November 2022
Published: 14 November 2022

1. Introduction

Titanium-based alloys have taken the frontline of reference biomaterials for orthopedic, maxillofacial, and dental therapeutic applications, aiming for bone healing and/or fixation, given the appropriate mechanical, chemical, and biological properties [1]. These include a high strength-to-weight ratio, high yield, and fatigue resistance, as well as an adequate biological response [2]. By and large, the major limitations of titanium-based materials for bone application rely on the potential feeble osseointegration—particularly in aged and disease-affected individuals—which may culminate into interfacial displacement between the implant and the adjacent bone; and the release of metallic cations with potential local and/or systemic toxicity [3,4]. To heighten the implants' functionality, coating applications—and specifically those with bioactive ceramic materials such as hydroxyapatite (HA)—have been developed, aiming for improved construct stability, long-term functionality, and decreased corrosion [5]. Bioceramic coatings show an effective osteoconduction and potential osteoinductive ability, translated into enhanced bioactivity with the human bone tissue [6].

The vast majority of clinically developed bioceramic coating strategies rely on plasma-spraying methodologies. Nevertheless, this coating approach may originate in structural and phase discrepancies—elapsing from the high processing temperature—that create a thick (30–100 μm), highly crystalline, non-uniform coating, and consequently, dissimilar surficial resorption and biofunctionality, as well as a reduction in the interfacial coating-substrate strength [7–9]. In the present study, an alternative coating methodology—the hydrothermal method—was used as a simple, scalable, cost-effective, environmentally friendly, and versatile process [9,10]. In addition, it can produce homogeneous coatings on complex-shaped substrates—such as threaded dental implants, with defined chemical composition and crystallinity similar to that of mineral bone tissue [11].

In previous reports, the research team established, characterized, and optimized the production of pure-phase HA through this methodology [11]. Aiming for improved bioactivity, the addition of fluoride (F^-) to the HA lattice, allowing the formation of fluorapatite via hydroxyl substitution, was innovatively assayed in the present study. The fluoride substitution was described to increase the density and reduce the solubility of the bioceramic [12], further improving the biological response within the bone tissue through the release of fluoride ions capable of increasing osteoblastic proliferation and differentiation [13,14]. Notwithstanding, fluoride content needs to be precisely balanced within the materials' composition, as a high fluoride release may be cytotoxic to bone cells and ultimately impair the bone healing/regeneration process [13,15].

Accordingly, this work aims at the preparation and characterization of fluorapatite coatings by the hydrothermal method, with distinct fluoride contents. Coatings were deposited over commercially available titanium implants and, as a proof-of-concept, the biological response of the constructs was evaluated upon surgical implantation within the rabbit proximal tibia and a microtomographic and histomorphometric analysis, at distinct time points.

2. Materials and Methods

2.1. Preparation and Characterization of Fluorapatite-Coated Titanium Implants

2.1.1. Implants

Commercially pure titanium Grade 4 implants, AnyRidge® 4 × 7 mm, kindly donated by MegaGen (Seoul, Korea), were used as substrates for coating deposition. Implants were cleaned with water and acetone and placed inside the autoclave until further preparation.

2.1.2. Synthesis of the Hydroxyapatite and Fluorapatite Coatings by the Hydrothermal Method

The hydroxyapatite and fluorapatite coating solutions were produced following the previously described precipitation method [11]. Briefly, a 0.6 M aqueous solution of citric acid ($C_6H_8O_7 \cdot H_2O$, 99.5%), with a pH of 8.0, was prepared with ammonium solution (NH_4, 25%). Then, a 0.2 M solution of calcium nitrate ($(CaNO_3)_2 \cdot 4H_2O$, 99%) was added to the citric acid solution (solution A). Finally, a 0.2 M, 0.1 M, or 0.01 M solution of ammonium hydrogen phosphate ($(NH_4)_2HPO_4$) was added, dropwise, to solution A, together with a 0.2 M, 0.1 M, or 0.01 M solution of ammonium fluoride, to obtain the hydroxyapatite coating and two coating solutions with different concentrations of F ions. Finally, the prepared solutions were immediately transferred to a Teflon vessel and placed in the autoclave. The sealed autoclave was set up to 180 °C for 4 h; and the coated implants were named HA, F 0.1, and F 0.01, respectively.

2.1.3. Physical and Chemical Characterization

The morphology of the developed coatings was evaluated using a scanning electron microscope JEOL-JSM7001F, at an operating voltage of 20 kV. The chemical composition of the coatings was determined using an X-ray energy dispersive spectrometer (EDS) analysis.

The particle size of the F 0.1 coating was studied using a Transmission Electron Microscope (TEM) (Hitachi H-8100-NA with an acceleration voltage of 200 kV). Before imaging, F 0.1 coating particles were detached from the titanium substrate and dispersed in ethanol. Then, the suspension particles were placed on the carbon-coated copper grid and dried at room temperature. Attenuated total reflectance (FTIR-ATR) spectroscopy using a Nicolet (Thermo Electron) was used to characterize the functional groups and chemical composition of the HA, F 0.01, and F 0.1 coating over a range of 650–4000 cm^{-1} and a resolution of 8 cm^{-1}. X-ray photoelectron spectroscopy (XPS; Kratos Axis Ultra HSA, Aluminum mono, Eo = 15 kV (90 W) 1 eV per step in a 300 μm × 700 μm area) was used for fluorine, calcium, and phosphorus content analysis at the surface of the F 0.1 coating.

2.2. *Biological Characterization—In Vivo Response to Bone Implantation*

2.2.1. Animals

In this study, 8 male New Zealand white rabbits (*Oryctolagus cuniculus*) weighing 2.9 ± 0.32 kg were acquired from a certified vendor. The sample size was calculated a priori using G power software (v.3.1.9.6) with the following parameters: significance level (α) was set at 0.05, statistical power (1-β) was set at 0.8, and the effect size (F) was set as at 0.5. All procedures were approved by the local Institutional Animal Care and Use Committee (IACUC), based on standard protocols, under national and European legislation for experimental animal research—European Directive 2010/63/EU.

Animals were acclimatized for 3 weeks before any experimental manipulation, and were housed in environmentally enriched individual cages, in a temperature-, humidity-, and air renewal-controlled room, in a 12 h light-dark cycle. Animals were fed a standard diet (Mucedola 2RB15) and water *ad libitum* and were monitored daily throughout the acclimatization and experimental period. All procedures were conducted in compliance with the ARRIVE guidelines.

2.2.2. Surgical Procedure

Each animal received a total of 6 implants, three on the proximal left tibia and three on the proximal right tibia, which were randomly distributed. Of the 8 animals, half were endorsed for each of the postoperative follow-up periods: 4 and 8 weeks (4 animals per time point, 8 implants per experimental group per time point; allowing the assessment of 8 samples for each group per time point—n = 8) [16].

Before the surgical implantation, animals were pre-medicated with intramuscular injections of 1 mg/kg midazolam. Buprenorphine (0.03 mg/kg), administered subcutaneously, was used for analgesia and continued for 5 days. General anesthesia was achieved upon the intraperitoneal administration of 25 mg/kg ketamine and 5 mg/kg xylazine. Throughout the surgical procedure, sterile saline was administered at 10 mL/kg/h while animals were maintained on a heated surface and carbomer eye gel was administered to prevent ocular lesions. O_2 was administered by a facial mask throughout the surgical procedure.

Following the validation of the anesthetic plane, trichotomy was conducted on both legs that were aseptically prepared for surgery upon chlorhexidine disinfection. Mepivacaine 3% (Scandinibsa, Inibsa) was infiltrated around the incision area and an anteromedial approach to the proximal tibia was conducted. Briefly, a 4 cm full-thickness incision was conducted and upon careful periosteum elevation, the tibial bone surface was exposed. Bone drilling protocol was established as recommended by the manufacturer—lance drill, followed by 2.0, 2.8, 3.2, and 3.8 mm in diameter drills, marked at 7 mm, with the recommended torque values. Implants were placed using the handpiece connector, at a 30 Ncm torque. The soft tissues were then closed in layers with absorbable sutures. During the postoperative recovery, animals were allowed to move freely and were routinely monitored for behavioral and physiological alterations. The biological response to the placed implants was evaluated through microtomography, focusing on the bone formation process in the vicinity of the implants (Figure 1).

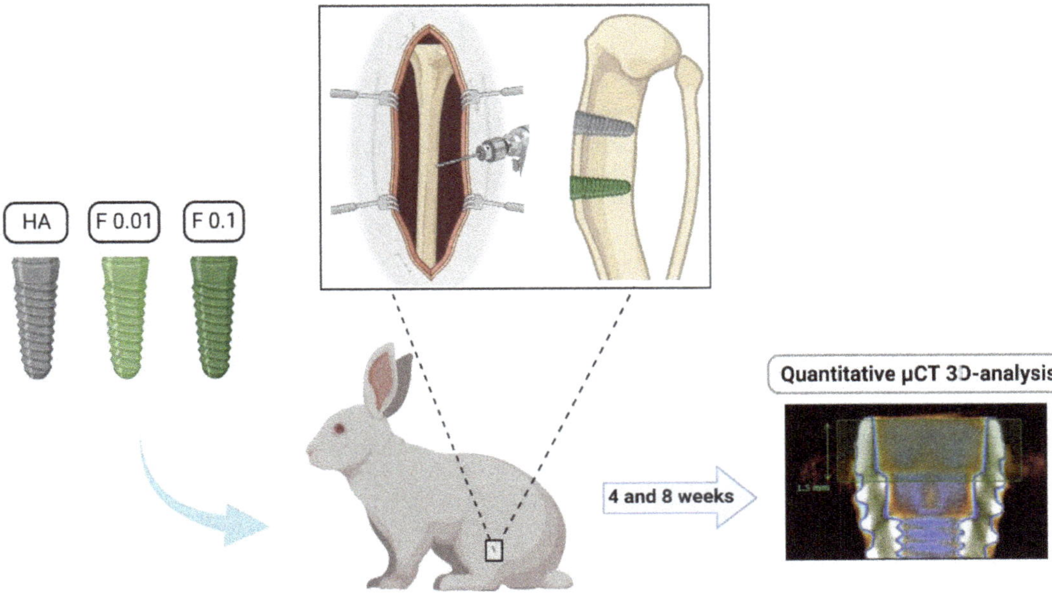

Figure 1. Study design overview.

2.2.3. Microtomographic Evaluation

At 4 and 8 weeks upon implantation, animals were euthanized. Tibiae were dissected, fixed in alcohol, and scanned using a Skyscan 1276 system (Bruker, Kontich, Belgium) at 100 kV, 200 uA, using an aluminum/copper filter and a resolution of 10 μm. The scans were performed with a 360° rotation, setting a rotation step of 0.2° and a framing averaging of eight.

The reconstruction of the obtained projection images was performed with the NRecon software (Bruker, version 1.7.4.2) with fixing parameters, such as bean hardening (16%), ring artifact reduction (0), and minimum/maximum for CS to image conversion of 0.0 to 0.07. Subsequently, implants were aligned along the coronal axis and isolated from each other using DataViewer software (Bruker, version 1.5.6.3). Three-dimensional images were obtained using CTVox software (Bruker, version 3.3.0).

Morphological analysis of the bone around implants was performed using the CT-Analyser software (Bruker, version 1.17.7.2) following the guidelines from Bruker [17]. Briefly, an anatomical reference was selected in the upper portion of the implant, and a fixed height of 1.5 mm was set. Then, the implants were isolated from the bone and other anatomical structures by the binary selection, and a ring of 20 pixels of thickness was drawn around the implant frame, to define the region of interest (ROI). Finally, the images were reloaded and binary thresholding was set to isolate the implant and the bone from the rest of the anatomical structures. The defined bone inside the ROI was analyzed three-dimensionally (bone volume (BV), bone volume fraction (BV/TV), bone surface (BS)), as well as in a 2D approach (bone-to-implant contact, calculated as the percent intersection surface (TIS/TS)—the ratio between total intersection surface (TIS), and total surface (TS)).

2.3. Statistical Analysis

Statistical analysis was conducted on the SPSS software (SPSS Statistics 27, Chicago, IL, USA). Quantitative data are expressed as mean ± standard deviation (SD). The Kruskal-Wallis nonparametric test was used and differences between groups were considered to be significant for $p < 0.05$.

3. Results and Discussion

3.1. Coating Preparation and Characterization

Figure 2 shows the morphology of HA-, F 0.01-, and F 0.1-coated titanium implants and an overall view of the uncoated implant is presented in (Figure S1). The HA coating presents a typical rod-like morphology (Figure 2a) [11]. On the other hand, F 0.01 reveals a uniform F-distributed coating with a "mud-like" morphology, completely covering the titanium surface, without evidence of porosity or discontinuities (Figure 2b). Comparatively, a morphology change was observed on the surface of F 0.1 (Figure 2c), in which individual and aggregated F particles were observed in a homogeneous "dumbbell-like" morphology, further aligned parallelly to the substrate surface (Figure 2d). The average length of the "dumbbell-like" particles determined by TEM was around 650 $\pm$ 20 nm by 250 $\pm$ 10 nm (inset Figure 2d). The observed morphology change can be attributed to a fractal growth of the fluorapatite particles, caused by the large dipolar field along the c axis of fluorapatite, provided by the presence of citrate ions in the precipitating medium [18,19]. Considering the obtained results, it is reasonable to surmise that citrate molecules have a strong interaction with the fluorapatite particles' surface, and conditionate the final morphology of the coating. Moreover, it is known that fluoride ions have a higher affinity to occupy positions on the hydroxyapatite lattice in comparison to hydroxyl ions, which enhances thermodynamic stability and decreases the solubility and degradation of the coating. These modifications are expected to lead to more gradual coating resorption, with the added reported benefit of increasing the differentiation behavior of osteoblastic cells and stimulating bone growth, when compared to HA coatings [12,20–22].

For a detailed elemental characterization, EDS mapping analyses were performed on the F 0.1 coating (Figure 2e–i) and EDS analyses on F 0.01 and HA coating. The obtained results reveal that fluorine (F) (Figure 2e), calcium (Ca) (Figure 2f), phosphorus (P) (Figure 2g), oxygen (O) (Figure 2h), and titanium (Ti) (Figure 2i) were homogeneously distributed over the implant surface. The relative fluoride content of each of the two coatings (F 0.01 and F 0.1), measured by EDS, revealed that the atomic percentage of fluorine ions increase from 0.7% on the F 0.01 coating, to 5.08% on the F 0.1. This result indicates the successful incorporation of F into both coatings. In principle, fluoride ions should be incorporated through the ionic substitution of hydroxyl ions [23].

Based on the previous hypothesis, ATR-FTIR analyses were performed. The notable typical absorption bands at 1053 and 1096 cm^{-1} assigned to stretching vibration of the phosphate groups [18,24] were detected in the ATR-FTIR spectrum of F 0.01 and F 0.1 (Figure 2j). The absence of the characteristic hydroxyl bands at 632 cm^{-1}, 3571 cm^{-1}, and the presence of one additional band (due to the high concentration of fluorine at ~740 cm^{-1}, usually attributed to the libration mode of the OH group connected with a fluoride ion by a hydrogen bond [12,25]) suggests that the formed coating is composed by fluorapatite [26–28]. In addition, the absorption peaks at 1402 and 1590 cm^{-1}, and a small absorption peak at 1459 cm^{-1}, observed in the F 0.1 spectrum revealed the presence of carboxylate groups in the F coating, most probably coming from the citrate species [11]. In addition, according to ATR-FTIR results (Figure 2j), the intensity of carboxylate and phosphates absorption peaks was increased by the F^- incorporation into the coating. In our previous results [11], it was found that only a weak band associated with carboxylic groups was observed on the HA coating, while for F-containing coatings, three bands were observed, which confirms that the coordination mode of the citrate species with the particles present in the titanium surface is dependent on the ionic composition of the precipitant medium. In this specific system, the presence of fluoride ions modifies the configuration of citrate species adsorbed, which could have a strong impact on the biological response [29]. Moreover, according to the literature, the presence of citrate in the precipitation medium is expected to enhance the substitution of OH^- by F^- and accelerate the crystallization process [22].

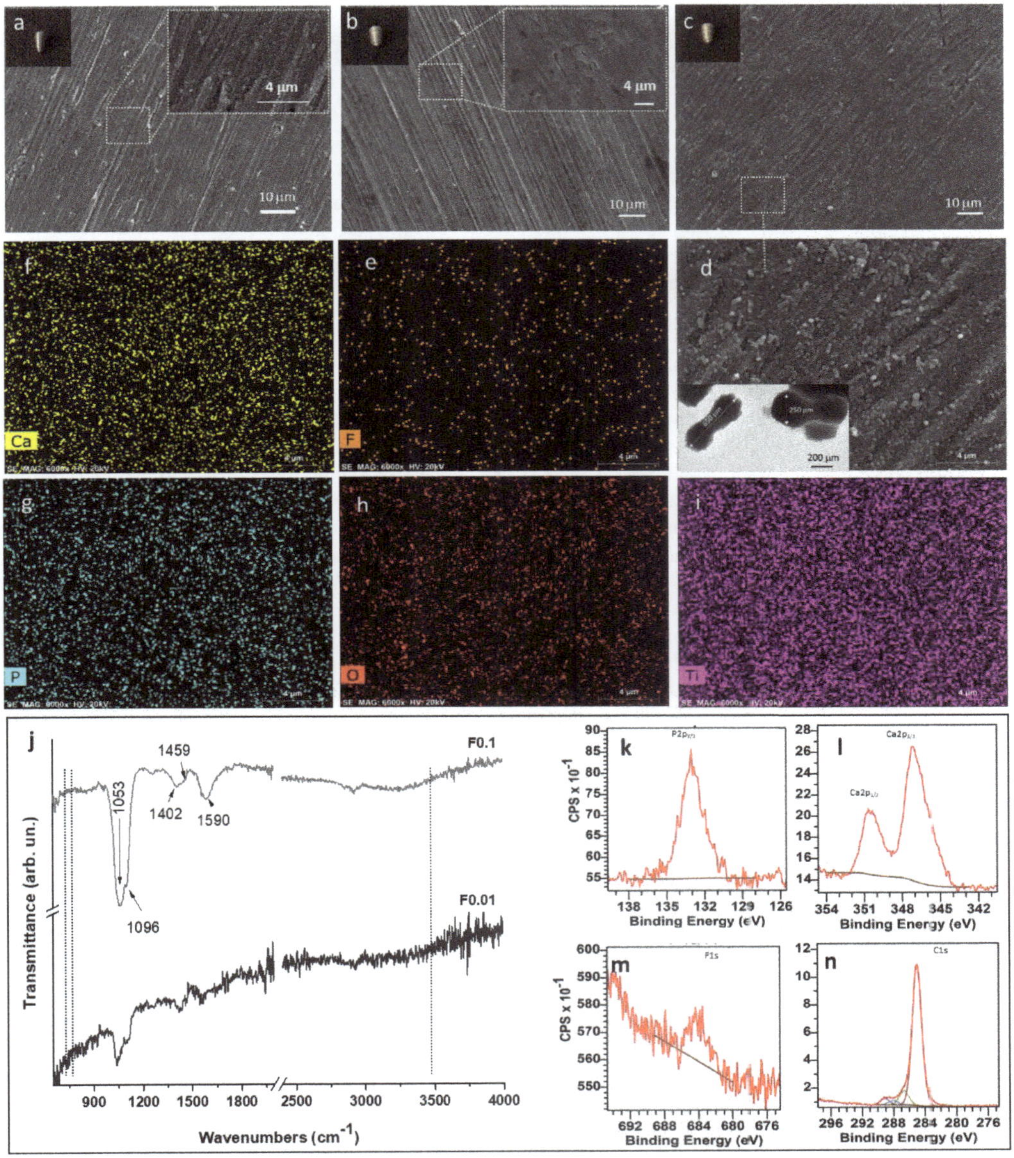

Figure 2. SEM images of the surfaces of the HA (**a**), F 0.01 (**b**). and F 0.1 (**c**) coatings obtained by the hydrothermal method. Magnified SEM image of F 0.1 (corresponding to the square area in (**c**)) and TEM image of F 0.1 coating particles (inset) (**d**). EDS elemental mapping images for F (**e**), Ca (**f**), P (**g**), O (**h**), and Ti (**i**) of the F 0.1 coating surface. ATR-FTIR spectra of the F 0.01 and F 0.1 coatings (**j**). XPS of the F 0.1 coating: (**k**) P 2p, (**l**) Ca 2p, (**m**) F 1s, (**n**) C 1s.

To validate the presence of the fingerprint peak of the fluorapatite structure, XPS analyses were performed and the results are shown in Figure 2k–n. It can be observed that all characteristic peaks of fluorapatite, P2p, and Ca2p, at 133.2 and 347.2 eV, respectively [30], were detected in the F 0.1 coating. Looking in detail, it can be observed one additional small peak at ~684.3 eV, belonging to F1s, indicating that F^- ions were incorporated into the fluorapatite lattice structure [31,32]. The F/Ca ratio, calculated directly from the XPS data, was F/Ca = 0.182. In stoichiometric fluorapatite $Ca_{10}(PO_4)_6F_2$, the F/Ca ratio should be, at maximum, 2:10 = 0.2. Considering the obtained ratio, it can be suggested that the produced fluorapatite is close to stoichiometry. Furthermore, looking at the calculated Ca/P ratio (~1.83), it turns out that the ratio is higher than the stoichiometric value Ca/P ratio = 1.67 [31], which may be due to the presence of citrate species on the F-containing coatings. The presence of the citrate species was confirmed by the three fitted peaks, corresponding to C1s photoelectrons from carbon bonded to other carbon and/or hydrogen atoms, carbon singly bonded to oxygen, and carbon in a carboxylate/carboxylic group [11]. Overall, from ATR-FTIR and XPS results, it can be confirmed that the developed F 0.1 coating is composed of fluorapatite.

3.2 Biological Evaluation

The biological characterization of the developed fluorapatite coatings was assayed in an in vivo translational model of the orthotopic implant placement, within the rabbit's proximal tibia. The rabbit has been a popular choice for the evaluation of biomaterials' biological response, reaching up to around one-third of the published literature on dental implant-related research [33]. Rabbits reach skeletal maturity at around 6 months of age, and given the fast bone turnover, allow an early evaluation of the bone tissue response [34]. Tibial implantation permits an adequate bone volume for the surgical placement of up to three clinically-relevant implants per side, within the range of 3 to 4 mm diameter and length up to 10 mm, allowing the use of routine characterization techniques to access osseointegration [35,36]. The thick cortical bone—broadly responsible for the primary fixation of the implants—also establishes a favorable environment for the early evaluation of the bone-to-implant interface [34].

The biological response to the developed implant-coated constructs was characterized through microtomography. This technique allows a nondestructive 3D imaging and morphometric analysis of the bone tissue with a very high resolution. Attained datasets can be used to reconstruct the implant and neighboring bone tissue, further characterizing tissue parameters in a given region of interest [37–39]. This allows the feasible analysis of bone tissue 3D parameters (e.g., bone volume (BV), bone volume ratio (BV/TV)), and 2D parameters (e.g., bone surface (BS)). In addition, information on the interaction between the implant and the bone tissue (e.g., bone-to-implant contact (BIC)), can also be estimated given proper data processing to minimize titanium-dependent imaging artifacts [40].

Biological outcomes were evaluated at two time points, 4 and 8 weeks. All animals recovered adequately during the postoperative period without any complications. At euthanasia, no signs of clinical alterations (i.e., ulceration, inflammation, infection, or abnormal tissue formation) were disclosed within the surgical area, with implants remaining integrated in situ. At 4 weeks, sectional reconstructions of the microtomographic data revealed an established cortical bone structure at the coronal aspect of the implant, with newly formed bone tissue growing along the threads (Figure 3), for all the constructs' compositions. Quantitative volumetric analysis revealed, as compared to HA (control), a significantly higher BV for F 0.01 and F 0.1, with the latter being significantly higher than that of F 0.01. Additionally, F 0.1 presented a significantly higher BV/TV ratio, as compared to HA and F 0.01. In regard to BS, both fluorapatite compositions presented a significantly increased level, a trend similarly verified for the BIC analysis.

The 3D reconstructions (Figure 4) substantiated the attained morphometric findings, presenting an increased bone volume for fluorapatite-containing substrates, with augmented bone surface and, as well, an increased contact area with the implant.

At 8 weeks, a more advanced bone formation process was attained for all conditions, with increased bone levels at the most coronal region of the implants, extending apically along the implant surface (Figure 5). Morphometric data revealed increased levels, as compared to data from the 4 weeks of implantation time (Figure 3). Compared to HA, fluorapatite coatings presented an increased BV and BS, and the F 0.1 formulation further presented an increased BV/TV. Additionally, the BIC was found to be significantly higher in both F 0.01 and F 0.1, despite the absence of differences between conditions.

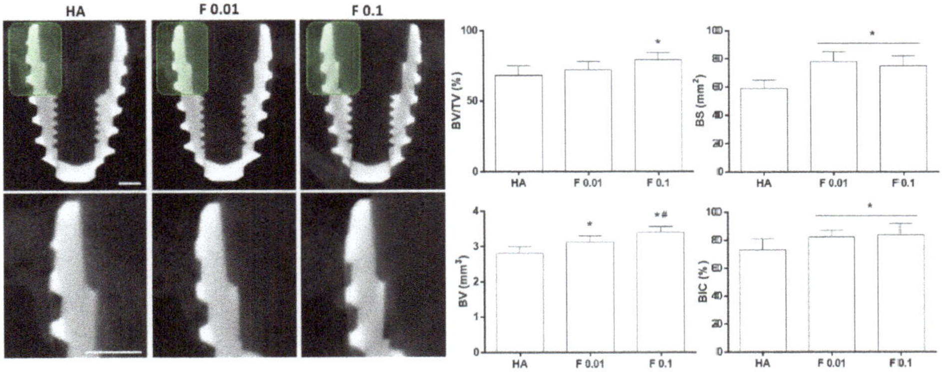

Figure 3. Bi-dimensional microtomographic images (left) and histomorphometric data of the coated constructs—HA, F 0.01, and F 0.1, at the 4 weeks timepoint. Scale bars correspond to 1 mm. $p < 0.05$; * significantly different from control; # significantly different from the other experimental group.

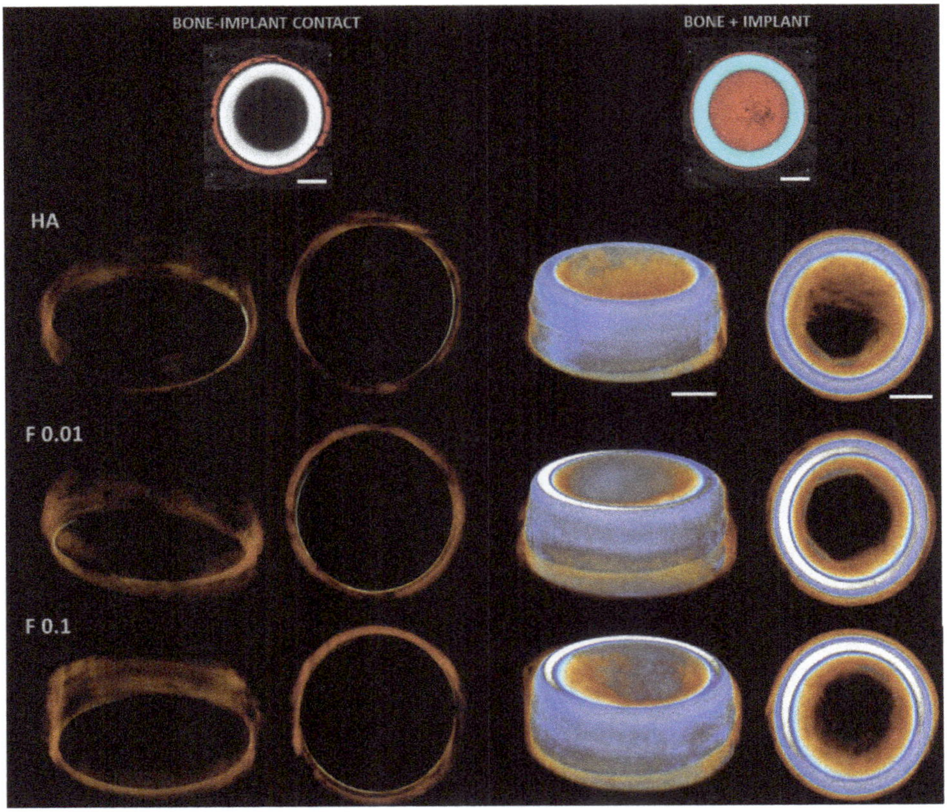

Figure 4. Representative three-dimensional microtomographic reconstructions of the coated constructs—HA, F 0.01, and F 0.1, at the 4 weeks timepoint. Scale bars correspond to 1 mm.

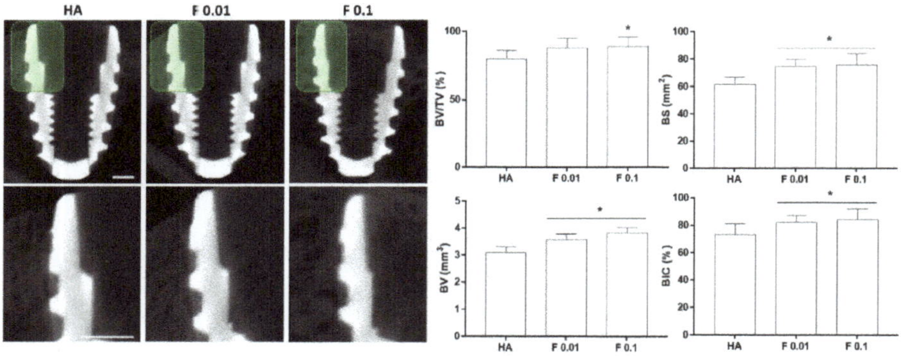

Figure 5. Bi-dimensional microtomographic images (**left**) and histomorphometric data (**right**) of the coated constructs—HA, F 0.01, and F 0.1, at the 8 weeks timepoint. Scale bars correspond to 1 mm. $p < 0.05$; * significantly different from control.

The 3D reconstruction of the peri-implant regions at 8 weeks of healing (Figure 6) corroborates the described findings, with increased bone formation at the coronal region, particularly within fluorapatite coatings, suggesting an increased mineralized tissue volume and increased surficial intersection with the implant surface.

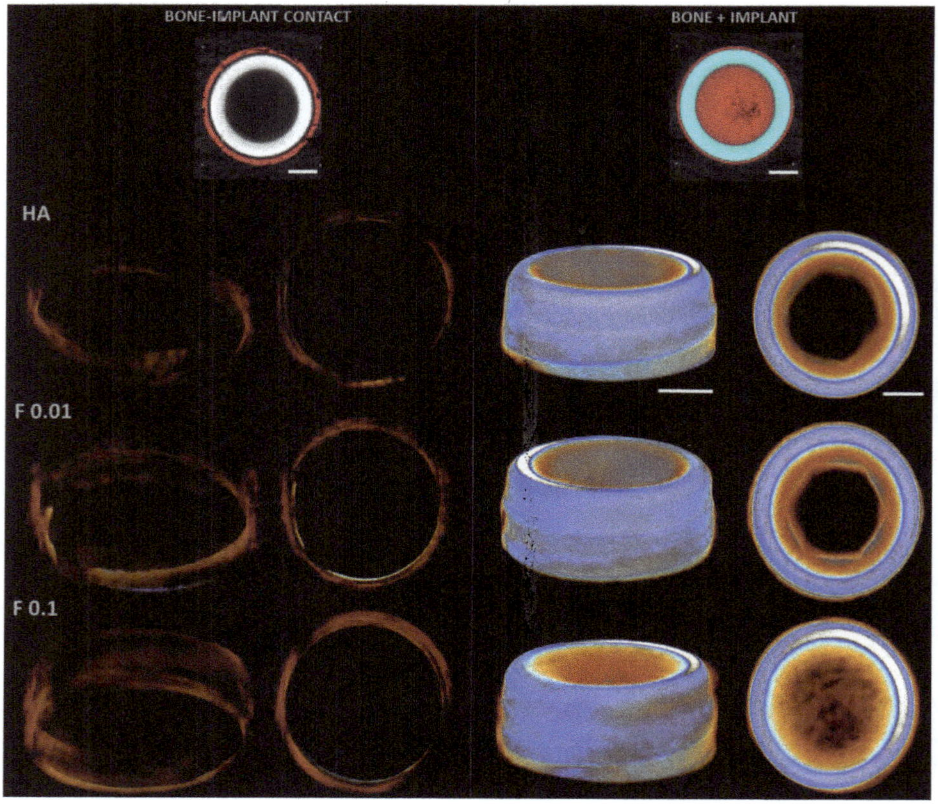

Figure 6. Representative three-dimensional microtomographic reconstructions of the coated constructs—HA, F 0.01, and F 0.1, at the 8 weeks timepoint. Scale bars correspond to 1 mm.

The biological assessment revealed the increased capability of fluorapatite coatings to enhance bone tissue formation in the vicinity of the implant and to increase the bone-to-implant contact, at both 4 and 8 weeks of implantation. F 0.1 coating was found to further induce bone tissue formation at the earliest time point, as compared to F 0.01, in line with the increased F content. Recently, fluoride-containing apatite coatings have become a topic of interest in implantology-related research [41]. Hydroxyapatite has long been considered the bioceramic of choice for bone-related applications, given its biocompatible response, high affinity to the bone tissue, and ability to induce early osseointegration [42]. Clinical applicability has, however, been limited by the reported coating delamination and dubious long-term success [43], associated with the plasma spraying coating technique. The alternative coating strategy currently employed, hydrothermal synthesis, is expected to surpass these limitations given the ability to control crystal structure, crystal morphology, and grain purity of the coating nanoparticles, by adjusting the reaction conditions. In addition, the coating thickness can reach tens of nanometers and uniform coatings can be prepared on complex surface shapes using the hydrothermal method. Furthermore, the prepared calcium phosphate-based coatings have high interfacial bonding strength and density, which can significantly improve the corrosion resistance of metallic substrates [44,45].

In addition, F-containing apatite coatings have demonstrated an enhanced biological response and bioactivity, as compared to hydroxyapatite, within distinct preparation and deposition methodologies [46–48]. Higher thermal stability and mechanical properties have also been recognized within F-containing apatites [49,50]. In the present study, F-containing surfaces enhanced bone formation and allowed an increased BIC, with the F 0.1 coating allowing for an enhanced biological outcome. Previous in vitro studies reported an increased osteoblastic proliferation within distinct F-substituted apatites, as compared to non-substituted ones [21]—a process that may be associated with the ability of fluorine to act on relevant cell signaling pathways, as the Jun N-terminal kinase (KNK) and p38 MAPK [51]. Similarly, F-containing substrates were also found to enhance osteoblastic differentiation, thus upregulating the expression of osteogenesis-related markers such as alkaline phosphatase and osteocalcin [52]. Mechanistically, this process may relate to the upregulation of the Wnt signaling pathway via the fluoride-mediated GSK-3β phosphorylation, or via BMP/Smad signaling, also modulated by fluoride [53,54]. On the other hand, F was found to diminish the osteoclastic functionality—either directly through the downregulation of a major transcription factor, NFATc1 [55]; or indirectly, increasing the expression of the osteoprotegerin decoy receptor, inhibiting the osteoblast-mediated osteoclast differentiation [56]. This regulation—a decreased bone resorption conjoined to an increased bone formation, further verified within in vivo models [57], is expected to accelerate the early osseointegration process and consequently, the overall implant success rate.

Nevertheless, in addition to the reported beneficial effects on bone metabolism/regeneration, fluoride may also elicit detrimental effects on bone tissue dynamics, altering cellular functionality and inducing structural damage, as verified in bone fluorosis [58]. The major factor determining fluoride-mediated biological outcomes seems to be the amount of bioavailable F^- within the microenvironment [59]. In accordance, both F 0.1 and F 0.01 formulations induced the osteogenic response, with the former outperforming the latter, demonstrating the adequacy of F levels in the coatings' composition [60].

Of additional relevance, the identified citrate species are further expected to tailor the biological outcomes. Citrate is a major component of the bone structure, distributed in two major pools: collagen-associated and HA-associated citrate [61]; and known to enhance the biomineralization process. Citrate also seems to play a chelating activity, binding to important ionic species, such as Mg, Zn, and Ca, constituting a major ionic store in the bone tissue [62]. Exogenous citrate supplementation has been further found to facilitate osteogenesis and cellular commitment of precursor populations, favoring the metabolic changes—switch from aerobic glycolysis to oxidative phosphorylation, needed to meet the increased energetic requirements determined by the osteogenic differentiation [29,62,63].

Citrate molecules may further indirectly induce the expression of osteogenic transcription factors (e.g., RUNX2 and downstream targets) by the stabilization of β-catenin [29,64]. Accordingly, the presence of citrate species on the produced coatings is expected to synergize with F to improve the osteogenic response.

Limitations of the present study are broadly related to the characteristics of the selected experimental model. While rabbits seem to be appealing models for bone research given the similarity in bone metabolism and Harversian remodeling capability, the anatomical and structural differences and scarcity of biomechanical data on skeletal functionality may hinder direct data translational application [65]. Moreover, future studies should further embrace longer time points of analysis, and embrace orthotopic similarity with functional biomechanical loading, reaching hand, for instance, of the oral implantation in a canine model.

4. Conclusions

In the present work, innovative thin ceramic coatings with F^- were prepared by the hydrothermal method and deposited over commercial titanium implants. The physicochemical characterization validated the incorporation of F^- into the HA lattice through OH^- substitution, leading to the formation of fluorapatite, in association with the presence of citrate species. Upon in vivo implantation in the bone tissue, fluorapatite-coated implants presented an enhanced bone formation process at the implant vicinity, with increased bone-to-implant contact, as compared to the control—HA-coated implants. The attained enhancement in osteogenesis is attributable to the conjoined modulatory activity of selected F^- and citrate levels, within the produced coatings. The production of fluorapatite coatings with citrate entails a promising approach for enhanced osseointegration in implant dentistry and orthopedic applications.

Supplementary Materials: The following supporting information can be downloaded at: https://www.mdpi.com/article/10.3390/jfb13040241/s1, Figure S1: SEM image of uncoated titanium implant.

Author Contributions: Conceptualization, P.S.G. and C.S.; Methodology, C.S., B.C. and P.S.G.; Validation, M.H.F.; Formal Analysis, V.M. and C.S.; Investigation, E.S. and V.M.; Resources, B.C., M.H.F. and C.S.; Writing—Original Draft Preparation, E.S.; Writing—Review and Editing, P.S.G., C.S. and M.H.F.; Visualization, V.M. and E.S.; Project Administration, P.S.G. and C.S.; Funding Acquisition, M.H.F. and C.S. All authors have read and agreed to the published version of the manuscript.

Funding: This work received financial support from PT national funds (FCT/MCTES, Fundação para a Ciência e Tecnologia and Ministério da Ciência, Tecnologia e Ensino Superior) through the projects UIDB/50006/2020, UIDP/50006/2020 and UIDB/CVT/00772/2020. VM thanks FCT through POCH for his Ph.D. grant, ref. 2020.04935.BD.

Institutional Review Board Statement: The animal study protocol was approved by the Direção-Geral de Alimentação e Veterinária (DGAV, Lisbon, Portugal) (nº010532/2018, 1 June 2018).

Informed Consent Statement: Not applicable.

Data Availability Statement: The data presented in this study are available on request from the corresponding author.

Acknowledgments: This work received support from PT national funds (FCT/MCTES, Fundação para a Ciência e Tecnologia and Ministério da Ciência, Tecnologia e Ensino Superior) through the projects UIDB/50006/2020, UIDP/50006/2020 and UIDB/CVT/00772/2020. The authors acknowledge the support of the i3S Scientific Platform Bioimaging, a member of the PPBI (PPBI-POCI-01-0145-FEDER-022122).

Conflicts of Interest: The authors declare no conflict of interest.

References

1. Zhang, L.; Chen, L. A Review on Biomedical Titanium Alloys: Recent Progress and Prospect. *Adv. Eng. Mater.* **2019**, *21*, 1801215. [CrossRef]
2. Kaur, M.; Singh, K. Review on titanium and titanium based alloys as biomaterials for orthopaedic applications. *Mater. Sci. Eng. C* **2019**, *102*, 844–862. [CrossRef] [PubMed]
3. Kim, K.T.; Eo, M.Y.; Nguyen, T.T.H.; Kim, S.M. General review of titanium toxicity. *Int. J. Implant Dent.* **2019**, *5*, 10. [CrossRef] [PubMed]
4. Mombelli, A.; Hashim, D.; Cionca, N. What is the impact of titanium particles and biocorrosion on implant survival and complications? A critical review. *Clin. Oral Implant. Res.* **2018**, *29*, 37–53. [CrossRef] [PubMed]
5. Heimann, R.B. Structure, properties, and biomedical performance of osteoconductive bioceramic coatings. *Surf. Coat. Technol.* **2013**, *233*, 27–38. [CrossRef]
6. Montazerian, M.; Hosseinzadeh, F.; Migneco, C.; Fook, M.V.L.; Baino, F. Bioceramic coatings on metallic implants: An overview. *Ceram. Int.* **2022**, *48*, 8987–9005. [CrossRef]
7. Vahabzadeh, S.; Roy, M.; Bandyopadhyay, A.; Bose, S. Phase stability and biological property evaluation of plasma sprayed hydroxyapatite coatings for orthopedic and dental applications. *Acta Biomater.* **2015**, *17*, 47–55. [CrossRef]
8. Mohseni, E.; Zalnezhad, E.; Bushroa, A.R. Comparative investigation on the adhesion of hydroxyapatite coating on Ti–6Al–4V implant: A review paper. *Int. J. Adhes. Adhes.* **2014**, *48*, 238–257. [CrossRef]
9. Moloodi, A.; Toraby, H.; Kahrobaee, S.; Razavi, M.K.; Salehi, A. Evaluation of fluorohydroxyapatite/strontium coating on titanium implants fabricated by hydrothermal treatment. *Prog. Biomater.* **2021**, *10*, 185–194. [CrossRef]
10. Kien, P.; Quan, T.; Tuyet Anh, L. Coating Characteristic of Hydroxyapatite on Titanium Substrates via Hydrothermal Treatment. *Coatings* **2021**, *11*, 1226. [CrossRef]
11. Arrés, M.; Salama, M.; Rechena, D.; Paradiso, P.; Reis, L.; Alves, M.M.; Botelho do Rego, A.M.; Carmezim, M.J.; Vaz, M.F.; Deus, A.M.; et al. Surface and mechanical properties of a nanostructured citrate hydroxyapatite coating on pure titanium. *J. Mech. Behav. Biomed. Mater.* **2020**, *108*, 103794. [CrossRef]
12. Borkowski, L.; Przekora, A.; Belcarz, A.; Palka, K.; Jozefaciuk, G.; Lübek, T.; Jojczuk, M.; Nogalski, A.; Ginalska, G. Fluorapatite ceramics for bone tissue regeneration: Synthesis, characterization and assessment of biomedical potential. *Mater. Sci. Eng. C* **2020**, *116*, 111211. [CrossRef]
13. Li, Z.; Huang, B.; Mai, S.; Wu, X.; Zhang, H.; Qiao, W.; Luo, X.; Chen, Z. Effects of fluoridation of porcine hydroxyapatite on osteoblastic activity of human MG63 cells. *Sci. Technol. Adv. Mater.* **2015**, *16*, 035006. [CrossRef]
14. Mansoorianfar, M.; Mansourianfar, M.; Fathi, M.; Bonakdar, S.; Ebrahimi, M.; Zahrani, E.M ; Hojjati-Najafabadi, A.; Li, D. Surface modification of orthopedic implants by optimized fluorine-substituted hydroxyapatite coating: Enhancing corrosion behavior and cell function. *Ceram. Int.* **2020**, *46*, 2139–2146. [CrossRef]
15. Rezaee, T.; Bouxsein, M.L.; Karim, L. Increasing fluoride content deteriorates rat bone mechanical properties. *Bone* **2020**, *136*, 115369. [CrossRef]
16. Charan, J.; Kantharia, N.D. How to calculate sample size in animal studies? *J. Pharmacol. Pharmacother.* **2013**, *4*, 303–306. [CrossRef]
17. Bruker-MicroCT MN074 Osteointegration: Analysis of Bone around a Metal Implant. In *Bruker-MicroCT Method Note*; Bruker Micro-Ct Academy: Kontich, Belgium, 2005; pp. 1–26.
18. Vidal, C.; Alves, P.; Alves, M.M.; Carmezim, M.J.; Fernandes, M.H.; Grenho, L.; Inácio, P.L.; Ferreira, F.B.; Santos, T.G.; Santos, C. Fabrication of a biodegradable and cytocompatible magnesium/nanohydroxyapatite/fluorapatite composite by upward friction stir processing for biomedical applications. *J. Mech. Behav. Biomed. Mater.* **2022**, *129*, 105137. [CrossRef]
19. Wu, Y.-J.; Tseng, Y.-H.; Chan, J.C.C. Morphology Control of Fluorapatite Crystallites by Citrate Ions. *Cryst. Growth Des.* **2010**, *10*, 4240–4242. [CrossRef]
20. Bhadang, K.A.; Gross, K.A. Influence of fluorapatite on the properties of thermally sprayed hydroxyapatite coatings. *Biomaterials* **2004**, *25*, 4935–4945. [CrossRef]
21. Qu, H.; Wei, M. The effect of fluoride contents in fluoridated hydroxyapatite on osteoblast behavior. *Acta Biomater.* **2006**, *2*, 113–119. [CrossRef]
22. Degli Esposti, L.; Adamiano, A.; Tampieri, A.; Ramirez-Rodriguez, G.B.; Siliqi, D.; Giannini, C.; Ivanchenko, P.; Martra, G.; Lin, F.-H.; Delgado-López, J.M.; et al. Combined Effect of Citrate and Fluoride Ions on Hydroxyapatite Nanoparticles. *Cryst. Growth Des.* **2020**, *20*, 3163–3172. [CrossRef]
23. Jha, L.J.; Best, S.M.; Knowles, J.C.; Rehman, I.; Santos, J.D.; Bonfield, W. Preparation and characterization of fluoride-substituted apatites. *J. Mater. Sci. Mater. Med.* **1997**, *8*, 185–191. [CrossRef] [PubMed]
24. Caligari Conti, M.; Xerri, G.; Peyrouzet, F.; Wismayer, P.S.; Sinagra, E.; Mantovani, D.; Vella, D.; Buhagiar, J. Optimisation of fluorapatite coating synthesis applied to a biodegradable substrate. *Surf. Eng.* **2019**, *35*, 255–265. [CrossRef]
25. Bulina, N.V.; Makarova, S.V.; Prosanov, I.Y.; Vinokurova, O.B.; Lyakhov, N.Z. Structure and thermal stability of fluorhydroxyapatite and fluorapatite obtained by mechanochemical method. *J. Solid State Chem.* **2020**, *282*, 121076. [CrossRef]
26. Slimen, J.B.; Hidouri, M.; Ghouma, M.; Salem, E.B.; Dorozhkin, S.V. Sintering of Potassium Doped Hydroxy-Fluorapatite Bioceramics. *Coatings* **2021**, *11*, 858. [CrossRef]
27. Wei, M.; Evans, J.H.; Bostrom, T.; Grøndahl, L. Synthesis and characterization of hydroxyapatite, fluoride-substituted hydroxyapatite and fluorapatite. *J. Mater. Sci. Mater. Med.* **2003**, *14*, 311–320. [CrossRef] [PubMed]

28. Alhilou, A.; Do, T.; Mizban, L.; Clarkson, B.H.; Wood, D.J.; Katsikogianni, M.G. Physicochemical and Antibacterial Characterization of a Novel Fluorapatite Coating. *ACS Omega* **2016**, *1*, 264–276. [CrossRef] [PubMed]
29. Fernandes, M.H.; Alves, M.M.; Cebotarenco, M.; Ribeiro, I.A.C.; Grenho, L.; Gomes, P.S.; Carmezim, M.J.; Santos, C.F. Citrate zinc hydroxyapatite nanorods with enhanced cytocompatibility and osteogenesis for bone regeneration. *Mater. Sci. Eng. C* **2020**, *115*, 111147. [CrossRef]
30. Deng, L.; Ou, J.; Yang, H.; Wen, G.; Huang, T. The surface regulation of calcite for defluoridation by fluorapatite-induced crystallization. *J. Water Process Eng.* **2021**, *41*, 102082. [CrossRef]
31. Joseph Nathanael, A.; Mangalaraj, D.; Hong, S.I.; Masuda, Y.; Rhee, Y.H.; Kim, H.W. Influence of fluorine substitution on the morphology and structure of hydroxyapatite nanocrystals prepared by hydrothermal method. *Mater. Chem. Phys.* **2013**, *137*, 967–976. [CrossRef]
32. Sun, J.; Wu, T.; Fan, Q.; Hu, Q.; Shi, B. Comparative study of hydroxyapatite, fluor-hydroxyapatite and Si-substituted hydroxyapatite nanoparticles on osteogenic, osteoclastic and antibacterial ability. *RSC Adv.* **2019**, *9*, 16106–16118. [CrossRef]
33. Blanc-Sylvestre, N.; Bouchard, P.; Chaussain, C.; Bardet, C. Pre-Clinical Models in Implant Dentistry: Past, Present, Future. *Biomedicines* **2021**, *9*, 1538. [CrossRef]
34. Stübinger, S.; Dard, M. The Rabbit as Experimental Model for Research in Implant Dentistry and Related Tissue Regeneration. *J. Investig. Surg.* **2013**, *26*, 266–282. [CrossRef]
35. Shin, D.; Blanchard, S.B.; Ito, M.; Chu, T.-M.G. Peripheral quantitative computer tomographic, histomorphometric, and removal torque analyses of two different non-coated implants in a rabbit model. *Clin. Oral Implant. Res.* **2011**, *22*, 242–250. [CrossRef]
36. Seong, W.-J.; Grami, S.; Jeong, S.C.; Conrad, H.J.; Hodges, J.S. Comparison of Push-In versus Pull-Out Tests on Bone-Implant Interfaces of Rabbit Tibia Dental Implant Healing Model. *Clin. Implant Dent. Relat. Res.* **2013**, *15*, 460–469. [CrossRef]
37. Garbieri, T.F.; Martin, V.; dos Santos, C.F.; Gomes, P.D.S.; Fernandes, M.H. The Embryonic Chick Femur Organotypic Model as a Tool to Analyze the Angiotensin II Axis on Bone Tissue. *Pharmaceuticals* **2021**, *14*, 469. [CrossRef]
38. Francisco, I.; Vale, F.; Martin, V.; Fernandes, M.H.; Gomes, P.S. From Blood to Bone—The Osteogenic Activity of L-PRF Membranes on the Ex Vivo Embryonic Chick Femur Development Model. *Materials* **2021**, *14*, 7830. [CrossRef]
39. Araújo, R.; Martin, V.; Ferreira, R.; Fernandes, M.H.; Gomes, P.S. A new ex vivo model of the bone tissue response to the hyperglycemic environment—The embryonic chicken femur organotypic culture in high glucose conditions. *Bone* **2022**, *158*, 116355. [CrossRef]
40. Liu, Y.; Xie, D.; Zhou, R.; Zhang, Y. 3D X-ray micro-computed tomography imaging for the microarchitecture evaluation of porous metallic implants and scaffolds. *Micron* **2021**, *142*, 102994. [CrossRef]
41. Pajor, K.; Pajchel, L.; Kolmas, J. Hydroxyapatite and Fluorapatite in Conservative Dentistry and Oral Implantology—A Review. *Materials* **2019**, *12*, 2683. [CrossRef]
42. Kattimani, V.S.; Kondaka, S.; Lingamaneni, K.P. Hydroxyapatite—Past, Present, and Future in Bone Regeneration. *Bone Tissue Regen. Insights* **2016**, *7*, BTRI.S36138. [CrossRef]
43. Alsabeeha, N.H.M.; Ma, S.; Atieh, M.A. Hydroxyapatite-coated oral implants: A systematic review and meta-analysis. *Int. J. Oral Maxillofac. Implants* **2012**, *27*, 1123–1130. [PubMed]
44. Yang, S.; Lee, S.; Bajpai, I.; Kim, S. Hydrothermal treatment of Ti surface to enhance the formation of low crystalline hydroxyl carbonate apatite. *Biomater. Res.* **2015**, *19*, 4. [CrossRef] [PubMed]
45. Lo, Y.-S.; Chang, C.-C.; Lin, P.-C.; Lin, S.-P.; Wang, C.-L. Direct growth of structurally controllable hydroxyapatite coating on Ti-6Al-4V through a rapid hydrothermal synthesis. *Appl. Surf. Sci.* **2021**, *556*, 149672. [CrossRef]
46. Bhadang, K.A.; Holding, C.A.; Thissen, H.; McLean, K.M.; Forsythe, J.S.; Haynes, D.R. Biological responses of human osteoblasts and osteoclasts to flame-sprayed coatings of hydroxyapatite and fluorapatite blends. *Acta Biomater.* **2010**, *6*, 1575–1583. [CrossRef]
47. Tredwin, C.J.; Young, A.M.; Abou Neel, E.A.; Georgiou, G.; Knowles, J.C. Hydroxyapatite, fluor-hydroxyapatite and fluorapatite produced via the sol–gel method: Dissolution behaviour and biological properties after crystallisation. *J. Mater. Sci. Mater. Med.* **2014**, *25*, 47–53. [CrossRef]
48. Kim, H.-W.; Kim, H.-E.; Knowles, J.C. Fluor-hydroxyapatite sol–gel coating on titanium substrate for hard tissue implants. *Biomaterials* **2004**, *25*, 3351–3358. [CrossRef]
49. Gross, K.A.; Rodríguez-Lorenzo, L.M. Sintered hydroxyfluorapatites. Part I: Sintering ability of precipitated solid solution powders. *Biomaterials* **2004**, *25*, 1375–1384. [CrossRef]
50. Gao, Y.; Karpukhina, N.; Law, R.V. Phase segregation in hydroxyfluorapatite solid solution at high temperatures studied by combined XRD/solid state NMR. *RSC Adv.* **2016**, *6*, 103782–103790. [CrossRef]
51. Everett, E.T. Fluoride's Effects on the Formation of Teeth and Bones, and the Influence of Genetics. *J. Dent. Res.* **2011**, *90*, 552–560. [CrossRef]
52. Wang, Y.; Zhang, S.; Zeng, X.; Ma, L.L.; Weng, W.; Yan, W.; Qian, M. Osteoblastic cell response on fluoridated hydroxyapatite coatings. *Acta Biomater.* **2007**, *3*, 191–197. [CrossRef]
53. Pan, L.; Shi, X.; Liu, S.; Guo, X.; Zhao, M.; Cai, R.; Sun, G. Fluoride promotes osteoblastic differentiation through canonical Wnt/β-catenin signaling pathway. *Toxicol. Lett.* **2014**, *225*, 34–42. [CrossRef]
54. Huo, L.; Liu, K.; Pei, J.; Yang, Y.; Ye, Y.; Liu, Y.; Sun, J.; Han, H.; Xu, W.; Gao, Y. Flouride Promotes Viability and Differentiation of Osteoblast-Like Saos-2 Cells Via BMP/Smads Signaling Pathway. *Biol. Trace Elem. Res.* **2013**, *155*, 142–149. [CrossRef]
55. Pei, J.; Li, B.; Gao, Y.; Wei, Y.; Zhou, L.; Yao, H.; Wang, J.; Sun, D. Fluoride decreased osteoclastic bone resorption through the inhibition of NFATc1 gene expression. *Environ. Toxicol.* **2014**, *29*, 588–595. [CrossRef]

56. Liu, X.; Song, J.; Liu, K.; Wang, W.; Xu, C.; Zhang, Y.; Liu, Y. Role of inhibition of osteogenesis function by Sema4D/Plexin-B1 signaling pathway in skeletal fluorosis in vitro. *J. Huazhong Univ. Sci. Technol. Med. Sci.* **2015**, *35*, 712–715. [CrossRef]
57. Zipkin, I.; Bernick, S.; Menczel, J. A Morphological Study of the Effect of Fluoride on the Periodontium of the Hydrocortisone-Treated Rat. *Periodontics* **1965**, *3*, 111–114.
58. Cook, F.J.; Seagrove-Guffey, M.; Mumm, S.; Veis, D.J.; McAlister, W.H.; Bijanki, V.N.; Wenkert, D.; Whyte, M.P. Non-endemic skeletal fluorosis: Causes and associated secondary hyperparathyroidism (case report and literature review). *Bone* **2021**, *145*, 115839. [CrossRef]
59. Wu, S.; Xia, B.; Mai, S.; Feng, Z.; Wang, X.; Liu, Y.; Liu, R.; Li, Z.; Xiao, Y.; Chen, Z.; et al. Sodium Fluoride under Dose Range of 2.4–24 μM, a Promising Osteoimmunomodulatory Agent for Vascularized Bone Formation. *ACS Biomater. Sci. Eng.* **2019**, *5*, 817–830. [CrossRef]
60. Nagendra, A.H.; Bose, B.; Shenoy, P.S. Recent advances in cellular effects of fluoride: An update on its signalling pathway and targeted therapeutic approaches. *Mol. Biol. Rep.* **2021**, *48*, 5661–5673. [CrossRef]
61. Costello, L.C.; Chellaiah, M.; Zou, J.; Franklin, R.B.; Reynolds, M.A. The status of citrate in the hydroxyapatite/collagen complex of bone; and Its role in bone formation. *J. Regen. Med. Tissue Eng.* **2014**, *3*, 4. [CrossRef]
62. Morganti, C.; Bonora, M.; Marchi, S.; Ferroni, L.; Gardin, C.; Wieckowski, M.R.; Giorgi, C.; Pinton, P.; Zavan, B. Citrate Mediates Crosstalk between Mitochondria and the Nucleus to Promote Human Mesenchymal Stem Cell In Vitro Osteogenesis. *Cells* **2020**, *9*, 1034. [CrossRef]
63. Ma, C.; Tian, X.; Kim, J.P.; Xie, D.; Ao, X.; Shan, D.; Lin, Q.; Hudock, M.R.; Bai, X.; Yang, J. Citrate-based materials fuel human stem cells by metabonegenic regulation. *Proc. Natl. Acad. Sci. USA* **2018**, *115*, E11741–E11750. [CrossRef] [PubMed]
64. Shares, B.H.; Busch, M.; White, N.; Shum, L.; Eliseev, R.A. Active mitochondria support osteogenic differentiation by stimulating β-catenin acetylation. *J. Biol. Chem.* **2018**, *293*, 16019–16027. [CrossRef] [PubMed]
65. Schafrum Macedo, A.; Cezaretti Feitosa, C.; Yoiti Kitamura Kawamoto, F.; Vinicius Tertuliano Marinho, P.; dos Santos Dal-Bó, Í.; Fiuza Monteiro, B.; Prado, L.; Bregadioli, T.; Antonio Covino Diamante, G.; Ricardo Auada Ferrigno, C. Animal modeling in bone research—Should we follow the White Rabbit? *Anim. Model. Exp. Med.* **2019**, *2*, 162–168. [CrossRef] [PubMed]

Journal of
Functional Biomaterials

MDPI

Article

Estrogen Deficiency Impairs Osseointegration in Hypertensive Rats Even Treated with Alendronate Coated on the Implant Surface

Gabriel Mulinari-Santos [1,*], Jaqueline Silva dos Santos [2], Igor Lebedenco Kitagawa [3], Fábio Roberto de Souza Batista [2], Paulo Roberto Botacin [2], Cristina Antoniali [2], Paulo Noronha Lisboa-Filho [4] and Roberta Okamoto [2,*]

1 Department of Diagnostic and Surgery, School of Dentistry, São Paulo State University (UNESP), Araçatuba, SP, Brazil
2 Department of Basic Science, School of Dentistry, São Paulo State University (UNESP), Araçatuba, SP, Brazil
3 Federal Institute of Education, Science and Technology of São Paulo (IFSP), Birigui, SP, Brazil
4 Department of Physics, School of Science, São Paulo State University (UNESP), Bauru, SP, Brazil
* Correspondence: gabriel.mulinari@unesp.br (G.M.-S.); roberta.okamoto@unesp.br (R.O.); Tel.: +55-18-3636-2787 (R.O.)

Abstract: Hypertension and estrogen deficiency can affect bone metabolism and therefore increase the risk of osseointegration. Antihypertensive drugs such as losartan not only control blood pressure but also enhance bone healing. In addition, alendronate sodium is widely used to treat postmenopausal osteoporosis. Hence, we evaluated the effect of systemic antihypertensive and local alendronate coted on implants on osseointegration under hypertensive and estrogen-deficiency conditions. A total of 64 spontaneously hypertensive rats (SHRs) treated with losartan were randomly divided according to the estrogen-deficiency induction by ovariectomy (OVX) or not (SHAM), and whether the implant surface was coated with sodium alendronate (ALE) or not, resulting in four groups: SHR SHAM, SHR SHAM ALE, SHR OVX, and SHR OVX ALE. The removal torque, microcomputed tomography, and epifluorescence microscopy were the adopted analyses. The hypertensive and estrogen-deficiency animals presented a lower removal torque even when treated with alendronate on implant surface. The microcomputed tomography revealed a higher bone volume and bone-to-implant contact in the SHRs than the SHR OVX rats. Epifluorescence showed a decreased mineral apposition ratio in the SHR OVX ALE group. The data presented indicate that estrogen deficiency impairs osseointegration in hypertensive rats; in addition, alendronate coated on the implant surface does not fully reverse this impaired condition caused by estrogen deficiency.

Keywords: bone; losartan; osseointegration; alendronate; spontaneously hypertensive rats

Citation: Mulinari-Santos, G.; dos Santos, J.S.; Kitagawa, I.L.; de Souza Batista, F.R.; Botacin, P.R.; Antoniali, C.; Lisboa-Filho, P.N.; Okamoto, R. Estrogen Deficiency Impairs Osseointegration in Hypertensive Rats Even Treated with Alendronate Coated on the Implant Surface. *J. Funct. Biomater.* **2023**, *14*, 471. https://doi.org/10.3390/jfb14090471

Academic Editor: Håvard J. Haugen

Received: 16 August 2023
Revised: 3 September 2023
Accepted: 11 September 2023
Published: 13 September 2023

1. Introduction

Hypertension and osteoporosis are the major causes of mortality and morbidity in public health, especially due to the increased risks for cardiovascular diseases and bone fractures [1]. These systemic alterations are more common in postmenopausal women; moreover, hypertensive females develop osteoporosis more frequently [2]. Osteoporotic patients are also more associated with hypertension, and there is a possible relation between hypertension and osteoporosis in diets with low calcium intake [2], mainly due to estrogen deficiency [3]. Hypertension and estrogen deficiency are responsible for promoting bone metabolism changes [4–7]. Thus, both of these associated conditions may be at high risk for impaired osseointegration. Despite the elevated frequency of postmenopausal hypertensive individuals, peri-implant bone healing under hypertension and estrogen deficiency remains uncertain. At least, therapeutic strategies must be adopted to improve this compromised bone metabolism.

Antihypertensive drugs act to suppress the renin–angiotensin system, prevent vasoconstriction, and control blood pressure [8]. Additionally, recent studies indicate antihypertensive drugs are capable of acting on the bone metabolism [9–11]. Components of the renin–angiotensin system play a role in bone [9,12]. Consequently, angiotensin II (Ang II) increases the bone resorption of osteoclasts by increasing the receptor activator of nuclear factor-κB ligand (RANKL) [9–11]. Furthermore, Ang II diminishes osteoblasts, reducing the expression of runt-related transcription factor 2 (RUNX2) and osteocalcin [13]. Conversely, losartan is an angiotensin II receptor type 1 antagonist widely used to control hypertension [14], which also impacts the bones [9–11]. Anticatabolic effects, such as improving the mechanical properties of the alveolar bone [15] and decreasing the risks of fractures from osteoporosis, were found [16,17]. In preclinical studies, losartan has improved bone mineralization and fracture healing [13,18], corroborating with clinical studies where antihypertensive drugs were stated as responsible for increasing the survival of osseointegrated implants [19,20]. Thereby, losartan may provide benefits for osseointegration in compromised bone conditions by hypertension and estrogen deficiency.

Alendronate sodium is a drug extensively used for osteoporosis; it consists of second-generation bisphosphonate, classified as an antiresorptive agent [21,22]. This drug can decrease bone fractures and improve alveolar bone healing [23]. Recent research has shown increased osseointegration parameters in osteoporosis models treated with alendronate [24]. However, the clinical use of alendronate is still imprecise, since there are side effects, such as nausea, vomiting, gastrointestinal pain [25], and osteonecrosis of the jaws, after long-term systemic use [26]. For reducing side effects, alendronate administrated locally has being one of the most promising therapies [23,24,27]. When it was applied on the implant surface, alendronate improved the values of bone volume and bone–implant contact [24], as well as in experimental osteoporosis [28]. Therefore, the local use of alendronate coated on the implant surface can be a beneficial treatment strategy for implant placement in hypertensive females with estrogen deficiency. Given the high incidence of hypertension and osteoporosis especially in females [4], the model of estrogen deficiency by ovariectomy in spontaneously hypertensive rats was proposed to simulate this challenging clinical condition for rehabilitation with osseointegrated implants.

2. Materials and Methods

2.1. Study Design and Ethics

This study was sent to the Animal Research Ethics Committee and obtained a favorable opinion under number 00332-2018. All research procedures have been established according to the ARRIVE guidelines [29]. A total of 64 female spontaneously hypertensive rats (SHRs), 6 months old, weighing around 300 g, were used. The animals were preserved in cages in a constant temperature environment (22 °C ± 2 °C; light control cycle of 12 light hours and 12 h dark) and provided a balanced diet (1.4% Ca and 0.8% Pe, Nuvilab, Curitiba, PR, Brazil). The rats were separated into four groups according to the estrogen-deficiency induction by ovariectomy surgery (OVX) or not (SHAM), and whether the implant surface was treated with sodium alendronate (ALE) or not, resulting in: SHR SHAM, SHR SHAM ALE, SHR OVX, and SHR OVX ALE. The randomization list was created with a computer-generated list in Stata 9.0 (StataCorp, College Station, TX, USA).

2.2. Strus Cycle and Ovariectomy Surgery

Before the ovariectomy surgery, it was checked whether the hypertensive rats were cycling normally. Daily, they were placed in individual cages and 1–2 drops of saline were introduced into the vagina, then extracted and positioned on a histology slide for microscopic reading to recognize the phases of the estrous cycle, using the Long and Evans technique [30]. Rats were used after obtaining 2 to 3 regular estrous cycles. Surgical removal of both ovaries was performed under anesthesia with 100 mg/kg of ketamine (Vetaset—Fort Dodge Saúde Animal Ltda., Campinas, São Paulo, Brazil) intramuscularly and 5 mg/kg of xylazine (Dopaser—Laboratório Calier do Brasil Ltda.—Osasco, São Paulo, Brazil) intraperitoneally.

2.3. Losartan Treatment and Systolic Blood Pressure

Losartan (Biosintetica, São Paulo, Brazil) was administered orally in all groups at concentrations of 30 mg/kg/day until euthanasia. The losartan treatment started 30 days after the ovariectomy and 7 days before the implant surgery, and continued after implant placement, extending until the euthanasia. Systolic blood pressure was checked immediately prior to the implant placement to ensure that the blood pressure was controlled below 150 mmHg. This verification was used to simulate a real control condition, since uncontrolled hypertension preoperatively contraindicates implant surgery. Here, we used the tail-cuff indirect plethysmography method with a Physiograph® MK-III-S (Narco Bio-Sistemas, Houston, TX, USA) according to a prior study [31].

2.4. Treatment of Implants with Sodium Alendronate

The implants were coated using the layer-by-layer technique, in which layers of polyelectrolyte were added via electrostatic interaction. For a monolayer of alendronate sodium coating on the implant's surface, implants were first immersed in a layer of sodium polystyrene sulfate. Then, the implants were immersed in a layer of titanium dioxide; finally, the implants were coated with a layer of sodium alendronate at a concentration of 1 g using electrostatic attraction. The suspension ingredients were blended in accurate quantities in water-cooled glass vials with a sonicator probe (model XLS-2015, Misonix, Inc., New York, NY, USA). Each step of the implant's coating layer followed a previously described method [32].

2.5. Implant Placement

The implant placement followed an earlier successful bicortical rat-tibia-implant model [33,34]. For the surgery to place the implants in the tibias, the animals underwent preoperative fasting for 8 h to be anesthetized with a combination of 100 mg/kg of ketamine (Vetaset—Fort Dodge Saúde Animal Ltda., Campinas, São Paulo, Brazil) intramuscularly and 5 mg/kg of xylazine (Dopaser—Laboratório Calier do Brasil Ltda.–Osasco, São Paulo, Brazil). Mepivacaine hydrochloride (0.3 mL/kg 2% Scandicaine, adrenaline 1: 100,000; Septodont, Saint-Maur-des-Fosse's, France) was administered for anesthesia and hemostasis. After anesthesia, the surgical site in the medial tibial portion was shaved and subjected to topical disinfection with degerming iodine (10% PVPI, Riodeine degermante; Rioquímica, São José do Rio Preto, SP, Brazil). A 1.5 cm incision was performed, followed by divulsion to expose the tibial metaphysis. A grade-four titanium implant treated with aluminum oxide and nitric acid (HNO3, Mendes, Itu, São Paulo, Brazil) was installed in the SHR SHAM and SHR OVX groups, with dimensions of 2 mm in diameter and 4 mm in length. The SHR SHAM ALE and SHR OVX ALE groups received the same implants with alendronate incorporated into their surfaces. With the 1.6 mm diameter reamer coupled to the electric motor (BLM 600, Driller, São Paulo, SP, Brazil) at a speed of 1000 RPM under isotonic irrigation of 0.9% sodium chloride (Fisiológico, Laboratórios Biosintética Ltda., Ribeirão Preto, SP, Brazil), the surgical defect was created for subsequent manual installation of the implant with a hexagonal digital driver. The wound was sutured in layers, with the deep layer using absorbable thread (Vycril 4.0, Ethicon, Johnson, São José dos Campos, Brazil) and the outer layer with monofilament (Nylon 5.0, Ethicon, Johnson, São José dos Campos, Brazil). In the immediate postoperative, the animals received an injection of intramuscular pentabiotic (0.1 mg/kg, Fort Dodge Saúde Animal Ltda., Campinas, São Paulo, Brazil) and sodium dipyrone (1 mg/kg/1 day $^{-1}$, Ariston Indústrias Químicas e Farmacêuticas Ltda., São Paulo, Brazil).

2.6. Fluorochrome Application

Twenty-one days after starting losartan treatment, fluorochrome calcein was administered intramuscularly (20 mg/kg, Sigma Chemical Company, St. Louis, MI, USA). In addition, alizarin red fluorochrome was used intramuscularly after 28 days (20 mg/kg, Sigma Chemical Company, St. Louis, MI, USA). The earliest infused calcein fluorochrome implied calcium deposition in the old bone, and later, the alizarin fluorochrome in the newly-formed bone, as per a previous study [31,33].

2.7. Euthanasia

All animals were euthanized 60 days after implant installation using a lethal dosage of anesthetic (60 mg/kg; Tiopental Cristália, Ltda., Itapira, SP, Brazil).

2.8. Analysis

2.8.1. Biomechanical Test

Removal torque was measured using an implant hexagon and a digital torque (torque meter, Conexão, São Paulo, Brazil). Then, the removal torque of eight tibias per subgroup was registered with an anticlockwise movement utilized by increasing the force for the removal torque until the implant turned in the bone at the highest torque point in Newton centimeters [34].

2.8.2. Computed Microtomography (MicroCT)

The tibias of eight rats per subgroup were removed after euthanasia, fixed in a 10% formalin solution (Analytical Reagents®, Dinâmica Odonto-Hospitalar Ltda., Catanduva, SP, Brazil) for 48 h, bathed in running water for 24 h, and stored in 70% alcohol to be transported until scanned. For scanning, SkyScan microtomography (SkyScan 1272 Bruker MicroCT, Aatselaar, Belgium, 2003) was used with 9 µm thick slices (50 Kv and 500 µ), copper and aluminum filters, and a rotation pitch of 0.3 mm. The sequence of images obtained was three-dimensionally reconstructed using the NRecon software (SkyScan, 2011; Version 1.6.6.0). In the Data Viewer software (SkyScan, Version 1.4.4 64-bit), the positioning of all samples in the transverse, longitudinal, and sagittal planes was standardized. Then, in the CTAnalyzer software (2003-11SkyScan, 2012 Bruker MicroCT Version 1.12.4.0), the analysis region around the implant was defined in the transverse plane. They were marked at 50 slices in a proximal direction and 50 slices in a distal direction in the bone in which the implant was installed, thus totaling a volume of 100 slices (874.3 µm). The CTAnalyzer software calculated the morphometric value based on the grayscale (threshold). The threshold used in the analysis was 25–90 shades of gray to delimit the bone formed around the implants. From the conversion, the software carried out the morphometric calculation applied to the implant in a transverse direction and provided the parameters of the bone volume percentage (BV/TV), the amount of bone in contact between the bone and implant (BIC), the trabecular thickness (Tb.Th), the number of trabeculae (Tb.N), the trabecular separation (Tb.Sp), and the total porosity (PO TOT). The adopted parameters followed the guidelines provided by the American Society of Bone and Mineral Research [35] and analyses carried out in previous studies [34]. The microtomographic reconstruction of each of the samples was performed using the CTvox software (SkyScan, Version 2.7).

2.8.3. Epifluorescence Microscopy

After fixation in formaldehyde 10% for 48 h, the pieces were washed for 24 h in continuously running water and dehydrated in a gradually increasing order of alcohols. The specimens were included in a solution at a 1:1 ratio of acetone and methylmethacrylate (Classical, Dental Articles Classical, São Paulo, SP, Brazil), followed by methylmethacrylate baths. In the final bath was added benzoyl peroxide (1%, Riedel-de Haen AG, Seelze-Hannover, Germany). Samples were kept in glass tubing at 37 °C for five days. After polymerization, resin blocks holding the samples were detached from the glass cylinders. Then, the blocks were cut down to the implant along the tibia longitudinal plane with a drill mounted on an electric motor (Strong 210, São Paulo, SP, Brazil). Later, the samples were ground on politriz (Ecomet 250 pro automate 250, Buchler, Lake Bluff, IL, USA) with sandpaper (120, 300, 400, 600, 800, and 1200 granules, Carbimet, Buchler, Lake Bluff, IL, USA) to the thickness of 80 μm, checked by a digital caliper (Mitutoyo, Pompeia, São Paulo, Brazil). The pieces were mounted on slides submerged in mineral oil (Petrolato liquid, Maantecor, Taquara, Rio de Janeiro, Brazil).

Later, once the slides were ready, it was possible to capture images of the peri-implant bone using the confocal laser microscope Leica CTR 4000 CS SPE (Leica Microsystems, Heidelberg, Germany), using a 10X objective (Original magnification of 100×). Figures were recreated in a microscope software (Leica CTR 4000 CS SPE, Leica Microsystems, Heidelberg, Germany). The blue filter was used to visualize the calcein fluorochrome, shown in green color. The alizarin fluorochrome was revealed in red by the green filter. The scanned images were analyzed with the Image J software version 1.53 (Image Analyzer Software, Ontario, ON, Canada). The tool "color selection" was used on each image for standardizing the hue, saturation, and brightness. The peri-implant bone was detected in the same location on a unique slide to measure the mineral apposition rate. The daily value of mineralization was calculated by the distance between the calcein and alizarin, divided by the 28 days among their applications, resulting in the mineral apposition rate, as previously reported [31].

2.9. Statistics

The statistics were performed in the GraphPad Prism 9 software (GraphPad Software; La Jolla, CA; USA). Having obtained all data, first, the homoscedasticity test (Shapiro–Wilk) was performed, considering $p < 0.05$ to confirm the distribution of the data in the normality curve. Thus, the two-way ANOVA test was applied for two independent variables, such as the systemic condition (SHAM vs. OVX) and the treatment surface with ALE or not, and one dependent variable, such as the removal torque in Newton per centimeter, as well as each microtomography parameter (BV/TV; BIC; Tb.Th; Tb.N; Tb.Sp; PO TOT), mineral apposition rate value in micrometers, and the systolic blood pressure. In case of a significant relational effect, Sidak's multiple comparisons test was directed to test all multiple-treatment groups.

3. Results

3.1 Systolic Blood Pressure

After 7 days of treatment with losartan, all animals obtained the systolic blood pressure below 150 mmHg preoperatively. This key feature of the target was determined since untreated hypertensive rats exhibit blood pressure levels above this threshold [31,36,37]. This adopted value reflects the effectiveness of the reduction in blood pressure by losartan with measurements based on prior studies [31,37]. Antihypertensive drugs have previously demonstrated to be capable of reducing the blood pressure of hypertensive animals from 171.4 ± 0.6 to 129.3 ± 0.9 mmHg when treated with losartan [31] and from 159.91 ± 7.09 to 114.07 ± 8.07 mmHg when treated with captopril [37]. Therefore, after confirmation that losartan effectively managed the systolic blood in the preoperative, all animals followed to the implant-placement surgery.

3.2. Removal Torque Values

Sixty days after implant installation, the torque was measured for all experimental groups (Figure 1). Before the removal torque test, it was observed clinically that all animals exhibited stable consolidation of the implants in their tibias. The successful installation of the implants was confirmed by the removal test and subsequent microtomography analysis, which showed that all implants demonstrated biomechanical resistance and peri-implant osteogenesis.

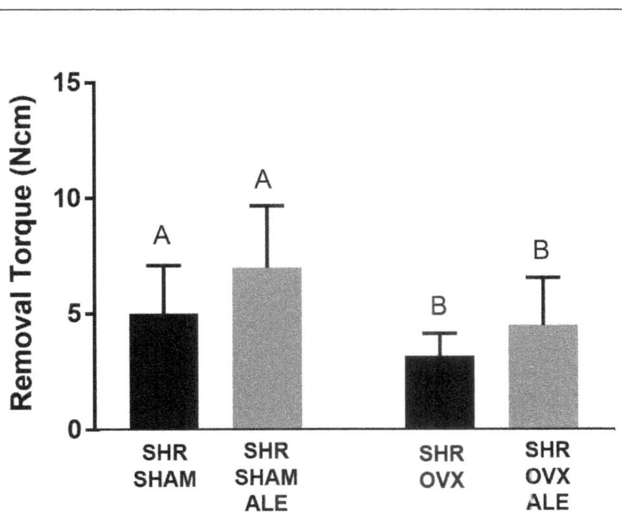

Figure 1. Biomechanical test of removal torque. Sixty days following implant installation, the removal torque for each experimental group was: SHR SHAM, SHR SHAM ALE, SHR OVX, and SHR OVX ALE. The removal torque was augmented until the implant rotated inside the bone and the highest torque in Newton centimeter (Ncm) was registered. The different uppercase letters indicate significant statistical differences in the systemic condition of the ovariectomy, comparing both SHAM groups to OVX groups ($p < 0.05$).

The removal torque increased until the implant turned within the bone, and the peak strength value was recorded. In evaluating the values, the SHR SHAM ALE group obtained the highest average of 7.2 Ncm. The SHR SHAM group achieved an average of 5 Ncm, and the SHR OVX ALE group was 4.7 Ncm. The SHR OVX group obtained the lowest value with 3.33 Ncm. Having these data, first, the homoscedasticity test (Shapiro–Wilk) was performed considering a $p < 0.05$. Thus, the two-way ANOVA test was applied for two independent variables, the systemic condition and treatment surface, and one dependent variable, the removal torque. A statistical difference was observed between the systemic condition in the SHAM vs. OVX groups ($p = 0.017$).

Thereby, it was first determined whether hypertension and estrogen deficiency negatively affected the resistance against removal by the torque wrench. In support of this hypothesis, the removal torque was significantly lower in both the SHR OVX and SHR OVX ALE groups compared to both the SHR SHAM and SHR SHAM ALE groups (3.33 ± 0.1 Ncm and 4.7 ± 0.3 Ncm vs. 5.0 ± 0.2 Ncm and 7.2 ± 0.3 Ncm; $p = 0.017$). In the SHR OVX ALE group, the local alendronate had a tendency toward an elevation in the removal torque; however, it did not achieve significance ($p = 0.55$). The biomechanical test exposed the negative impacts of estrogen deficiency caused by ovariectomy on implant biomechanical stability in hypertensive rats (Figure 1).

3.3. Microcomputed Tomography Results

The effects of untreated hypertension on the analyzed parameters of microtomography are already reported in the literature [34]. Untreated hypertension has shown a peri-implant bone thickness of 0.096 ± 0.003 mm and a BV/TV of about 45% in hypertensive rats [34]. These data were used as a starting point for the present experimental design.

Consistent with the biomechanical outcomes, the systemic condition of estrogen deficiency by ovariectomy significantly decreased the mean BV/TV in the hypertensive and ovariectomized rats of both the SHR OVX and SHR OVX ALE groups compared to the SHR SHAM and SHR SHAM ALE groups ($45.5\% \pm 0.6\%$ and $47.9\% \pm 0.4\%$ vs. $61.5\% \pm 0.6\%$ and $64.8\% \pm 0.6\%$; $p < 0.0001$; two-way ANOVA). Also, a statistical difference was noted between the alendronate groups of the SHR SHAM ALE and SHR OVX ALE groups (p = 0018; Sidak's test). Additionally, the SHR SHAM and SHR OVX groups with conventional surfaces were statically different (p = 0.0039; Sidak's test), reinforcing the systemic condition, and not the implant's alendronate surface, as the major cause of the difference when comparing the groups (Figure 2a).

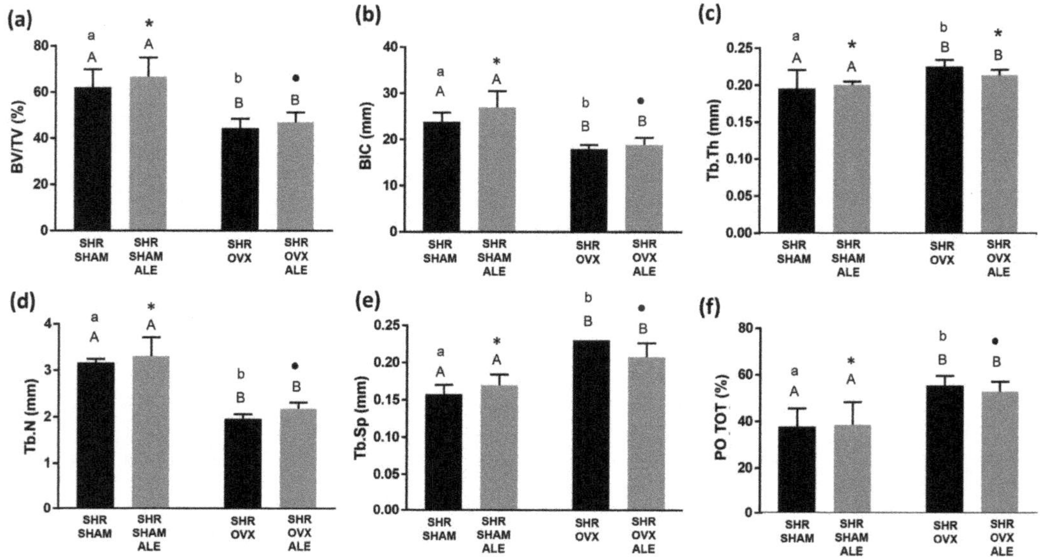

Figure 2. Microcomputerized tomography of peri-implant bone. Morphologic parameters were calculated and reported as: (**a**) Bone volume per tissue volume (BV/TV); (**b**) Bone-to-implant contact; (**c**) Trabecular thickness (Tb.Th); (**d**) Trabecular number (Tb.N.); (**e**) Trabecular separation (Tb.Sp); (**f**) Total porosity. The different uppercase letters indicate a significant statistical difference ($p < 0.05$) between systemic conditions. The different lowercase letters denote significant statistical differences ($p < 0.05$) between the SHR SHAM and SHR OVX groups with no alendronate surface. The alteration of symbols "*" and dot represents statistical differences ($p < 0.05$) between the SHR SHAM ALE and SHR OVX ALE groups.

Accordingly, the BIC demonstrated the highest value for the SHR SHAM ALE group, followed by the SHR SHAM group (25 mm ± 0.6 and 22 mm ± 0.4), and both were statistically different with both OVX groups ($p < 0.0001$; two-way ANOVA). A statistical difference between the alendronate groups of SHR SHAM ALE and SHR OVX ALE was obtained (p = 0.004; Sidak's test). Also, the SHR SHAM and SHR OVX groups with conventional surfaces were statically different (p = 0005; Sidak's test; Figure 2b).

Possibly, the catabolic effect of estrogen deficiency on BV/TV obtained a trend to recompense by increasing the thickness of the bone trabeculae in the SHR OVX and SHR OVX ALE animals compared to the SHR SHAM and SHR SHAM ALE controls (0.22 ± 0.149 mm and 0.21 ± 0.198 mm vs. 0.019 ± 0.003 mm and 0.019 ± 0.003 mm; *, p = 0.01; two-way ANOVA). Interestingly, the SHR OVX group obtained a statistically different greater thickness in trabeculae than the SHR SHAM group with conventional surfaces (p = 0.02; Sidak's test; Figure 2c).

Nevertheless, the number of bone trabeculae decreased in both OVX groups compared to both SHAM groups ($p < 0.0001$; two-way ANOVA; Figure 2d). The catabolic changes by estrogen deficiency caused a higher trabecular separation in both OVX groups ($p < 0.05$; two-way ANOVA; Figure 2e) and elevated the porosity (p = 0.0006; two-way ANOVA; Figure 2f) when compared only with the hypertensive animals. The number of trabeculae and the total porosity were statistically different between the groups with implant surfaces with no alendronate ($p < 0.05$; Sidak's test) and with an alendronate surface ($p < 0.05$; Sidak's test). Figure 3 contains the microtomography volumetric images of the following groups of samples: SHR SHAM, SHR SHAM ALE, SHR OVX, and SHR OVX ALE. The microtomography structural analysis indicates that estrogen deficiency added to hypertension exerts catabolic effects in the peri-implant bone of hypertensive rats, even with systemic losartan and local alendronate coated on the implant surface.

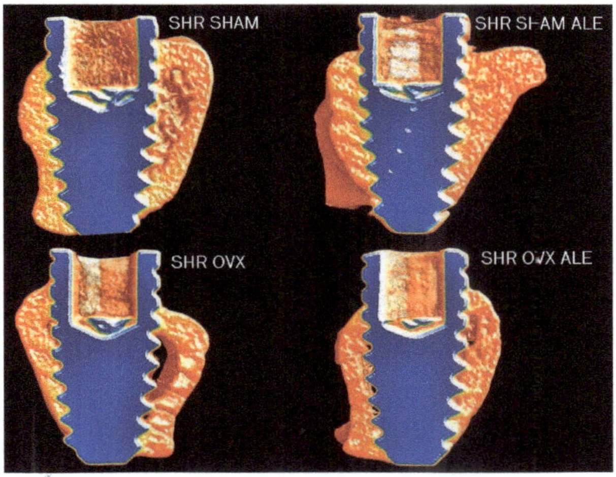

Figure 3. The microtomographic reconstruction of the bicortical implant in the tibia of each group. The microtomography images are representative from all four groups: SHR SHAM, SHR SHAM ALE, SHR OVX, and SHR OVX ALE, respectively. The microtomography images evidenced that estrogen deficiency reduces the peri-implant bone volume in hypertensive ovariectomized rats from the SHR OVX group. The microtomography was executed applying the CTvox software (SkyScan, Version 2.7).

3.4. *Epifluorescence Results*

In the epifluorescence analysis, it is possible to measure the amount of calcium and alizarin precipitation at the time of the fluorochrome injection (Figure 4). The graph further down shows precipitation at the time of the injection of calcein and alizarin fluorochrome. The daily mineral apposition showed a balance between all evaluated groups. The synergistic action of systemic losartan and local alendronate contributed to the fact that daily calcium precipitation on the peri-implant bone occurred in all groups. It is worth highlighting a slight statistically different reduction in this rate in rats with the implant's alendronate surface for the SHR SHAM ALE and SHR OVX ALE groups compared to the conventional implant of the SHR SHAM and SHR OVX groups ($p < 0.0001$; two-way ANOVA). Curiously, the trabecular thickness data supported the mineral apposition rate, where the higher MAR value was obtained in the SHR OVX group rather than in the SHR SHAM group, since the SHR OVX group had thicker bone trabeculae than the other groups in a probably compensatory mechanism. The lowest MAR achieved in the SHR OVX ALE group corroborates the smallest trabeculae thickness of them.

The descriptive evaluation of Figure 5 shows the peri-implant bone dynamics through the precipitation of fluorochromes in all groups. The fluorochromes are tied to the calcium during their precipitation in the organic bone matrix. Thus, there is an overlapping of green calcein, indicating the previous bone formation, and red alizarin, indicating the new bone formation. The first fluorochrome applied was calcein; therefore, it was the green biomarker of the old bone. Calcein biomarkers were present in all groups, with higher intensity in both SHAM groups, mainly in SHR SHAM, and the lowest in the SHR OVX

ALE group. The following administrated fluorochrome was alizarin, so the biomarker in red denoted the fresh newly formed bone. It was also slightly present in all groups.

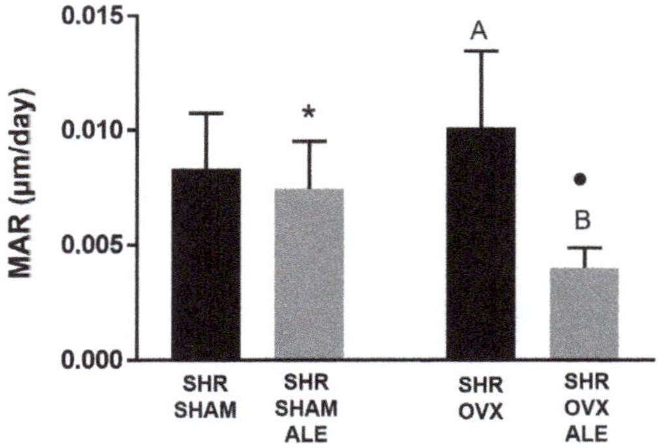

Figure 4. Parameters of epifluorescence evaluated in the peri-implant bone. It was quantified by the daily mineral apposition rate. The different uppercase letters indicate a significant statistical difference between the SHR OVX and SHR OVX ALE groups ($p < 0.05$). In addition, the different symbols "*" and dot demonstrate statistical differences when comparing the SHR SHAM ALE to SHR OVX ALE groups ($p < 0.05$).

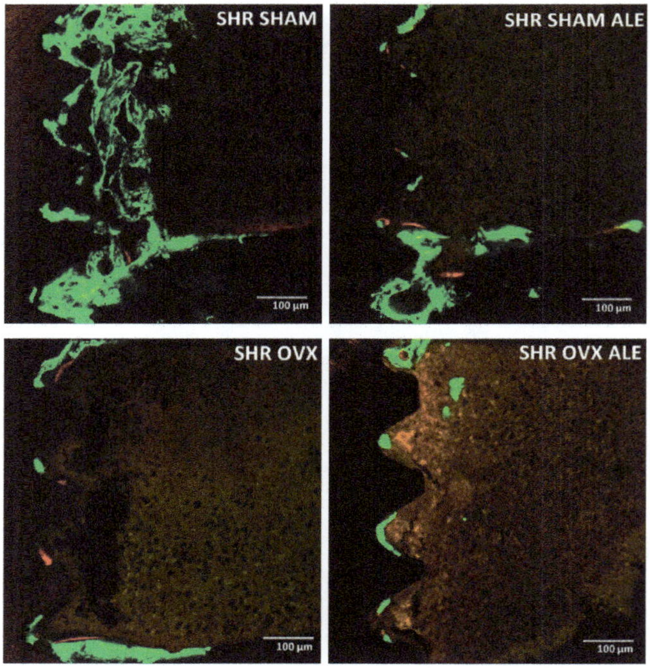

Figure 5. Peri-implant bone dynamics by epifluorescence analysis. Representative figure of the biomarking of the fluorochromes in the peri-implant bone around the implant assessed by the "Color Selection" instrument in the Image J software version 1.53 (Processing Software and Image Analysis, Ontario, ON, Canada). The image represents the overlying of both fluorochromes characterizing the peri-implant bone dynamics by green biomarkers of calcein fluorochrome on the old bone and red biomarkers of the alizarin fluorochrome on the fresh new bone in the SHR SHAM, SHR SHAM ALE, SHR OVX, and SHR OVX ALE groups, respectively.

4. Discussion

Untreated hypertension delays alveolar bone healing [36], though antihypertensive drugs can improve bone healing and metabolism [34,37] Besides hypertension, osteoporosis by estrogen deficiency also impairs alveolar bone healing [33]. Antiresorptive drugs administrated locally, such as sodium alendronate, can improve bone healing during osseointegration [23]. Considering the benefits of antihypertensive and antiresorptive drugs, the hypothesis of the mutually beneficial use of both drugs on postmenopausal bones under hypertensive conditions arise. Interestingly, a prior study showed a higher rate of alveolar bone apposition in hypertensive animals treated with the antihypertensive losartan [31]. Furthermore, untreated hypertensive animals revealed impaired osseointegration [34]. Taking these early results, and considering the replication of a real clinical setting, all hypertensive animals in this study were treated with losartan, since uncontrolled hypertension contraindicates the surgery of implant placement. Together with these propositions, the outcomes revealed a greater reverse torque and higher quality of bone microarchitecture in hypertensive animals than hypertensive and ovariectomized rats, even when treated with alendronate coated on the implant surface and with systemic losartan.

Hypertensive and estrogen-deficient rats confirmed the compromised peri-implant bone microarchitecture established by both conditions. The local administration of alendronate was ineffective in entirely reversing the unfavorable outcome. Also, the antihypertensive treatment did not demonstrate overcoming the consequences of estrogen deficiency on bone, despite its previously established osteoprotective properties in an osteoporotic model [17]. Possibly, the high osteoclastic activity from hypertension [10] and estrogen deficiency rationalizes the nonreversion [38]. The synergism of the drugs worked out in the exclusive presence of hypertension and the absence of estrogen deficiency, where the highest torque was registered. Hypertension leads to an increase in RANKL expression and reduces RUNX2, with a consequent imbalance in the bone metabolism [36]. Osteoporosis also acts by increasing the RANKL and the number of osteoclasts [39]. Therefore, both conditions can decrease calcium deposition in the mineral apposition rate, as shown by epifluorescence. Although alendronate has an affinity to hydroxyapatite crystal and tends to inhibit osteoclast activity [33], the use of ALE was not able to enhance calcium precipitation in hypertensive and ovariectomized animals. By these means, the mineral-apposition-rate values were lower in the SHR OVX ALE group than without ALE. Supporting our results, the mineral apposition rate was lower in animals treated with systemic alendronate through an increase via RANKL and osteoprotegerin [33].

The lower mineral apposition rate observed by fluorochromes in hypertensive and ovariectomized animals can be explained by changes in the calcium metabolism. Increased inflammatory mediators, such as interleukin-6 and tumor necrosis factor, are correlated with the onset of cardiovascular disease in hypertension and osteoporosis [40]. Arteriosclerosis is common in these disorders, causing an intravascular mineralization by a shift of vascular cells into osteoblasts and osteoblast-like cells [1]. Inherently, another study has shown a negative interference in osteogenesis directly by the blood flow [41], which clarifies the lesser percentage of bone volume present in both SHR OVX groups. Insufficient blood supply due to hypertension [42] or estrogen deficiency [41] can be inferred in the presented data. A beneficial effect was revealed on the bone endothelium of elderly mice treated with alendronate [41]. In this previous research, alendronate improved the blood flow and increased endothelial cells, type H osteoprogenitor, and osterix as zinc-finger transcription factor [41]. Feasibly, the alterations in bone vascularization could not be totally reversed with the losartan via the renin–angiotensin system or with the local alendronate in hypertensive and ovariectomized rats.

Postmenopausal women with estrogen deficiency show a rise the expression of type 1 angiotensin II receptors and a decrease estradiol levels, consequently increasing vasoconstriction and blood pressure [5]. Additionally, angiotensin II is considerably associated with an increase in bone resorption [12,18]. However, losartan, as a blocker of angiotensin II, was revealed to improve the mass and strength of bones in osteoporotic rats [43]. Furthermore, this drug increased peri-implant bone thickness in hypertensive rats [34]. The same anabolic effect was found in the alveolar bone mineralization of normotensive rats [31]. In addition, this antihypertensive reduced bone loss in a model of experimental periodontitis [44]. Thus, besides the benefits of losartan in controlling blood pressure and reducing bone loss, it can avoid bone resorption. Although the Tb.Th and calcium precipitation were not compromised in the hypertensive ovariectomized rats, the BV/TV of both SHR OVX groups were lower. The data imply a reduced BV/TV due to decreases in Tb.N and increases in Tb.Sp; the increased Tb.Th can indicate a compensatory mechanism. These results revealed an increased bone loss related to estrogen deficiency. However, the beneficial influence of the local application of alendronate should be stated, since the SHR SHAM ALE and SHR OVX ALE groups showed a slightly higher, but not statistically significant, torque value and bone volume. Thus, alendronate coted on the implant surface may improve, but not totally reverse, this impaired condition by estrogen deficiency.

Previous studies reported inadequate osseointegration in spontaneously hypertensive animals, where losartan proved to be efficient in overcoming this condition [34]. In the present study, losartan used in all groups demonstrated to be ineffective in revering the impaired osseointegration caused by estrogen deficiency in the ovariectomized groups. Hence, the effect of losartan on bone and local alendronate did not effectively reverse the impaired osseointegration in the presence of estrogen deficiency caused by ovariectomy in the hypertension model. Likewise, the use of losartan in synergy with alendronate on the implant surface seems to be a therapeutic strategy for avoiding side effects, such as osteonecrosis. Additionally, new investigations need to consider the local application of alendronate on implants as a treatment strategy for osseointegration in hypertensive females. In this way, the outcomes can also be partially extrapolated to the clinical relevance, where further studies need to elucidate the risk of implants in patients with controlled hypertension and estrogen deficiency. The findings suggest the lack of combined effect between systemic losartan and local alendronate in these simultaneous conditions. In further studies, the concentrations can be adjusted to evaluate new cellular responses in this experimental model. Innovative studies to shift compromised bone healing caused by hypertension and osteoporosis are crucial. Moreover, it is still necessary to verify whether the effect of losartan and alendronate confirmed in hypertensive animals occurs through the vascularization or induction of bone cells.

5. Conclusions

In conclusion, the impairment of osseointegration caused by estrogen deficiency in hypertensive rats is characterized by altering biomechanical resistance and peri-implant bone microarchitecture and mineralization. Furthermore, the application of alendronate on the surface of the implants does not fully reverse this impaired condition affected by estrogen deficiency in hypertensive rats.

Author Contributions: Conceptualization, R.O., P.R.B., G.M.-S. and P.N.L.-F.; methodology, J.S.d.S., C.A., F.R.d.S.B. and G.M.-S.; software, I.L.K. and G.M.-S.; validation, I.L.K., J.S.d.S., C.A. and P.R.B.; formal analysis, G.M.-S.; investigation, G.M.-S. and J.S.d.S.; resources, G.M.-S. and J.S.d.S.; data curation, F.R.d.S.B. and C.A.; writing—original draft preparation, G.M.-S. and R.O.; writing—review and editing, G.M.-S. and R.O.; visualization, R.O.; supervision, R.O. and P.R.B.; project administration, P.N.L.-F. and R.O.; funding acquisition, R.O. All authors have read and agreed to the published version of the manuscript.

Funding: This research was supported by grant #2017/23082-3, São Paulo Research Foundation (FAPESP) and grant #2017/16912-0, São Paulo Research Foundation (FAPESP). In addition, Roberta Okamoto is affiliated with Research productivity scholarship (process: 309408/2020-2).

Acknowledgments: The authors are grateful for the dental implants provided by the "Medens Indústria e Comércio". In addition, the authors thank the Multi-User Laboratory of FOA-UNESP and FINEP (FINEP/CT-INFRA—Agreement FINEP: 01.12.0530.00-PROINFRA 01/2011) for providing the high-resolution computed microtomography system for the analysis of this study. Finally, the authors are grateful for the Multiuser Laboratory of Confocal Microscopy, Department of Cell and Molecular Biology and Pathogenic Bioagents, Ribeirão Preto Medical School, University of São Paulo (FMRP/USP) and Elisabete Rosa Milani for assisting the epifluorescence microscopy analysis.

Conflicts of Interest: The authors declare no conflict of interest. Additionally, the funders had no role in the design of the study; in the collection, analyses, or interpretation of data; in the writing of the manuscript; or in the decision to publish the results.

References

1. Warburton, D.E.; Nicol, C.W.; Gatto, S.N.; Bredin, S.S. Cardiovascular Disease and Osteoporosis: Balancing Risk Management. *Vasc. Health Risk Manag.* **2007**, *3*, 673–689. [PubMed]
2. Varenna, M.; Manara, M.; Galli, L.; Binelli, L.; Zucchi, F.; Sinigaglia, L. The Association between Osteoporosis and Hypertension: The Role of a Low Dairy Intake. *Calcif. Tissue Int.* **2013**, *93*, 86–92. [CrossRef] [PubMed]
3. Barton, M.; Meyer, M.R. Postmenopausal Hypertension: Mechanisms and Therapy. *Hypertension* **2009**, *54*, 11–18. [CrossRef] [PubMed]
4. Chai, H.; Ge, J.; Li, L.; Li, J.; Ye, Y. Hypertension Is Associated with Osteoporosis: A Case-Control Study in Chinese Postmenopausal Women. *BMC Musculoskelet. Disord.* **2021**, *22*, 253. [CrossRef]
5. Khosla, S.; Hofbauer, L.C. Osteoporosis Treatment: Recent Developments and Ongoing Challenges. *Lancet Diabetes Endocrinol.* **2017**, *5*, 898–907. [CrossRef]
6. McCarron, D.A.; Yung, N.N.; Ugoretz, B.A.; Krutzik, S. Disturbances of Calcium Metabolism in the Spontaneously Hypertensive Rat. *Hypertension* **1981**, *3*, I162–I167. [CrossRef]
7. Wright, G.L.; Rankin, G.O. Concentrations of Ionic and Total Calcium in Plasma of Four Models of Hypertension. *Am. J. Physiol.* **1982**, *243*, H365–H370. [CrossRef]
8. Musini, V.M.; Gueyffier, F.; Puil, L.; Salzwedel, D.M.; Wright, J.M. Pharmacotherapy for Hypertension in Adults Aged 18 to 59 Years. *Cochrane Database Syst. Rev.* **2017**, *8*, CD008276. [CrossRef]
9. Santos, C.F.; Morandini, A.C.; Dionisio, T.J.; Faria, F.A.; Lima, M.C.; Figueiredo, C.M.; Colombini-Ishikiriama, B.L.; Sipert, C.R.; Maciel, R.P.; Akashi, A.P.; et al. Functional Local Renin-Angiotensin System in Human and Rat Periodontal Tissue. *PLoS ONE* **2015**, *10*, e0134601. [CrossRef]
10. Donmez, B.O.; Unal, M.; Ozdemir, S.; Ozturk, N.; Oguz, N.; Akkus, O. Effects of losartan treatment on the physicochemical properties of diabetic rat bone. *J. Bone Miner. Metab.* **2017**, *35*, 161–170. [CrossRef]
11. Moura, A.P.; Montalvany-Antonucci, C.C.; Rodrigues de Albuquerque Taddei, S.; Queiroz-Junior, C.M.; Biguetti, C.C.; Garlet, G.P.; Ferreira, A.J.; Teixeira, M.M.; Silva, T.A.; Andrade, I., Jr. Effects of Angiotensin II Type I Receptor Blocker Losartan on Orthodontic Tooth Movement. *Am. J. Orthod. Dentofac. Orthop.* **2016**, *149*, 358–365. [CrossRef] [PubMed]
12. Shimizu, H.; Nakagami, H.; Osako, M.K.; Hanayama, R.; Kunugiza, Y.; Kizawa, T.; Tomita, T.; Yoshikawa, H.; Ogihara, T.; Morishita, R. Angiotensin II Accelerates Osteoporosis by Activating Osteoclasts. *FASEB J.* **2008**, *22*, 2465–2475. [CrossRef] [PubMed]
13. Nakai, K.; Kawato, T.; Morita, T.; Yamazaki, Y.; Tanaka, H.; Tonogi, M.; Oki, H.; Maeno, M. Angiotensin II suppresses osteoblastic differentiation and mineralized nodule formation via AT1 receptor in ROS17/2.8 cells. *Arch. Med. Sci. AMS* **2015**, *11*, 628–637. [CrossRef] [PubMed]
14. Al-Majed, A.R.; Assiri, E.; Khalil, N.Y.; Abdel-Aziz, H.A. Losartan: Comprehensive Profile. *Profiles Drug Subst. Excip. Relat. Methodol.* **2015**, *40*, 159–194. [CrossRef]
15. Bastos, M.F.; Brilhante, F.V.; Goncalves, T.E.; Pires, A.G.; Napimoga, M.H.; Marques, M.R.; Duarte, P.M. Hypertension May Affect Tooth-Supporting Alveolar Bone Quality: A Study in Rats. *J. Periodontol.* **2010**, *81*, 1075–1083. [CrossRef]
16. Kunutsor, S.K.; Blom, A.W.; Whitehouse, M.R.; Kehoe, P.G.; Laukkanen, J.A. Renin-Angiotensin System Inhibitors and Risk of Fractures: A Prospective Cohort Study and Meta-Analysis of Published Observational Cohort Studies. *Eur. J. Epidemiol.* **2017**, *32*, 947–959. [CrossRef] [PubMed]
17. Rajkumar, D.S.; Faitelson, A.V.; Gudyrev, O.S.; Dubrovin, G.M.; Pokrovski, M.V.; Ivanov, A.V. Comparative Evaluation of Enalapril and Losartan in Pharmacological Correction of Experimental Osteoporosis and Fractures of Its Background. *J. Osteoporos.* **2013**, *2013*, 325693. [CrossRef]
18. Wiens, M.; Etminan, M.; Gill, S.S.; Takkouche, B. Effects of Antihypertensive Drug Treatments on Fracture Outcomes: A Meta-Analysis of Observational Studies. *J. Intern. Med.* **2006**, *260*, 350–362. [CrossRef]
19. Lee, J.; Adachi, K.; Gionhaku, N.; Fujita, S.; Uchida, T.; Gerstner, G.E.; Koshikawa, N. Evidence That Angiotensin II Enhances Apomorphine-Induced Jaw Movements via AT1 Receptors in the Ventrolateral Striatum: Studies by Magnet-Sensing System in Freely Moving Rats. *Methods Find. Exp. Clin. Pharmacol.* **2004**, *26*, 195–199. [CrossRef]
20. Garcia-Denche, J.T.; Wu, X.; Martinez, P.P.; Eimar, H.; Ikbal, D.J.; Hernandez, G.; Lopez-Cabarcos, E.; Fernandez-Tresguerres, I.; Tamimi, F. Membranes over the Lateral Window in Sinus Augmentation Procedures: A Two-Arm and Split-Mouth Randomized Clinical Trials. *J. Clin. Periodontol.* **2013**, *40*, 1043–1051. [CrossRef]
21. Jeal, W.; Barradell, L.B.; McTavish, D. Alendronate. A Review of Its Pharmacological Properties and Therapeutic Efficacy in Postmenopausal Osteoporosis. *Drugs* **1997**, *53*, 415–434. [CrossRef] [PubMed]

22. Tanaka, S.; Yamamoto, T.; Oda, E.; Nakamura, M.; Fujiwara, S. Real-World Evidence of Raloxifene versus Alendronate in Preventing Non-Vertebral Fractures in Japanese Women with Osteoporosis: Retrospective Analysis of a Hospital Claims Database. *J. Bone Miner. Metab.* **2016**, *36*, 87–94. [CrossRef] [PubMed]
23. Kim, J.H.; Park, Y.B.; Li, Z.; Shim, J.S.; Moon, H.S.; Jung, H.S.; Chung, M.K. Effect of Alendronate on Healing of Extraction Sockets and Healing around Implants. *Oral Dis.* **2011**, *17*, 705–711. [CrossRef] [PubMed]
24. Kellesarian, S.V.; Abduljabbar, T.; Vohra, F.; Malignaggi, V.R.; Malmstrom, H.; Romanos, G.E.; Javed, F. Role of Local Alendronate Delivery on the Osseointegration of Implants: A Systematic Review and Meta-Analysis. *Int. J. Oral Maxillofac. Surg.* **2017**, *46*, 912–921. [CrossRef]
25. Kennel, K.A.; Drake, M.T. Adverse Effects of Bisphosphonates: Implications for Osteoporosis Management. *Mayo Clin. Proc.* **2009**, *84*, 632–638. [CrossRef]
26. He, L.; Sun, X.; Liu, Z.; Qiu, Y.; Niu, Y. Pathogenesis and Multidisciplinary Management of Medication-Related Osteonecrosis of the Jaw. *Int. J. Oral Sci.* **2020**, *12*, 30. [CrossRef]
27. Jakobsen, T.; Kold, S.; Bechtold, J.E.; Elmengaard, B.; Søballe, K. Effect of Topical Alendronate Treatment on Fixation of Implants Inserted with Bone Compaction. *Clin. Orthop. Relat. Res.* **2006**, *444*, 229–234. [CrossRef]
28. Vohra, F.; Al-Rifaiy, M.Q.; Almas, K.; Javed, F. Efficacy of Systemic Bisphosphonate Delivery on Osseointegration of Implants under Osteoporotic Conditions: Lessons from Animal Studies. *Arch. Oral Biol.* **2014**, *59*, 912–920. [CrossRef]
29. Percie du Sert, N.; Hurst, V.; Ahluwalia, A.; Alam, S.; Avey, M.T.; Baker, M.; Browne, W.J.; Clark, A.; Cuthill, I.C.; Dirnagl, U.; et al. The ARRIVE Guidelines 2.0: Updated Guidelines for Reporting Animal Research. *PLoS Biol.* **2020**, *18*, e3000410. [CrossRef]
30. Long, J.A.; Evans, H.M.L. *The Oestrous Cycle in the Rat and It's Associated Phenomena;* Memoirs of the University of California, University of California Press: Berkeley, CA, USA, 1922.
31. Mulinari-Santos, G.; Santos, J.S.D.; Palin, L.P.; da Silva, A.C.E.; Antoniali, C.; Faverani, L.P.; Okamoto, R. Losartan Improves Alveolar Bone Dynamics in Normotensive Rats but Not in Hypertensive Rats. *J. Appl. Oral Sci.* **2019**, *27*, e20180574. [CrossRef]
32. Kitagawa, I.L.; Miyazaki, C.M.; Pitol-Palin, L.; Okamoto, R.; de Vasconcellos, L.M.R.; Constantino, C.J.L.; Lisboa-Filho, P.N. Titanium-Based Alloy Surface Modification with TiO2 and Poly(Sodium 4-Styrenesulfonate) Multilayers for Dental Implants. *ACS Appl. Bio Mater.* **2021**, *4*, 3055–3066. [CrossRef]
33. Ramalho-Ferreira, G.; Faverani, L.P.; Grossi-Oliveira, G.A.; Okamoto, T.; Okamoto, R. Alveolar Bone Dynamics in Osteoporotic Rats Treated with Raloxifene or Alendronate: Confocal Microscopy Analysis. *J. Biomed. Opt.* **2015**, *20*, 038003. [CrossRef] [PubMed]
34. Mulinari-Santos, G.; de Souza Batista, F.R.; Kirchweger, F.; Tangl, S.; Gruber, R.; Okamoto, R. Losartan Reverses Impaired Osseointegration in Spontaneously Hypertensive Rats. *Clin. Oral Implant. Res.* **2018**, *29*, 1126–1134. [CrossRef] [PubMed]
35. Dempster, D.W.; Compston, J.E.; Drezner, M.K.; Glorieux, F.H.; Kanis, J.A.; Malluche, H.; Meunier, P.J.; Ott, S.M.; Recker, R.R.; Parfitt, A.M. Standardized Nomenclature, Symbols, and Units for Bone Histomorphometry: A 2012 Update of the Report of the ASBMR Histomorphometry Nomenclature Committee. *J. Bone Miner. Res. Off. J. Am. Soc. Bone Miner. Res.* **2013**, *28*, 2–17. [CrossRef] [PubMed]
36. Okamoto, K.; Aoki, K. Development of a strain of spontaneously hypertensive rats. *Jpn. Circ. J.* **1963**, *27*, 282–293. [CrossRef] [PubMed]
37. Zhang, N.; Huo, Y.; Yao, C.; Sun, J.; Zhang, Y. The Effect of the Angiotensin-Converting Enzyme Inhibitor on Bone Health in Castrated Hypertensive Rats Is Mediated via the Kinin-Kallikrein System. *J. Renin. Angiotensin Aldosterone Syst.* **2022**, *2022*, 9067167. [CrossRef] [PubMed]
38. Gruber, R. Molecular and Cellular Basis of Bone Resorption. *Wien. Med. Wochenschr.* **2015**, *165*, 48–53. [CrossRef]
39. Teitelbaum, S.L. Bone Resorption by Osteoclasts. *Science* **2000**, *289*, 1504–1508. [CrossRef]
40. Mantovani, A.; Sozzani, S.; Introna, M. Endothelial Activation by Cytokines. *Ann. N. Y. Acad. Sci.* **1997**, *832*, 93–116. [CrossRef]
41. Ramasamy, S.K.; Kusumbe, A.P.; Schiller, M.; Zeuschner, D.; Bixel, M.G.; Milia, C.; Gamrekelashvili, J.; Limbourg, A.; Medvinsky, A.; Santoro, M.M.; et al. Blood Flow Controls Bone Vascular Function and Osteogenesis. *Nat. Commun.* **2016**, *7*, 13601. [CrossRef]
42. Rizzoni, D.; Pasini, E.; Flati, V.; Rodella, L.F.; Paiardi, S.; Assanelli, D.; De Ciuceis, C.; Porteri, E.; Boari, G.E.; Rezzani, R.; et al. Angiotensin Receptor Blockers Improve Insulin Signaling and Prevent Microvascular Rarefaction in the Skeletal Muscle of Spontaneously Hypertensive Rats. *J. Hypertens.* **2008**, *26*, 1595–1601. [CrossRef] [PubMed]
43. Donmez, B.O.; Ozdemir, S.; Sarikanat, M.; Yaras, N.; Koc, P.; Demir, N.; Karayalcin, B.; Oguz, N. Effect of Angiotensin II Type 1 Receptor Blocker on Osteoporotic Rat Femurs. *Pharmacol. Rep. PR* **2012**, *64*, 878–888. [CrossRef] [PubMed]
44. Dionísio, T.J.; Souza, G.P.; Colombini-Ishikiriama, B.L.; Garbieri, T.F.; Parisi, V.A.; Oliveira, G.M.; Cano, I.P.; Rodini, C.O.; Oliveira, S.H.P.; Greene, A.S.; et al. AT1 Receptor Antagonism Promotes Bone Loss Attenuation in Experimental Periodontitis, Block Inflammatory Mediators, Upregulate Antioxidant Enzymes and Bone Formation Markers. *J. Periodontol.* **2019**, *91*, 533–544. [CrossRef] [PubMed]

Journal of
Functional Biomaterials

Article

Adipogenesis-Related Metabolic Condition Affects Shear-Stressed Endothelial Cells Activity Responding to Titanium

Thaís Silva Pinto, Anderson Moreira Gomes, Paula Bertin de Morais and Willian F. Zambuzzi *

Lab. of Bioassays and Cellular Dynamics, Department of Chemical and Biological Sciences, Institute of Biosciences, UNESP—São Paulo State University, Botucatu 18618-970, SP, Brazil
* Correspondence: w.zambuzzi@unesp.br

Abstract: Purpose: Obesity has increased around the world. Obese individuals need to be better assisted, with special attention given to dental and medical specialties. Among obesity-related complications, the osseointegration of dental implants has raised concerns. This mechanism depends on healthy angiogenesis surrounding the implanted devices. As an experimental analysis able to mimic this issue is currently lacking, we address this issue by proposing an in vitro high-adipogenesis model using differentiated adipocytes to further investigate their endocrine and synergic effect in endothelial cells responding to titanium. Materials and methods: Firstly, adipocytes (3T3-L1 cell line) were differentiated under two experimental conditions: Ctrl (normal glucose concentration) and High-Glucose Medium (50 mM of glucose), which was validated using Oil Red O Staining and inflammatory markers gene expression by qPCR. Further, the adipocyte-conditioned medium was enriched by two types of titanium-related surfaces: Dual Acid-Etching (DAE) and Nano-Hydroxyapatite blasted surfaces (nHA) for up to 24 h. Finally, the endothelial cells (ECs) were exposed in those conditioned media under shear stress mimicking blood flow. Important genes related to angiogenesis were then evaluated by using RT-qPCR and Western blot. Results: Firstly, the high-adipogenicity model using 3T3-L1 adipocytes was validated presenting an increase in the oxidative stress markers, concomitantly with an increase in intracellular fat droplets, pro-inflammatory-related gene expressions, and also the ECM remodeling, as well as modulating mitogen-activated protein kinases (MAPKs). Additionally, Src was evaluated by Western blot, and its modulation can be related to EC survival signaling. Conclusion: Our study provides an experimental model of high adipogenesis in vitro by establishing a pro-inflammatory environment and intracellular fat droplets. Additionally, the efficacy of this model to evaluate the EC response to titanium-enriched mediums under adipogenicity-related metabolic conditions was analyzed, revealing significant interference with EC performance. Altogether, these data gather valuable findings on understanding the reasons for the higher percentage of implant failures in obese individuals.

Keywords: bone; wound healing; adipogenesis; obese; dental implants; titanium; failure; angiogenesis

Citation: Pinto, T.S.; Gomes, A.M.; de Morais, P.B.; Zambuzzi, W.F. Adipogenesis-Related Metabolic Condition Affects Shear-Stressed Endothelial Cells Activity Responding to Titanium. *J. Funct. Biomater.* **2023**, *14*, 162. https://doi.org/10.3390/jfb14030162

Academic Editor: Masaaki Nakai

Received: 19 January 2023
Revised: 14 March 2023
Accepted: 16 March 2023
Published: 17 March 2023

1. Introduction

Physiological changes in obese patients are widely studied due to their systemic complications. The comorbidities known to be associated with obesity include cardiovascular, pulmonary, renal, endocrine, and musculoskeletal problems (e.g., arthrosis, osteoarthritis, back and joint pain), as well as impaired wound healing [1]. It is known that the adipose tissue develops endocrine activity by secreting bioactive substances, named adipokines, that can reach all systems and develop important roles in metabolism [2]. In obese individuals there is a dysfunction of the adipose tissue, presenting higher adipogenesis (high number and size of adipocytes, as well as intracellular fat droplets) that directly affects the profiles of released adipokines, including an increase in proinflammatory interleukin secretion, which triggers different responses systemically, including in the cardiovascular system and vasculature [3].

The vasculature exerts an important function in supplying blood to the whole body, delivering oxygen, nutrients, and essential factors to ensure adequate physiologic and metabolic functioning, also guaranteeing adequate tissue regeneration and repair in cases of injuries, when the vasculature also provides undifferentiated cells [4]. In this context, the endothelial cells (ECs), present along the luminal surface of the blood vessels, develop a central role in vascular activity by perceiving and responding to the changes in chemical and mechanical factors in the blood [5–8]. It is important to mention that angiogenesis is a decisive event during tissue healing [9] and it has been shown that ECs differently respond to metallic implantable medical devices, such as titanium alloys [10–12].

Titanium is commonly used as a biomaterial in several dental and medical fields, e.g., implantology, orthopedics, cardiology, and gastroenterology [13]. This metallic alloy is considered the gold standard for these purposes because it possesses important properties such as mechanical and corrosion resistance and proper biocompatibility [14]. After biomaterial implantation, the reactional tissue surrounding the implants suffers adaptation processes that require blood supplements through the blood vessels, highlighting the importance of EC activities and angiogenesis [15]. In addition, studies have revealed that angiogenesis and osteogenesis are coupled processes, and shown that better osseointegration occurs by interfering with the process of appositional new bone growth [16]. This coupling between cells seems to be affected in specific metabolic conditions, such as diabetes and obesity.

In fact, obesity is related to severe complications in the process of bone healing and osseointegration, likely because it provokes a burst of pro-inflammatory involvement, and an increased risk of bone loss [17]. The explanation for the obesity-related tissue regeneration complications lies in the effects of the unbalanced adipose-derived proinflammatory cytokines and adipokines [18], however, the biology involved in this context is barely known.

Thus, studies exploring the metabolic effects of high adipogenicity on different systems are needed, as well as strategies and alternative technologies and methodologies to study these conditions and predict biological responses. In this context, in vitro methods and analyses, such as cell culture, have the relevant advantages, despite presenting a huge set of limitations in comparison to in vivo protocols arising from age, genetics, environments, habits, hormones, etc. Furthermore, many in vitro methodologies capable of mimicking or closely addressing scenarios in biological systems, as well as in the presence of pathologies, have been proposed and used to guide preclinical experiments.

Using in vitro methodologies, we investigated the potential crosstalk between adipogenicity-related metabolic conditions and ECs responding to a titanium-enriched medium. Specifically, we demonstrate the proof of concept of the creation of the high-adipogenicity model using an adipocyte cell line, and observed its interference with EC activity. Altogether, these data gather new findings to understand the higher rate of failure of implants in obese individuals.

2. Methods

2.1. Implants

This experimental workflow was performed using two titanium surfaces, as follows: Dual Acid-Etched (DAE) and nanohydroxyapatite-coated surfaces (nHA). Both set of titanium discs were kindly provided by S.I.N.—Sistema Nacional de Implantes (Sao Paulo, SP, Brazil). The nHA titanium surface is described in more detail elsewhere by Gottlander et al. (1997) and Meirelles et al. (2008) [19,20].

2.2. Reagents

Dulbecco's modified Eagle's medium (DMEM), Fetal Bovine Serum (FBS), trypsin, penicillin, and streptomycin (antibiotics) were obtained from Nutricell (Campinas, Sao Paulo, Brazil). Trypan Blue (T6146), acetic acid glacial (695092), (3-(4,5-dimethylthiazol-2-yl)-2,5-diphenyltetrazolium bromide) (MTT) (M2128), ethanol (459844), Oil Red O (O0625), 3-isobutyl-1-methylxanthine (IBMX) (I5879), Dexamethasone (D1756), β-glycerophosphate (G9422), glucose (G-5400) and insulin (I2643) were obtained from Sigma Chemical Co. (St. Louis, MO, USA). The antibodies #9102, #8690, #3700, and #2109 were obtained from Cell Signaling Technology (Beverly, MA, USA). TRIzol™ reagent (15596026), DNase I (18068015), and the High-Capacity cDNA Reverse Transcription Kit (4368314) were purchased from Thermo Fisher Scientific Inc. (Waltham, MA, USA). GoTaq qPCR Master Mix (A6002) was obtained from PROMEGA (Madison, WI, USA). Oligonucleotides for gene expression were purchased from Exxtend (Campinas, São Paulo, Brazil).

2.3. Cell Culture

Two cell lines were used in this study: both 3T3-L1 preadipocytes (passage < 10) and Human Umbilical Vein Endothelial Cells (HUVECs; ECs) (ATCC; CRL1730; passage < 10) were cultured in Dulbecco's modified Eagle's medium (DMEM—Nutricell, Campinas, Brazil) containing penicillin 100 U/mL, streptomycin 100 mg/mL, and 10% fetal bovine serum (FBS). The adipocyte differentiation is described below. Importantly, ECs were cultured subjected to a shear stress model, as is described in detail below. In all experiments, the cells were maintained at 37 °C with a 5% CO_2 and 95% humidity environment.

2.4. Adipocyte Differentiation

The 3T3-L1 adipocyte cultures were maintained in 100 mm culture dishes in DMEM supplemented with 10% FBS and 1% penicillin/streptomycin at 37 °C, 5% CO_2, and 95% humidity until reaching proper confluence. Then, adipocyte differentiation was induced by supplementing the cell culture medium with insulin (1 mg/mL), dexamethasone (1 mM), and 3-isobutyl-1-methylxanthine (IBMX) (0.5 mM), which was used to expose the cells for 3 days. The medium was changed to maintenance DMEM containing 10% FBS and insulin (1 mg/mL) for an additional 7 days (completing 10 days in toto), being changed every 2 days in the meantime. This differentiation process was followed under 2 separate experimental conditions; one under normal conditions (Ctrl), and the other subjected to a high-glucose medium (HGM) to induce higher adipogenesis, reaching a final glucose concentration of 50 mM. The cultures were maintained up to 10 days, at which point the cells were harvested and the conditioned medium collected to further expose ECs.

2.5. Shear Stress Model

ECs were seeded in the peripheral area of previously modified 100 mm culture dishes (Figure 1). The modification in the 100 mm culture dishes was made by using medical silicone to bond a 60 mm culture dish at the center-bottom of the 100 mm culture dishes. Thereafter, the modified dishes were sterilized using UV light for 30 min. To perform the model, the ECs were exposed to the tension forces induced by the circuit of shear stress triggered by the rotations of an orbital shaker (Scilogex, Rocky Hill, CT, USA) placed in the cell culture incubator, as in other experiments [21,22]. We calculated the maximal wall shear stress of ~3 Pa (physiological arterial shear stress = ~1–4 Pa) by using this equation: $\tau max = \alpha\sqrt{\rho\eta}(2\pi f)3$. This equation encompasses the τmax that is the shear stress (Pascal), α is the radius of orbital rotation (12 cm), ρ is the density of the cell culture medium (937.5 kg/m^3), η is the viscosity of the cell culture medium (7.5×10^{-4} Pa s), and f is the frequency of rotation.

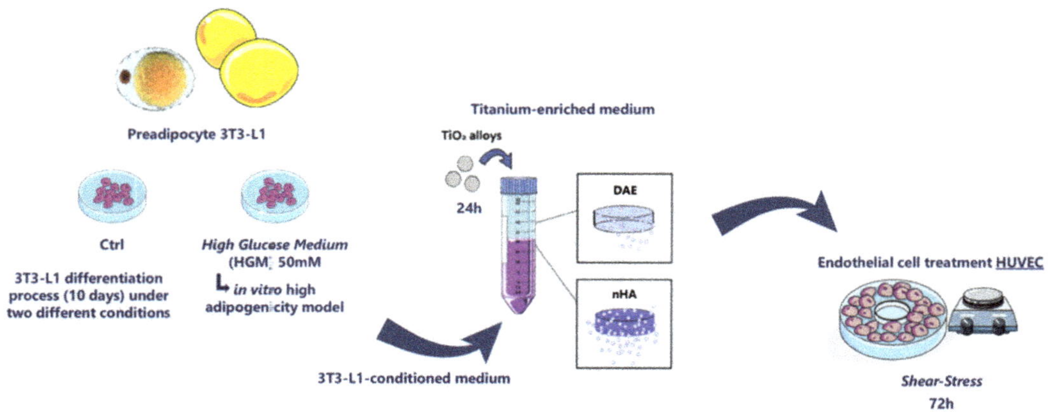

Figure 1. Outline and workflow. To evaluate the ECs behavior in response to titanium-enriched medium in concomitance with high adipogenesis conditions arising from 3T3-L1 adipocytes, we collected the medium conditioned by the adipocytes during their differentiation, which was later enriched with titanium for up to 24 h, as recommended by ISO 10993:2016. This conditioned medium was also further used to expose ECs for 72 h under shear stress mimicking blood flow, at which point the samples were collected to allow the molecular analysis.

2.6. Cell Viability Assay

The high-glucose medium was prepared previously at 50 mM final concentration. Concomitantly, the 3T3-L1 adipocytes were maintained on 96-well plates (5×10^4 cells/mL) and later incubated for up to 24 h. One group of cells was treated with the HGM to evaluate whether the high glucose concentration would affect cell viability. The control group contained cells under normal cell culture conditions. The cells were maintained in both treatments (Ctrl and HGM 50 mM) for up to 72 h, after which the cell viability was measured by adding 1 mg/mL of 3-(4,5-dimethyl-2-thiazolyl)-2,5-diphenyl-2H-tetrazolium bromide (MTT) to evaluate the mitochondrial dehydrogenase activity through the MTT reduction reaction after 3 h in a CO_2 incubator. During this reaction, the yellow-colored water-soluble tetrazolium salt MTT becomes the purple-colored soluble compound formazan proportional to mitochondrial activity. The dye of formazan was later dissolved in DMSO, and the absorbance was measured at 570 nm (Synergy II; BioTek Instruments, Winooski, VT, USA).

2.7. Oil Red O Staining

Previously, Oil Red O solution was obtained by dissolving the Oil Red O dye in propylene glycol (0.5%; w/v) in a heater at 95 °C. By using a 0.45 µm syringe filter the Oil Red O solution was filtered to eliminate residual particulates from the solution and later used to stain the adipocytes. Adipocytes were differentiated for 10 days using a 24-well plate. At the end of differentiation, the medium was removed, the cells were washed with warm PBS, fixed in 4% paraformaldehyde for 10 min at room temperature, washed twice in deionized water, and then maintained in absolute propylene glycol for 5 min. The cells were stained in Oil Red O solution 0.5% up to 30 min at room temperature, then washed in 85% propylene glycol solution for 3 min. Finally, 3T3-L1 was washed twice in deionized water. Images were acquired using an inverted microscope (Axio Vert.A1, Carl Zeiss microscopy GMBH, Göttingen, Germany).

2.8. Titanium-Enriched Medium Obtaining

Adipocyte-related conditioned medium was later used to incubate both types of titanium disc for 24 h: DAE and nHA, in accordance with ISO 10993:2016 (0.2 g/mL w/v) with slight modification as suggested by Zambuzzi et al. [23–26]. The final conditioned medium was later used to expose ECs for 3 days under shear stress to mimic blood flow.

2.9. Western Blot

Both adipocytes and ECs were harvested using lysis buffer [Lysis Cocktail (50 mM Tris [tris(hydroxymethyl)aminomethane]–HCl [pH 7.4], 1% Tween 20, 0.25% sodium deoxycholate, 150 mM NaCl, 1 mM EGTA (ethylene glycol tetraacetic acid), 1 mM *O*-Vanadate, 1 mM NaF, and protease inhibitors [1 µg/mL aprotinin, 10 µg/mL leupeptin, and 1 mM 4-(2-amino-ethyl)-benzolsulfonylfluorid-hydrochloride])] for 2 h, after which the samples were cleared by centrifugation, and the protein concentration was measured using the Lowry method [27]. An equal volume of 2x sodium dodecyl sulfate (SDS) gel loading buffer (100 mM Tris-HCl [pH 6.8], 200 mM dithiothreitol [DTT], 4% SDS, 0.1% bromophenol blue, and 20% glycerol) was added to the samples and boiled for 5 min. Aliquots of protein extracts were resolved into SDS-PAGE (10 or 12%) and transferred to PVDF membranes (Millipore, Burlington, MA, USA). Membranes were blocked with either 5% fat-free dried milk dissolved in Tris-buffered saline (TBS)–Tween 20 (0.05%) and incubated overnight at 4° C with appropriate primary antibody at 1:1000 dilutions. After washing 3x TBS-Tween 20 (0.05%), those membranes were incubated with horseradish peroxidase-conjugated secondary IgGs antibodies, at 1:5000 dilutions, in a blocking buffer for 1 h. Thereafter, the bands were detected by enhanced chemiluminescence (ECL), or by fluorescence (ODYSSEY® CLx Infrared Imaging System).

2.10. Quantitative PCR Assay (qPCR)

The same experimental workflow was performed and the cells were harvested now in Ambion TRIzol Reagent (Life Sciences—Fisher Scientific Inc, Waltham, MA, USA), and treated with DNase I (Invitrogen, Carlsband, CA, USA). cDNA synthesis was performed using High-Capacity cDNA Reverse Transcription Kit (Applied Biosystems, Foster City, CA, USA) following the manufacturer's instructions. qPCR was performed on a total of 10 µL, containing PowerUp™ SYBR™ Green Master Mix 2x (5 µL) (Applied Biosystems, Foster City, CA, USA), 0.4 µM of each primer, and 50 ng of cDNA and nuclease-free H_2O. Data were expressed as relative amounts of each target gene normalized considering the expression of 18SrRNA and Gapdh genes, here used as housekeeping genes, using the cycle threshold (Ct) method. Specific primers and running details are described in Table 1.

Table 1. Sequences of the primers and conditions of the quantitative polymerase chain reaction cycle.

Genes	Primers	5′-3′ Sequences	Reaction's Conditions
H-AKT	Forward	CAG CGC GGC CCG AAG GAC	95 °C, 3 s; 55 °C, 8 s; 72 °C, 20 s
	Reverse	GAC GCT CAC GCG CTC CTC TC	
H-CDK2	Forward	CTT TGC TGA GAT GGT GAC TCG	95 °C, 3 s; 55 °C, 8 s; 72 °C, 20 s
	Reverse	GCC TCC CAG ATT CCT CAT GC	
H-CDK4	Forward	CTC TCT AGC TTG CGG CCT G	95 °C, 3 s; 55 °C, 8 s; 72 °C, 20 s
	Reverse	GCA GGG ATA CAT CTC GAG GC	
H-ERK	Forward	GCA GCG CCT CCC TTG CTA GA	95 °C, 3 s; 55 °C, 8 s; 72 °C, 20 s
	Reverse	AAC AGC CTC TGG CCC ACC CAT	

Table 1. *Cont.*

Genes	Primers	5′-3′ Sequences	Reaction's Conditions
H-IL1B	Forward	GGA GAA TGA CCT GAG CAC CT	95 °C, 3 s; 55 °C, 8 s; 72 °C, 20 s
	Reverse	GGA GGT GGA GAG CTT TCA GT	
H-IL6	Forward	AGT CCT GAT CCA GTT CCT GC	95 °C, 3 s; 55 °C, 8 s; 72 °C, 20 s
	Reverse	CTA CAT TTG CCG AAG AGC CC	
H-JNK	Forward	AAA GGT GGT GTT TTG TTC CCA GGT	95 °C, 3 s; 55 °C, 8 s; 72 °C, 20 s
	Reverse	TGA TGA TGG ATG CTG AGA GCC ATT G	
H-P38	Forward	GAG AAC TGC GGT TAC TTA	95 °C, 3 s; 55 °C, 8 s; 72 °C, 20 s
	Reverse	ATG GGT CAC CAG ATA CAC AT	
H-VEGF	Forward	TGC AGA TTA TGC GGA TCA AAC C	95 °C, 3 s; 55 °C, 8 s; 72 °C, 20 s
	Reverse	TGC ATT CAC ATT TGT TGT GCT GTA G	
H-VEGFr1	Forward	CAG GCC CAG TTT CTG CCA TT	95 °C, 3 s; 55 °C, 8 s; 72 °C, 20 s
	Reverse	TTC CAG CTC AGC GTG GTC GTA	
M-Gapdh	Forward	AGG CCG GTG CTG AGT ATG TC	95 °C, 3 s; 55 °C, 8 s; 72 °C, 20 s
	Reverse	TGC CTG CTT CAC CAC CTT CT	
M-Il13	Forward	CAG TCC TGG CTC TTG CTT G	95 °C, 3 s; 55 °C, 8 s; 72 °C, 20 s
	Reverse	CCA GGT CCA CAC TCC ATA CC	
M-Il18	Forward	ACT TTG GCC GAC TTC ACT GT	95 °C, 3 s; 55 °C, 8 s; 72 °C, 20 s
	Reverse	GGG TTC ACT GGC ACT TTG AT	
M-Il1b	Forward	GAC CTT CCA GGA TGA GGA CA	95 °C, 3 s; 55 °C, 8 s; 72 °C, 20 s
	Reverse	AGC TCA TAT GGG TCC GAC AG	
M-Il1r	Forward	ACC CCC ATA TCA GCG GAG CG	95 °C, 3 s; 55 °C, 8 s; 72 °C, 20 s
	Reverse	TTG CTT CCC CCG GAA CGT AT	
M-Il33	Forward	CCT TCT CGC TGA TTT CCA AG	95 °C, 3 s; 55 °C, 8 s; 72 °C, 20 s
	Reverse	CCG TTA CGG ATA TGG TGG TC	
M-Il6	Forward	AGT TGC CTT CTT GGG ACT GA	95 °C, 3 s; 55 °C, 8 s; 72 °C, 20 s
	Reverse	CAG AAT TGC CAT TGC ACA AC	
M-Myd88	Forward	ATG GTG GTG GTT GTT TCT GAC GA	95 °C, 3 s; 55 °C, 8 s; 72 °C, 20 s
	Reverse	GCA AGG GTT GGT ATA GTC GCA TAT A	
M-Nfkb	Forward	CAC CTG TTC CAA AGA GCA CC	95 °C, 3 s; 55 °C, 8 s; 72 °C, 20 s
	Reverse	GGT TCA GGA GCT GCT GAA AC	
M-Pparg	Forward	TTT TCA AGG GTG CCA GTT TC	95 °C, 3 s; 55 °C, 8 s; 72 °C, 20 s
	Reverse	AAT CCT TGG CCC TCT GAG AT	
M-Tnf	Forward	CCA CAT CTC CCT CCA GAA AA	95 °C, 3 s; 55 °C, 8 s; 72 °C, 20 s
	Reverse	AGG GTC TGG GCC ATA GAA CT	

Note H = Human; M = Mice.

2.11 Oxidative Stress Markers

After obtaining adipocytes, the cells were harvested in PBS and sonicated. Protein carbonylation (CBO) was measured by using DNPH (2,4-dinitrophenyl hydrazine) as derivatizing agent [28]. The experiment was performed in a dark chamber to prevent the light. Firstly, the samples (10 µL of the lysed cells) were incubated with DNPH 10 mM (100 µL) for 10 min, after which 50 µL of 6 M NaOH (sodium hydroxide) was added. The reaction was interrupted after 10 min. The CBO was estimated by reading the final solution coloration spectrophotometrically at 450 nm. Finally, the results were calculated using the molar extinction coefficient (22,000 M^{-1} cm^{-1}) of DNPH and expressed as nmol/mg protein.

2.12. Matrix Metalloproteinases (MMPs) Activities by Zymography

Differentiating adipocyte cell culture medium was collected to measure the activity of matrix metalloproteinases (MMPs). The conditioned medium was centrifuged at 14,000 rpm for 15 min to avoid cell debris, and the protein concentration was determined using the Lowry method [27]. The same concentration of protein was resolved into a 12% polyacrylamide gel containing 4% gelatin. The gelatinolytic activity of MMPs was determined in the resolved proteins (bands). The proteins' structures were renatured in Triton X-100 aqueous solution (2% *w*/*v*) for 40 min, followed by incubation for 18 h in proteolysis buffer (Tris-$CaCl_2$) at 37 °C, when the gels were stained using Coomassie Blue R-250 dye solution 0.05% for 3 h. Thereafter, the stained gels were washed in a 30% methanol (*v*/*v*) and 10% glacial acetic acid solution (*v*/*v*). The opposite staining (clear bands) was obtained in the gels exactly where there was the gelatinolytic activity (bands) of MMP2 (~62 kDa) and MMP9 (~84 kDa), and then they were analyzed using the software ImageJ (Bethesda, MD, USA), as previously proposed [29].

2.13. Statistical Analysis

Data were expressed as mean ± standard error of the mean (SEM) of the replicates of each experiment (n = 3). The samples assumed a normal distribution, and they were subjected to Student's *t*-test (two-tailed) with $p < 0.05$ considered statistically significant. In the experiment where there were more than two groups, the statistical analyses were performed using either analysis of variance (one-way ANOVA) combined with appropriate Bonferroni's correction post-test, or nonparametric analysis. A $p < 0.05$ was considered to be statistically significant. The software used was GraphPad Prism 7 (GraphPad Software, La Jolla, CA, USA).

3. Results

To better evaluate the molecular mechanism underlying EC response to high-adipogenesis conditions concomitant to a titanium-enriched medium, we proposed an in vitro experimental model subjecting adipocyte cells to differentiation under two conditions: Control cultures (Ctrl) and high-glucose treated cells (H_Adip). Furthermore, the adipocyte-obtained medium was enriched by titanium and then used to expose semiconfluent EC cultures dynamically responding to mechanotransduction mimicking blood flow [22]. Additionally, the titanium-enriched medium was obtained using two different titanium-modified surfaces: DAE, in which the discs were subjected to dual acid-etching, and nHA, in which DAE surfaces were covered by nano-hydroxyapatite [11,30,31], as we have shown previously.

3.1. Validation of the High-Adipogenesis Model

To validate the proposed high-adipogenesis model, we first analyzed classical biomarkers of adipocyte phenotype in cells responding to high-glucose exposition, which did not affect their viability (Figure 2A). Oxidative stress markers and fat droplet staining were analyzed and correlated with inflammatory profile and adipogenesis. Reactive oxygen species and concomitant oxidative stress in adipocytes was better investigated by evaluating the protein carbonylation profile. Our data show that this parameter was significantly affected in cells responding to the high-glucose exposure, with higher values than the conventional condition (Ctrl) (Figure 2B). Images acquired by using light microscopy show that adipocytes responding to the high-glucose medium presented a higher number of bigger-sized intracellular fat droplets stained by Oil Red O dye (Figure 2C).

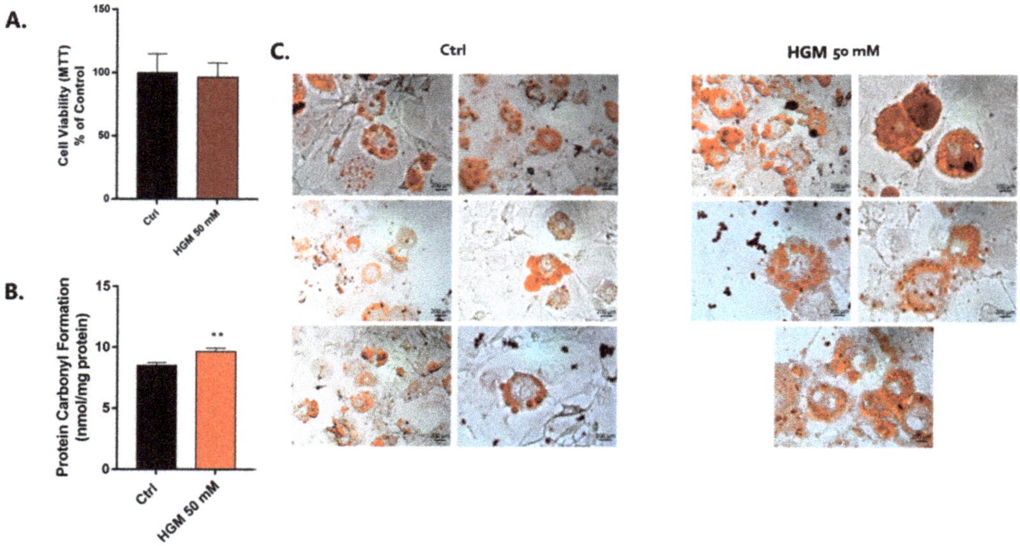

Figure 2. High-glucose medium affects adipocyte differentiation. The 50 mM high-glucose medium was used to expose pre-adipocyte cells for stimulation to differentiation. Firstly, cytotoxicity was measured by using an MTT assay (**A**). Thereafter, oxidative stress was measured by evaluating the protein carbonylation (nmol/mg protein) by performing a method using DNPH (2,4-dinitrophenylhydrazine derivatizing agent) (**B**). The data are plotted respecting mean ± SD (n = 3), and the significance was shown using Student's *t*-test, ** $p = 0.0058$. Intracellular fat droplets of the pre-adipocytes were acquired using a light microscope (40× magnification) thereafter stained by using Oil Red O Staining (**C**). HGM: high-glucose medium.

Adipocyte differentiation and inflammation signaling pathways are widely studied regarding obesity concerns. Herein, the gene expression was evaluated by investigating the behavior of the interleukins IL-1β, IL-6, IL-13, IL-18, and IL-33 genes. They were significantly higher in cells responding to the adipogenesis model where pre-adipocytes were exposed to the high-glucose medium (HGM) (Figure 3A–E), as well as considering TNF-α gene expression (Figure 3G). Importantly, the expression activity of the PPAR-γ gene was also investigated in this study. It was significantly higher in cells responding to HGM (Figure 3J), while both Myd88 and IL1 receptor genes expression were lower in the HGM group (Figure 3F,I), and NFkB remains unchanged (Figure 3H).

We also investigated the behavior of mitogen-activated protein kinase (MAPKs) genes to infer about cell survival signaling in differentiated adipocytes. Our data show that there is a higher profile of MAPK-ERK proteins in cells responding to HGM (Figure 4A,B), while the MAPK-P38 protein remains unchanged (Figure 4C,D). Finally, the perspective of extracellular matrix (ECM) remodeling was investigated by analyzing the activities of matrix metalloproteinases (MMPs) through gelatin-based proteolysis assay. Our data show that there is higher activity of both MMP2 and 9 in differentiated adipocytes (Figure 4E–J).

3.2. *Angiogenesis-Related Genes Were Evaluated in ECs Responding to High Adipogenesis and Titanium*

To analyze the behavior of ECs responding to high-adipogenesis and titanium-enriched mediums, we first evaluated angiogenesis-related genes and viability. The Ctrl group now refers to the adipocyte-conditioned medium subjected to normal conditions, while the H_Adip group refers to the adipocyte-conditioned medium chronically responding to high glucose concentrations (50 mM); furthermore, the DAE and nHA refer to the adipocyte-conditioned medium previously used to incubate titanium discs with respect to the difference on their surfaces: H_Adip + DAE and H_Adip + nHA. The VEGF gene remains unchanged when compared to Ctrl or when the cells were treated with either of the titanium-enriched mediums (Figure 5A). In this way, the VEGFr1 gene remains unchanged when compared to the Ctrl group, but significantly decreases in response to both nHA and H_Adip + nHA groups when compared to H_Adip (Figure 5B).

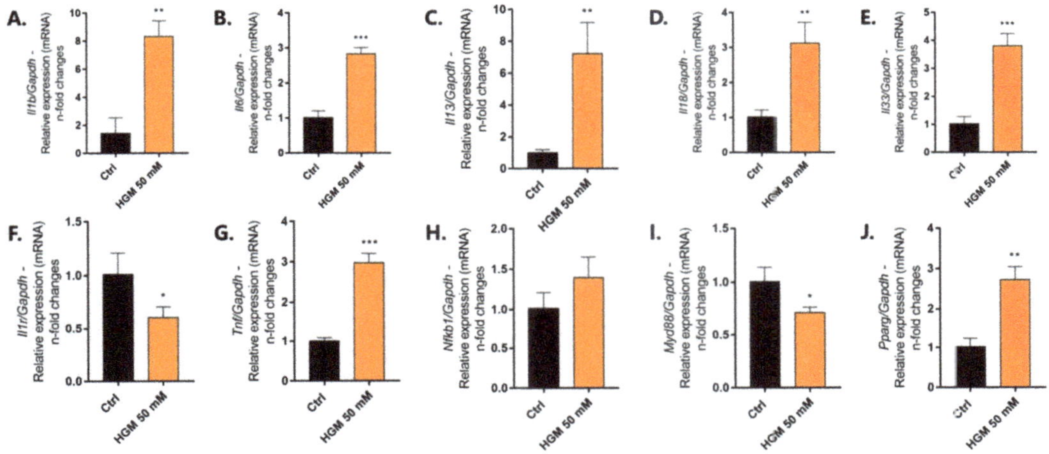

Figure 3. Adipogenesis model recapitulates the inflammation microenvironment and requires PPAR-γ. Pre-adipocytes were differentiated for 10 days using classical model when the cells were harvested and the biological samples forwarded to perform the qPCR technology. A significantly higher expression of IL-1β (**A**), IL-6 (**B**), IL-13 (**C**), IL-18 (**D**), IL-33 (**E**), TNF-α (**G**), and PPAR-γ (**J**) genes were observed in the adipocytes responding to HGM (50 mM). The graphs bring the n-fold change of the profile of gene expression normalized to the GAPDH gene (housekeeping gene). Significant differences were considered when * $p < 0.05$, and ** $p < 0.01$, *** $p < 0.001$. HGM: high glucose medium.

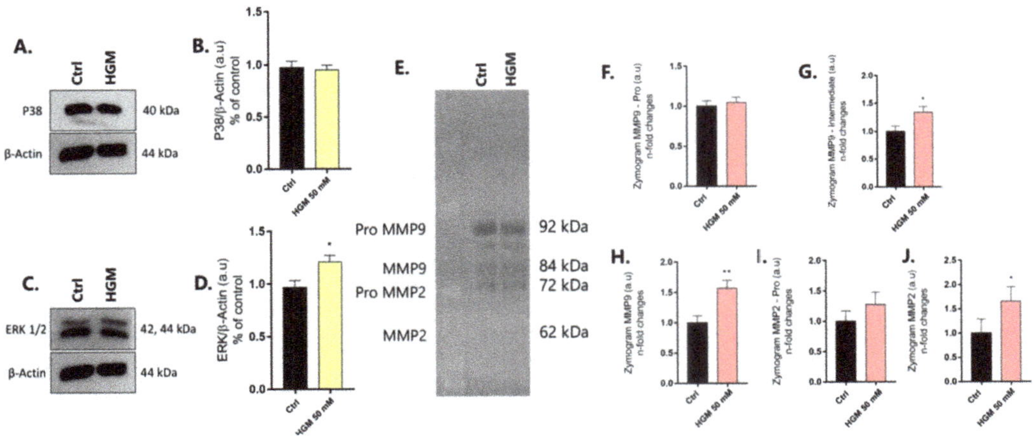

Figure 4. Adipocytes require MAPK and MMP activity. Adipocytes require MAPK-ERK (**A**,**B**), while the MAPK-P38 protein remains unchanged (**C**,**D**). β-Actin was used as the protein loading control. Additionally, higher activities of MMP9 (**F–H**) and MMP2 (**I**,**J**) were found in adipocytes responding to HGM. Data are plotted as means ± standard deviations (n = 3). Significant differences were considered when * $p < 0.05$, and ** $p < 0.01$. HGM: high glucose medium.

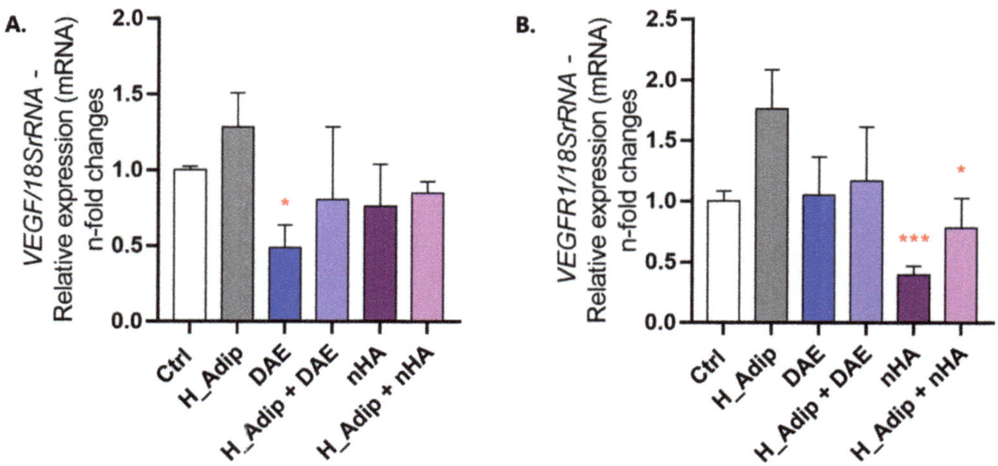

Figure 5. VEGF and VEGFR1 genes are modulated in response to titanium-based surfaces. Both genes are related to EC phenotype as well as to angiogenesis. Their response to H_Adip and the titanium-enriched medium seems to be relevant to the lower angiogenesis profile in adipogenesis. The VEGF gene presented a low expression profile in ECs responding to DAE, while there was no significance when considering the other groups (**A**). Additionally, the VEGFR1 gene presented a low profile of expression in ECs responding to nHA (**B**). The data show the n-fold changes in the profile of transcripts normalized to the 18 S gene (housekeeping gene). Differences were considered statistically significant when * $p < 0.05$, and *** $p < 0.001$, represented by red * when compared to the H_Adip group.

3.3. Proliferation and Survival-Related Genes in ECs

The protein kinase B (AKT), cyclin-dependent kinase 2 (CDK2), and CDK4 genes were significantly higher in ECs exposed to adipocyte-conditioned medium (H_Adip) (Figure 6A–C). Thereafter, our data show that there is a significant involvement of the AKT gene in the coupling of adipogenicity and titanium-enriched medium (Figure 6A). Thereafter, the CDK genes presented a very similar profile between DAE and nHA, with CDK2 being higher in nHA (Figure 6B), and CDK4 expressing a very similar profile (Figure 6C).

MAPK genes were also investigated in ECs. Figure 7 shows there is a significant modulation in response to the coupling between high-adipogenicity and titanium-enriched medium. The titanium DAE-enriched medium increased the expression of the MAPK-ERK gene independently of normal or high-adipogenicity conditions (Figure 7A). ECs required MAPK-JNK gene expression when they were exposed to the DAE-enriched medium under normal adipogenesis conditions (DAE) when compared to H_Adip group, but without difference in the H_Adip + DAE group, while cells responding to nHA and H_Adip + nHA remain unchanged (Figure 7B). The MAPK-P38 gene showed involvement but without statistically significant changes when the groups were compared to the Ctrl group, showing an increase only in the DAE group when compared to H_Adip (Figure 7C).

As c-Src kinase is related to the survival mechanism governing the viability of eukaryotic cells, we decided to evaluate whether this protein was involved in response to the titanium-enriched medium under normal and high-adipogenesis conditions. Additionally, our data show that c-Src seems to be required in high-adipogenesis conditions regardless of the presence of the titanium DAE-enriched medium as it remained higher in the group H_Adip + DAE (Figure 8A,B). Moreover, in the titanium nHA-enriched medium, the cells under high adipogenesis showed a significant difference when compared with Ctrl (Figure 8).

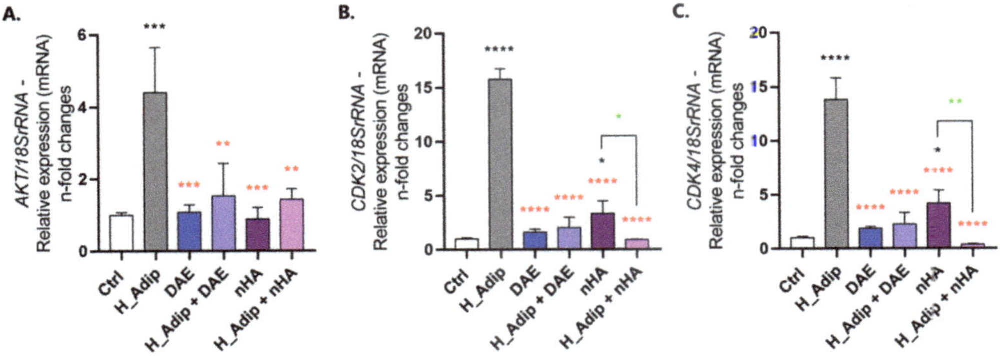

Figure 6. Both survival and cell proliferation progress were investigated in ECs. ECs require an increase in survival and cell cycle-related gene expression even more responding to high-adipogenesis condition, observing the behavior of AKT (**A**), CDK2 (**B**), and CDK4 (**C**). Differences were considered statistically when * $p < 0.05$, ** $p < 0.01$, *** $p < 0.001$, and **** $p < 0.0001$, represented by black * when compared to the Ctrl group, by red * when compared to the H_Adip group, and by green * when compared between the groups with nHA.

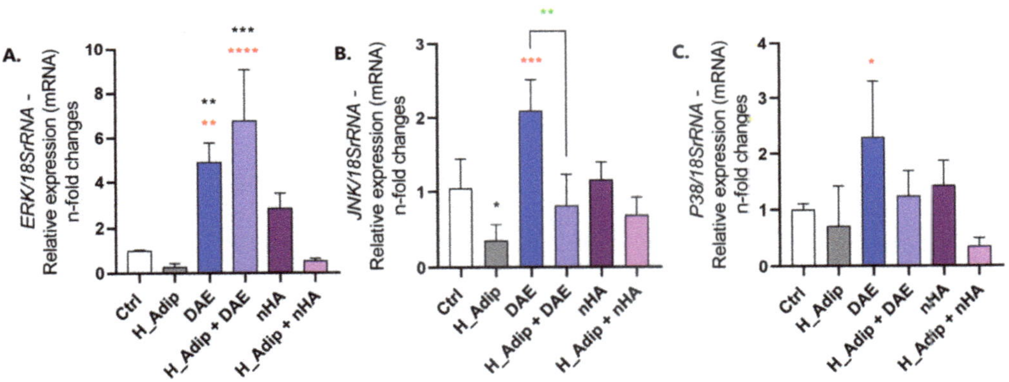

Figure 7. MAPK-related genes were modulated in EC responding to adipogenesis. qPCR shows different modulations of the MAPKs genes in ECs: the MAPK-ERK gene was higher in cells responding to the DAE group, and even higher in the H_Adip + DAE group. However, there was no difference in the H_Adip + nHA group when compared to Ctrl (**A**). The JNK was down-regulated in the H_Adip group when compared to Ctrl, and higher in response to DAE treatment when compared to H_Adip. Data are reported as means ± standard deviations (n = 3). Comparison by one-way ANOVA. Statistical differences were considered when * $p < 0.05$, ** $p < 0.01$, *** $p < 0.001$, and **** $p < 0.0001$, represented by black * when compared to the Ctrl group, by red * when compared to the H_Adip group, and by green * when compared between the groups with DAE.

3.4. Endothelial Cell Appears to Be Important in Inflammatory Gene Microenvironment

The interleukins IL-6 and IL1-β genes were evaluated using RT-qPCR technology. This revealed an important modulation in ECs responding to different treatments. The IL-6 gene was significantly higher when ECs were treated with titanium DAE and nHA under the high-adipogenesis condition (Figure 9A). The IL-1β gene was also significantly higher in the high-adipogenesis condition, a situation that seems to be controlled in the presence of the titanium-enriched medium, whether considering DAE or nHA (Figure 9B).

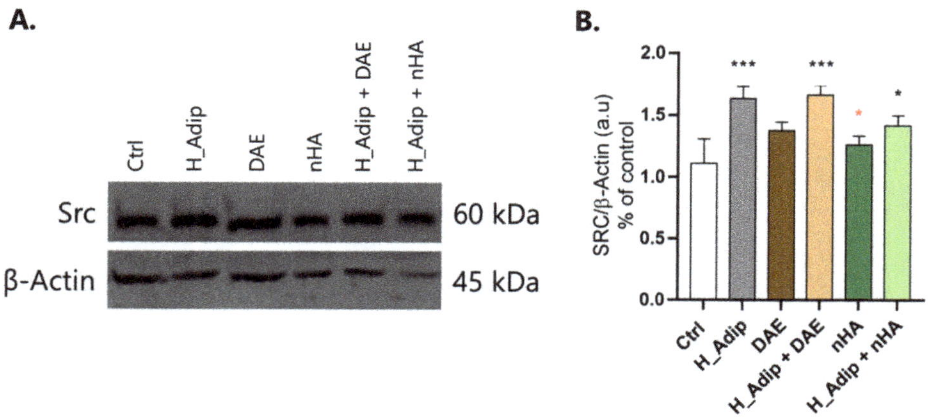

Figure 8. c-Src involvement. After exposing the ECs to different shear-stress treatments for 72 h, the samples were obtained to perform the Western blotting assay and evaluate SRC protein content (**A**,**B**). The high adipogenesis condition increased the content of SRC in ECs, significantly in the H_Adip + DAE group, and less significantly in the H_Adip + nHA group. β-Actin was considered the protein loading control. Differences were considered significant when * $p < 0.05$, and *** $p < 0.001$, represented by black * when compared to the Ctrl group, and by red * when compared to the H_Adip group.

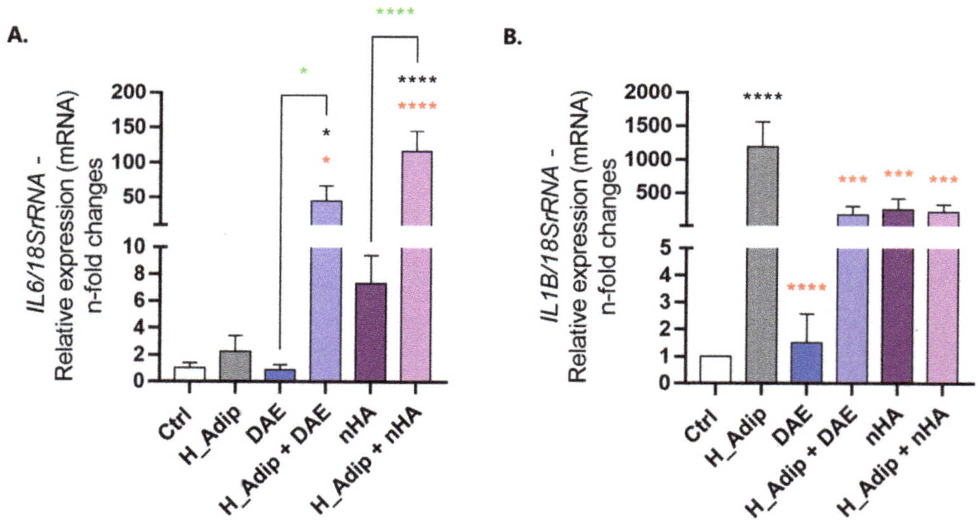

Figure 9. IL6 and IL1B gene expression changed in response to conditioned mediums. The samples were harvested as described earlier and the gene expression was measured using qPCR. To address the inflammatory effect of different conditions on ECs, we evaluated IL-6 (**A**) and IL-1β (**B**) genes. The 18SrRNA gene was considered the housekeeping gene and used to normalize the expression values. Data are reported in means ± standard deviations (n = 3). Differences were considered significant when * $p < 0.05$, *** $p < 0.001$, and **** $p < 0.0001$, represented by black * when compared to the Ctrl group, by red * when compared to the H_Adip group, and by green * when compared between the groups with DAE or between the groups with nHA. DAE = titanium with Dual Acid-Etching; nHA = titanium with nano-Hydroxyapatite-coated surface.

4. Discussion

The experimental model proposed in this study permits the evaluation of the effect of high adipogenesis in the ECs responding to a titanium-enriched medium. The adipocyte differentiation under high-glucose conditions has been used as a tool in vitro to understand obesity-related metabolic dysfunctions [32], once the glucose enhances lipid accumulation and adipogenesis [33]. Regarding the validation of this alternative model, the adipocyte exposed to high glucose concentrations in our model significantly modulated specific interleukin gene expression as well as the PPAR-γ gene activation. These genes are related to adipocyte differentiation [34]. Taken together, this validates our biological model for obtaining functional adipocytes. Importantly, the increase in protein carbonylation in obtained adipocytes can be correlated with the high oxidative stress expected in differentiated adipocytes. In general, our proposed adipogenesis model promotes an increase in the intracellular fat droplets during the pre-adipocyte differentiation concomitantly with the increase in the pro-inflammatory profile, as expected in adipocytes [32,35]. Additionally, MAPK and PPAR are also both involved with adipocyte metabolism [33], which corroborates with our findings. Additionally, we have also shown significant morphological changes in adipocytes and which can explain the higher activities of MMPs and is expected to modulate the ECM remodeling. Considering in vitro studies, this experimental model presents limitations, such as evaluating the crosstalk between cells of different origins, however, it can be overcome by the evolutionarily conserved structure and functions of proteins and genes over 80%.

Although some progress has been made on the way to understanding the etiology of systemic and metabolic dysfunctions such as diabetes and obesity, their relevance to bone-healing peri-implants, which might explain the higher failure of implants in obese patients [36], is barely understood. In fact, it has been hypothesized that angiogenesis is compromised in obese and diabetic individuals. Thus, we have applied an experimental model to better evaluate the impact on ECs responding to a titanium-enriched medium which plays crucial roles during the osseointegration mechanism of angiogenesis, such as interacting with osteoblasts [37,38]. There are important similarities between osseointegration and wound healing, a situation that is harmful in patients with compromised metabolism [39]. To advance with this proposal, we have investigated the effect of two titanium-modified surfaces. Firstly, genes related to the phenotype of endothelial cells were investigated. While VEGF seemed not to be affected, its receptor, VEGFR1, was higher in ECs responding to the titanium-enriched medium. This might be explained by a correlation with its ligand and suggests an autocrine loop in this condition. The increase in the intracellular signaling upon VEGFR1 activation requires the involvement of MAPK upstream, modulating cell survival and proliferative phenotype. This explains the capacity of those cells to involve p38 and CDKs [40]. Additionally, the VEGFR1-related intracellular cascade seems to require Src kinase in this context, and also might be correlated with ECM remodeling by regulating MMP activity.

Lastly, the gene expression of interleukins IL-6 and IL-1β was shown to be sensitive to the response to titanium in an adipogenesis-related metabolic condition once both DAE and nHA promoted their higher expression. It is important that the IL-related cascade also requires the activation of MAPKs and Src. An important aspect is that IL-1β effectively and rapidly induces human mesenchymal stem cells differentiation into osteoblasts [41]. This might be an important axis coupling angiogenesis and osteogenesis during the osseointegration mechanism, meaning nHA is able to improve the capacity of ECs to drive bone healing in obese patients.

5. Conclusions

Taking our data into consideration, it is possible to suggest that nHA-coated surface favor biological events related to angiogenesis and might be an alternative strategy in adipogenesis-related metabolic conditions where usually the percentage of dental implant failure is higher. Altogether, this study gathers valuable information on understanding the higher failure of dental implants in obese individuals.

Author Contributions: Methodology, T.S.P., A.M.G. and P.B.d.M.; Validation, T.S.P.; Formal analysis, A.M.G. and P.B.d.M.; Investigation, T.S.P., A.M.G. and P.B.d.M.; Data curation, P.B.d.M.; Writing—original draft, A.M.G. and W.F.Z.; Writing—review & editing, T.S.P. and P.B.d.M.; Supervision, W.F.Z.; Project administration, W.F.Z.; Funding acquisition, W.F.Z. All authors have read and agreed to the published version of the manuscript.

Funding: Fundação de Amparo à Pesquisa do Estado de São Paulo-FAPESP (FAPESP: 2019/26854-2; 2014/22689-3), Conselho Nacional de Desenvolvimento Científico e Tecnológico (CNPq–Bolsa Produtividade em Pesquisa, 314166/2021-1), and Coordenação de Aperfeiçoamento de Pessoal de Nível Superior (Capes), code 1.

Institutional Review Board Statement: Not applicable.

Informed Consent Statement: Not applicable.

Data Availability Statement: The data that support the findings of this study are available from the corresponding author upon reasonable request.

Conflicts of Interest: The authors declare no conflict of interest.

References

1. Stevens, S.M.; O'Connell, B.P.; Meyer, T.A. Obesity related complications in surgery. *Curr. Opin. Otolaryngol. Head Neck Surg.* **2015**, *23*, 341–347. [CrossRef]
2. Adamczak, M.; Wiecek, A. The adipose tissue as an endocrine organ. *Semin. Nephrol.* **2013**, *33*, 2–13. [CrossRef]
3. Fuster, J.J.; Ouchi, N.; Gokce, N.; Walsh, K. Obesity-Induced Changes in Adipose Tissue Microenvironment and Their Impact on Cardiovascular Disease. *Circ. Res.* **2016**, *118*, 1786–1807. [CrossRef] [PubMed]
4. Wuche, C. The cardiovascular system and associated disorders. *Br. J. Nurs.* **2022**, *31*, 886–892. [CrossRef]
5. Givens, C.; Tzima, E. Endothelial Mechanosignaling: Does One Sensor Fit All? *Antioxid. Redox Signal.* **2016**, *25*, 373–388. [CrossRef]
6. da Silva, R.A.; Ferreira, M.R.; Gomes, A.M.; Zambuzzi, W.F. LncRNA HOTAIR is a novel endothelial mechanosensitive gene. *J. Cell. Physiol.* **2020**, *235*, 4631–4642. [CrossRef]
7. Gomes, A.M.; Pinto, T.S.; da Costa Fernandes, C.J.; da Silva, R.A.; Zambuzzi, W.F. Wortmannin targeting phosphatidylinositol 3-kinase suppresses angiogenic factors in shear-stressed endothelial cells. *J. Cell. Physiol.* **2020**, *235*, 5256–5269. [CrossRef] [PubMed]
8. da Silva, R.A.; Fernandes, C.J.d.C.; Feltran, G.d.S.; Gomes, A.M.; de Camargo Andrade, A.F.; Andia, D.C.; Peppelenbosch, M.P.; Zambuzzi, W.F. Laminar shear stress-provoked cytoskeletal changes are mediated by epigenetic reprogramming of TIMP1 in human primary smooth muscle cells. *J. Cell. Physiol.* **2019**, *234*, 6382–6396. [CrossRef]
9. Johnson, K.E.; Wilgus, T.A. Vascular Endothelial Growth Factor and Angiogenesis in the Regulation of Cutaneous Wound Repair. *Adv. Wound Care* **2014**, *3*, 647–661. [CrossRef] [PubMed]
10. Martins, B.R.; Pinto, T.S.; da Costa Fernandes, C.J.; Bezerra, F.; Zambuzzi, W.F. PI3K/AKT signaling drives titanium-induced angiogenic stimulus. *J. Mater. Sci. Mater. Med.* **2021**, *32*, 18. [CrossRef]
11. Pinto, T.S.; Martins, B.R.; Ferreira, M.R.; Bezerra, F.; Zambuzzi, W.F. Nanohydroxyapatite-Blasted Bioactive Surface Drives Shear-Stressed Endothelial Cell Growth and Angiogenesis. *BioMed Res. Int.* **2022**, *2022*, 1433221. [CrossRef]
12. Chen, W.; Xu, K.; Tao, B.; Dai, L.; Yu, Y.; Mu, C.; Shen, X.; Hu, Y.; He, Y.; Cai, K. Multilayered coating of titanium implants promotes coupled osteogenesis and angiogenesis in vitro and in vivo. *Acta Biomater.* **2018**, *74*, 489–504. [CrossRef] [PubMed]
13. Prasad, K.; Bazaka, O.; Chua, M.; Rochford, M.; Fedrick, L.; Spoor, J.; Symes, R.; Tieppo, M.; Collins, C.; Cao, A.; et al. Metallic Biomaterials: Current Challenges and Opportunities. *Materials* **2017**, *10*, 884. [CrossRef] [PubMed]
14. Coelho, P.G.; Granjeiro, J.M.; Romanos, G.E.; Suzuki, M.; Silva, N.R.F.; Cardaropoli, G.; Thompson, V.P.; Lemons, J.E. Basic research methods and current trends of dental implant surfaces. *J. Biomed. Mater. Res. B Appl. Biomater.* **2009**, *88*, 579–596. [CrossRef] [PubMed]
15. Rademakers, T.; Horvath, J.M.; van Blitterswijk, C.A.; LaPointe, V.L.S. Oxygen and nutrient delivery in tissue engineering: Approaches to graft vascularization. *J. Tissue Eng. Regen. Med.* **2019**, *13*, 1815–1829. [CrossRef]
16. Roseti, L.; Parisi, V.; Petretta, M.; Cavallo, C.; Desando, G.; Bartolotti, I.; Grigolo, B. Scaffolds for Bone Tissue Engineering: State of the art and new perspectives. *Mater. Sci. Eng. C Mater. Biol. Appl.* **2017**, *78*, 1246–1262. [CrossRef] [PubMed]

17. Coelho, P.G.; Pippenger, B.; Tovar, N.; Koopmans, S.-J.; Plana, N.M.; Graves, D.T.; Engebretson, S.; van Beusekom, H.M.M.; Oliveira, P.G.F.P.; Dard, M. Effect of Obesity or Metabolic Syndrome and Diabetes on Osseointegration of Dental Implants in a Miniature Swine Model: A Pilot Study. *J. Oral Maxillofac. Surg.* **2018**, *76*, 1677–1687. [CrossRef]
18. Di Benedetto, A.; Gigante, I.; Colucci, S.; Grano, M. Periodontal disease: Linking the primary inflammation to bone loss. *Clin. Dev. Immunol.* **2013**, *2013*, 503754. [CrossRef]
19. Gottlander, M.; Johansson, C.B.; Wennerberg, A.; Albrektsson, T.; Radin, S.; Ducheyne, P. Bone tissue reactions to an electrophoretically applied calcium phosphate coating. *Biomaterials* **1997**, *18*, 551–557. [CrossRef]
20. Meirelles, L.; Arvidsson, A.; Andersson, M.; Kjellin, P.; Albrektsson, T.; Wennerberg, A. Nano hydroxyapatite structures influence early bone formation. *J. Biomed. Mater. Res. A* **2008**, *87*, 299–307. [CrossRef]
21. dela Paz, N.G.; Walshe, T.E.; Leach, L.L.; Saint-Geniez, M.; D'Amore, P.A. Role of shear-stress-induced VEGF expression in endothelial cell survival. *J. Cell Sci.* **2012**, *125*, 831–843. [CrossRef] [PubMed]
22. Pinto, T.S.; Fernandes, C.J.d.C.; da Silva, R.A.; Gomes, A.M.; Vieira, J.C.S.; Padilha, P.d.M.; Zambuzzi, W.F. c-Src kinase contributes on endothelial cells mechanotransduction in a heat shock protein 70-dependent turnover manner. *J. Cell. Physiol.* **2019**, *234*, 11287–11303. [CrossRef]
23. Zambuzzi, W.F.; Bonfante, E.A.; Jimbo, R.; Hayashi, M.; Andersson, M.; Alves, G.; Takamori, E.R.; Beltrao, P.J.; Coelho, P.G.; Granjeiro, J.M. Nanometer scale titanium surface texturing are detected by signaling pathways involving transient FAK and Src activations. *PLoS ONE* **2014**, *9*, e95662. [CrossRef]
24. Machado, M.I.P.; Gomes, A.M.; Rodrigues, M.F.; Silva Pinto, T.; da Costa Fernandes, C.J.; Bezerra, F.J.; Zambuzzi, W.F. Cobalt-chromium-enriched medium ameliorates shear-stressed endothelial cell performance. *J. Trace Elem. Med. Biol.* **2019**, *54*, 163–171. [CrossRef]
25. da Costa Fernandes, C.J.; Bezerra, F.J.B.; de Campos Souza, B.; Campos, M.A.; Zambuzzi, W.F. Titanium-enriched medium drives low profile of ECM remodeling as a pre-requisite to pre-osteoblast viability and proliferative phenotype. *J. Trace Elem. Med. Biol.* **2018**, *50*, 339–346. [CrossRef] [PubMed]
26. Fernandes, C.J.C.; Bezerra, F.; das D do Carmo, M.; Feltran, G.S.; Rossi, M.C.; da Silva, R.A.; Padilha, P.; Zambuzzi, W. CoCr-enriched medium modulates integrin-based downstream signaling and requires a set of inflammatory genes reprograming in vitro. *J. Biomed. Mater. Res. A* **2018**, *106*, 839–849. [CrossRef]
27. Hartree, E.F. Determination of protein: A modification of the Lowry method that gives a linear photometric response. *Anal. Biochem.* **1972**, *48*, 422–427. [CrossRef] [PubMed]
28. Mesquita, C.S.; Oliveira, R.; Bento, F.; Geraldo, D.; Rodrigues, J.V.; Marcos, J.C. Simplified 2,4-dinitrophenylhydrazine spectrophotometric assay for quantification of carbonyls in oxidized proteins. *Anal. Biochem.* **2014**, *458*, 69–71. [CrossRef] [PubMed]
29. Lefebvre, V.; Peeters-Joris, C.; Vaes, G. Production of gelatin-degrading matrix metalloproteinases ('type IV collagenases') and inhibitors by articular chondrocytes during their dedifferentiation by serial subcultures and under stimulation by interleukin-1 and tumor necrosis factor alpha. *Biochim. Biophys. Acta* **1991**, *1094*, 8–18. [CrossRef]
30. Bezerra, F.; Ferreira, M.R.; Fontes, G.N.; da Costa Fernandes, C.J.; Andia, D.C.; Cruz, N.C.; da Silva, R.A.; Zambuzzi, W.F. Nano hydroxyapatite-blasted titanium surface affects pre-osteoblast morphology by modulating critical intracellular pathways. *Biotechnol. Bioeng.* **2017**, *114*, 1888–1898. [CrossRef]
31. Fernandes, C.J.C.; Bezerra, F.; Ferreira, M.R.; Andrade, A.F.C.; Pinto, T.S.; Zambuzzi, W.F. Nano hydroxyapatite-blasted titanium surface creates a biointerface able to govern Src-dependent osteoblast metabolism as prerequisite to ECM remodeling. *Colloids Surf. B Biointerfaces* **2018**, *163*, 321–328. [CrossRef] [PubMed]
32. Palacios-Ortega, S.; Varela-Guruceaga, M.; Martínez, J.A.; de Miguel, C.; Milagro, F.I. Effects of high glucose on caveolin-1 and insulin signaling in 3T3-L1 adipocytes. *Adipocyte* **2016**, *5*, 65–80. [CrossRef] [PubMed]
33. Chuang, C.C.; Yang, R.S.; Tsai, K.S.; Ho, F.M.; Liu, S.H. Hyperglycemia Enhances Adipogenic Induction of Lipid Accumulation: Involvement of Extracellular Signal-Regulated Protein Kinase 1/2, Phosphoinositide 3-Kinase/Akt, and Peroxisome Proliferator-Activated Receptor γ Signaling. *Endocrinology* **2007**, *148*, 4267–4275. [CrossRef] [PubMed]
34. Lefterova, M.I.; Haakonsson, A.K.; Lazar, M.A.; Mandrup, S. PPARγ and the global map of adipogenesis and beyond. *Trends Endocrinol. Metab.* **2014**, *25*, 293–302. [CrossRef] [PubMed]
35. Shilpa, K.; Dinesh, T.; Lakshmi, B.S. An In Vitro Model to Probe the Regulation of Adipocyte Differentiation under Hyperglycemia. *Diabetes Metab. J.* **2013**, *37*, 176–180. [CrossRef]
36. von Wilmowsky, C.; Stockmann, P.; Harsch, I.; Amann, K.; Metzler, P.; Lutz, R.; Moest, T.; Neukam, F.W.; Schlegel, K.A. Diabetes mellitus negatively affects peri-implant bone formation in the diabetic domestic pig. *J. Clin. Periodontol.* **2011**, *38*, 771–779. [CrossRef]
37. Kusumbe, A.P.; Ramasamy, S.K.; Adams, R.H. Coupling of angiogenesis and osteogenesis by a specific vessel subtype in bone. *Nature* **2014**, *507*, 323–328. [CrossRef] [PubMed]
38. Ramasamy, S.K.; Kusumbe, A.P.; Schiller, M.; Zeuschner, D.; Bixel, M.G.; Milia, C.; Gamrekelashvili, J.; Limbourg, A.; Medvinsky, A.; Santoro, M.M.; et al. Blood flow controls bone vascular function and osteogenesis. *Nat. Commun.* **2016**, *7*, 13601. [CrossRef]
39. de Oliveira, P.G.F.P.; Bonfante, E.A.; Bergamo, E.T.P.; de Souza, S.L.S.; Riella, L.; Torroni, A.; Benalcazar Jalkh, E.B.; Witek, L.; Lopez, C.D.; Zambuzzi, W.F.; et al. Obesity/Metabolic Syndrome and Diabetes Mellitus on Peri-implantitis. *Trends Endocrinol. Metab.* **2020**, *31*, 596–610. [CrossRef]

40. Zhang, X.; Simons, M. Receptor tyrosine kinases endocytosis in endothelium: Biology and signaling. *Arterioscler. Thromb. Vasc. Biol.* **2014**, *34*, 1831–1837. [CrossRef]
41. Sonomoto, K.; Yamaoka, K.; Oshita, K.; Fukuyo, S.; Zhang, X.; Nakano, K.; Okada, Y.; Tanaka, Y. Interleukin-1β induces differentiation of human mesenchymal stem cells into osteoblasts via the Wnt-5a/receptor tyrosine kinase-like orphan receptor 2 pathway. *Arthritis Rheum.* **2012**, *64*, 3355–3363. [CrossRef] [PubMed]

Journal of *Functional Biomaterials*

MDPI

Article

Laser-Modified Ti Surface Improves Paracrine Osteogenesis by Modulating the Expression of DKK1 in Osteoblasts

Jorge Felipe Lima Teixeira [1], João Antônio Chaves de Souza [2], Fernando Augusto Cintra Magalhães [3], Guilherme José Pimentel Lopes de Oliveira [4], José Bernardo de Santis [5], Carlos Alberto de Souza Costa [1] and Pedro Paulo Chaves de Souza [6,*]

1 Department of Physiology and Pathology, School of Dentistry, São Paulo State University, Araraquara 14801-385, Brazil
2 School of Dentistry, Federal University of Goiás, Goiânia 74605-020, Brazil
3 Nursing Department, Federal University of Maranhão, Imperatriz 65905-240, Brazil
4 Department of Periodontology and Implant Dentistry, School of Dentistry, Federal University of Uberlândia, Uberlândia 38405-320, Brazil
5 Department of Basic and Oral Biology, Bone Research Lab, School of Dentistry of Ribeirão Preto, University of São Paulo, Ribeirão Preto 14040-904, Brazil
6 Innovation in Biomaterials Laboratory (iBioM), School of Dentistry, Federal University of Goiás, Goiânia 74605-020, Brazil
* Correspondence: pedrosouza@ufg.br

Abstract: Titanium surface modifications are widely used to modulate cellular behavior by recognition of topographical cues. However, how those modifications affect the expression of mediators that will influence neighboring cells is still elusive. This study aimed to evaluate the effects of conditioned media from osteoblasts cultured on laser-modified titanium surfaces on the differentiation of bone marrow cells in a paracrine manner and to analyze the expression of Wnt pathway inhibitors. Mice calvarial osteoblasts were seeded on polished (P) and Yb:YAG laser-irradiated (L) Ti surfaces. Osteoblast culture media were collected and filtered on alternate days to stimulate mice BMCs. Resazurin assay was performed every other day for 20 days to check BMC viability and proliferation. After 7 and 14 days of BMCs maintained with osteoblasts P and L-conditioned media, alkaline phosphatase activity, Alizarin Red staining, and RT-qPCR were performed. ELISA of conditioned media was conducted to investigate the expression of Wnt inhibitors Dickkopf-1 (DKK1) and Sclerostin (SOST). BMCs showed increased mineralized nodule formation and alkaline phosphatase activity. The L-conditioned media enhanced the BMC mRNA expression of bone-related markers *Bglap*, *Alpl*, and *Sp7*. L-conditioned media decreased the expression of DKK1 compared with P-conditioned media. The contact of osteoblasts with Yb:YAG laser-modified Ti surfaces induces the regulation of the expression of mediators that affect the osteoblastic differentiation of neighboring cells. DKK1 is among these regulated mediators.

Keywords: laser ablation; Ti; nanotopography; osteoblast; osseointegration

Citation: Teixeira, J.F.L.; de Souza, J.A.C.; Magalhães, F.A.C.; de Oliveira, G.J.P.L.; de Santis, J.B.; de Souza Costa, C.A.; de Souza, P.P.C. Laser-Modified Ti Surface Improves Paracrine Osteogenesis by Modulating the Expression of DKK1 in Osteoblasts. *J. Funct. Biomater.* **2023**, *14*, 224. https://doi.org/10.3390/jfb14040224

Academic Editors: Marco Tatullo and Jarosław Jakubowicz

Received: 8 February 2023
Revised: 4 April 2023
Accepted: 12 April 2023
Published: 16 April 2023

1. Introduction

Osseointegrated implants are widely used with high clinical success rates and predictability in oral rehabilitation [1,2]. Since the mid-1960s, titanium (Ti) has been considered a reliable material for bone implantations due to its excellent biocompatibility, corrosion resistance, and mechanical properties, providing stable implant anchorage in most cases [3–5].

The clinical success of a dental implant depends on many factors, including surgical procedure, bone quality, and how fast and uniform the osseointegration occurs [6,7]. Faster osseointegration allows an earlier loading and increases success rates even in patients with poor bone quality or systemic conditions [8].

Disordered microstructures are common cues that influence cell morphology and behavior [9–11]. Thus, the specialized industry widely investigates Ti surface modification protocols and alternatives to increase the quality of dental implant osseointegration [12]. These modifications usually involve chemical treatment or particle incrustation, raising the product's final cost and adding potential contaminants to its surface [13,14].

High-power laser irradiation is a cheaper, faster, and reproducible alternative for Ti surface modifications that can eliminate chemicals and residual particles adhered to the substrate. This method creates different roughness and topographical patterns with a superficial oxide layer of nanometer thickness that improves corrosion resistance and biocompatibility and plays a critical role in osseointegration [15–18]. Laser-irradiated Ti surfaces exhibit high wettability, increased cell spreading, and adhesion to the material [10,19,20]. Previous in vivo studies have demonstrated that high-power laser Ti modifications produced nano- and microstructures capable of accelerating the implant osseointegration [16,18], which is a complex process that includes the recruitment and differentiation of osteoblasts, mainly controlled by the early maturation runt-related transcription factor 2 (Runx2) and the downstream transcriptional regulator Osterix (Sp7). Differentiated osteoblasts secrete bone matrix proteins, including collagen type 1 alpha 1 (Col1a1), osteocalcin (OC), and alkaline phosphatase (Alp), that will regulate mineralized matrix deposition and new bone formation [21,22].

For effective implant integration, osteogenesis must occur in two different sites: in contact with the Ti implant surface and at a distance, on the host bone surface [3,23]. The distance osteogenesis will connect the host bone to the implant surface, while the contact osteogenesis will enable an effective bone apposition. Although contact osteogenesis seems to require other factors to be triggered, histological evidence demonstrates that implant surface modifications play a critical role in paracrine osteogenesis and might accelerate the final implant osseointegration [23–25].

To date, in vitro studies using the contact osteogenesis approach showed that modifications on the Ti surface can accelerate and improve implant/bone interface consolidation [5,10,21]. It has been demonstrated that micro- and nanopatterns on surface topography influence cell morphology, increasing the chemotactic signal of cell adhesion and affecting integrin receptors and the subsequent secretion of cytokines that may modulate an osteogenic differentiation [26]. Our group has previously shown that irradiation of Ti with high-power laser creates a surface that can induce osteoblastic differentiation, increasing the mRNA expression of osteoblast transcription factors (*Runx2* and *Sp7*), matrix proteins (*Col1a1*, *Spp1*, and *Bglap*), and enzymes (*Alpl*) [10]. However, whether and how this microenvironment may influence bone formation in a paracrine manner has yet to be investigated. Understanding how Ti surface modifications affect the expression of mediators of osteoblastic differentiation that may act in a paracrine manner will shed a new light on the comprehension of the mechanisms of osseointegration.

The Wnt/β-catenin signaling pathway is essential to embryonic skeleton development; it modulates bone formation and resorption by promoting osteoblast and osteocyte differentiation, maturation, and survival and inhibiting osteoclastogenesis [27]. Its activation by binding Wnt proteins to a Frizzled (Fzd) receptor and co-receptor (LDL receptor-related protein (Lrp5 or Lpr6)) results in the accumulation of non-phosphorylated β-catenin in the cytoplasm that will eventually be translocated into the nucleus and regulates gene transcription [28]. The Dickkopf (Dkk) family is a well-known group of proteins that modulate the Wnt signaling pathway; they prevent the formation of the Wnt–Fzd–Lrp complex binding to co-receptor Lpr5/6 [29]. Even though Wnt proteins and their downstream signaling components have been widely studied and targeted as bone formation modulators on modified surfaces [30,31], the role of Wnt signaling inhibitors on distance bone formation within the implant/bone interface microenvironment remains unclear.

Here, we report that conditioned culture media from osteoblasts cultured on laser-modified Ti surface stimulates bone marrow cell (BMC) osteoblastic differentiation in a paracrine manner. A possible mechanism is the decrease in the expression of DKK1 by osteoblasts cultured on this surface.

2. Materials and Methods

The in vitro study was approved by the local Ethical Committee for Animal Experimentation from the Faculty of Dentistry of Araraquara (protocol # 23/2013, approved on March 2016) and conducted according to the ARRIVE guidelines. The experiments were carried out in triplicate, with three independent experiments.

2.1. Ti Disks Preparation and Characterization

Commercially pure (grade 4) Ti disks (175 mm^2 of area and 2 mm thickness) were sequentially polished under water-cooling with a 320- to 1200-grit grinding paper. They were immersed in acetone PA ACS (Qhemis, Indaiatuba, SP, Brazil) and ultrasonically cleaned for 15 min and washed with deionized water and 100% ethanol (Qhemis, Indaiatuba, SP, Brazil). The disks were finally rinsed with distilled water for 15 min and dried for 90 min in an incubator at 60 °C (Orion 515, Fanem, São Paulo, SP, Brazil). The polished samples were divided into two experimental groups, polished Ti (P Ti) and laser Ti (L Ti). The laser surface was obtained by laser ablation with a pulsed Yb:YAG laser (Omnimark 20F, Ominitek Tecnologia Ltda, São Paulo, Brazil) as previously described [10]. Briefly, the surfaces were irradiated in an ambient atmosphere, and the laser beam was perpendicular to the surface with a fixed focal length of 170 mm. The laser wavelength was 1064 nm, scanning speed 100 mm/s, pulse duration 100 ns, average pulse power 10 kW, spot diameter 50 µm and cutoff distance 0.1 mm.

To access the surface topography, the specimens were characterized by scanning electron microscopy (SEM) (JMS-T33A, JEOL, Tokyo, Japan). The average surface roughness measured in our previous publication [10] was 0.3151 ± 0.0069 µm for the P surface and 10.57 µm ± 0.39 for the L surface.

2.2. Isolation and Culture of Mouse Calvarial Osteoblasts

Primary osteoblasts were isolated from neonatal calvarial bones obtained from 2- to 3-day-old mice (Mus musculus C57Bl/6) using sequential enzymatic digestion [32]. The cells obtained were seeded in plain minimum essential medium α (α-MEM, Invitrogen, Carlsbad, CA, USA) supplemented with penicillin/streptomycin (Pen-Strep, Sigma Aldrich, St. Louis, MO, USA) and 10% fetal bovine serum (FBS, Sigma Aldrich, St. Louis, MO, USA). After three days of incubation, 20,000 cells were seeded on Ti disks with L or P surface (24-well plates) in osteogenic differentiation media and α-MEM-supplemented with 10 mmol/L β-glycerophosphate and 2 mmol/L ascorbic acid (Sigma Aldrich, Carlsbad, CA, USA). The OM was collected and filtered every other day to stimulate the bone marrow cell culture and analyze bone metabolism, as described below.

2.3. Isolation and Culture of Mouse Bone Marrow Cells

BMCs were isolated from the tibia and femur of 4-week-old mice as previously described [33]. A total of 5×10^5 cells were seeded in 96-well plates in complete α-MEM. After 24 h, the complete medium was replaced for the OM medium obtained from calvarial osteoblast-like cultures seeded over P or L Ti surfaces; every 48 h, the fresh conditioned medium was collected from calvarial osteoblasts cultured in parallel to the BMC cultures, filtered and used to stimulate BMCs.

2.4. Alamar Blue

Resazurin (Alamar Blue, Sigma Aldrich, St. Louis, MO, USA) assay was performed every other day after the first day of culture for 20 days, according to the manufacturer's protocol, to analyze the BMC's viability and proliferation. The 1:10 Alamar Blue solution was directly added into the culture wells, and the plates were incubated for 1 h at 37 °C. Cell viability was assessed by the proportional absorbance at 570 nm in a microplate reader (Power Wave XS, BioTek Instruments, Winooski, VT, USA).

2.5. Alkaline Phosphatase Activity

After 7 and 14 days, the osteoblasts were lysed using 500 μL of a 50 mmol/L Tris–HCl solution (pH 7.4) containing 0.5% of Nonidet P40. The total protein content was quantified using a bicinchoninic acid assay (BCA Protein Assay Kit, Thermo Scientific, Rockford, IL, USA). ALPase activity was assessed by measuring the release of thymolphthalein from thymolphthalein monophosphate using a commercial test kit (Labtest Diagnostica SA, Belo Horizonte MG, Brazil). According to the estimated protein amount, 20 μg of total protein was added to a solution of thymolphthalein monophosphate (22 mmol/L) and diethanolamine buffer solution (0.3 mmol/mL, pH 10.1) and incubated for 40 min at 37 °C. About 200 μL of a Na_2CO_3 (0.09 mmol/mL) and NaOH (0.25 mmol/mL) solution was added and incubated at 37 °C for additional 10 min; the color gradient was measured by the absorbance at 590 nm in a microplate reader (Power Wave XS, BioTek Instruments, USA). ALPase activity was calculated from a standard curve using alkaline phosphatase as the standard in a range varying from 22.5 to 900 μU.

2.6. Alizarin Red Staining

The mineral deposition of osteoblast-like cells treated with OM was accessed by Alizarin Red staining at 14 and 21 days after the cells were cultured in 96-well plates. The cells were rinsed twice with PBS at 37 °C and then fixed with 150 μL of 4% paraformaldehyde for 10 min. After rinsing the cells again with PBS, 50 μL of 10% *w/v* Alizarin Red (Sigma Aldrich, St. Louis, MO, USA) staining solution was added to each well at room temperature for 10 min, the staining solution was removed, and the wells were rinsed twice with PBS. The mineralization was quantified by dissolving stained nodules with 200 μL of 10% *w/v* cetylpyridinium chloride (Sigma Aldrich, St. Louis, MO, USA) solution. The dissolved nodules were transferred to 96-well plates and read at 562 nm in a microplate reader (Power Wave XS, BioTek Instruments, USA).

2.7. Real-Time PCR

Total RNA was extracted from seeded BMCs treated with OM for 14 days using an affinity column method according to the manufacturer's protocol (RNAqueous-micro Kit, Thermo Scientific, Rockford, IL, USA). The RNA from each well was quantified by UV absorbance (Eppendorf, Hamburg, Germany) and considered acceptable when the 260/280 nm absorbance ratio was above 1.8. Three hundred nanograms of total RNA was converted into cDNA using random hexamer primers and reverse transcriptase in a total reaction volume of 20 μL (high-capacity cDNA synthesis kit, Applied Biosystems).

Quantitative real-time PCRs (qPCRs) were performed to assess the expression levels of the genes encoding the transcription factors *Runx2* (forward primer, 5′-CCTGAACTCTGCA CCAAGTCCT-3′; reverse primer, 5′-TCATCTGGCTCAGATAGGAGGG-3′) and Osterix (*Sp7*) (forward primer, 5′-TGCTTGAGGAGGAAGTTCAC-3′; reverse primer, 5′-AGGTCAC TGCCCACAGAGTA-3′), and bone-specific proteins osteopontin/secreted phosphoprotein (*Spp1*) (forward primer, 5′-TCACCATTCGGATGAGTCTG-3′; reverse primer, 5′-ACTTGTG GCTCTGATGTTCC-3′), alkaline phosphatase (*Alpl*), α1 chain of collagen type 1 (*Col1a1*) (forward primer, 5′-CCTCAGGGTATTGCTGGACAAC-3′; reverse primer, 5′-CAGAAGGA CCTTGTTTGCCAGG-3′) and osteocalcin/bone gamma-carboxy glutamic acid protein (*Bglap*) (forward primer, 5′-GCAATAAGGTAGTGAACAGACTCC-3′; reverse primer, 5′-CCATAGATGCGTTTGTAGGCGG-3′). The 20 μL total volume reactions included a TaqMan qPCR master mix (Applied Biosystems), cDNA, deionized water, and species-specific predesigned and optimized pairs of primers and probes (TaqMan gene expression assays).

The relative levels of gene expression were determined using the quantification cycle by the standard curve method using the gene encoding to β-actin (*Actb*) as the reference gene. Results were expressed as fold change over the levels of expression of the normalized target gene determined in cDNA prepared from the P Ti sample group.

2.8. ELISA

The ELISA tests for Dkk-1 (# MKK100) and Sclerostin (# MSST00) were performed following the manufacturer's recommended protocol (R&D Systems, Minneapolis, MN, USA). For these assays, aliquots of the calvarial osteoblast culture supernatants (OM) were collected at 0, 2, 4, 6, 8, 10, and 12 days and were mixed to obtain pooled samples. Six different pools were evaluated per group (n = 6).

2.9. Statistics

Data from Alamar Blue, RT-qPCR, alkaline phosphatase activity, Alizarin red, and ELISA assays were submitted to t-tests followed by Welch's correction since they adhered to the normal distribution (Shapiro–Wilk, $p > 0.05$). The level of significance was 95% ($p < 0.05$). Results were represented as mean + SD.

3. Results

3.1. Ti Surface Characterization

The results of the SEM analysis revealed topographic differences between the P and L Ti surfaces. The polished surface was homogeneous with minor residual roughness from the manufacturing process (Figure 1a–c). In contrast, laser surfaces generally presented an irregular surface, characterized by the presence of spheres and protrusions of different diameters due to the irradiation pattern (Figure 1d–f). Longitudinal channel-shaped depressions with different depths and widths were observed. At high magnification, it was possible to detect in the L surface the presence of secondary nano-roughness inside the spheres caused by the Ti ablation. Therefore, laser irradiation created a nano-to-micro hybrid surface (Figure 1f).

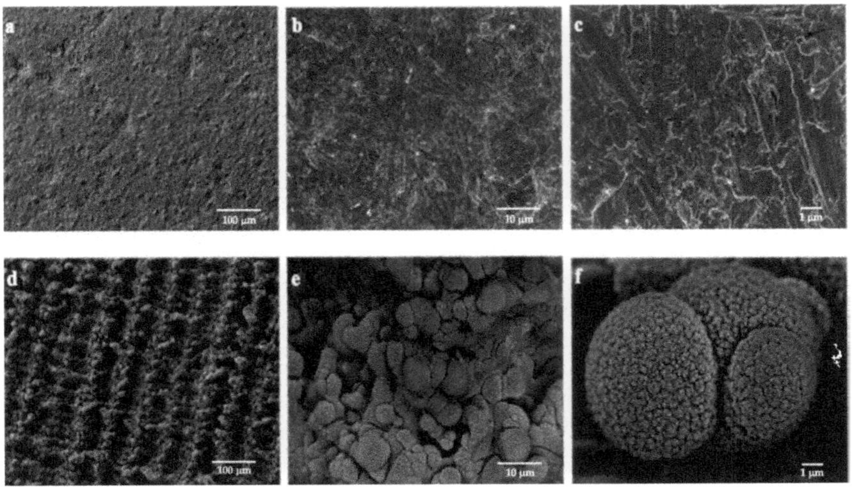

Figure 1. Representative SEM images of polished Ti surface in ×200 (**a**), ×2000 (**b**), and ×10,000 (**c**); and laser Ti surface in ×200 (**d**), ×2000 (**e**), and ×10,000 (**f**).

3.2. Cell Viability and Proliferation

The effect of surface topography on the osteoblastic expression of proliferation/differentiation factors was first investigated on the ability of the osteoblastic conditioned medium to induce BMC viability and proliferation. The cells were treated with the conditioned medium from osteoblasts cultured in P and L Ti surfaces. For all the experimentation periods, no significant difference between the groups was observed (Figure 2) ($p > 0.05$).

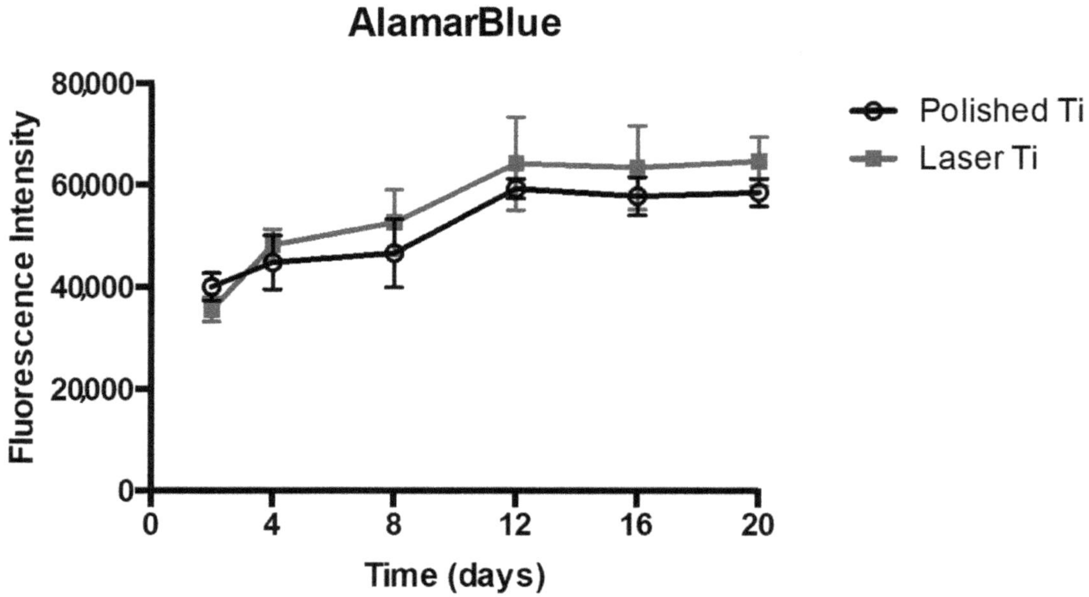

Figure 2. BMC proliferation and viability assessed by Alamar Blue assay. Mean (±SD) of fluorescence intensity at 2, 4, 8, 12, 16, and 20 days (n = 8) ($p > 0.05$).

3.3. Conditioned Medium from Osteoblasts Cultured on Laser-Modified Surface Affects Osteoblastic Genes Expression in BMCs

The effects of the laser-modified Ti surface on the osteoblast production of differentiation factors were then investigated on the expression of mRNAs, encoding for transcription factors (*Runx2* and *Sp7*) and osteoblast phenotypic markers (*Alpl*, *Col1a1*, and *Bglap*), by BMCs treated with the conditioned medium obtained from osteoblasts cultured on P and L surfaces. After 14 days, the L surface-conditioned medium upregulated the mRNA expression of the transcription factor *Sp7* and the phenotypic markers *Bglap* and *Alpl* (Figure 3).

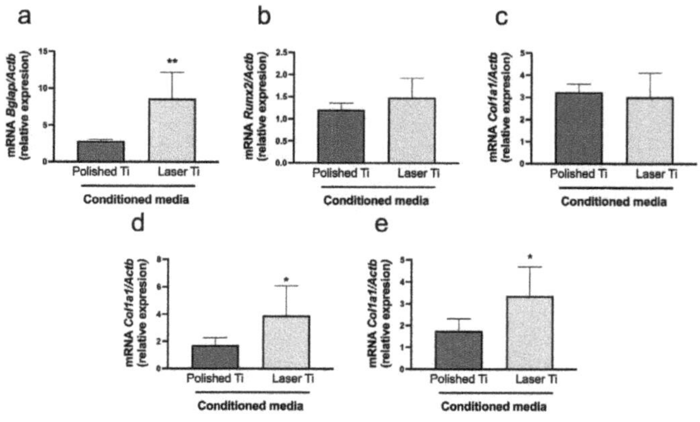

Figure 3. mRNA relative expression of osteoblast phenotypical marks after 14 days of culture of BMCs treated with P and L Ti surface-conditioned media. Mean (±SD) of mRNA expression of *Bglap* (**a**), *Runx2* (**b**), *Col1a1* (**c**), *Alpl* (**d**), and *Sp7* (**e**), (n = 8). Real-time PCR (* $p < 0.05$; ** $p < 0.01$).

3.4. Laser Ti Surface Stimulates ALPase Activity in BMC

After 7 and 14 days, the activity of ALPase was increased in BMCs treated with the conditioned medium obtained from the culture of osteoblasts grown on L Ti-irradiated surfaces, compared with BMCs treated with the medium obtained from osteoblasts grown on the P surfaces (Figure 4).

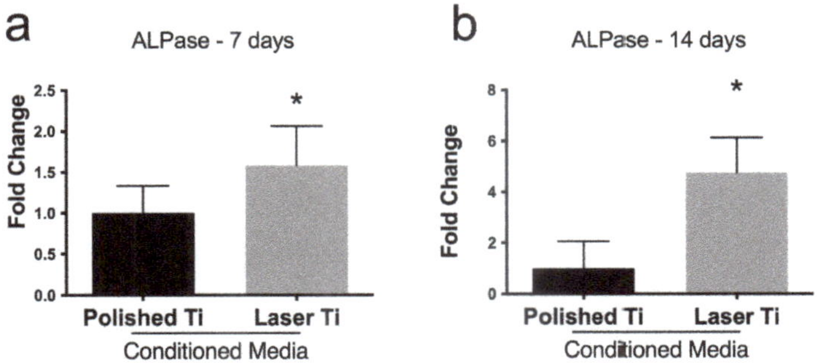

Figure 4. ALPase activity of BMCs treated with P and L Ti surface-conditioned medium at the 7th and 14th day of culture. Mean (±SD) of spectrophotometric quantification of ALPase activity in BMC cultures at (**a**) 7 and (**b**) 14 days (n = 8) (* $p < 0.05$).

3.5. Laser Ti Surface Stimulates Mineralized Nodule Formation

The formation of a mineralized extracellular matrix is what defines fully differentiated and active osteoblasts. To investigate how the conditioned medium from osteoblasts cultured on different Ti surfaces affected BMC differentiation on fully differentiated osteoblasts, we quantified the calcified nodules formed in BMC cultures treated with a P or L Ti surface-conditioned medium after 14 and 21 days of culture. Active formation of mineralized nodules was observed in both groups and experimental periods. After 14- and 21-day cultures, BMCs treated with the laser Ti surface-conditioned medium showed higher calcified matrix formation compared with cells treated with the polished Ti surface-conditioned medium (Figure 5, Supplemental Figure S1) ($p < 0.05$).

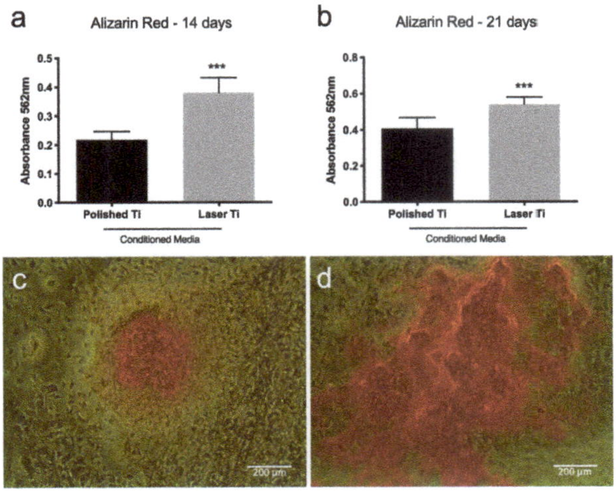

Figure 5. Mineralized matrix deposition of BMCs maintained in P or L Ti-conditioned medium after 14 (**a**) and 21 (**b**) days of culture. Mean (±SD) of absorbance after mineralized nodule dissolution (n = 8) (*** $p < 0.001$). Representative images of the nodules formed by BMCs cultured on P-conditioned media (**c**) or L-conditioned media (**d**) at 21 days.

3.6. Laser Ti Surface Downregulates Wnt Signaling Inhibitor in Osteoblasts

To investigate the impact of the laser-modified surface on the expression of Wnt signaling inhibitors, we performed an ELISA for Dkk-1 and Sclerostin using the supernatant of osteoblasts cultured on P and L surfaces. The expression of Dkk-1 was decreased in the L surface group compared with that in the polished P surface group (Figure 6). The protein Sclerostin was not detected in all groups.

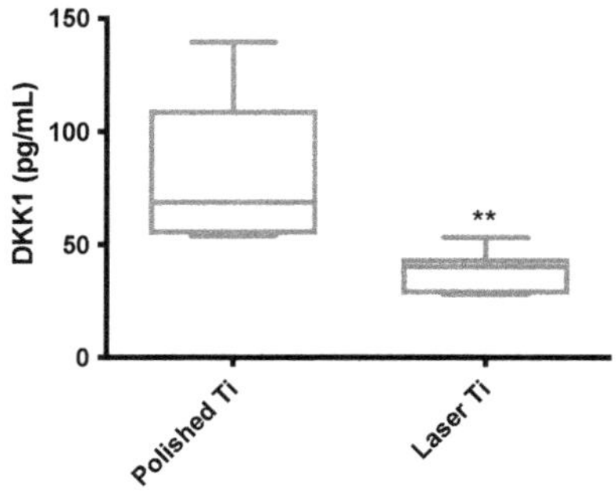

Figure 6. Dkk-1 protein measured by ELISA is downregulated in laser Ti surface samples. Dkk-1 was measured by ELISA in osteoblasts cultured on P and L Ti surfaces. Mean (±SD) expression of Dkk-1, OPN, and TGF-beta proteins (n = 6) (** $p < 0.01$).

4. Discussion

The implant osseointegration is environment-dependent and largely influenced by Ti surface topographical properties [5,34]. Several techniques aiming for micro- and nanostructured surface modifications have been proposed to modulate this biological process [35]. Laser surface treatments have often been described as a highly controllable approach to improving the cellular response to the implant surface due to the creation of micro- and nanometric features [16,18]. Our group has previously fabricated and characterized a nano-to-micro roughened surface produced by Yb:YAG laser ablation [10]. Besides changing the surface roughness, the melting caused by laser ablation led to the incorporation of atmospheric gases and the formation of a TiO_2 layer containing both the rutile and anatase phases [10]. In agreement with previous publications showing that the anatase phase of titanium plays an important role in cell adhesion, proliferation, and differentiation [36,37], our surface increased the expression of osteoblastic genes and their differentiation, as shown by the increased ALPase activity and mineral nodule deposition [10]. Besides the well-documented direct effect on bone cells in contact with the implant surface, laser ablation might influence cytokine and growth factor expression and release patterns that dictate paracrine osteogenesis in the bone/implant microenvironment [38–40]. Distance osteogenesis occurs on the host bone at the peri-implant area, and it is an important event for complete osseointegration and endosseous healing [3,23]. Here, we report that the laser-ablated surface decreased the expression of Wnt modulator Dkk-1 in osteoblast cultures. The conditioned medium obtained from these cultures induced the osteogenic differentiation and activity of BMCs in a paracrine fashion.

Although the laser Ti surface-conditioned medium did not affect cell proliferation and viability, it enhanced BMC ALPase activity and mineralized nodule formation compared with the P-conditioned medium group, suggesting that the ALPase activity of bone marrow stromal cells present in the host bone plays an important role in the bone matrix calcification of the peri-implant area. It has been demonstrated that the Ti-conditioned medium enhanced pre-osteoblast adhesion and activated key signaling proteins related to bone metabolism [38]. The laser surface-conditioned medium also increased the mRNA expression of *Sp7*, *Bglap*, and *Alpl*. It was recently reported that when stimulated with a conditioned medium obtained from RAW cells cultured on microstructured Ti topographies, bone marrow stem cells exhibited nearly 1.5-fold higher expression of osteogenic genes and production of mineralized nodules compared with the direct stimulation of Ti surfaces [40]. Recently, laser-induced micro–nano patterned structures were proven to overlay isotropic and anisotropic cues that could influence cell shape and induce nuclear orientation and activation of the Wnt/β-catenin pathway, signaling to increase the expression and activation of integrin α5, integrin β1, cadherin 2, Runx2, OPN, and OCN [30]. The decreased expression of the Wnt/β-catenin inhibitor Dkk-1, allied to the subsequent increased release of bone-related proteins induced by laser-irradiated Ti surface, may explain the modulation of BMC differentiation.

The Wnt/β-catenin signaling pathway is an important mechanism that interferes in cell proliferation, cell polarity, and cell fate determination during embryonic and postnatal development [24]. Bone diseases, such as osteoporosis, pseudoglioma syndrome, sclerosteosis, and van Buchem's disease, have been associated with atypical Wnt signaling [41]. The Wnt signaling pathway is one of the most critical regulators of bone tissue formation and homeostasis. The Wnt/β-catenin inhibitor Dkk-1 negatively regulates bone mass, and the reduction of this molecule results in increased bone mass in mice [42]. In the present study, the expression of Dkk-1 decreased on osteoblasts cultured on the L surface group, suggesting that besides the direct mechanical stimulation on cellular cytoskeleton and its integrin receptors [43], the laser treatment might modulate osteogenesis occurring within the implant/bone microenvironment by downregulating the Wnt canonical signaling inhibitor. We also analyzed the expression of another important inhibitor of Wnt signaling, sclerostin, an important factor regulating bone mass [44]. However, sclerostin (SOST) mRNA was not detected in any experimental groups. Since osteocytes are the leading producers of SOST, several growth factors and bone-related proteins have been deposited in ECM during mineralization [45,46]. We believe that the levels of SOST present in the osteoblast culture supernatant were insufficient to be reactive in ELISA measurement.

As demonstrated, the laser-modified Ti surface can affect bone cells at a distance, enhancing paracrine mineralized tissue formation by downregulating Dkk-1 and modulating osteoblastic lineage transcription factors and key protein expression. Additionally, the conditioned medium is a worthy tool for stimulating bone regeneration. Cultured MSCs secrete several growth factors and cytokines into the medium with the potential to stimulate tissue regeneration [47]. Pioneering studies have reported that the conditioned medium from mesenchymal stem cells improved bone regeneration in critical-sized calvarial defects in rats [48]. Moreover, the bone-conditioned medium enhanced osteoblastic differentiation on collagen membranes [49] and inorganic bovine bone grafts [50]. Thus, understanding how cells interact with different surfaces will provide new avenues for bone engineering. There are a plethora of surface modifications nowadays [51], and they can be used not only to accelerate the osseointegration of implanted devices but also to serve as bioreactors for the production of "bone growth media". However, further molecular investigations, gene editing tools, and in vivo studies are still needed to extend the understanding of the underlying mechanisms governing surface-induced paracrine osteogenesis.

5. Conclusions

Considering the methodology used in the present study, one may conclude that modifying the Ti surface with Yb:YAG laser positively impacts osteoblast behavior, enhancing the expression of key osteoblastic markers and downregulating the Wnt signaling inhibitor Dkk-1. The pre-conditioned medium stimulated the bone marrow cells into an osteogenic phenotype showing that laser-ablated surfaces might affect the bone tissue regeneration process occurring at a distance in the bone/implant microenvironment.

Supplementary Materials: The following supporting information can be downloaded at: https://www.mdpi.com/article/10.3390/jfb14040224/s1, Figure S1: Representative images of BMCs cultured in P-conditioned or L-conditioned media for 7 and 21 days.

Author Contributions: Conceptualization, J.A.C.d.S., C.A.d.S.C. and P.P.C.d.S.; methodology, J.F.L.T., J.A.C.d.S., F.A.C.M., G.J.P.L.d.O. and J.B.d.S.; formal analysis, J.A.C.d.S. and P.P.C.d.S.; data curation, J.F.L.T., J.A.C.d.S. and P.P.C.d.S.; writing—original draft preparation, J.F.L.T., J.A.C.d.S. and P.P.C.d.S.; writing—review and editing, J.F.L.T., J.A.C.d.S., P.P.C.d.S. and C.A.d.S.C.; supervision, P.P.C.d.S., J.A.C.d.S. and C.A.d.S.C.; project administration, C.A.d.S.C.; funding acquisition, C.A.d.S.C. and P.P.C.d.S. All authors have read and agreed to the published version of the manuscript.

Funding: This research was funded by the São Paulo Research Foundation—FAPESP (grants #2017/22739-9, #2014/05283-3) and the National Council for Scientific and Technological Development—CNPq (grant #303599/2014-6).

Data Availability Statement: Not applicable.

Conflicts of Interest: The authors declare no conflict of interest.

References

1. Moraschini, V.; Poubel, L.D.C.; Ferreira, V.; dos Sp Barboza, E. Evaluation of survival and success rates of dental implants reported in longitudinal studies with a follow-up period of at least 10 years: A systematic review. *Int. J. Oral Maxillofac. Surg.* **2015**, *44*, 377–388. [CrossRef]
2. Sousa, V.; Mardas, N.; Farias, B.; Petrie, A.; Needleman, I.; Spratt, D.; Donos, N. A systematic review of implant outcomes in treated periodontitis patients. *Clin. Oral Implants Res.* **2015**, *27*, 787–844. [CrossRef]
3. Berglundh, T.; Abrahamsson, I.; Lang, N.P.; Lindhe, J. *De novo* alveolar bone formation adjacent to endosseous implants: A model study in the dog. *Clin. Oral Implants Res.* **2003**, *14*, 251–262. [CrossRef] [PubMed]
4. Olivares-Navarrete, R.; Hyzy, S.L.; Hutton, D.L.; Erdman, C.P.; Wieland, M.; Boyan, B.D.; Schwartz, Z. Direct and indirect effects of microstructured titanium substrates on the induction of mesenchymal stem cell differentiation towards the osteoblast lineage. *Biomaterials* **2010**, *31*, 2728–2735. [CrossRef]
5. Wall, I.; Donos, N.; Carlqvist, K.; Jones, F.; Brett, P. Modified titanium surfaces promote accelerated osteogenic differentiation of mesenchymal stromal cells in vitro. *Bone* **2009**, *45*, 17–26. [CrossRef] [PubMed]
6. Jimbo, R.D.; Anchieta, R.D.; Baldassarri, M.; Granato, R.D.; Marin, C.D.; Teixeira, H.S.D.; Tovar, N.; Vandeweghe, S.D.; Janal, M.N.; Coelho, P.G.D. Histomorphometry and Bone Mechanical Property Evolution Around Different Implant Systems at Early Healing Stages. *Implants Dent.* **2013**, *22*, 596–603. [CrossRef]
7. De Bruyn, H.; Christiaens, V.; Doornewaard, R.; Jacobsson, M.; Cosyn, J.; Jacquet, W.; Vervaeke, S. Implant surface roughness and patient factors on long-term peri-implant bone loss. *Periodontology 2000* **2016**, *73*, 218–227. [CrossRef] [PubMed]
8. Khandelwal, N.; Oates, T.W.; Vargas, A.; Alexander, P.P.; Schoolfield, J.D.; McMahan, C.A. Conventional SLA and chemically modified SLA implants in patients with poorly controlled type 2 Diabetes mellitus—A randomized controlled trial. *Clin. Oral Implants Res.* **2011**, *24*, 13–19. [CrossRef]
9. Gittens, R.A.; Olivares-Navarrete, R.; McLachlan, T.; Cai, Y.; Hyzy, S.L.; Schneider, J.M.; Schwartz, Z.; Sandhage, K.H.; Boyan, B.D. Differential responses of osteoblast lineage cells to nanotopographically-modified, microroughened titanium–aluminum–vanadium alloy surfaces. *Biomaterials* **2012**, *33*, 8986–8994. [CrossRef]
10. Mariscal-Muñoz, E.; Costa, C.A.S.; Tavares, H.S.; Bianchi, J.; Hebling, J.; Machado, J.P.; Lerner, U.; Souza, P.P.C. Osteoblast differentiation is enhanced by a nano-to-micro hybrid titanium surface created by Yb:YAG laser irradiation. *Clin. Oral Investig.* **2015**, *20*, 503–511. [CrossRef]
11. Meirelles, L.; Arvidsson, A.; Albrektsson, T.; Wennerberg, A. Increased bone formation to unstable nano rough titanium implants. *Clin. Oral Implants Res.* **2007**, *18*, 326–332. [CrossRef]
12. Du, Z.; Xiao, Y.; Hashimi, S.; Hamlet, S.M.; Ivanovski, S. The effects of implant topography on osseointegration under estrogen deficiency induced osteoporotic conditions: Histomorphometric, transcriptional and ultrastructural analysis. *Acta Biomater.* **2016**, *42*, 351–363. [CrossRef] [PubMed]
13. Bonsignore, L.A.; Colbrunn, R.W.; Tatro, J.M.; Messerschmitt, P.J.; Hernandez, C.J.; Goldberg, V.M.; Stewart, M.C.; Greenfield, E.M. Surface contaminants inhibit osseointegration in a novel murine model. *Bone* **2011**, *49*, 923–930. [CrossRef]
14. Bosshardt, D.D.; Chappuis, V.; Buser, D. Osseointegration of titanium, titanium alloy and zirconia dental implants: Current knowledge and open questions. *Periodontology 2000* **2016**, *73*, 22–40. [CrossRef] [PubMed]
15. Miyauchi, T.; Yamada, M.; Yamamoto, A.; Iwasa, F.; Suzawa, T.; Kamijo, R.; Baba, K.; Ogawa, T. The enhanced characteristics of osteoblast adhesion to photofunctionalized nanoscale TiO2 layers on biomaterials surfaces. *Biomaterials* **2010**, *31*, 3827–3839. [CrossRef] [PubMed]

16. Queiroz, T.P.; de Molon, R.S.; Souza, F.Á.; Margonar, R.; Thomazini, A.H.A.; Guastaldi, A.C.; Hochuli-Vieira, E. In vivo evaluation of cp Ti implants with modified surfaces by laser beam with and without hydroxyapatite chemical deposition and without and with thermal treatment: Topographic characterization and histomorphometric analysis in rabbits. *Clin. Oral Investig.* **2016**, *21*, 685–699. [CrossRef]
17. Sawase, T.; Jimbo, R.; Baba, K.; Shibata, Y.; Ikeda, T.; Atsuta, M. Photo-induced hydrophilicity enhances initial cell behavior and early bone apposition. *Clin. Oral Implants Res.* **2008**, *19*, 491–496. [CrossRef]
18. Souza, F.Á.; Queiroz, T.P.; Sonoda, C.K.; Okamoto, R.; Margonar, R.; Guastaldi, A.C.; Nishioka, R.S.; Júnior, I.R.G. Histometric analysis and topographic characterization of *cp Ti* implants with surfaces modified by laser with and without silica deposition. *J. Biomed. Mater. Res. Part B Appl. Biomater.* **2014**, *102*, 1677–1688. [CrossRef]
19. Oliveira, N.T.; Guastaldi, F.P.; Perrotti, V.; Hochuli-Vieira, E.; Guastaldi, A.C.; Piattelli, A.; Iezzi, G. Biomedical Ti-Mo Alloys with Surface Machined and Modified by Laser Beam: Biomechanical, Histological, and Histometric Analysis in Rabbits. *Clin. Implants Dent. Relat. Res.* **2011**, *15*, 427–437. [CrossRef]
20. Queiroz, T.P.; Souza, F.Á.; Guastaldi, A.C.; Margonar, R.; Garcia-Júnior, I.R.; Hochuli-Vieira, E. Commercially pure titanium implants with surfaces modified by laser beam with and without chemical deposition of apatite. Biomechanical and topographical analysis in rabbits. *Clin. Oral Implants Res.* **2012**, *24*, 896–903. [CrossRef]
21. Calciolari, E.; Hamlet, S.; Ivanovski, S.; Donos, N. Pro-osteogenic properties of hydrophilic and hydrophobic titanium surfaces: Crosstalk between signalling pathways in in vivo models. *J. Periodontal Res.* **2018**, *53*, 598–609. [CrossRef]
22. Rutkovskiy, A.; Stensløkken, K.-O.; Vaage, I.J. Osteoblast Differentiation at a Glance. *Med. Sci. Monit. Basic Res.* **2016**, *22*, 95–106. [CrossRef]
23. Offermanns, V.; Andersen, O.Z.; Sillassen, M.; Almtoft, K.P.; Andersen, I.H.; Kloss, F.; Foss, M. A comparative in vivo study of strontium-functionalized and SLActive™ implant surfaces in early bone healing. *Int. J. Nanomed.* **2018**, *13*, 2189–2197. [CrossRef]
24. Choi, J.Y.; Lai, J.K.; Xiong, Z.; Ren, M.; Moorer, M.C.; Stains, J.P.; Cao, K. Diminished Canonical β-Catenin Signaling During Osteoblast Differentiation Contributes to Osteopenia in Progeria. *J. Bone Miner. Res.* **2018**, *33*, 2059–2070. [CrossRef]
25. Schouten, C.; Meijer, G.J.; Beucken, J.J.V.D.; Spauwen, P.H.; Jansen, J.A. The quantitative assessment of peri-implant bone responses using histomorphometry and micro-computed tomography. *Biomaterials* **2009**, *30*, 4539–4549. [CrossRef]
26. Yang, L.; Gao, Q.; Ge, L.; Zhou, Q.; Warszawik, E.M.; Bron, R.; Lai, K.W.C.; Van Rijn, P. Topography induced stiffness alteration of stem cells influences osteogenic differentiation. *Biomater. Sci.* **2020**, *8*, 2638–2652. [CrossRef] [PubMed]
27. Delgado-Calle, J.; Sato, A.Y.; Bellido, T. Role and mechanism of action of sclerostin in bone. *Bone* **2017**, *96*, 29–37. [CrossRef]
28. Baron, R.; Rawadi, G. Targeting the Wnt/β-Catenin Pathway to Regulate Bone Formation in the Adult Skeleton. *Endocrinology* **2007**, *148*, 2635–2643. [CrossRef] [PubMed]
29. Li, X.; Zhang, Y.; Kang, H.; Liu, W.; Liu, P.; Zhang, J.; Harris, S.E.; Wu, D. Sclerostin Binds to LRP5/6 and Antagonizes Canonical Wnt Signaling. *J. Biol. Chem.* **2005**, *280*, 19883–19887. [CrossRef] [PubMed]
30. Fu, J.; Liu, X.; Tan, L.; Cui, Z.; Liang, Y.; Li, Z.; Zhu, S.; Zheng, Y.; Yeung, K.; Chu, P.K.; et al. Modulation of the mechanosensing of mesenchymal stem cells by laser-induced patterning for the acceleration of tissue reconstruction through the Wnt/β-catenin signaling pathway activation. *Acta Biomater.* **2020**, *101*, 152–167. [CrossRef]
31. Olivares-Navarrete, R.; Hyzy, S.; Wieland, M.; Boyan, B.D.; Schwartz, Z. The roles of Wnt signaling modulators Dickkopf-1 (Dkk1) and Dickkopf-2 (Dkk2) and cell maturation state in osteogenesis on microstructured titanium surfaces. *Biomaterials* **2010**, *31*, 2015–2024. [CrossRef] [PubMed]
32. Granholm, S.; Henning, P.; Lindholm, C.; Lerner, U.H. Osteoclast progenitor cells present in significant amounts in mouse calvarial osteoblast isolations and osteoclastogenesis increased by BMP-2. *Bone* **2013**, *52*, 83–92. [CrossRef] [PubMed]
33. Takeshita, S.; Kaji, K.; Kudo, A. Identification and Characterization of the New Osteoclast Progenitor with Macrophage Phenotypes Being Able to Differentiate into Mature Osteoclasts. *J. Bone Miner. Res.* **2000**, *15*, 1477–1488. [CrossRef] [PubMed]
34. Khang, D.; Choi, J.; Im, Y.-M.; Kim, Y.-J.; Jang, J.-H.; Kang, S.S.; Nam, T.-H.; Song, J.; Park, J.-W. Role of subnano-, nano- and submicron-surface features on osteoblast differentiation of bone marrow mesenchymal stem cells. *Biomaterials* **2012**, *33*, 5997–6007. [CrossRef]
35. Jemat, A.; Ghazali, M.J.; Razali, M.; Otsuka, Y. Surface Modifications and Their Effects on Titanium Dental Implants. *BioMed Res. Int.* **2015**, *2015*, 791725. [CrossRef]
36. Yu, W.-Q.; Jiang, X.-Q.; Zhang, F.-Q.; Xu, L. The effect of anatase TiO2 nanotube layers on MC3T3-E1 preosteoblast adhesion, proliferation, and differentiation. *J. Biomed. Mater. Res. Part A* **2010**, *94*, 1012–1022. [CrossRef] [PubMed]
37. He, J.; Zhou, W.; Zhou, X.; Zhong, X.; Zhang, X.; Wan, P.; Zhu, B.; Chen, W. The anatase phase of nanotopography titania plays an important role on osteoblast cell morphology and proliferation. *J. Mater. Sci. Mater. Med.* **2008**, *19*, 3465–3472. [CrossRef]
38. Rossi, M.C.; Bezerra, F.J.B.; Silva, R.A.; Crulhas, B.P.; Fernandes, C.J.C.; Nascimento, A.S.; Pedrosa, V.A.; Padilha, P.; Zambuzzi, W.F. Titanium-released from dental implant enhances pre-osteoblast adhesion by ROS modulating crucial intracellular pathways. *J. Biomed. Mater. Res. Part A* **2017**, *105*, 2968–2976. [CrossRef] [PubMed]
39. Wang, X.; Wang, Y.; Bosshardt, D.D.; Miron, R.J.; Zhang, Y. The role of macrophage polarization on fibroblast behavior-an in vitro investigation on titanium surfaces. *Clin. Oral Investig.* **2017**, *22*, 847–857. [CrossRef]
40. Zhang, Z.; Xie, Y.; Pan, H.; Huang, L.; Zheng, X. Influence of patterned titanium coatings on polarization of macrophage and osteogenic differentiation of bone marrow stem cells. *J. Biomater. Appl.* **2017**, *32*, 977–986. [CrossRef]

41. Monroe, D.G.; McGee-Lawrence, M.E.; Oursler, M.J.; Westendorf, J.J. Update on Wnt signaling in bone cell biology and bone disease. *Gene* **2012**, *492*, 1–18. [CrossRef]
42. Morvan, F.; Boulukos, K.; Clément-Lacroix, P.; Roman-Roman, S.; Suc-Royer, I.; Vayssière, B.; Ammann, P.; Martin, P.; Pinho, S.; Pognonec, P.; et al. Deletion of a Single Allele of the Dkk1 Gene Leads to an Increase in Bone Formation and Bone Mass. *J. Bone Miner. Res.* **2006**, *21*, 934–945. [CrossRef] [PubMed]
43. Mastrangelo, F.; Quaresima, R.; Canullo, L.; Scarano, A.; Muzio, L.; Piattelli, A. Effects of Novel Laser Dental Implant Microtopography on Human Osteoblast Proliferation and Bone Deposition. *Int. J. Oral Maxillofac. Implants* **2020**, *35*, 320–329. [CrossRef] [PubMed]
44. Marini, F.; Giusti, F.; Palmini, G.; Brandi, M.L. Role of Wnt signaling and sclerostin in bone and as therapeutic targets in skeletal disorders. *Osteoporos. Int.* **2023**, *34*, 213–238. [CrossRef] [PubMed]
45. Brie, I.-C.; Soritau, O.; Dirzu, N.; Berce, C.; Vulpoi, A.; Popa, C.; Todea, M.; Simon, S.; Perde-Schrepler, M.; Virag, P.; et al. Comparative in vitro study regarding the biocompatibility of titanium-base composites infiltrated with hydroxyapatite or silicatitanate. *J. Biol. Eng.* **2014**, *8*, 14. [CrossRef]
46. Crane, J.L.; Cao, X. Bone marrow mesenchymal stem cells and TGF-β signaling in bone remodeling. *J. Clin. Investig.* **2014**, *124*, 466–472. [CrossRef]
47. Leuning, D.G.; Beijer, N.R.M.; du Fossé, N.A.; Vermeulen, S.; Lievers, E.; van Kooten, C.; Rabelink, T.J.; de Boer, J. The cytokine secretion profile of mesenchymal stromal cells is determined by surface structure of the microenvironment. *Sci. Rep.* **2018**, *8*, 7716. [CrossRef]
48. Osugi, M.; Katagiri, W.; Yoshimi, R.; Inukai, T.; Hibi, H.; Ueda, M. Conditioned media from mesenchymal stem cells enhanced bone regeneration in rat calvarial bone defects. *Tissue Eng. Part A* **2012**, *18*, 1479–1489. [CrossRef]
49. Fujioka-Kobayashi, M.; Caballé-Serrano, J.; Bosshardt, D.D.; Gruber, R.; Buser, D.; Miron, R.J. Bone conditioned media (BCM) improves osteoblast adhesion and differentiation on collagen barrier membranes. *BMC Oral Health* **2016**, *17*, 7. [CrossRef]
50. Caballé-Serrano, J.; Fujioka-Kobayashi, M.; Bosshardt, D.D.; Gruber, R.; Buser, D.; Miron, R.J. Pre-coating deproteinized bovine bone mineral (DBBM) with bone-conditioned medium (BCM) improves osteoblast migration, adhesion, and differentiation in vitro. *Clin. Oral Investig.* **2016**, *20*, 2507–2513. [CrossRef]
51. Yeo, I.-S.L. Modifications of Dental Implant Surfaces at the Micro- and Nano-Level for Enhanced Osseointegration. *Materials* **2019**, *13*, 89. [CrossRef] [PubMed]

Journal of *Functional Biomaterials*

MDPI

Article

Effects of Modulation of the Hedgehog and Notch Signaling Pathways on Osteoblast Differentiation Induced by Titanium with Nanotopography

Paola Gomes Souza, Leticia Faustino Adolpho, Helena Bacha Lopes, Denise Weffort, Alann Thaffarell Portilho Souza, Fabiola Singaretti Oliveira, Adalberto Luiz Rosa and Marcio Mateus Beloti *

Bone Research Lab, School of Dentistry of Ribeirão Preto, University of São Paulo, Ribeirão Preto 14040-904, SP, Brazil

* Correspondence: mmbeloti@usp.br; Tel.: +55-16-3315-4785

Abstract: Background: The events of bone formation and osteoblast/titanium (Ti) interactions may be affected by Hedgehog and Notch signalling pathways. Herein, we investigated the effects of modulation of these signalling pathways on osteoblast differentiation caused by the nanostructured Ti (Ti-Nano) generated by H_2SO_4/H_2O_2. Methods: Osteoblasts from newborn rat calvariae were cultured on Ti-Control and Ti-Nano in the presence of the Hedgehog agonist purmorphamine or antagonist cyclopamine and of the Notch antagonist N-(3,5-Difluorophenacetyl)-L-alanyl]-S-phenylglycine t-butyl ester (DAPT) or agonist bexarotene. Osteoblast differentiation was evaluated by alkaline phosphatase activity and mineralization, and the expression of Hedgehog and Notch receptors was also evaluated. Results: In general, purmorphamine and DAPT increased while cyclopamine and bexarotene decreased osteoblast differentiation and regulated the receptor expression on both Ti surfaces, with more prominent effects on Ti-Nano. The purmorphamine and DAPT combination exhibited synergistic effects on osteoblast differentiation that was more intense on Ti-Nano. Conclusion: Our results indicated that the Hedgehog and Notch signalling pathways drive osteoblast/Ti interactions more intensely on nanotopography. We also demonstrated that combining Hedgehog activation with Notch inhibition exhibits synergistic effects on osteoblast differentiation, especially on Ti-Nano. The uncovering of these cellular mechanisms contributes to create strategies to control the process of osseointegration based on the development of nanostructured surfaces.

Keywords: hedgehog; nanotopography; notch; osteoblast; titanium

Citation: Souza, P.G.; Adolpho, L.F.; Lopes, H.B.; Weffort, D.; Souza, A.T.P.; Oliveira, F.S.; Rosa, A.L.; Beloti, M.M. Effects of Modulation of the Hedgehog and Notch Signaling Pathways on Osteoblast Differentiation Induced by Titanium with Nanotopography. *J. Funct. Biomater.* **2023**, *14*, 79. https://doi.org/10.3390/jfb14020079

Academic Editor: Dusan Losic

Received: 16 December 2022
Revised: 14 January 2023
Accepted: 28 January 2023
Published: 30 January 2023

1. Introduction

Regenerative dentistry is based on developing strategies for inducting and maintaining cellular functions to promote oral tissue's structural and functional reestablishment [1–3]. In this context, osteogenesis is characterized by the sequential steps of cell adhesion, proliferation, differentiation, extracellular matrix apposition, and mineralization dependent on cellular signalling pathways [4]. Among these pathways, the Hedgehog and Notch act in several cellular processes, including osteoblast differentiation [5,6].

The Hedgehog signalling pathway acts on cell proliferation during embryonic development, stem cell maintenance, tissue repair, and regeneration [7–10]. When the Hedgehog precursor protein binds to the patched-1 (PTCH1) receptor, the smoothened (SMO) protein becomes constitutively active. It induces the signalling cascade, stabilizing the family zinc finger 2 transcription factor (GLI2) [11]. GLI2 is translocated to the nucleus and induces transcription of target genes, such as *Ptch1*, *Gli1*, and *Gli2*, and bone morphogenetic proteins (BMPs) [12]. In the absence of Hedgehog-PTCH1 binding, the SMO protein remains inactive, and the suppressor of fused protein is activated, which downregulates the pathway and leads to the production of Gli3 [13]. Purmorphamine, an agonist of the Hedgehog pathway, increases the expression of the runt-related transcription factor 2 (Runx2) in osteoprogenitor cells, favours the differentiation of osteoblasts and the formation of mineralized extracellular matrix [14–17]. The effects of purmorphamine are induced by the stimulation of the signalling cascade and an increase in the expression of the genes *Gli1*, *Gli2*, *Ptch1*, and *Ptch2* [18,19]. Cyclopamine, an antagonist of the Hedgehog pathway, binds to the SMO protein and inhibits its activity by changing the protein conformation, inhibiting GLI1 and GLI2 [20,21]. Cyclopamine reduces the expression of Ptch1 and alters the pattern of the BMP-2 expression in osteoblasts [22,23].

The Notch signalling pathway is involved in cellular proliferation and differentiation processes, mainly during embryonic development [24,25]. The interaction of the ligand with the receptor releases the intracellular domain of the Notch receptor (NICD), which is translocated from the membrane to the nucleus, acting as a transcriptional coactivator. This domain heterodimerizes with a protein complex containing the DNA-binding protein, called recombination signal sequence-binding protein Jk. This interaction results in the removal of corepressors, recruitment of co-activators, and, consequently, in the transcription of target genes from the hairy/enhancer of split (*Hes*) and hairy-related transcription factor (*Hey*) families [26]. While the HES1 subtype favours, the HEY1 and HEY2 subtypes strongly inhibit RUNX2 activity [27–31]. Bexarotene is a Notch agonist that exerts its biological action by binding to the gamma-secretase protein complex, inducing activation of the pathway, which inhibits cell growth and differentiation [32]. In contrast, the Notch antagonist N-(3,5-Difluorophenacetyl)-L-alanyl]-S-phenylglycine t-butyl ester (DAPT), an inhibitor of the gamma-secretase protein complex, promotes osteogenesis [33,34].

The modulation of these cellular mechanisms using agonists and antagonists can affect the bone tissue and titanium (Ti) implants interaction, which also depends on surface characteristics. The mechanisms of favouring or inducing osteoblast differentiation by topographical and chemical characteristics of Ti surfaces have been widely discussed in the literature [35–38]. The investigation of the behaviour of cells cultured on Ti with nanotopography (Ti-Nano) obtained through chemical conditioning with H_2SO_4/H_2O_2 showed that nanotopography induces osteoblast differentiation in osteogenic and non-osteogenic environments [39–41]. To date, we have shown the participation of integrin, BMP, and Wnt signalling pathways in the osteogenic potential of this nanotopography [39,42–46]. As Hedgehog and Notch regulate osteogenesis and their possible involvement with the osteogenic potential of Ti-Nano has not been investigated yet, we hypothesized that Hedgehog and Notch signalling pathways participate in the nanotopography-induced osteoblast differentiation. Thus, this study aimed to investigate the effects of agonists and antagonists of these signalling pathways on the osteoblast differentiation of cells grown on Ti-Nano.

2. Materials and Methods

2.1. Ti Surface Modification and Characterization

All reagents were laboratory grade. There were two commercially pure grade Ti discs (13 × 2 mm, Realum, São Paulo, SP, Brazil) conditioned with a 10 N H_2SO_4 (Merck Millipore, Darmstadt, Hesse, Germany) and 30% H_2O_2 (Merck Millipore) solution for 4 h to create nanotopography (Ti-Nano), as previously described [40]. The control samples were non-treated Ti discs (Ti-Control). To characterize the surface topography, the Ti discs were examined under field emission scanning electron microscopy (SEM) operated at 5 kV (Inspect S50, FEI Company, Hillsboro, OR, USA).

2.2. Selection of the Concentrations of the Hedgehog and Notch Agonists and Antagonists

2.2.1. Preparation of the Hedgehog and Notch Agonists and Antagonists

The Hedgehog agonist purmorphamine [2-(1-Naphthoxy)-6-(4-morpholinoanilino)-9-cyclohexylpurin] (Sigma-Aldrich, Saint Louis, MO, USA) was prepared at different concentrations of 0.5, 1, and 2 μM. The antagonist cyclopamine-KAAD [3-keto-N-aminoethyl-N′-aminocaproyldihydrocinnamoyl cyclopamine] (Calbiochem, Gibbstown, NJ, USA) was prepared at different concentrations of 10, 100, and 1000 nM. The Notch antagonist DAPT [GSI-IX, LY-374973, N-[N-(3,5-Difluorophenacetyl)-L-alanyl]-S-phenylglycine t-butyl ester] (Sigma-Aldrich) was prepared at different concentrations of 10, 15, and 20 μM, and the antagonist bexarotene [4-[1-(5,6,7,8-Tetrahydro-3,5,5,8,8-pentamethyl-2-naphthalenyl)ethenyl)benzoic acid] (Santa Cruz Biotechnology, Dallas, TX, USA) was prepared at different concentrations of 0.1, 0.5, and 1 μM. All agonists and antagonists were dissolved in the vehicle dimethylsulfoxide (DMSO, Sigma-Aldrich) and diluted in a culture medium, and the concentrations were selected based on the literature [18,19,32,34].

2.2.2. Isolation and Culture of Osteoblasts

After approval of the Ethics Committee on the Use of Animals of the School of Dentistry of Ribeirão Preto, University of São Paulo (Protocol # 2019.5.309.58.0), the osteoblasts were isolated from calvariae of newborn male Sprague-Dawley rats aged 2–4 days, as previously described [47,48]. Osteoblasts (2×10^4 cells per well) were cultured in a minimum essential medium, alpha modification (α-MEM, (Gibco-Life Technologies, Waltham, MA, USA)) supplemented with 10% foetal bovine serum (Gibco-Life Technologies), 5 μg/mL ascorbic acid (Gibco-Life Technologies), 7 mM β-glycerophosphate (Sigma-Aldrich), and 50 μg/mL gentamicin in 24-well polystyrene culture plates (Corning Life Sciences, Corning, NY, USA). The cultures were kept for up to 17 days at 37 °C in a humidified atmosphere (5% CO_2 and 95% atmospheric air) in the presence of either the vehicle (DMSO), the Hedgehog, or Notch agonists and antagonists.

2.2.3. Analysis of Gene Expression by Real-Time Polymerase Chain Reaction (RT-qPCR)

The osteoblastic marker osteopontin (*Opn*) gene expression was evaluated on day 10 by RT-qPCR. The total RNA was extracted using the SV Total RNA Isolation System kit (Promega, Fitchburg, WI, USA) and reverse transcription reaction was carried out to synthesize the complementary DNA (cDNA) using the High-Capacity cDNA Reverse Transcription Kit (Thermo Fisher Scientific, Waltham, MA, USA). For RT-qPCR, the SYBR Green system, and primers (Table 1) were used in the QuantStudio™ 7 Flex System device (Applied Biosystems, Waltham, MA, USA). The reactions were done (n = 4) and the data analysed using the cycle threshold value (Ct). The expression of the constitutive gene eukaryotic translation initiation factor 2B, subunit 1 alpha (*Eif2b1*) was evaluated, and the 2^{-ddCt} method was used to compare the gene expression of the experimental groups [49,50].

Table 1. Primer sequences for RT-qPCR.

Gene	Forward	Reverse
Opn	GAAGCCTGACCCATCTCAGAA	GTTGCTTGGAAGAGTTTCTTGCTT
Runx2	CGTATTTCAGATGATGACACTGCC	AAATGCCTGGGAACTGCCTG
Alp	TACTGCTGATCACTCCCACG	ACCGTCCACCACCTTGTAAC
Gli1	ACCTGCAAACCGTAATCCGT	TCCTAAAGAAGGGCTCATGGTG
Gli2	CCAACCAGAATAAGCAGAACAGC	TGAGATCAGCCAGTTGCTCC
Gli3	AGTCAGCCCTGCGGAATACT	GGGAAATCTGGTGCTGTCCAT
Hes1	ACGACACCGGACAAACCAAA	CGGGAGCTATCTTTCTTAAGTGCAT
Hey1	GCCGACGAGACCGAATCAAT	ATAGTCCATAGCCAGGGCGT
Hey2	CGTGGGGAGCGAGAACAATTA	ATTTATTCGATCCCGACGCCT
Eif2β	ACCTCCCTGGAATACTCTGACT	TCGCCCCGTCTTTGATGAAT

2.2.4. Analysis of the ALP Activity by Fast Red Staining

The ALP activity was evaluated on day 7 using Fast red staining, as previously described [51]. The cultures were incubated with 1 mL of a solution containing 1.8 mM Fast Red-TR Salt (Sigma-Aldrich), 0.9 mM Naphthol AS-MX phosphate (Sigma-Aldrich), and 4 mg/mL dimethylformamide (Sigma-Aldrich) for 30 min. The wells were dried and macroscopic images of the whole wells with the stained cultures were obtained with a high-resolution camera (Canon EOS Digital Rebel Camera, Canon, Lake Success, NY, USA). The stained areas of the whole wells were quantified by counting the pixels using the

ImageJ 1.52 software (National Institute of Mental Health, Bethesda, MD, USA). The data (n = 5) were expressed as a percentage of area.

2.2.5. Analysis of the Extracellular MATRIX Mineralization by Alizarin Red Staining

The formation of a mineralized extracellular matrix was evaluated on day 17 by Alizarin red staining. The cultures were fixed with 10% formalin at 4 °C for 24 h, dehydrated with alcohol, dried, and stained with 2% Alizarin red (Sigma-Aldrich), and the quantification was performed according to what was previously described [52]. The absorbance was measured in a spectrophotometer (BioTek Instruments Inc., Winooski, VT, USA) using a wavelength of 405 nm, and the data (n = 5) were expressed as absorbance.

2.3. Effects of the Hedgehog Signalling Modulation on Osteoblast Differentiation and Expression of Hedgehog Receptors in Cells Grown on Ti-Control and Ti-Nano

The osteoblasts were cultured on Ti-Control and Ti-Nano discs in 24-well polystyrene culture plates (Corning Life Sciences) at a density of 2×10^4 cells per disc in the presence of either vehicle, the agonist purmorphamine (Sigma-Aldrich) or the antagonist cyclopamine (Calbiochem) at the previously selected concentrations. The analyses of the gene expression of *Runx2*, *Opn*, *Alp*, *Gli1*, *Gli2*, and *Gli3* using the primers presented in Table 1, ALP activity, and extracellular matrix mineralization were performed as already detailed here, at the same time points. Additionally, the expression of RUNX2 and GLI1 proteins was evaluated by western blot on day 10.

Analysis of the Protein Expression by Western Blot

Western blot detected the expression of RUNX2 and GLI1 proteins on day 10 as previously described [40]. The cells were lysed and 25 µg of the total protein was denatured, separated in SDS polyacrylamide electrophoresis gel, and transferred to a PVDF membrane (Bio-Rad Laboratories, Hercules, CA, USA). The antibodies used were primary antibody either anti-RUNX2 (8486, 1:1000; Cell Signaling Technology, Danvers, MA, USA), anti-GLI1 (ab273018-1:1000; Abcam, Cambridge, UK) or anti-GAPDH (sc-25778, 1:1000; Santa Cruz Biotechnology, Dallas, TX, USA), and secondary antibody goat anti-rabbit IgG (7074, 1:3000, Cell Signaling Technology). The proteins were revealed with Clarity™ Western ECL Substrate (PerkinElmer Life Sciences, Waltham, MA, USA), and the images were obtained in a G: BOX device (Syngene, Cambridge, UK). The RUNX2 and GLI1 expressions were quantified (n = 3) using ImageJ Software (NIH, Bethesda, MD, USA) and normalized to GAPDH.

2.4. Effects of the Notch Signalling Modulation on Osteoblast Differentiation and the Expression of Notch Receptors in Cells Grown on Ti-Control and Ti-Nano

Osteoblasts were cultured on Ti-Control and Ti-Nano discs in 24-well polystyrene culture plates (Corning Life Sciences) at a density of 2×10^4 cells per disc in the presence of either vehicle, the antagonist DAPT (Sigma-Aldrich) or the agonist bexarotene (Santa Cruz Biotechnology) at the previously selected concentrations. The analyses of the gene expression of *Runx2*, *Opn*, *Alp*, *Hes1*, *Hey2*, and *Hey3* using the primers presented in Table 1, the protein expression of RUNX2 and HES1, ALP activity, and extracellular matrix mineralization were done as already detailed here, at the same time points. The antibody used to detect HES1 by western blot was anti-HES1 (11,988, 1:1000; Cell Signalling Technology), and the secondary antibody was goat anti-rabbit IgG (7074, 1:3000, Cell Signaling Technology).

2.5. Effects of the Combination of the Hedgehog and Notch Signalling Modulation on the Gene Expression of Bone Markers in Cells Grown on Ti-Control and Ti-Nano

To evaluate the effects of combining the modulation of both Hedgehog and Notch signalling, osteoblasts were cultured on Ti-Control and Ti-Nano discs in 24-well polystyrene culture plates (Corning Life Sciences) at a density of 2×10^4 cells per disc in the presence of either vehicle, the association of the Hedgehog agonist purmorphamine (Sigma-Aldrich) with the Notch antagonist DAPT (Sigma-Aldrich) or the association of the Hedgehog antagonist cyclopamine (Calbiochem) with the Notch agonist bexarotene (Santa Cruz Biotechnology) at the previously selected concentrations. The analyses of the gene expression of *Runx2*, *Opn*, and *Alp* using the primers presented in Table 1 were done, as already detailed here, at the same time.

2.6. Statistical Analysis

The software SigmaPlot free trial version 15.0 (Systat Software Inc., San Jose, CA, USA) was used to analyse the data. The data of concentrations' selection of the Hedgehog and Notch agonists and antagonists were analysed by one-way ANOVA, followed by Tukey's post-test. The data of the effects of the Hedgehog and Notch signalling modulation on osteoblast differentiation and the expression of Hedgehog and Notch receptors in the cells grown on Ti-Control and Ti-Nano were analysed using two-way ANOVA, followed by the Tukey's post-test. The results were expressed as the mean ± standard deviation (SD), and the significance level was established at $p \leq 0.05$.

3. Results

3.1. Ti-Control and Ti-Nano Surfaces

The SEM demonstrated that Ti-Control presents a polished surface (Figure 1A), and the Ti-Nano produced by H_2SO_4/H_2O_2 treatment exhibited nanopores over the entire surface (Figure 1B).

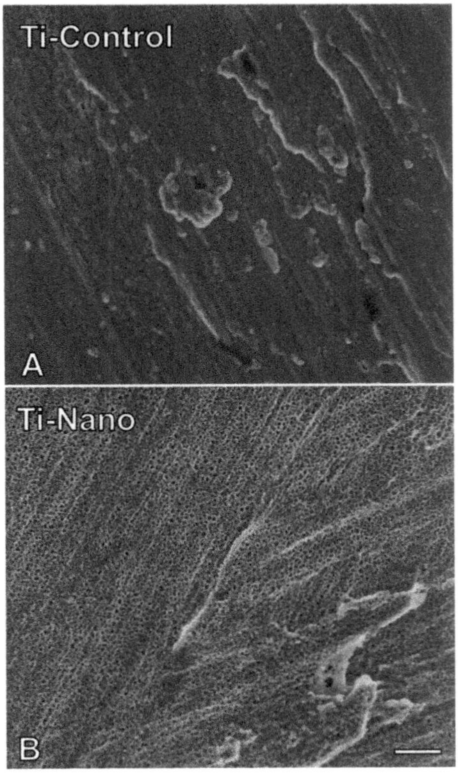

Figure 1. Surface topography of the titanium (Ti) discs. Images generated by scanning electron micrography of polished Ti ((**A**), Ti-Control) and nanostructured Ti ((**B**), Ti-Nano). Scale bar (**A**,**B**): 200 nm. Original magnification: 100,000×.

3.2. Selection of the Concentration of the Hedgehog and Notch Agonists and Antagonists

The selection of the concentration of the Hedgehog and Notch agonists and antagonists was based on the *Opn* gene expression, ALP activity, and extracellular matrix mineralization of cells grown on polystyrene. Higher osteoblast differentiation induced by purmorphamine was observed at a concentration of 2 M (Figure 2A–C). The *Opn* gene expression was higher at a concentration of 2 μM compared with 1 μM ($p < 0.001$), which was higher than 0.5 μM ($p < 0.001$) and vehicle ($p < 0.001$), and there was no statistically significant difference between 0.5 μM and vehicle ($p = 0.992$, Figure 2A). The ALP activity was higher at the concentration of 2 μM compared with 1 μM ($p < 0.001$), 0.5 μM ($p < 0.001$), and vehicle ($p < 0.001$), which was lower in vehicle compared with 0.5 μM ($p < 0.001$) and 1 μM ($p < 0.001$), without statistically significant difference between them ($p = 0.520$, Figure 2B). The extracellular matrix mineralization was greater at the concentration of 2 μM compared with 1 μM ($p < 0.001$), which was higher than 0.5 μM ($p < 0.001$) and vehicle ($p < 0.001$), and there was no statistically significant difference between the vehicle and 0.5 μM ($p = 1.000$, Figure 2C). The best effect of cyclopamine in decreasing osteoblast differentiation was observed at the concentration of 1000 nM (Figure 2D–F); however, as cyclopamine 1000 nM exhibited some cytotoxic effect, we selected 10 nM for further experiments. The *Opn* gene expression was lower at the concentration of 1000 nM compared with 10 nM ($p < 0.001$), which was lower than 100 nM ($p < 0.001$) and vehicle ($p < 0.001$), and there was no statistically significant difference between them ($p = 0.549$, Figure 2D). The ALP activity was lower at the concentration of 1000 nM compared with 10 nM ($p = 0.001$), which was lower than 100 nM ($p < 0.001$) that was lower than the vehicle ($p < 0.001$, Figure 2E). The extracellular matrix mineralization was lower at the concentrations of 1000 nM and 10 nM compared with 100 nM ($p = 0.002$ and $p = 0.009$) and vehicle ($p = 0.001$ and $p = 0.004$), without statistically significant differences between 1000 nm and 10 nM ($p = 0.553$), and 100 nM and vehicle ($p = 0.937$, Figure 2F). The best effect of DAPT in increasing osteoblast differentiation was observed at the concentration of 20 μM (Figure 2G–I). The *Opn* gene expression was higher at the concentration of 20 μM compared with 15 μM ($p = 0.028$), 10 μM ($p = 0.005$), and vehicle ($p = 0.002$), and there were no statistically significant differences among vehicle, 10 μM and 15 μM (($p = 0.384$, $p = 0.775$, and $p = 0.894$, Figure 2G). The ALP activity was higher at the concentrations of 20 μM and 15 μM compared with 10 μM ($p < 0.001$ and $p < 0.001$) and vehicle ($p < 0.001$ and $p < 0.001$), without statistically significant differences between 20 μM and 15 μM ($p = 0.831$), and 10 μM and vehicle ($p = 0.871$, Figure 2H). The extracellular matrix mineralization was higher at the concentration of 20 μM compared with 15 μM ($p = 0.041$), 10 μM ($p = 0.027$), and vehicle ($p = 0.014$), and there were no statistically significant differences among vehicle, 10 μM and 15 μM ($p = 0.866$, $p = 0.990$, and $p = 0.964$, Figure 2I). The best effect of bexarotene in decreasing osteoblast differentiation was observed at the concentration of 0.1 μM (Figure 2J–L). The *Opn* gene expression was lower at 0.1 μM compared with 0.5 μM ($p = 0.013$) and 1 μM ($p = 0.003$), which were lower than vehicle ($p < 0.001$ and $p < 0.001$), and there was no statistically significant difference between 0.5 μM and 1 μM ($p = 0.809$, Figure 2J). The ALP activity was lower at the concentrations of 0.1 μM and 0.5 μM compared with 1 μM ($p < 0.001$, $p < 0.001$), which was lower than vehicle ($p = 0.007$), and there was no statistically significant difference between 0.1 μM and 0.5 μM ($p = 0.918$, Figure 2K). The extracellular matrix mineralization was lower at the concentrations of 0.1 μM and 1 μM compared with 0.5 μM ($p = 0.024$ and $p = 0.034$) and vehicle ($p = 0.012$ and $p = 0.016$), without statistically significant differences between 0.1 μM and 1 μM ($p = 0.994$), and 0.5 μM and vehicle ($p = 0.948$, Figure 2L). Based on these results, we selected the following concentrations for the further experiments of this study: purmorphamine 2 μM, cyclopamine 10 nM, DAPT 20 μM, and bexarotene 0.1 μM.

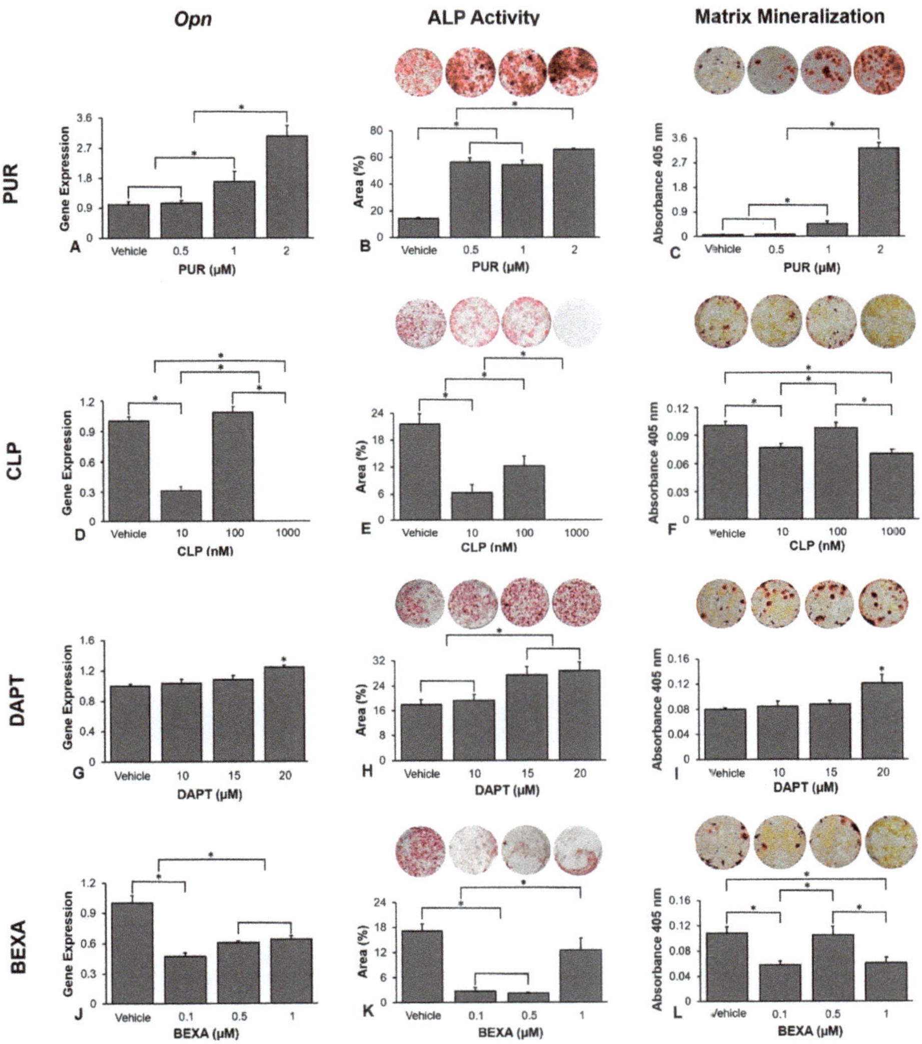

Figure 2. Selection of the concentrations of the Hedgehog and Notch agonists and antagonists based on their effects on osteoblast differentiation. Osteopontin (*Opn*) gene expression on day 10 (**A**), alkaline phosphatase (ALP) activity on day 7 (**B**), and extracellular matrix mineralization on day 17 (**C**) in osteoblasts cultured on polystyrene with either vehicle or the Hedgehog agonist Purmorphamine (PUR) at the concentrations of 0.5, 1, and 2 μM of (PUR). *Opn* gene expression on day 10 (**D**), alkaline phosphatase (ALP) activity on day 7 (**E**), and extracellular matrix mineralization on day 17 (**F**) in osteoblasts cultured on polystyrene with either vehicle or with cyclopamine (CLP) at the concentrations of 10, 100, and 1000 nM. *Opn* gene expression on day 10 (**G**), alkaline phosphatase (ALP) activity on day 7 (**H**), and extracellular matrix mineralization on day 17 (**I**) in osteoblasts cultured on polystyrene with either vehicle or with DAPT at the concentrations of 10, 15, and 20 μM. *Opn* gene expression on day 10 (**J**), alkaline phosphatase (ALP) activity on day 7 (**K**), and extracellular matrix mineralization on day 17 (**L**) in osteoblasts cultured on polystyrene with either vehicle or with bexarotene (BEXA) at the concentrations of 0.1, 0.5, and 1 μM. The original diameter of the bottom of the polystyrene wells presented in (**B**,**C**,**E**,**F**,**H**,**I**,**K**,**L**) is 15.62 mm. The data are presented as the mean ± SD, and the asterisks (*) indicate a statistically significant difference ($p \leq 0.05$).

3.3. Effects of the Hedgehog Signalling Modulation on Osteoblast Differentiation and the Expression of Hedgehog Receptors in Cells Grown on Ti-Control and Ti-Nano

The interaction between Ti surfaces and purmorphamine treatment affected the gene expression of *Runx2* ($p \leq 0.001$), *Alp* ($p \leq 0.001$), and *Opn* ($p = 0.043$, Figure 3A). Purmorphamine upregulated the gene expression of *Runx2* ($p < 0.001$ and $p < 0.001$), *Alp* ($p = 0.001$ and $p = 0.004$), and *Opn* ($p < 0.001$and $p < 0.001$) in cells grown on both Ti-Control and Ti-Nano (Figure 3A). In the presence of vehicle, the gene expression of *Runx2* ($p < 0.001$) was lower, while *Alp* ($p < 0.001$) and *Opn* ($p = 0.008$) was higher in cells grown on Ti-Nano than on Ti-Control (Figure 3A). In the presence of purmorphamine, the gene expression of *Runx2* ($p < 0.001$), *Alp* ($p < 0.001$), and *Opn* ($p < 0.001$) was higher in cells grown on Ti-Nano than on Ti-Control (Figure 3A). The interaction between Ti surfaces and purmorphamine treatment also affected the RUNX2 protein expression ($p \leq 0.001$), Figure 3B). Purmorphamine increased the RUNX2 protein expression ($p = 0.003$ and $p = 0.016$) in the cells grown on both Ti-Control and Ti-Nano (Figure 3B). In the presence of vehicle or purmorphamine, the RUNX2 protein expression ($p < 0.001$ and $p < 0.001$) was higher in cells grown on Ti-Nano than on Ti-Control (Figure 3B). The interaction between Ti surfaces and purmorphamine treatment did not affect the expression of ALP activity ($p = 0.270$, Figure 3C). Purmorphamine increased the ALP activity ($p < 0.001$ and $p < 0.001$) in cells grown on both Ti-Control and Ti-Nano (Figure 3C). In the presence of vehicle, the ALP activity ($p = 0.431$) was not affected by Ti surfaces (Figure 3C). In the presence of purmorphamine, the ALP activity (0.028) was higher on Ti-Nano than Ti-Control (Figure 3C). The interaction between Ti surfaces and purmorphamine treatment did not affect the extracellular matrix mineralization ($p = 0.111$, Figure 3D). Purmorphamine increased the extracellular matrix mineralization ($p < 0.001$ and $p < 0.001$) in the cells grown on both Ti-Control and Ti-Nano (Figure 3D). In the presence of vehicle or purmorphamine, the extracellular matrix mineralization ($p = 0.284$ and $p = 0.221$) was not affected by Ti surfaces (Figure 3D). The interaction between Ti surfaces and purmorphamine treatment affected the expression of *Gli1* ($p \leq 0.001$), *Gli2* ($p \leq 0.001$), and *Gli3* ($p \leq 0.001$, Figure 3E). Purmorphamine upregulated the gene expression of *Gli1* ($p < 0.001$ and $p < 0.001$), *Gli2* ($p < 0.001$ and $p < 0.001$), and downregulated *Gli3* ($p < 0.001$and $p < 0.001$) in cells grown on both Ti-Control and Ti-Nano (Figure 3E). In the presence of vehicle, the gene expression of *Gli1* ($p = 0.018$) was higher, while *Gli2* ($p < 0.001$) and *Gli3* ($p < 0.001$) were lower in cells grown on Ti-Nano than on Ti-Control (Figure 3E). In the presence of purmorphamine, the gene expression of *Gli1* ($p = 0.001$) and *Gli2* ($p < 0.001$) was higher, while *Gli3* ($p < 0.001$) was lower in cells grown on Ti-Nano than on Ti-Control (Figure 3E). The interaction between Ti surfaces and purmorphamine treatment also affected the GLI1 protein expression ($p < 0.001$, Figure 3F). Purmorphamine increased the GLI1 protein expression ($p = 0.003$ and $p = 0.016$) in cells grown on Ti-Control and Ti-Nano (Figure 3F). In the presence of vehicle or purmorphamine, the GLI1 protein expression ($p < 0.001$ and $p < 0.001$) was higher in cells grown on Ti-Nano than on Ti-Control (Figure 3F).

The interaction between Ti surfaces and cyclopamine treatment affected the gene expression of *Runx2* ($p \leq 0.001$), *Alp* ($p = 0.001$), and *Opn* ($p \leq 0.001$, Figure 4A). Cyclopamine downregulated the gene expression of *Runx2* ($p < 0.001$ and $p < 0.001$), *Alp* ($p < 0.001$ and $p < 0.001$), and *Opn* ($p < 0.001$ and $p < 0.001$) in cells grown on both Ti-Control and Ti-Nano (Figure 3B). In the presence of vehicle, the gene expression of *Runx2* ($p < 0.001$) was lower, while *Alp* ($p = 0.020$) and *Opn* ($p < 0.001$) were higher in cells grown on Ti-Nano than on Ti-Control (Figure 4A). In the presence of cyclopamine, the gene expression of *Runx2* ($p = 0.038$), *Alp* ($p = 0.009$), and *Opn* ($p < 0.001$) was lower in cells grown on Ti-Nano than on Ti-Control (Figure 4A). The interaction between Ti surfaces and cyclopamine treatment also affected the RUNX2 protein expression (($p \leq 0.001$), Figure 4B). Cyclopamine decreased the RUNX2 protein expression ($p < 0.001$ and $p < 0.001$) in cells grown on both Ti-Control and Ti-Nano (Figure 4B). In the presence of vehicle, the RUNX2 protein expression ($p < 0.001$ and $p < 0.001$) was higher, while in the presence of cyclopamine, it was lower in cells grown on Ti-Nano than on Ti-Control (Figure 4B). The interaction between Ti surfaces and cyclopamine treatment did not affect the ALP activity ($p = 0.647$, Figure 4C) and extracellular matrix mineralization ($p = 0.653$, Figure 4D). Cyclopamine decreased the ALP activity ($p < 0.001$ and $p < 0.001$, Figure 4C) and extracellular matrix mineralization ($p < 0.001$ and $p < 0.001$, Figure 4D) in cells grown on both Ti-Control and Ti-Nano. In the presence of vehicle or cyclopamine, the ALP activity ($p = 0.513$ and $p = 0.993$, Figure 4C) and extracellular matrix mineralization ($p = 0.185$ and $p = 0.059$, Figure 4D) were not affected by Ti surfaces. The interaction between Ti surfaces and cyclopamine treatment affected the expression of *Gli1* ($p \leq 0.001$), *Gli2* ($p \leq 0.001$), and *Gli3* ($p = 0.001$, Figure 4E). Cyclopamine downregulated the gene expression of *Gli1* ($p < 0.001$ and $p < 0.001$) and *Gli2* ($p < 0.001$ and $p < 0.001$) and upregulated *Gli3* ($p < 0.001$ and $p < 0.001$) in cells grown on both Ti-Control and Ti-Nano (Figure 4E). In the presence of vehicle, the gene expression of *Gli1* ($p = 0.346$) was not affected, while *Gli2* ($p < 0.001$) was higher and *Gli3* ($p < 0.001$) was lower in cells grown on Ti-Nano than on Ti-Control (Figure 4E). In the presence of cyclopamine, the gene expression of *Gli1* ($p = 0.006$) and *Gli2* ($p < 0.001$) was lower, while *Gli3* ($p < 0.001$) was higher in cells grown on Ti-Nano than on Ti-Control (Figure 4E). The interaction between Ti surfaces and cyclopamine treatment also affected the GLI1 protein expression ($p \leq 0.001$, Figure 4F). Cyclopamine decreased the GLI1 protein expression ($p < 0.001$, $p < 0.001$) in cells grown on both Ti-Control and Ti-Nano (Figure 4F). In the presence of vehicle or cyclopamine, the GLI1 protein expression ($p = 0.006$ and $p < 0.001$) was higher in cells grown on Ti-Nano than on Ti-Control (Figure 4F).

3.4. Effects of the Notch Signalling Modulation on Osteoblast Differentiation and the Expression of Notch Receptors in the Cells Grown on Ti-Control and Ti-Nano

The interaction between Ti surfaces and DAPT treatment did not affect the gene expression of *Runx2* ($p = 0.757$) but affected *Alp* ($p = 0.043$) and *Opn* ($p = 0.026$, Figure 5A). DAPT upregulated the gene expression of *Runx2* ($p < 0.001$ and $p < 0.001$), *Alp* ($p < 0.001$ and $p = 0.008$), and *Opn* ($p < 0.001$ and $p < 0.001$) in cells grown on both Ti-Control and Ti-Nano (Figure 5A). In the presence of vehicle, the gene expression of *Runx2* ($p = 0.114$), *Alp* ($p = 0.863$), and *Opn* ($p = 0.938$) was not affected by Ti surfaces (Figure 5A). In the presence of DAPT, the gene expression of *Runx2* ($p = 0.233$) were not affected, while *Alp* ($p = 0.011$) and *Opn* ($p = 0.005$) were higher in cells grown on Ti-Nano than on Ti-Control (Figure 5A). The interaction between Ti surfaces and DAPT treatment also affected the RUNX2 protein expression ($p \leq 0.001$, Figure 5B). DAPT increased the RUNX2 protein expression ($p < 0.001$ and $p < 0.001$) in cells grown on both Ti-Control and Ti-Nano (Figure 5B). In the presence of vehicle, the RUNX2 protein expression was not affected ($p = 0.571$), while in the presence of DAPT, it was higher ($p = 0.016$) in cells grown on Ti-Nano than on Ti-Control (Figure 5B). The interaction between Ti surfaces and DAPT treatment affected the ALP activity ($p \leq 0.001$, Figure 5C) and extracellular matrix mineralization ($p = 0.014$, Figure 5D). DAPT increased the ALP activity ($p < 0.001$ and $p < 0.001$, Figure 5C) and extracellular matrix mineralization ($p < 0.001$ and $p < 0.001$, Figure 5D) in cells grown on both Ti-Control and Ti-Nano. In the presence of either vehicle or DAPT, the ALP activity ($p = 0.007$ and $p = 0.023$) was greater in cells grown on Ti-Nano than on Ti-Control (Figure 5C). In the presence of vehicle, the extracellular matrix mineralization was not affected ($p = 0.527$), while in the presence of DAPT, it was greater ($p = 0.005$) in cells grown on Ti-Nano than on Ti-Control (Figure 5D). The interaction between Ti surfaces and DAPT treatment affected the gene expression of *Hes1* ($p = 0.002$) but not of *Hey1* ($p = 0.869$) and *Hey2* ($p = 0.681$, Figure 5E). DAPT upregulated the gene expression of *Hes1* ($p < 0.001$ and $p < 0.001$) and downregulated *Hey1* ($p < 0.001$ and $p < 0.001$), and *Hey2* ($p < 0.001$ and $p < 0.001$) in cells grown on both Ti-Control and Ti-Nano (Figure 5E). In the presence of vehicle, the gene expression of *Hes1* ($p = 0.202$), *Hey1* ($p = 0.975$), and *Hey2* ($p = 0.771$) was not affected by Ti surfaces (Figure 5E). In the presence of DAPT, the gene expression of *Hes1* ($p = 0.001$) was higher in cells grown on Ti-Nano than on Ti-Control, while *Hey1* ($p = 0.840$) and *Hey2* ($p = 0.772$) were not affected by Ti surfaces (Figure 3E). The interaction between Ti surfaces and DAPT treatment also affected the HES1 protein expression ($p \leq 0.001$, Figure 5F). DAPT increased the HES1 protein expression ($p < 0.001$ and $p < 0.001$) in cells grown on both Ti-Control and Ti-Nano (Figure 5F). In the presence of vehicle, the HES1 protein expression was not affected ($p = 0.274$), while in the presence of DAPT, it was higher ($p = 0.003$) in cells grown on Ti-Nano than on Ti-Control (Figure 5F).

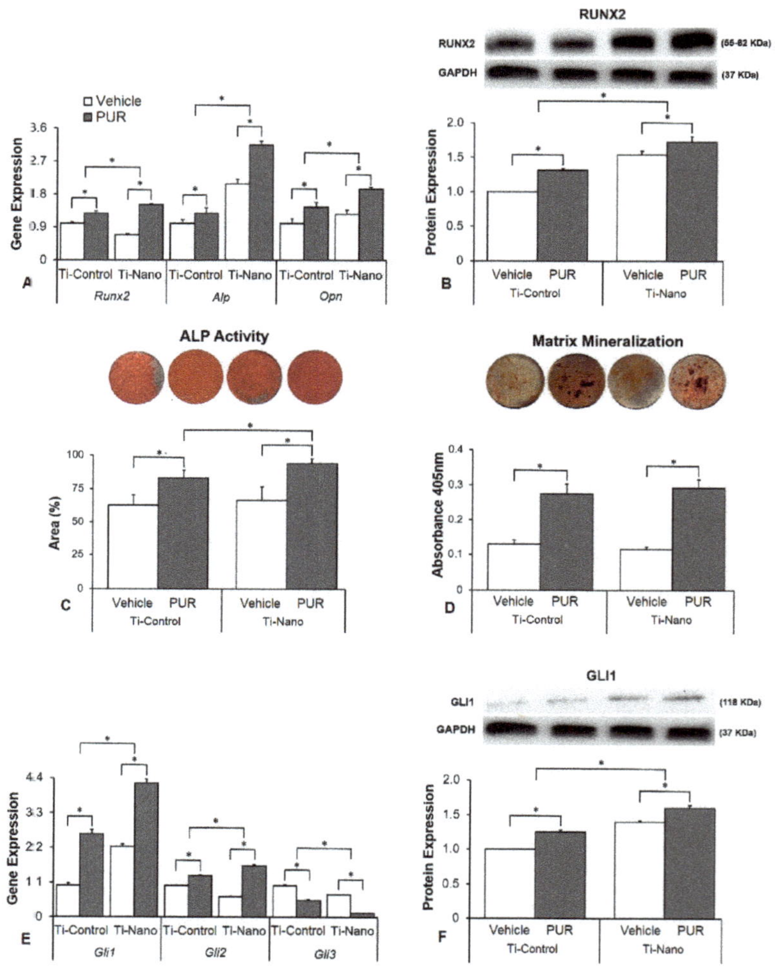

Figure 3. Effects of the Hedgehog agonist purmorphamine (PUR) on osteoblast differentiation and the expression of Hedgehog receptors in osteoblasts grown on polished Ti (Ti-Control) and Ti with nanotopography (Ti-Nano). The gene expression of the osteoblastic markers runt-related transcription factor 2 (*Runx2*), osteopontin (*Opn*), and alkaline phosphatase activity (*Alp*) on day 10 (**A**), RUNX2 protein expression on day 10 (**B**), ALP activity on day 7 (**C**), extracellular matrix mineralization on day 17 (**D**), gene expression of the Hedgehog receptors zinc finger 1, 2 and 3 transcription factors (*Gli1*, *Gli2*, and *Gli3*) on day 10 (**E**) and GLI1 protein expression on day 10 (**F**) in osteoblasts cultured on Ti-Control and Ti-Nano with either vehicle or PUR 2 μM. The original diameter of the Ti discs presented in C and D is 13 mm. The data of gene expression (n = 4), protein expression (n = 3), ALP activity (n = 5), and extracellular matrix mineralization (n = 5) are presented as mean ± SD, and * indicate statistically significant differences ($p \leq 0.05$).

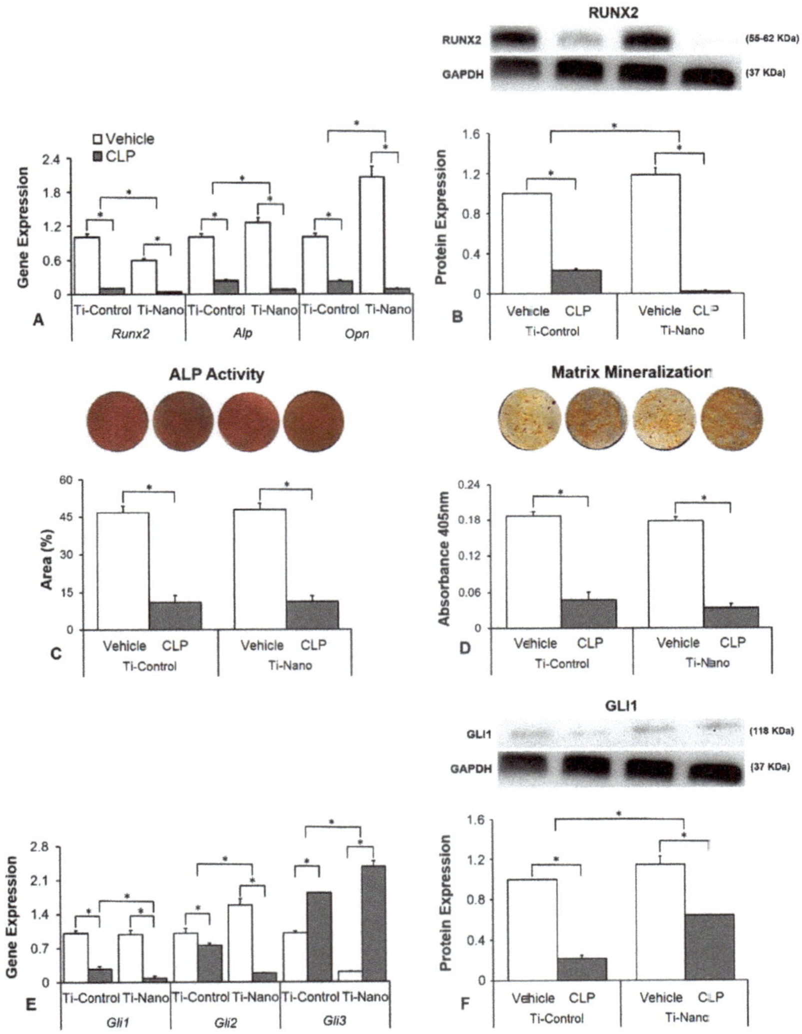

Figure 4. Effects of the Hedgehog antagonist cyclopamine (CLP) on osteoblast differentiation and the expression of Hedgehog receptors in osteoblasts grown on polished Ti (Ti-Control) and Ti with nanotopography (Ti-Nano). The gene expression of the osteoblastic markers runt-related transcription factor 2 (*Runx2*), osteopontin (*Opn*), and alkaline phosphatase activity (*Alp*) on day 10 (**A**), RUNX2 protein expression on day 10 (**B**), ALP activity on day 7 (**C**), extracellular matrix mineralization on day 17 (**D**), gene expression of the Hedgehog receptors zinc finger 1, 2 and 3 transcription factors (*Gli1*, *Gli2*, and *Gli3*) on day 10 (**E**) and GLI1 protein expression on day 10 (**F**) in osteoblasts cultured on Ti-Control and Ti-Nano with either vehicle or CLP 10 nM. The original diameter of the Ti discs presented in C and D is 13 mm. The data of gene expression (n = 4), protein expression (n = 3), ALP activity (n = 5), and extracellular matrix mineralization (n = 5) are presented as the mean ± SD, and * indicates statistically significant differences ($p \leq 0.05$).

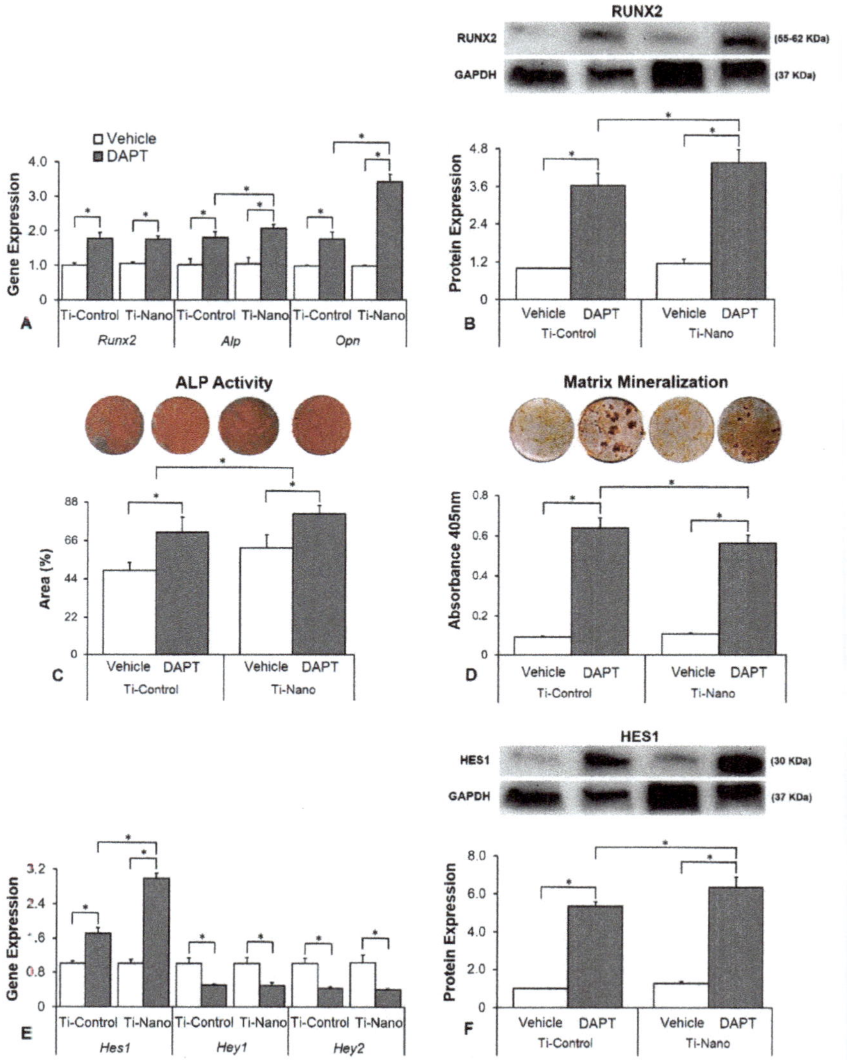

Figure 5. Effects of the Notch antagonist DAPT on osteoblast differentiation and the expression of Notch receptors in osteoblasts grown on polished Ti (Ti-Control) and Ti with nanotopography (Ti-Nano). The gene expression of the osteoblastic markers runt-related transcription factor 2 (*Runx2*), osteopontin (*Opn*), and alkaline phosphatase activity (*Alp*) on day 10 (**A**), RUNX2 protein expression on day 10 (**B**), ALP activity on day 7 (**C**), extracellular matrix mineralization on day 17 (**D**), gene expression of the Notch receptors hairy/enhancer of split 1 (*Hes1*) and hairy-related transcription factors 1 and 2 (*Hey1* and *Hey2*) on day 10 (**E**) and HES1 protein expression on day 10 (**F**) in osteoblasts cultured on Ti-Control and Ti-Nano with either vehicle or DAPT 20 μM. The original diameter of the Ti discs presented in C and D is 13 mm. The data of gene expression (n = 4), protein expression (n = 3), ALP activity (n = 5), and extracellular matrix mineralization (n = 5) are presented as the mean ± SD, and * indicates statistically significant differences ($p \leq 0.05$).

The interaction between Ti surfaces and bexarotene treatment did not affect the gene expression of *Runx2* ($p = 0.061$) but affected *Alp* ($p = 0.014$) and *Opn* ($p = 0.019$, Figure 6A). Bexarotene downregulated the gene expression of *Runx2* ($p < 0.001$ and $p < 0.001$), *Alp*

($p = 0.004$ and $p < 0.001$), and *Opn* ($p < 0.001$ and $p < 0.001$) in cells grown on both Ti-Control and Ti-Nano (Figure 6A). In the presence of vehicle, the gene expression of *Runx2* ($p = 0.024$) and *Opn* ($p < 0.001$) was lower, while *Alp* ($p = 0.007$) and was higher in cells grown on Ti-Nano than on Ti-Control (Figure 6A). In the presence of bexarotene, the gene expression of *Runx2* ($p = 0.746$) and *Alp* ($p = 0.439$) was not affected, while *Opn* ($p = 0.041$) was lower in cells grown on Ti-Nano than on Ti-Control (Figure 6A). The interaction between Ti surfaces and bexarotene treatment also affected the RUNX2 protein expression ($p \leq 0.001$, Figure 6B). Bexarotene decreased the RUNX2 protein expression ($p < 0.001$ and $p < 0.001$) in cells grown on both Ti-Control and Ti-Nano (Figure 6B). In the presence of vehicle, the RUNX2 protein expression was higher ($p < 0.001$) in cells grown on Ti-Nano than on Ti-Control, while in the presence of bexarotene, it was not affected ($p = 0.731$) by Ti surfaces (Figure 6B). The interaction between Ti surfaces and bexarotene treatment affected the ALP activity ($p \leq 0.001$, Figure 6C) and extracellular matrix mineralization ($p \leq 0.001$, Figure 6D). Bexarotene decreased the ALP activity ($p < 0.001$ and $p < 0.001$, Figure 6C) and extracellular matrix mineralization ($p = 0.002$ and $p = 0.001$, Figure 6D) in cells grown on both Ti-Control and Ti-Nano. In the presence of vehicle or bexarotene, the ALP activity ($p = 0.001$ and $p = 0.011$) was higher in cells grown on Ti-Nano than on Ti-Control (Figure 6C). In the presence of vehicle or bexarotene, the extracellular matrix mineralization ($p < 0.001$ and $p < 0.001$) was lower in cells grown on Ti-Nano than on Ti-Control (Figure 6D). The interaction between Ti surfaces and bexarotene treatment affected the gene expression of *Hes1* ($p \leq 0.001$), *Hey1* ($p = 0.004$), and *Hey2* ($p = 0.003$, Figure 6E). Bexarotene downregulated the gene expression of *Hes1* ($p < 0.001$ and $p < 0.001$) and upregulated *Hey1* ($p < 0.001$ and $p < 0.001$) and *Hey2* ($p < 0.001$ and $p = 0.004$) in cells grown on both Ti-Control and Ti-Nano (Figure 6E). In the presence of vehicle, the gene expression *Hey1* ($p = 0.740$) was not affected, while *Hes1* ($p = 0.004$) and *Hey2* ($p = 0.002$) were higher in cells grown on Ti-Control than on Ti-Nano (Figure 6E). In the presence of bexarotene, the gene expression of *Hes1* ($p < 0.001$), *Hey1* ($p < 0.001$), and *Hey2* ($p < 0.001$) was lower in cells grown on Ti-Nano than on Ti-Control (Figure 6E). The interaction between Ti surfaces and bexarotene treatment also affected the HES1 protein expression ($p = 0.001$, Figure 6F). Bexarotene decreased the HES1 protein expression ($p < 0.001$ and $p < 0.001$) in cells grown on both Ti-Control and Ti-Nano (Figure 4F). In the presence of vehicle, the HES1 protein expression was higher ($p = 0.021$), while, in the presence of bexarotene, it was lower ($p = 0.005$) in cells grown on Ti-Nano than on Ti-Control (Figure 6F).

3.5. Effects of the Combination of the Hedgehog and Notch Signalling Modulation on the Gene Expression of Bone Markers in the Cells Grown on Ti-Control and Ti-Nano

The interaction between Ti surfaces and the treatment with the combination of purmorphamine and DAPT affected the expression of *Runx2* ($p \leq 0.001$), *Alp* ($p \leq 0.001$), and *Opn* ($p \leq 0.001$, Figure 7). The combination of purmorphamine and DAPT upregulated the gene expression of *Runx2* ($p < 0.001$ and $p < 0.001$), *Alp* ($p < 0.001$ and $p < 0.001$), and *Opn* ($p < 0.001$ and $p < 0.001$) in cells grown on both Ti-Control and Ti-Nano (Figure 7A). In the presence of vehicle, the gene expression of *Runx2* ($p = 0.707$), *Alp* ($p = 0.345$), and *Opn* ($p = 0.969$) was not affected by Ti surfaces (Figure 7A). In the presence of the combination of purmorphamine and DAPT, the gene expression of *Runx2* ($p < 0.001$), *Alp* ($p < 0.001$), and *Opn* ($p < 0.001$) was higher in cells grown on Ti-Nano than Ti-Control (Figure 7A). The interaction between Ti surfaces and the treatment with the combination of cyclopamine and bexarotene affected the expression of *Runx2* ($p \leq 0.001$), *Alp* ($p = 0.010$), and *Opn* ($p \leq 0.001$, Figure 7B). The combination of cyclopamine and bexarotene downregulated the gene expression of *Runx2* ($p = 0.001$ and $p < 0.001$), *Alp* ($p = 0.001$ and $p < 0.001$), and *Opn* ($p = 0.001$ and $p < 0.001$) in cells grown on both Ti-Control and Ti-Nano (Figure 7B). In the presence of vehicle, the gene expression of *Runx2* ($p = 0.248$), *Alp* ($p = 0.416$), and *Opn* ($p = 0.874$) were not affected by Ti surfaces (Figure 7B). In the presence of the combination of cyclopamine and bexarotene, the gene expression of *Runx2* ($p < 0.001$), *Alp* ($p = 0.005$), and *Opn* ($p < 0.001$) was lower in cells grown on Ti-Nano than on Ti-Control (Figure 7B).

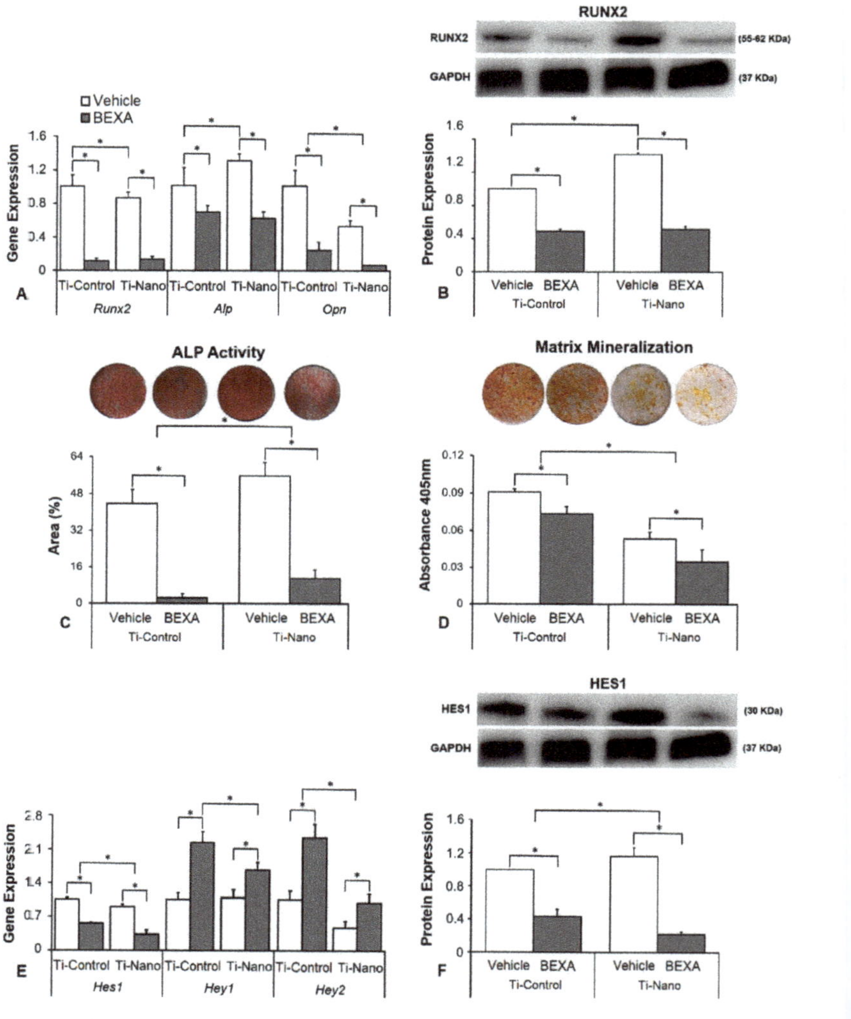

Figure 6. Effects of the Notch agonist bexarotene (BEXA) on osteoblast differentiation and the expression of Notch receptors in osteoblasts grown on polished Ti (Ti-Control) and Ti with nanotopography (Ti-Nano). The gene expression of the osteoblastic markers runt-related transcription factor 2 (*Runx2*), osteopontin (*Opn*), and alkaline phosphatase activity (*Alp*) on day 10 (**A**), RUNX2 protein expression on day 10 (**B**), ALP activity on day 7 (**C**), extracellular matrix mineralization on day 17 (**D**), gene expression of the Notch receptors hairy/enhancer of split 1 (*Hes1*) and hairy-related transcription factors 1 and 2 (*Hey1* and *Hey2*) on day 10 (**E**), and HES1 protein expression on day 10 (**F**), in osteoblasts cultured on Ti-Control and Ti-Nano with either vehicle or BEXA 0.1 μM. The original diameter of the Ti discs presented in C and D is 13 mm. The data of gene expression ($n = 4$), protein expression ($n = 3$), ALP activity ($n = 5$), and extracellular matrix mineralization ($n = 5$) are presented as the mean ± SD, and * indicates statistically significant differences ($p \leq 0.05$).

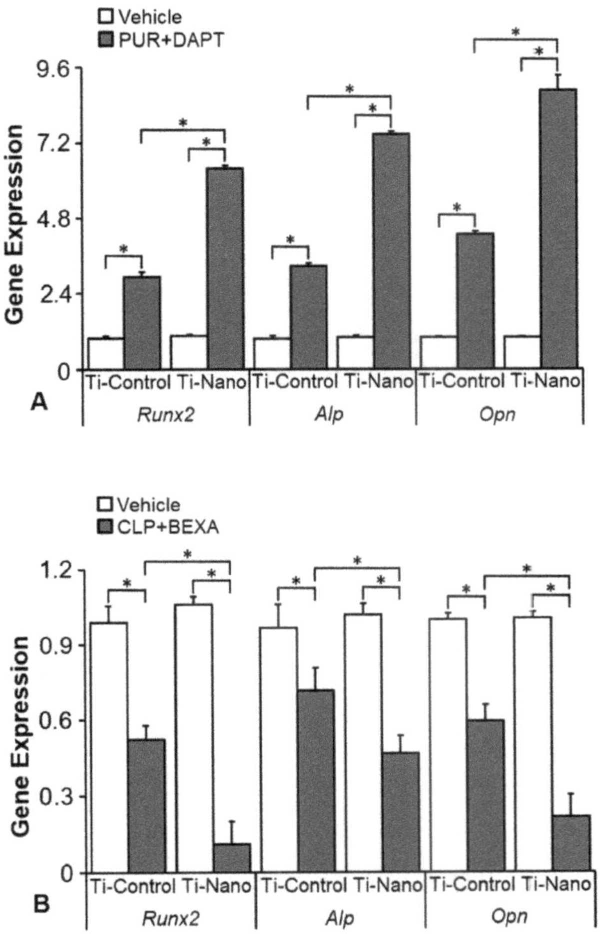

Figure 7. Effects of the combination of the Hedgehog agonist purmorphamine (PUR) with Notch antagonist DAPT and Hedgehog antagonist cyclopamine (CLP) with Notch agonist bexarotene (BEXA) on the gene expression of bone markers in osteoblasts grown on polished Ti (Ti-Control) and Ti with nanotopography (Ti-Nano). The gene expression of runt-related transcription factor 2 (*Runx2*), osteopontin (*Opn*), and alkaline phosphatase activity (*Alp*) on day 10 in osteoblasts cultured on Ti-Control and Ti-Nano with either vehicle or the combination of purmorphamine 2 μM with DAPT 20 μM (PUR+DAPT, (**A**)) or with either vehicle or the combination of cyclopamine 10 nM with bexarotene 0.1 μM (CLP+BEXA) (**B**). The data (n = 4) are presented as the mean ± SD, and * indicates statistically significant differences ($p \leq 0.05$).

4. Discussion

The modulation of the cell signalling involved in osteogenesis impacts the interaction between osteoblasts and Ti surfaces [40,42,45,53,54]. This study showed that agonists and antagonists of the Hedgehog and Notch signalling pathways affect osteoblast differentiation. Using either the Hedgehog agonist purmorphamine or the Notch antagonist DAPT increased while the Hedgehog antagonist cyclopamine or the Notch agonist bexarotene decreased the osteoblast differentiation of cells cultured on Ti-Control and Ti-Nano. Additionally, the association between purmorphamine and DAPT seems to have a synergistic effect in increasing the osteoblast differentiation of cells grown on both Ti surfaces, especially on Ti-Nano.

To select the concentration of the Hedgehog and Notch agonists and antagonists, we tested three doses of each, based on data from the literature, and evaluated three critical parameters of osteoblast differentiation; *Opn* gene expression, ALP activity, and extracellular matrix mineralization [18,19,32,34]. As observed in other studies, the Hedgehog agonist purmorphamine was more efficient in inducing osteoblast differentiation at a concentration of 2 μM [14,15,55]. The Hedgehog antagonist cyclopamine inhibited the osteoblast differentiation and because the concentration of 1000 nM seems to induce some toxicity specifically based on its effect on *Opn* gene expression and ALP activity, we selected 10 nM as it was more efficient than 100 nM in inhibiting osteoblast differentiation [56–58]. In agreement with previous studies, the Notch antagonist DAPT at the concentration of 20 μM was more osteogenic [33,59,60]. The Notch agonist bexarotene was more efficient in reducing osteoblast differentiation at the concentration of 0.1 μM, which agrees with previous studies [61–63].

Hedgehog agonist purmorphamine being used to enhance the osteoblast differentiation of cells grown on Ti has already been investigated [15,64]. Herein, we showed that the osteogenic effects of purmorphamine were more prominent on Ti-Nano, as evidenced by an increase in gene expression of *Runx2*, *Opn*, and *Alp*, RUNX2 protein expression, and ALP activity, which were more evident in the cells grown on Ti-Nano than on Ti-Control. The Hedgehog antagonist cyclopamine inhibited the osteoblast differentiation in a more pronounced way in cells grown on Ti-Nano compared with Ti-Control, as shown by the gene expression of *Runx2*, *Opn*, and *Alp*, and RUNX2 protein expression. This higher susceptibility to the Hedgehog agonist and antagonist of cells grown on Ti-Nano in osteoblast differentiation could be related to the modulation of the Hedgehog signalling pathway being more intense in cells grown on this surface. Indeed, the gene expression of the Hedgehog receptors *Gli1*, *Gli2*, and *Gli3*, and the GLI1 protein expression were more modulated by purmorphamine and cyclopamine except for GLI1 protein expression, with the expected opposite effects of the agonist and antagonist. Together, these results suggest that the Hedgehog signalling pathway is more relevant to the osteogenic potential of the Ti-Nano than of Ti-Control and that this nanotopography can regulate this cellular mechanism by itself. Despite few information on this subject is available in the literature, it was demonstrated that Ti with micro-/nanotextured topography, either with or without TiO_2 nanotubes, enhances osteoblast differentiation of MG63 cell lineage by activating Hedgehog-Gli1 signalling, which is inhibited by cyclopamine [65].

The involvement of the Notch signalling pathway in the osteoblast-Ti interaction is underexplored, despite its well-known participation in osteogenesis [31,66,67]. The Ti surface hydrophilicity was observed to favour bone formation by acting on several signalling pathways involved in proliferation and osteoblast precursor differentiation, including Notch signalling [68]. Additionally, the inhibition of the Notch signalling enhances the osteoblast differentiation of mesenchymal stem cells cultured on Ti substrates [69]. In keeping with this, we demonstrated that the Notch antagonist DAPT enhanced the osteoblast differentiation of cells grown on both Ti-Control and Ti-Nano, with more pronounced effects on Ti-Nano as noticed by the gene expression of *Opn* and *Alp*, RUNX2 protein expression, and ALP activity. Corroborating these data, the Notch agonist bexarotene inhibited the osteoblast differentiation more intensely in the cells grown on Ti-Nano than on Ti-Control, by reducing the same parameters and the extracellular matrix mineralization. As for the Hedgehog signalling, the higher responsiveness to the Notch antagonist and agonist of cells grown on Ti-Nano regarding osteoblast differentiation could be attributed to the higher intensity of the regulation of the Notch signalling in cells grown on this surface, specifically through the regulation of the Notch receptor Hes1. Indeed, the gene and protein expression of Hes1 was more modulated by DAPT and Bexarotene in cells grown on Ti-Nano than on Ti-Control while the gene expression of *Hey1* and *Hey2* was not affected by surface topography. Collectively, these data suggest that the Notch signalling pathway is more important to the osteogenic potential of the Ti-Nano than Ti-Control and that this nanotopography can regulate this signal. Although the participation of the Notch signalling in the osteoblast differentiation of cells grown on Ti surfaces has already been described, to the best of our knowledge, this is the first evidence that the distinct effects elicited by different Ti surface topographies on osteoblast differentiation involve the regulation of this signalling pathway [68,69].

As our results showed more prominent effects of the modulation of the Hedgehog and Notch signalling in osteoblasts grown on Ti-Nano, we started an investigation on the possible synergistic effect of the combination of the modulation of both signalling pathways. Although the effect of this combination on osteoblast differentiation was not previously evaluated, the only study presented in the literature is not related to bone tissue and demonstrated that the concomitant regulation of the Hedgehog and Notch signalling pathways potentiates the anti-leukemic effects of the Notch modulation alone [70]. Here, the association of the Hedgehog agonist purmorphamine with the Notch antagonist DAPT increased the upregulation of the *Runx2*, *Opn*, and *Alp* gene expression compared with the use of either purmorphamine or DAPT alone in osteoblasts grown on both Ti surfaces with more intense effects on Ti-Nano compared with Ti-Control. Although combining the Hedgehog antagonist cyclopamine with the Notch agonist bexarotene downregulated these gene expressions, the synergistic effect was not as evident as we observed when purmorphamine and DAPT were combined. Thus, despite further studies are needed to confirm the synergism, it is possible to suggest that the activation of Hedgehog along with the inhibition of Notch signalling may favour the osteoblast differentiation of cells grown on Ti, especially with nanostructured surfaces.

In conclusion, our results indicate that the Hedgehog and Notch signalling pathways are involved in the responses of osteoblasts to Ti surfaces, with more relevant effects on osteoblast differentiation of cells grown on the nanostructured surface, which may regulate these signals by itself. We also demonstrated that the concomitant activation of Hedgehog and inhibition of Notch might synergistically affect osteoblast differentiation, especially in cells grown on nanotopography. These cellular mechanisms may explain, at least in part, the higher osteogenic potential of this nanostructured Ti surface, which opens windows to develop strategies to drive the process of osseointegration.

Author Contributions: Conceptualization: A.L.R. and M.M.B.; Methodology: P.G.S., L.F.A., H.B.L., D.W., A.T.P.S. and F.S.O.; Software: P.G.S., L.F.A., H.B.L., D.W., A.T.P.S. and F.S.O.; Validation: P.G.S., L.F.A., H.B.L., D.W., A.T.P.S. and F.S.O.; Formal analysis: P.G.S., L.F.A., H.B.L., F.S.O., A.L.R. and M.M.B.; Investigation: P.G.S., L.F.A., H.B.L., D.W., A.T.P.S. and F.S.O.; Data curation: P.G.S., L.F.A., H.B.L., F.S.O., A.L.R. and M.M.B.; Writing—original draft preparation: P.G.S. and M.M.B.; Writing—review and editing: P.G.S., L.F.A., H.B.L., D.W., A.T.P.S., F.S.O., A.L.R. and M.M.B.; Supervision: A.L.R. and M.M.B.; Project administration: M.M.B.; Funding acquisition: M.M.B. All authors have read and agreed to the published version of the manuscript.

Funding: This work was supported by the São Paulo Research Foundation (FAPESP, grants # 2018/17356-6 and 2019/15531-8) and the National Council for Scientific and Technological Development (CNPq, grant # 303115/2019-8).

Data Availability Statement: The datasets used and/or analysed during the current study are available from the corresponding author on reasonable request.

Acknowledgments: The authors would like to thank Roger Rodrigo Fernandes for their technical assistance during the execution of the study.

Conflicts of Interest: The authors declare no conflict of interest.

References

1. Adolpho, L.F.; Lopes, H.B.; Freitas, G.P.; Weffort, D.; Campos Totoli, G.G.; Loyola Barbosa, A.C.; Freire Assis, R.I.; Silverio Ruiz, K.G.; Andia, D.C.; Rosa, A.L.; et al. Human periodontal ligament stem cells with distinct osteogenic potential induce bone formation in rat calvaria defects. *Regen. Med.* **2022**, *17*, 341–353. [CrossRef] [PubMed]
2. Chen, Q.; Liu, W.; Sinha, K.M.; Yasuda, H.; de Crombrugghe, B. Identification and characterization of microRNAs controlled by the osteoblast-specific transcription factor Osterix. *PLoS ONE* **2013**, *8*, e58104. [CrossRef] [PubMed]
3. Souza, A.T.P.; Lopes, H.B.; Oliveira, F.S.; Weffort, D.; Freitas, G.P.; Adolpho, L.F.; Fernandes, R.R.; Rosa, A.L.; Beloti, M.M. The extracellular matrix protein Agrin is expressed by osteoblasts and contributes to their differentiation. *Cell Tissue Res.* **2021**, *386*, 335–347. [CrossRef] [PubMed]
4. Yang, J.; Andre, P.; Ye, L.; Yang, Y.-Z. The Hedgehog signalling pathway in bone formation. *Int. J. Oral Sci.* **2015**, *7*, 73–79. [CrossRef] [PubMed]
5. Levi, B.; James, A.W.; Nelson, E.R.; Li, S.; Peng, M.; Commons, G.W.; Lee, M.; Wu, B.; Longaker, M.T. Human Adipose-Derived Stromal Cells Stimulate Autogenous Skeletal Repair via Paracrine Hedgehog Signaling with Calvarial Osteoblasts. *Stem Cells Dev.* **2011**, *20*, 243–257. [CrossRef] [PubMed]
6. Yan, X.; Yang, Z.; Chen, Y.; Li, N.; Wang, L.; Dou, G.; Liu, Y.; Duan, J.; Feng, L.; Deng, S.; et al. Endothelial cells-targeted soluble human Delta-like 4 suppresses both physiological and pathological ocular angiogenesis. *Sci. China Life Sci.* **2015**, *58*, 425–431. [CrossRef]
7. Heretsch, P.; Tzagkaroulaki, L.; Giannis, A. Cyclopamine and Hedgehog Signaling: Chemistry, Biology, Medical Perspectives. *Angew. Chem. Int. Ed.* **2010**, *49*, 3418–3427. [CrossRef] [PubMed]
8. Roberge, L.; Origa-Alves, A.C.; Rebelatto, C.L.K.; Dallagiovanna, B.; Shigunov, P. Inhibition of Hedgehog signaling pathway affects the expression of miR-20a and miR-3. *J. Biotechnol. Biodivers.* **2013**, *4*, 322–332.
9. Ullah, A.; Ullah, N.; Nawaz, T.; Aziz, T. Molecular mechanisms of Sanguinarine in cancer prevention and treatment. *Anti-Cancer Agents Med. Chem.* **2022**, *online ahead of print*. [CrossRef]
10. Wang, Q.; Huang, C.; Zeng, F.; Xue, M.; Zhang, X. Activation of the Hh pathway in periosteum-derived mesenchymal stem cells induces bone formation in vivo: Implication for postnatal bone repair. *Am. J. Pathol.* **2010**, *177*, 3100–3111. [CrossRef]
11. Evangelista, M.; Tian, H.; de Sauvage, F.J. The hedgehog signaling pathway in cancer. *Clin. Cancer Res. Off. J. Am. Assoc. Cancer Res.* **2006**, *12*, 5924–5928. [CrossRef] [PubMed]
12. Ingham, P.W.; McMahon, A.P. Hedgehog signaling in animal development: Paradigms and principles. *Genes Dev.* **2001**, *15*, 3059–3087. [CrossRef] [PubMed]
13. Plaisant, M.; Giorgetti-Peraldi, S.; Gabrielson, M.; Loubat, A.; Dani, C.; Peraldi, P. Inhibition of hedgehog signaling decreases proliferation and clonogenicity of human mesenchymal stem cells. *PLoS ONE* **2011**, *6*, e16798. [CrossRef] [PubMed]
14. Beloti, M.M.; Bellesini, L.S.; Rosa, A.L. Purmorphamine enhances osteogenic activity of human osteoblasts derived from bone marrow mesenchymal cells. *Cell Biol. Int.* **2005**, *29*, 537–541. [CrossRef]
15. Beloti, M.M.; Bellesini, L.S.; Rosa, A.L. The effect of purmorphamine on osteoblast phenotype expression of human bone marrow mesenchymal cells cultured on titanium. *Biomaterials* **2005**, *26*, 4245–4248. [CrossRef] [PubMed]
16. Oliveira, F.S.; Bellesini, L.S.; Defino, H.L.A.; da Silva Herrero, C.F.; Beloti, M.M.; Rosa, A.L. Hedgehog signaling and osteoblast gene expression are regulated by purmorphamine in human mesenchymal stem cells. *J. Cell. Biochem.* **2012**, *113*, 204–208. [CrossRef]
17. Wu, X.; Ding, S.; Ding, Q.; Gray, N.S.; Schultz, P.G. A small molecule with osteogenesis-inducing activity in multipotent mesenchymal progenitor cells. *J. Am. Chem. Soc.* **2002**, *124*, 14520–14521. [CrossRef] [PubMed]
18. Sinha, S.; Chen, J.K. Purmorphamine activates the Hedgehog pathway by targeting Smoothened. *Nat. Chem. Biol.* **2006**, *2*, 29–30. [CrossRef]
19. Wu, X.; Walker, J.; Zhang, J.; Ding, S.; Schultz, P.G. Purmorphamine induces osteogenesis by activation of the hedgehog signaling pathway. *Chem. Biol.* **2004**, *11*, 1229–1238. [CrossRef]
20. Heretsch, P.; Tzagkaroulaki, L.; Giannis, A. Modulators of the hedgehog signaling pathway. *Bioorganic Med. Chem.* **2010**, *18*, 6613–6624. [CrossRef]
21. Katoh, Y.; Katoh, M. Hedgehog target genes: Mechanisms of carcinogenesis induced by aberrant hedgehog signaling activation. *Curr. Mol. Med.* **2009**, *9*, 873–886. [CrossRef] [PubMed]
22. Quint, E.; Smith, A.; Avaron, F.; Laforest, L.; Miles, J.; Gaffield, W.; Akimenko, M.-A. Bone patterning is altered in the regenerating zebrafish caudal fin after ectopic expression of sonic hedgehog and bmp2b or exposure to cyclopamine. *Proc. Natl. Acad. Sci. USA* **2002**, *99*, 8713–8718. [CrossRef] [PubMed]
23. Suh, J.M.; Gao, X.; McKay, J.; McKay, R.; Salo, Z.; Graff, J.M. Hedgehog signaling plays a conserved role in inhibiting fat formation. *Cell Metab.* **2006**, *3*, 25–34. [CrossRef] [PubMed]
24. Panepucci, R.A.; Oliveira, L.H.B.; Zanette, D.L.; Viu Carrara, R.d.C.; Araujo, A.G.; Orellana, M.D.; Bonini de Palma, P.V.; Menezes, C.C.B.O.; Covas, D.T.; Zago, M.A. Increased levels of NOTCH1, NF-kappaB, and other interconnected transcription factors characterize primitive sets of hematopoietic stem cells. *Stem Cells Dev.* **2010**, *19*, 321–332. [CrossRef]
25. Radtke, F.; Schweisguth, F.; Pear, W. The Notch "gospel". *EMBO Rep.* **2005**, *6*, 1120–1125. [CrossRef]
26. High, F.; Epstein, J.A. Signalling Pathways Regulating Cardiac Neural Crest Migration and Differentiation. In *Vascular Development*; John Wiley & Sons, Ltd.: New York, NY, USA, 2007; pp. 152–164, ISBN 978-0-470-31941-3.

27. Fischer, A.; Gessler, M. Delta-Notch—And then? Protein interactions and proposed modes of repression by Hes and Hey bHLH factors. *Nucleic Acids Res.* **2007**, *35*, 4583–4596. [CrossRef]
28. Garg, V.; Muth, A.N.; Ransom, J.F.; Schluterman, M.K.; Barnes, R.; King, I.N.; Grossfeld, P.D.; Srivastava, D. Mutations in NOTCH1 cause aortic valve disease. *Nature* **2005**, *437*, 270–274. [CrossRef]
29. McLarren, K.W.; Lo, R.; Grbavec, D.; Thirunavukkarasu, K.; Karsenty, G.; Stifani, S. The mammalian basic helix loop helix protein HES-1 binds to and modulates the transactivating function of the runt-related factor Cbfa1. *J. Biol. Chem.* **2000**, *275*, 530–538. [CrossRef]
30. Shen, Q.; Christakos, S. The vitamin D receptor, Runx2, and the Notch signaling pathway cooperate in the transcriptional regulation of osteopontin. *J. Biol. Chem.* **2005**, *280*, 40589–40598. [CrossRef]
31. Zamurovic, N.; Cappellen, D.; Rohner, D.; Susa, M. Coordinated activation of notch, Wnt, and transforming growth factor-beta signaling pathways in bone morphogenic protein 2-induced osteogenesis. Notch target gene Hey1 inhibits mineralization and Runx2 transcriptional activity. *J. Biol. Chem.* **2004**, *279*, 37704–37715. [CrossRef]
32. Chitranshi, N.; Dheer, Y.; Kumar, S.; Graham, S.L.; Gupta, V. Molecular docking, dynamics, and pharmacology studies on bexarotene as an agonist of ligand-activated transcription factors, retinoid X receptors. *J. Cell. Biochem.* **2019**, *120*, 11745–11760. [CrossRef] [PubMed]
33. Dishowitz, M.I.; Terkhorn, S.P.; Bostic, S.A.; Hankenson, K.D. Notch signaling components are upregulated during both endochondral and intramembranous bone regeneration. *J. Orthop. Res.* **2012**, *30*, 296–303. [CrossRef] [PubMed]
34. Jing, W.; Xiong, Z.; Cai, X.; Huang, Y.; Li, X.; Yang, X.; Liu, L.; Tang, W.; Lin, Y.; Tian, W. Effects of γ-secretase inhibition on the proliferation and vitamin D3 induced osteogenesis in adipose derived stem cells. *Biochem. Biophys. Res. Commun.* **2010**, *392*, 442–447. [CrossRef] [PubMed]
35. Costa, D.G.; Ferraz, E.P.; Abuna, R.P.F.; de Oliveira, P.T.; Morra, M.; Beloti, M.M.; Rosa, A.L. The effect of collagen coating on titanium with nanotopography on in vitro osteogenesis. *J. Biomed. Mater. Res. A* **2017**, *105*, 2783–2788. [CrossRef] [PubMed]
36. Martin, J.Y.; Schwartz, Z.; Hummert, T.W.; Schraub, D.M.; Simpson, J.; Lankford, J., Jr.; Dean, D.D.; Cochran, D.L.; Boyan, B.D. Effect of titanium surface roughness on proliferation, differentiation, and protein synthesis of human osteoblast-like cells (MG63). *J. Biomed. Mater. Res.* **1995**, *29*, 389–401. [CrossRef]
37. Mendonça, G.; Mendonça, D.B.S.; Simões, L.G.P.; Araújo, A.L.; Golin, A.L.; Duarte, W.R.; Cooper, L.F.; Aragão, F.J.L. Efeito de superfícies de implantes nano-estruturadas na expressão de genes de osteoblastos e no contato osso-implante in vivo. *Rev. Odontológica Bras. Cent.* **2010**, *19*, 196–204. [CrossRef]
38. Silverwood, R.K.; Fairhurst, P.G.; Sjöström, T.; Welsh, F.; Sun, Y.; Li, G.; Yu, B.; Young, P.S.; Su, B.; Meek, R.M.D.; et al. Analysis of Osteoclastogenesis/Osteoblastogenesis on Nanotopographical Titania Surfaces. *Adv. Healthc. Mater.* **2016**, *5*, 947–955. [CrossRef] [PubMed]
39. Rosa, A.; Kato, R.; Castro-Raucci, L.; Teixeira, L.; de Oliveira, F.; Bellesini, L.; de Oliveira, PT.; Hassan, M.; Beloti, M. Nanotopography drives stem cell fate toward osteoblast differentiation through α1β1 integrin signaling pathway. *J. Cell. Biochem.* **2014**, *115*, 540–548. [CrossRef]
40. Bighetti-Trevisan, R.L.; Almeida, L.O.; Castro-Raucci, L.M.S.; Gordon, J.A.R.; Tye, C.E.; Stein, G.S.; Lian, J.B.; Stein, J.L.; Rosa, A.L.; Beloti, M.M. Titanium with nanotopography attenuates the osteoclast-induced disruption of osteoblast differentiation by regulating histone methylation. *Biomater. Adv.* **2022**, *134*, 112548. [CrossRef]
41. de Oliveira, P.T.; Zalzal, S.F.; Beloti, M.M.; Rosa, A.L.; Nanci, A. Enhancement of in vitro osteogenesis on titanium by chemically produced nanotopography. *J. Biomed. Mater. Res. A* **2007**, *80*, 554–564. [CrossRef]
42. Abuna, R.P.F.; Oliveira, F.S.; Adolpho, L.F.; Fernandes, R.R.; Rosa, A.L.; Beloti, M.M. Frizzled 6 disruption suppresses osteoblast differentiation induced by nanotopography through the canonical Wnt signaling pathway. *J. Cell. Physiol.* **2020**, *235*, 8293–8303. [CrossRef] [PubMed]
43. Kato, R.B.; Roy, B.; De Oliveira, F.S.; Ferraz, E.P.; De Oliveira, P.T.; Kemper, A.G.; Hassan, M.Q.; Rosa, A L.; Beloti, M.M. Nanotopography Directs Mesenchymal Stem Cells to Osteoblast Lineage through Regulation of microRNA-SMAD-BMP-2 Circuit. *J. Cell. Physiol.* **2014**, *229*, 1690–1696. [CrossRef] [PubMed]
44. Lopes, H.B.; Freitas, G.P.; Elias, C.N.; Tye, C.; Stein, J.L.; Stein, G.S.; Lian, J.B.; Rosa, A.L.; Beloti, M.M. Participation of integrin β3 in osteoblast differentiation induced by titanium with nano or microtopography. *J. Biomed. Mater. Res. A* **2019**, *107*, 1303–1313. [CrossRef]
45. Lopes, H.B.; Souza, A.T.P.; Freitas, G.P.; Elias, C.N.; Rosa, A.L.; Beloti, M.M. Effect of focal adhesion kinase inhibition on osteoblastic cells grown on titanium with different topographies. *J. Appl. Oral Sci. Rev. FOB* **2020**, *28*, e20190156. [CrossRef] [PubMed]
46. Castro-Raucci, L.M.S.; Francischini, M.S.; Teixeira, L.N.; Ferraz, E.P.; Lopes, H.B.; de Oliveira, P.T.; Hassan, M.Q.; Rosa, A.L.; Beloti, M.M. Titanium With Nanotopography Induces Osteoblast Differentiation by Regulating Endogenous Bone Morphogenetic Protein Expression and Signaling Pathway. *J. Cell. Biochem.* **2016**, *117*, 1718–1726. [CrossRef] [PubMed]
47. de Oliveira, P.T.; Zalzal, S.F.; Irie, K.; Nanci, A. Early Expression of Bone Matrix Proteins in Osteogenic Cell Cultures. *J. Histochem. Cytochem.* **2003**, *51*, 633–641. [CrossRef]
48. Tambasco de Oliveira, P.; Nanci, A. Nanotexturing of titanium-based surfaces upregulates expression of bone sialoprotein and osteopontin by cultured osteogenic cells. *Biomaterials* **2004**, *25*, 403–413. [CrossRef]

49. Abuna, R.P.F.; Oliveira, F.S.; Ramos, J.I.R.; Lopes, H.B.; Freitas, G.P.; Souza, A.T.P.; Beloti, M.M.; Rosa, A.L. Selection of reference genes for quantitative real-time polymerase chain reaction studies in rat osteoblasts. *J. Cell. Physiol.* **2018**, *234*, 749–756. [CrossRef]
50. Livak, K.J.; Schmittgen, T.D. Analysis of relative gene expression data using real-time quantitative PCR and the 2(-Delta Delta C(T)) Method. *Methods* **2001**, *25*, 402–408. [CrossRef]
51. Majors, A.K.; Boehm, C.A.; Nitto, H.; Midura, R.J.; Muschler, G.F. Characterization of human bone marrow stromal cells with respect to osteoblastic differentiation. *J. Orthop. Res.* **1997**, *15*, 546–557. [CrossRef]
52. Gregory, C.A.; Gunn, W.G.; Peister, A.; Prockop, D.J. An Alizarin red-based assay of mineralization by adherent cells in culture: Comparison with cetylpyridinium chloride extraction. *Anal. Biochem.* **2004**, *329*, 77–84. [CrossRef] [PubMed]
53. Boyan, B.D.; Berger, M.B.; Nelson, F.R.; Donahue, H.J.; Schwartz, Z. The Biological Basis for Surface-dependent Regulation of Osteogenesis and Implant Osseointegration. *JAAOS-J. Am. Acad. Orthop. Surg.* **2022**, *30*, e894. [CrossRef] [PubMed]
54. Zhao, Y.; Bai, L.; Zhang, Y.; Yao, R.; Sun, Y.; Hang, R.; Chen, X.; Wang, H.; Yao, X.; Xiao, Y.; et al. Type I collagen decorated nanoporous network on titanium implant surface promotes osseointegration through mediating immunomodulation, angiogenesis, and osteogenesis. *Biomaterials* **2022**, *288*, 121684. [CrossRef] [PubMed]
55. van der Horst, G.; Farih-Sips, H.; Löwik, C.W.G.M.; Karperien, M. Hedgehog stimulates only osteoblastic differentiation of undifferentiated KS483 cells. *Bone* **2003**, *33*, 899–910. [CrossRef] [PubMed]
56. Chen, J.K.; Taipale, J.; Cooper, M.K.; Beachy, P.A. Inhibition of Hedgehog signaling by direct binding of cyclopamine to Smoothened. *Genes Dev.* **2002**, *16*, 2743–2748. [CrossRef]
57. Stanton, B.Z.; Peng, L.F. Small-molecule modulators of the Sonic Hedgehog signaling pathway. *Mol. Biosyst.* **2009**, *6*, 44–54. [CrossRef]
58. Yuan, Y.-F.; Zhu, W.-X.; Liu, T.; He, J.-Q.; Zhou, Q.; Zhou, X.; Zhang, X.; Yang, J. Cyclopamine functions as a suppressor of benign prostatic hyperplasia by inhibiting epithelial and stromal cell proliferation via suppression of the Hedgehog signaling pathway. *Int. J. Mol. Med.* **2020**, *46*, 311–319. [CrossRef] [PubMed]
59. Androutsellis-Theotokis, A.; Leker, R.R.; Soldner, F.; Hoeppner, D.J.; Ravin, R.; Poser, S.W.; Rueger, M.A.; Bae, S.-K.; Kittappa, R.; McKay, R.D.G. Notch signalling regulates stem cell numbers in vitro and in vivo. *Nature* **2006**, *442*, 823–826. [CrossRef]
60. Canalis, E. Notch Signaling in Osteoblasts. *Sci. Signal.* **2008**, *1*, pe17. [CrossRef]
61. Ai, X.; Mao, F.; Shen, S.; Shentu, Y.; Wang, J.; Lu, S. Bexarotene inhibits the viability of non-small cell lung cancer cells via slc10a2/PPARγ/PTEN/mTOR signaling pathway. *BMC Cancer* **2018**, *18*, 407. [CrossRef]
62. Fantini, J.; Di Scala, C.; Yahi, N.; Troadec, J.-D.; Sadelli, K.; Chahinian, H.; Garmy, N. Bexarotene blocks calcium-permeable ion channels formed by neurotoxic Alzheimer's β-amyloid peptides. *ACS Chem. Neurosci.* **2014**, *5*, 216–224. [CrossRef] [PubMed]
63. Kamp, F.; Scheidt, H.A.; Winkler, E.; Basset, G.; Heinel, H.; Hutchison, J.M.; LaPointe, L.M.; Sanders, C.R.; Steiner, H.; Huster, D. Bexarotene Binds to the Amyloid Precursor Protein Transmembrane Domain, Alters Its α-Helical Conformation, and Inhibits γ-Secretase Nonselectively in Liposomes. *ACS Chem. Neurosci.* **2018**, *9*, 1702–1713. [CrossRef] [PubMed]
64. Lin, Y.; Huang, Y.; He, J.; Chen, F.; He, Y.; Zhang, W. Role of Hedgehog-Gli1 signaling in the enhanced proliferation and differentiation of MG63 cells enabled by hierarchical micro-/nanotextured topography. *Int. J. Nanomed.* **2017**, *12*, 3267–3280. [CrossRef] [PubMed]
65. Xie, Y.; Chen, X.; Zheng, X.; Li, L.; Li, J.; Xu, Y.; He, J.; Lin, Y. Beta1-integrin/Hedgehog-Gli1 signaling pathway fuels the diameter-dependent osteoblast differentiation on different TiO2 nanotubes: The optimal-diameter nanotubes for osteoblast differentiation. *Int. J. Biochem. Cell Biol.* **2021**, *137*, 106026. [CrossRef] [PubMed]
66. Delgado-Calle, J.; McAndrews, K.; Wu, G.; Orr, A.L.; Ferrari, A.; Tu, X.; Srinivasan, V.; Roodman, G.D.; Ebetino, F.H.; Boeckman, R.K., Jr.; et al. The Notch pathway regulates the bone gain induced by PTH anabolic signaling. *FASEB J.* **2022**, *36*, e22196. [CrossRef] [PubMed]
67. Zanotti, S.; Smerdel-Ramoya, A.; Stadmeyer, L.; Durant, D.; Radtke, F.; Canalis, E. Notch inhibits osteoblast differentiation and causes osteopenia. *Endocrinology* **2008**, *149*, 3890–3899. [CrossRef]
68. Calciolari, E.; Hamlet, S.; Ivanovski, S.; Donos, N. Pro-osteogenic properties of hydrophilic and hydrophobic titanium surfaces: Crosstalk between signalling pathways in in vivo models. *J. Periodontal Res.* **2018**, *53*, 598–609. [CrossRef]
69. Wang, H.; Jiang, Z.; Zhang, J.; Xie, Z.; Wang, Y.; Yang, G. Enhanced osteogenic differentiation of rat bone marrow mesenchymal stem cells on titanium substrates by inhibiting Notch3. *Arch. Oral Biol.* **2017**, *80*, 34–40. [CrossRef]
70. Okuhashi, Y.; Itoh, M.; Nara, N.; Tohda, S. Effects of combination of notch inhibitor plus hedgehog inhibitor or Wnt inhibitor on growth of leukemia cells. *Anti-Cancer Res.* **2011**, *31*, 893–896.

Journal of
*Functional
Biomaterials*

MDPI

Review

Wood as Possible Renewable Material for Bone Implants—Literature Review

Vadims Nefjodovs [1,2,*], Laura Andze [3], Martins Andzs [3], Inese Filipova [3], Ramunas Tupciauskas [3], Linda Vecbiskena [3] and Martins Kapickis [2]

1 Faculty of Residency, Riga Stradins University, Dzirciema iela 16, LV-1007 Riga, Latvia
2 Microsurgery Centre of Latvia, Brivibas Gatve 410, LV-1024 Riga, Latvia
3 Latvian State Institute of Wood Chemistry, Dzerbenes Street 27, LV-1006 Riga, Latvia; linda.vecbiskena@gmail.com (L.V.)
* Correspondence: vadims.nefjodovs@gmail.com

Abstract: Bone fractures and bone defects affect millions of people every year. Metal implants for bone fracture fixation and autologous bone for defect reconstruction are used extensively in treatment of these pathologies. Simultaneously, alternative, sustainable, and biocompatible materials are being researched to improve existing practice. Wood as a biomaterial for bone repair has not been considered until the last 50 years. Even nowadays there is not much research on solid wood as a biomaterial in bone implants. A few species of wood have been investigated. Different techniques of wood preparation have been proposed. Simple pre-treatments such as boiling in water or preheating of ash, birch and juniper woods have been used initially. Later researchers have tried using carbonized wood and wood derived cellulose scaffold. Manufacturing implants from carbonized wood and cellulose requires more extensive wood processing—heat above 800 °C and chemicals to extract cellulose. Carbonized wood and cellulose scaffolds can be combined with other materials, such as silicon carbide, hydroxyapatite, and bioactive glass to improve biocompatibility and mechanical durability. Throughout the publications wood implants have provided good biocompatibility and osteoconductivity thanks to wood's porous structure.

Keywords: wood implants; bone repair; biocomposites; osteosynthesis

Citation: Nefjodovs, V.; Andze, L.; Andzs, M.; Filipova, I.; Tupciauskas, R.; Vecbiskena, L.; Kapickis, M. Wood as Possible Renewable Material for Bone Implants—Literature Review. *J. Funct. Biomater.* **2023**, *14*, 266. https://doi.org/10.3390/jfb14050266

Academic Editors: Xiaodu Wang, Adalberto Luiz Rosa and Marcio Mateus Beloti

Received: 25 February 2023
Revised: 25 April 2023
Accepted: 6 May 2023
Published: 10 May 2023

1. Introduction

Bone fractures first drew prehistoric humans' attention up to 46 thousand years ago during the Early Upper Paleolithic age, the period from which the first healed bone fractures were found by archeologists [1]. After thousands of years of using traction and immobilization as the only treatment for bone fractures, the first true external fixation was applied only 120 years ago. That was developed by a Belgian surgeon Albin Lambotte. Lambotte who also introduced the term "osteosynthesis"—fixation of bone by using mechanical devices [2]. At the beginning of the 20th century, with the development of antiseptics, anesthesiology, and bone imaging possibilities, the modern principles of the internal fixation of fractures were developed. The first material for osteosynthesis implants was nickel-coated steel, developed in the 19th century [3,4]. Other metals such as silver [5], aluminium, and brass [6] have been used to produce different bone implants. Nevertheless, these materials were found not to be fully suitable due to inadequate mechanical properties and corrosion. The first successful material was stainless steel, later joined by titanium and cobalt-chromium alloys [7]. Although the problem with obvious and quick corrosion was resolved, there are still a few debatable issues. First, the density of a metal alloys is up to three times higher than cancellous bone [8,9]. Thus, aseptic loosening of the metallic implants is considered a possible complication within 15 years after surgery [10]. Second, bio-corrosion of stainless steel [11] and titanium [12] alloys is being investigated as well. Demand for non-metallic implant materials is growing, not only because of bone damage over time due to the loosening and biocorrosion of implant material, but also because of the increased use of modern medical diagnostic systems, e.g., nuclear magnetic resonance [13]. Metal implants cause significant artifacts in computer tomography and magnetic resonance images. The lower image quality of artifacts cause blurring. In the last decades, numerous studies have been published about reducing the effects of artifacts. However, the issue is still present in everyday clinical practice [14,15].

Aside from bone fractures, bone defects are a common issue in orthopaedic and reconstructive surgery. Bone defects can be caused by severe injuries, congenital anomalies, and tissue resection due to oncological masses. Although bone has great capabilities for rejuvenation, the healing of large defects is challenging. Treatment with bone xenografts (grafts from animals) from dogs and goats for cranioplasty was first described more than 500 years ago by Ottoman empire surgeon Ibrahim Bin Abdullah [16]. Nowadays, bone defect reconstruction still relies mostly on autologous (from the patient), allogeneic (from another human donor), and xenogeneic (animal-derived) bone grafts. For very extensive defects, vascularized bone flaps are harvested from the patient. Harvesting bone tissues from the patient adds additional surgical sites, with possible complications. Using allogeneic and xenogeneic grafts posts immunological challenges, as well as logistical and ethical issues [17]. In attempts to improve bone defect reconstruction, different biomaterials have been widely investigated—calcium phosphates [18,19], bioactive glass [20], collagen [21], silk fibroin [22], etc.—for potential use in clinical practise [23].

Great interest has been observed among industry, researchers, and society towards new materials produced from natural sources, due to several reasons, e.g., topicality of environmental protection issues, new regulations, unsustainability of fossil fuels and their reserve, increasing plastic pollution, and global concepts of circular bio-economy [24]. Reducing emissions in the next decade is crucial for Europe, which intends to become the world's first climate-neutral continent by 2050 by making the European Green Deal a reality [25]. Concerns about environmental issues have encouraged research on biomaterials, including bone implants [26]. Over the last few years, comparative studies have been carried out on the ecological footprints of raw materials for bone implants [26,27], as well as studies on the use of natural or bio-based polymers in bone tissues [27–29], the synthesis of hydroxyapatite from sustainable natural raw materials [30,31], and the use of new technologies such as 3D printing as a solution for a sustainable and circular economy [32]. Materials obtained directly from nature are being studied as well, e.g., corals [27]. However, the materials developed thus far do not provide sufficient mechanical strength for osteosynthesis implants compared with metal implants. These issues promote further investigations for alternative implants' material.

2. Similarities between Wood and Bone

Humanity has known about wood as a biomaterial since the Stone Age, with wood has played a major role in humanity's greatest achievements—from discovering fire to creating transport. Wood is an anisotropic natural material usually obtained from the trunk of a tree. It can be defined as a heterogeneous composite that consists mainly of natural polymers such as cellulose (40–50%), hemicelluloses (15–25%), and lignin (15–30%) [33,34]. Tree cross-sections can be distinguished into three components—the bark, cambium, and wood parts—xylem. The bark consists of a cork layer on the outside and a phloem layer on the inside. The cambium, located between the bark and the xylem, consists of living cells that form the new xylem and phloem layers. Xylem has two wood parts—sapwood and heartwood. The sapwood consists of dead cells and a small number of living parenchymal cells. It acts as food storage, as a water and nutrients transporter, and as mechanical support for the tree. The heartwood consists entirely of dead cells and provides only a support function for the tree [35].

Similar to wood, bone is also an anisotropic heterogeneous composite material, as it consists of about 60% inorganic material (calcium and phosphate in a form of natural or calcium-deficient hydroxyapatite), 30% organic material (collagen) and 10% water [36–38].

Similar to wood, which fulfills the function of support in a tree, the main function of bone is to support the static and movement functions of the body. Bone also acts as storage for minerals such as calcium and phosphate, as well as in the maintenance of homeostasis [39]. Three parts of bone can be observed in the cross-section—cortical bone, also called dense or compact bone, trabecular bone, also called spongy or cancellous bone, and bone marrow cavity. Bone tissue contains three main cell types: osteoblasts, osteoclasts, and osteocytes. Osteoblasts are responsible for bone formation, while osteoclasts are cells that resorb bone. Bone homeostasis is maintained by the connection between bone formation and bone resorption (bone turnover). Osteocytes are cells that are found in fully formed bone and make up most of the bone [40]. Similarities between wood and bone have been observed and were described as early as the invention of the microscopic magnification itself. The pioneer of microscopy, Antonie van Leeuwenhoek, described the analogy between the osteoid bone structure and the fibre structure of wood [41]. Since then, several authors have continued researching similarities between wood and bone, revealing the hierarchical macroscopic and microscopic structures of both, as well as functional similarities such as biomechanical characteristics, remodeling, and liquid transportation abilities [42–46]. Figure 1 shows the structural similarity between cortical bone and wood at the micro and nano level.

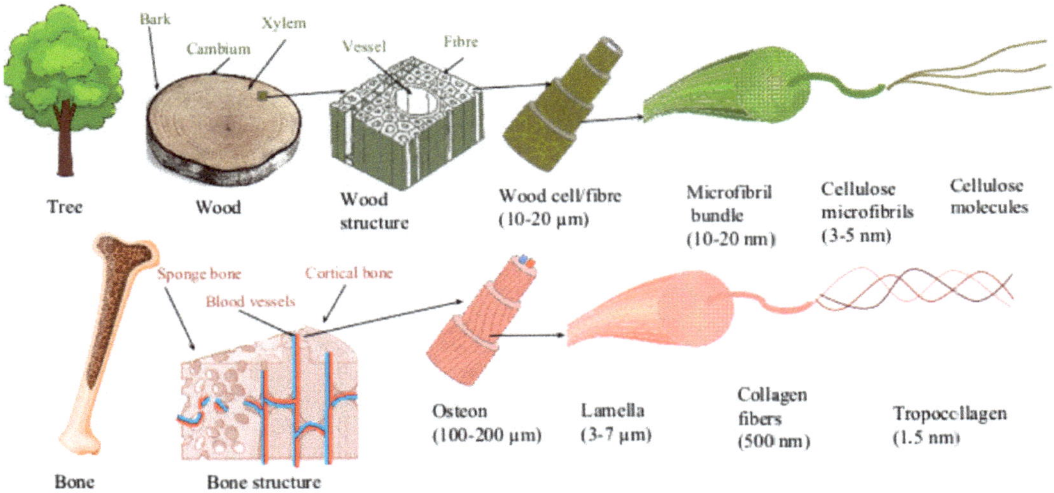

Figure 1. Schematical structure of bone and wood in macro, micro and nano scale.

The strength of both cortical bone and wood is ensured by its structural construction. The cortical bone base is formed by osteons, while wood consists of wood cells. Both the osteon and the tree cell are oriented in the direction of the long axis of the bone and wood, respectively, and are composed of several concentric layers of parallel fibers or fibrils. Each layer is oriented in different directions, thus providing mechanical strength. In bone, the layers are formed of collagen fibers, while wood cells are composed of bundles of cellulose micro fibrils. Collagen fiber consists of collagen fibrils constructed from triple helix collagen molecules and mineral nanocrystals, while cellulose microfibril is constructed from amorphous and crystalline parts of parallel cellulose molecules [36,47,48]. Based on similarities between wood and bone, wood has been used as a testing model for orthopedic implants [49]. Despite the structural similarity between wood and bone, solid wood has not been amply considered as a possible biomaterial for bone implants. Only in recent decades has wood been studied by a few authors as a possible implant material.

Considering the mentioned topical problems, this review presents a brief overview of the use of wood as a biomaterial for bone implants, emphasizing the wood processing and current research methodologies.

3. Wood Species for Bone Implants

Individual species of trees have been studied as a source of biomaterial. Most of the published studies have main purpose of creating biomaterials for bone defect substitution and rarely for creating wood-based orthopedic implants.

3.1. Birch

One of the first studies regarding wood as a possible implant material was performed by Kristen, Bösch et al. using birch wood [50]. Birch is one of the most widespread and economically important species of deciduous trees in Europe and Scandinavia. Silver birch (*Betula pendula*) and European white birch (*Betula pubescens*) are among the most common birch species found in most of Europe, up to Central Siberia. *Betula pubescens*, which is the northernmost tree species, is more common in the Northern and Eastern parts of Europe. *Betula pendula* is more common in the southern regions of Europe, such as the Iberian Peninsula, southern Italy, and Greece [51]. Birch wood has an average density of 600–650 kg/m^3 and a high Jank hardness of 4000–5000 N, and contains little extractive material, which makes this wood well suited for bone implants [52,53]. The early studies were conducted in vivo using rabbits. Birch implants were pre-treated with ethanol and placed transcortically into rabbit tibias. Evaluation was done after 3, 5, 14, and 32 weeks. Although the tissues produced a foreign body reaction, a new bone formed around the wood implants. Additionally, bone ingrowth into the implant's pores was recorded [54]. Similar ethanol pre-treated birch implants were implanted into a rabbit's soft tissues. After controls within 2, 6, 12, and 30 weeks, it was concluded that ethanol pre-treatment was not sufficient to prevent a foreign body reaction [50]. Years later, Rekola, Aho et al. published a novel wood pre-treatment method—preheating of birch implants at different temperatures—140 °C, 200 °C, and 220 °C for 2 h. The implants were placed into the drilled cavities of rabbit femurs and observed after 4, 8 and 20 weeks detecting the bone ingrowth. Preheated birch implants showed better osteoconductivity compared to untreated implants. However, when applying the highest temperature of 220 °C, the biomechanical characteristics of the implants were decreased [55]. In vitro studies were performed by immersing the birch implants into simulated body fluid (SBF) for 63 days at 37 °C. It was documented that immersion in the SBF significantly decreases the biomechanical properties of the untreated implants, while heat pre-treated implants preserve these properties [56,57].

3.2. Ash

Ash is a tree of the olive family that is widespread in Europe, Asia, Canada, and North America. As a hardwood with a low content of extractive substances, a high density of 600–680 kg/m^3, strength, and flexibility, it is suitable for bone implant materials [58]. The in vivo study with ash implants was conducted simultaneously with early birch studies. Ash specimens were ethanol-pretreated and fixed in rabbit calcaneus bones with Achilles tendons reattached and analyzed after 5 and 14 weeks. The ethanol pre-treatment of ash resulted not only in bone ingrowth, but the tendons' tissues grew into the wood pores as well, along with moderate foreign body reaction [59].

3.3. Lime, Willow, and Fir

Spruce wood is widespread in Scandinavia, Northern Eastern Europe, North America, Canada, and Japan, and is one of the most economically important coniferous wood species [60]. The white willow (*Salix alba*) is the most well-known of the willows, widely distributed throughout Europe except for the most northern regions. The northern part of Europe, where willow is common, includes the British Isles, the Netherlands, and the Baltic coast (Latvia and Lithuania). Willow is also found in Mediterranean regions, as well as in

North Africa (Morocco and Algeria) [61]. Lime trees are common in Eastern North America (*Tilia americana*) and Europe (*Tilia Europen*; hybrid wood). All named wood species have a low density—400–450 kg/m^3 for fir and willow, 450–550 kg/m^3 for lime wood. These wood species, together with birch and ash, were used to fix fractures in rabbit femurs by Horsky, Huraj, Paukovic. Implants were untreated before implanting in vivo. Birch, ash, and fir were well tolerated, while lime and willow caused acute inflammatory reaction, indicating that differences in wood species meant that not all species would be suitable for bone implants [62]. Although all the mentioned wood species contain a high extractive content, they differ in their composition. Fir extract contains the most lignans [63], while willow extracts contain a large amount of salicylic compounds, flavonoids, and tannins. These substances are bioactive compounds characterized by antipyretic, analgesic, anti-inflammatory, antirheumatic, and anticoagulant properties. As with all bioactive substances, they can be toxic at certain levels [64,65].

3.4. Juniper

Juniper is the world's most widespread and northernmost coniferous tree. It is common both in Europe and Asia, as well as in North America and Japan. Juniper can be found both in the farthest North areas of Scandinavia and in the mountain areas of the warmer regions of Southern Europe. The density of juniper is 450–600 kg/m^3 [66]. Juniper has long been studied for its antibacterial properties, but not for use in bone implants. The essential oils in juniper wood can also be toxic at high dosages; therefore, pre-treatment is required [67]. A unique in vivo study considered juniper wood for potential orthopedic hardware. Hip prostheses were crafted and pre-treated in boiling water for 10 min. The proximal part of rabbit femurs were resected and hemiarthroplasty with the juniper prostheses was performed. Rabbits were allowed to bear weight with no restrictions. Histological analysis was done after 3, 6, 18, and 36 months. No foreign body reaction was documented in any specimens. Initial bone ingrowth was detected after 6 months. After 3 years, wood implants were fully integrated with bone tissues (Figure 2). Essential oils from juniper were tested for their capacity to induce a toxic response in rats and was demonstrated to be well tolerated, especially when released slowly [68]. Almost 20 years later, preliminary studies have been carried out for the possible development of bone implants from partially delignified and compressed solid juniper wood, thus improving the mechanical properties of the implant. A compressed wood density of 1170 kg/m^3 was achieved (100% increase compared to natural juniper wood). The modulus of rupture was increased by 85%, reaching 174 MPa, and the modulus of elasticity by 620%, reaching 12,500 MPa [69].

3.5. Carbonized Wood

Another trial for the development of bone implants has been proposed by pre-treating wood at high temperatures to create a charcoal-type material. In one of the earliest studies, wood from clematis was carbonized at 850 °C for 5 h. Samples were implanted in vivo into rabbit bone, whose tissue was able to grow into the carbonized wood [70]. A similar in vivo study was performed with bamboo charcoal. The results showed that charcoal bamboo as a bone substitute has good biocompatibility and osteoconductivity [71]. Although pure carbonized wood had good biocompatibility and osteoconductivity, the complete loss of its mechanical properties made it an impractical material. Years later, the mechanical properties of pure carbonized wood were improved by an impregnation with silicon carbide (SiC) to produce a biomaterial called ecoceramics. In the preparation process, natural wood was pyrolyzed at 1000 °C using argon gas; the natural wood lost around 75% of its weight and 60% of its volume as a result of the treatment. The remaining scaffold was infiltrated with melted Si at 1550 °C. Si reacts with carbon in pyrolyzed wood to form SiC. Different wood species have been used to produce wood-based ecoceramics, for example, maple [72], eucalyptus [73], mango [74], oak [75], beech [76], pine [77], and others [78]. The technique preserved the porous structure of the wood while adding the rigidity of SiC. It is also a light-weight material, with density around 1100–2300 kg/cm^3, depending on the selected wood [74–76]. In addition, ecoceramics have great heat and electric resistance [79,80]. Due to various properties, ecoceramics have attracted more interest of researchers in civil [81], aeronautical [82] and electronic [83] engineering, and only a small number of studies consider ecoceramics as a material for medical applications. One in vivo study has been done with SiC scaffolds that were implanted in sheep metatarsal bones. Histological analysis was performed after 4, 8, 12, and 48 weeks. Analysis revealed good scaffold-to-bone adhesion, and new bone ingrowth inside the scaffolds was documented as well [84]. Since then, few authors have proposed combining wooden scaffolds with other biomaterials. In one study, SiC scaffolds derived from beech, eucalyptus, and sapele were combined with bioactive glass. An in vitro study with MG-63 osteoblasts showed good cellular attachment to both coated and uncoated SiC scaffolds. Additionally, the osteoblasts proliferated equally in standardized environments and on the surface of bioactive-glass-coated SiC scaffolds [85]. In another study, carbonized wood scaffolds derived from cane and pine [86,87] or rattan [88] were combined with hydroxyapatite (HA). The obtained samples showed the preserved porous structure and improved mechanical properties; the compressive strength reached 0.4 MPa and the tensile modulus increased 2–3 times [86–88].

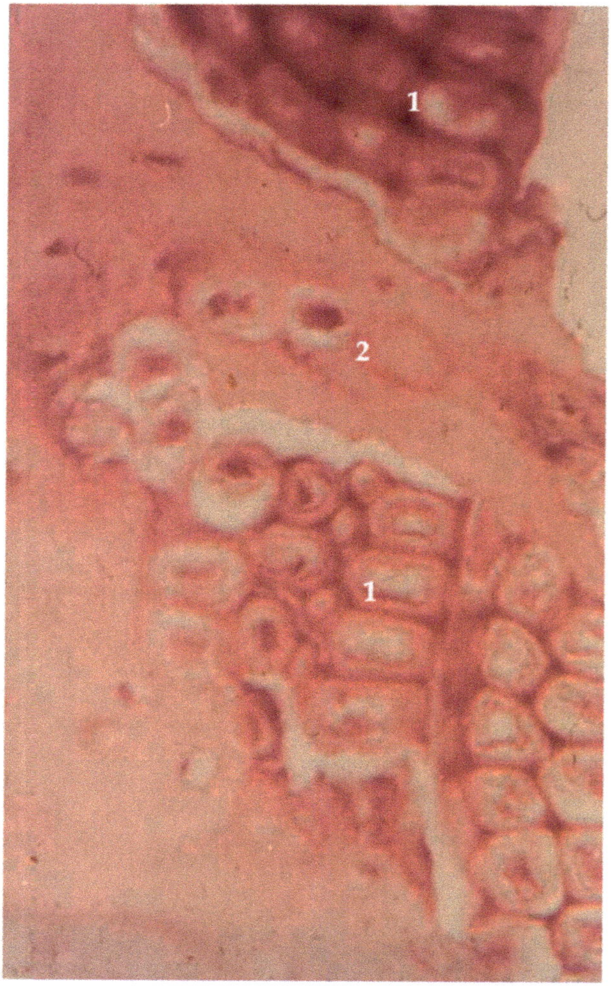

Figure 2. Juniper implant in the in vivo model. 1—juniper implant; 2—bone tissue ingrowth.

3.6. Cellulose-Based Scaffold

Few authors have considered wood as a base for cellulose-based scaffolds. To create such scaffolds, more extensive wood processing is required. Firstly, wood is processed into cellulose. Cellulose is a natural linear cell polysaccharide consisting of glucose $(C_6H_{10}O_5)_n$. (Figure 3). It is the main component of cell wall in green plants and algae, and bacteria produce cellulose to form a biofilm as well. While the purest natural form of cellulose is cotton, where cellulose comprises about 90% of cotton's mass, wood is made of around 57% cellulose and remains the main source for producing cellulose [89]. As a natural raw material, cellulose has been used for fabrics and papers for hundreds of years, but only 185 years ago, in 1838, the chemical structure of cellulose was discovered and described by French chemist Anselme Payen [90]. Since then, production of cellulose from wood stock is performed by chemically dissolving unwanted components such as lignin, short-chained polymers, etc. Cellulose is widely used for its porous structure and insolubility in water and organic substances in medical filters [91], pharmacy [92], and wound dressings [93]. In the last few decades, cellulose has also attracted researchers' attention as a potential biomaterial for medical applications, similar to using wood as an implant material. In the 1960s, implantation of cellulose sponges was used to study tissue inflammation and granulation formation shortly after implantation [94]. Years later, in the 1990s, researchers began to investigate the long-term effects of cellulose implantation. Märtson, Viljanto et al. used industrial soft cellulose sponges derived from eucalyptus, birch, or oak. Cellulose sponges were tested in vivo in rat soft tissue, and histological examinations were performed consecutively after 1–60 weeks. Histological evaluation revealed that the inflammatory response of the surrounding tissues subsided after 4–6 weeks and revealed good connective tissue ingrowth into the cellulose sponges. The researchers also detected a slow resorption and degradation of the pure cellulose sponges [95,96]. In addition, the biocompatibility of cellulose sponges with bone tissue was investigated in vivo; cellulose sponges were tested into the femoral bone cavity of rats. Bone ingrowth into cellulose sponges was recorded after 4–6 weeks [97]. Later researchers started combining cellulose fibers with other biomaterials. An in vitro study was performed with chondrocytes from the bovine knee joint. Cellulose scaffolds were exposed to saturated calcium hydroxide $(Ca(OH)_2)$ solution, then immersed in supersaturated simulated body fluid (SBF). Thus, a calcium phosphate coating was created. Although the cellulose and calcium phosphate scaffolds caused an acidic reaction in solution and the pH had to be adjusted with calcium hydroxide [98], better cellular adhesion was detected compared to untreated cellulose scaffolds. In another study, cellulose fibers were impregnated with hydroxyapatite particles. Tomilla, Ekholm et al. published two studies on cellulose coating with hydroxyapatite derived from bioactive glass. Bioactive glass S53P4 (Abmin Technologies Ltd., Turku, Finland) was dissolved in SBF, and cellulose sponges were immersed in the SBF solution at 37 °C for 24 h. After 24 h of immersion in SBF, calcium hydroxyapatite was formed on the surface of the scaffold. In an in vivo study with scaffolds implanted in rat soft tissue, an acute inflammatory response was reported on the first day after implantation. More extensive connective tissue formation was observed in the hydroxyapatite layer, while the inflammatory response disappeared within 14 days [99]. In another in vivo study, biomimetically coated cellulose sponges with silica-rich apatite were implanted into femoral bone defects in rats. After 12 weeks post-implantation, apatite-coated cellulose sponges did not significantly improve bone ingrowth compared to uncoated cellulose sponges. [100]. Later, Daugela, Pranskunas et al. investigated cellulose-based scaffolds substituted with micro- and nano-hydroxyapatite particles. An in vitro study was performed on human-like osteoblastic cells (Mg-63) to determine cytotoxicity and cell adhesion. According to the results, cell adhesion was improved by hydroxyapatite nanoparticles compared to cellulose-based scaffolds substituted with hydroxyapatite microparticles, and no cytotoxic response was detected. Similar to the in vitro results, scaffolds with hydroxyapatite nanoparticles significantly improved bone tissue ingrowth in rabbit calvaria bones [101]. Further studies also involved cellulose derivatives; carboxymethyl cellulose scaffolds were prepared using a freeze-drying process. An in vitro study showed that prepared scaffolds supported the proliferation and differentiation of Saos-2 cells, and extensive tissue proliferation was detected in rat subcutaneous tissues in vivo [102]. Cellulose scaffolds have also attracted interested as a localized drug-delivery system in damaged bone. Different studies have been published on the delivery of growth factors, bioactive proteins, antibiotics, and anti-inflammatory drugs [103].

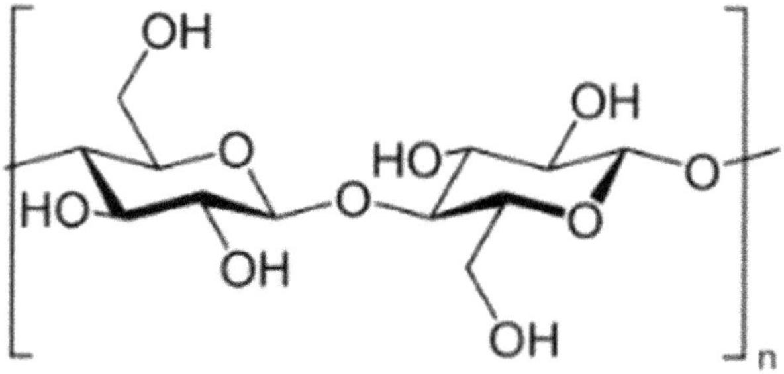

Figure 3. Cellulose molecule.

Table 1 summarizes brief facts about the wood species studied in the literature.

Table 1. Summary of potential wood species for bone implants.

References	Wood Species	Study Design	Results
Kristen, Bosch et al., 1977 [54]	Birch	In vivo	Foreign body reaction after ethanol pre-treatment
Aho, Rekola et al., 2007 [55]		In vivo	After heat pre-treatment good biocompatibility and osteoconductivity
Bosch, Kristen et al., 1979 [59]	Ash	In vivo	After ethanol pre-treatment good biocompatibility and osteoconductivity
Horsky, Huraj et al., 1987 [62]	Fir	In vivo	With no pre-treatment good biocompatibility
	Lime	In vivo	With no pre-treatment acute foreign body reaction
	Willow	In vivo	With no pre-treatment acute foreign body reaction
Gross and Ezerietis, 2003 [68]	Juniper	In vivo	After heat pre-treatment good biocompatibility and osteoconductivity
Andze, Andzs et al., 2022 [69]		Mechanical studies	Partial delignification of wood and subsequent densification showed improved mechanical properties comparable to bone
Kosuwon, Laupattarakasem et al., 1994 [71]	Bamboo	In vivo	Carbonized charcoal showed good biocompatibility and osteoconductivity

Table 1. *Cont.*

References	Wood Species	Study Design	Results
de Carlos, Borrajo et al., 2006 [85]	Beech	In vitro	SiC scaffolds allowed good cell proliferation
	Sapele		
Qian, Kang et al., 2008 [86] Tampieri, Sprio et al., 2009 [87]	Cane	Mechanical studies	SiC scaffolds combined with other biomaterials provided significantly improved mechanical resistance
	Pine		
Finardi and Sprio, 2012 [88]	Rattan		

4. Discussion

4.1. Advantages and Disadvantages of Wood as Bone Implants

The most important advantage of wood as a bone implant is the structural similarities between bone and wood, described by the pioneer of microscopy, Anton van Leeuwenhoek, back in 1693. Since then, the structural and functional similarities have been described by many other authors [41–44,104]. This was discussed in the previous sections.

Wood is a natural composite material consisting of three main components—cellulose, hemicelluloses and lignin [105]. Simply, each of the components gives the wood specific mechanical properties—strength, flexibility, and stiffness, respectively. The chemical composition of wood is variable and depends on the wood species, age, genetic factors, and growing conditions. [106–108] In principle, this could be considered a disadvantage, but this property of wood as a biomaterial is also an advantage. Knowledge of these effects on wood properties can help in finding suitable wood materials for a specific application. Furthermore, chemical [109,110], thermal [111,112], or enzymatic [113,114] treatment can change the chemical composition of wood, thus affecting its mechanical properties.

Porosity is another advantage of wood as a biomaterial. Pore size distribution in the wood varies from 1 nm to 100 μm and can be classified into macro-, meso- and micropores. Porosity is inversely proportional to wood density [110,115]. Osteoblasts are approximately 10–50 μm in size and require 100–200 nm pores for ingrowth and bone regeneration. In the pores with a smaller size, the formation of osteoid and fibrous tissue occurs [116]. It can be concluded that the ingrowth of bone cells will occur more easily in less dense wood. At the same time, wood density directly affects the mechanical properties of wood. The porous structure of wood easily lends to impregnation and can be used to introduce bioactive substances or drugs for bone regeneration.

The biodegradation of wood is one of the possible reasons why wood as an implant material causes skeptical reaction. To justify this property of wood, it is important to consider that wood decay occurs at a certain humidity and in the presence of oxygen. In a highly wet and oxygen-free environment, wood biodegradation occurs very slowly and wood can be preserved for hundreds of years [117,118]. Water content in human body lean mass (or fat-free mass) is around 70–75%; thus, it is considered a non-oxygen wet environment when inserting a wood implant [119].

Shrinking and swelling in water is a typical characteristic of wood. The amount of water in the wood is significantly affected by the humidity and temperature of the environment. Accordingly, in a dry environment, the wood dries quickly and cracks form; conversely, in a water environment, the wood absorbs moisture and swells. This characteristic complicates the use of wood materials and requires evaluation of the preparation process of wood samples. The water content of natural green wood is approximately 60–70%. The moisture content of dry wood is approximately 7%. Considering that the water content in human muscle mass is 70–75%, it is recommended to keep the moisture content in wood above 50% when developing a wood implant, thus preventing possible problems that may arise due to shrinking and swelling. It should be noted that the water content in solid wood affects its mechanical properties. The elasticity of wood is directly proportional, while the strength is inversely proportional to the water content of the wood [120,121].

Other disadvantages of wood as a biomaterial for bone substitution are its uneven and different properties, variable density and chemical composition depending on the wood species, growing conditions, genetic aspects, and age [105,107,108]. These makes it difficult to obtain materials with the same properties. One of the solutions is to use wood from plantations or to obtain pre-treated wood.

Apart from the basic components of wood—cellulose, hemicellulose, and lignin—it also contains a small number of inorganic compounds (up to 1%) and soluble organic compounds called extractives. The presence of extractives in wood samples can be critical for their use in bone implants. The extractives consist of mixtures of various components, from relatively low-molar-mass molecules to the higher molar-mass substances [122] such as fats, fatty acids, waxes, sterols, terpenic compounds, phenolic compounds, pectins, flavonoids, stilbenes, tannins, etc. [120] Some of the compounds are bioactive substances and, depending on the concentration, can be either therapeutic or toxic to the human body. Extractives can be divided into groups based on their chemical type—lipophilic or non-polar and hydrophilic or polar compounds. Each of the mentioned groups can be dissolved in different solvents—organic solvents or water. To eliminate all extractives, consecutive extraction is performed using different solvents of increasing polarity, e.g., dichloromethane, acetone, ethanol, and water [123]. Since the human body comprises 70–75% water, separation of the water-soluble extractives from the wood before implantation is critical. Pre-treatment of wood with both ethanol and water is necessary so that the extractives do not cause toxic reactions.

As a biological material, wood also provides a habitat for various microorganisms, such as fungi and bacteria, that are not desirable in bone implant material. Considering the above, special attention should be paid to the chosen sterilization methods. Not all popular bone implant sterilization methods are applicable to wood samples. Such classical methods as UV, ethanol, or ethylene oxide treatment [124], autoclaving, or steam treatment [125] cannot be used for wood materials. Wood is destroyed under the influence of UV [126], swelling occurs during water vapor treatment, and wood hydrolysis begins at elevated temperatures above 140 degrees [111]; but in the case of ethanol or ethylene oxide (toxic) treatment [124], it could be problematic to ensure the removal of all substances from the sample due to its porous structure. It is possible to use gamma irradiation or microwave treatment [127], but in this case, a suitable processing time should be chosen, as the wood may be destroyed due to heating (more than 140 degree) [128,129]. Sterilization with gaseous phase compounds such as supercritical CO_2, hydroxyl peroxide, or peracetic acid is applicable to porous fibrous materials, including wood [124,130,131].

Additionally, greater mechanical properties and density are needed to use wood in osteosynthesis implants. Densification of wood increases the mechanical properties and density of wood. Chemical pretreatment makes it possible to reduce the variability of the chemical composition. Chemical pretreatment of wood and subsequent densification is a promising method for wood processing to obtain implants with density and mechanical strength suitable for osteosynthesis biomaterials [69].

4.2. Mechanical Properties of Wood as Bone Implant Compared with Other Implant Materials

Natural wood has a density from 450 up to 700 kg/m^3, depending on species. The higher density is observed in hardwood species, whereas softwoods have lower densities. The modulus of elasticity (MOE) reaches 1550–13,500 MPa and the modulus of rupture (MOR) reaches 60–100 MPa [132,133]. Mechanical properties can be altered by various methods such as applying heat and chemical treatment. Applying lower heat, up to 200 °C, can increase the MOE, and rupture can be increased by up to 50% [134]. If wood is heated over 800 °C, it loses up to 80% of its mass and 60% of its volume. The obtained carbonized wood has poor mechanical properties, with a density around 200–400 kg/m^3, although biocompatibility and osteoconductivity is preserved [71,135,136]. Another technique for wood processing is densification, which involves partial delignification and compression. Authors Andze L. et al., in a mechanical study, increased juniper's density by 100%, reaching almost 1200 kg/m^3. The MOR and MOE were increased by 85% and 620%, accordingly [69]. Studies of wood's mechanical properties show that natural wood is not strong enough to produce durable orthopedic implants for bone fracture fixation. Nevertheless, with certain processing mechanical properties of wood can be improved to fit requirements for orthopedic implants. The most common material for orthopedic implants is still titanium and its alloys. Titanium has a density of 4500 kg/m^3, MOR of 45,000 MPa and MOE of 120,000 MPa [137]. Obviously, titanium's mechanical strength is multiple times higher than any biological material, including bone—its density is up to 1200 kg/m^3 [138–140], its MOE varies from 10–3000 MPa, and its MOR is 150–180 MPa. [141,142] This significant disparity in mechanical properties allows stable fixation for fractures, but can also cause complications such as aseptic loosening [143–148].

Since other materials used in bone repair are dedicated to bone defect substitution, their mechanical properties are unessential, as they are not supposed to provide mechanical support to the bone [149]. Among the investigated biomaterials are tricalcium phosphate bioceramics (density 3070 kg/m^3, MOR 1.3 MPa, MOE 49 MPa [150]), hydroxyapatite bioceramics (density 3050 kg/m^3, MOR 18 MPa, MOE 174 MPa [151]) bioactive glass 45S5 (density 2850 kg/m^3, MOR 45 MPa, MOE 60 MPa [152]), collagen (density 2700 kg/m^3, MOR 2 MPa, MOE 46 MPa [153]), and silk fibrion (density 1400 kg/m^3, MOR 5 MPa, MOE 100 MPa [154]). Mechanical properties of different materials are summarized in Table 2.

Table 2. Summary of mechanical properties for different materials.

Material	Density, kg/m^3	MOE, MPa	MOR, MPa
Human bone [140,141]	up to 1200	10–3000	150–180
Natural wood [132]	450–700	1550–13,500	60–100
Carbonized wood [135]	200–400	15–140	11–53
Densified wood [69]	1170	12,500	174
Titanium [137]	4500	120,000	45,000
Calcium phosphate bioceramics [150]	3070	49	1.3
Hydroxyapatite bioceramics [151]	3050	174	18
Bioactive glass 45S5 [152]	2850	60	45
Collagen [153]	2700	46	2
Silk fibrion [154]	1400	100	5

4.3. *Summary for Further Investigation*

The interest in biomaterials has been constantly growing in the past decades. As environmental issues are a growing concern, an alternative to unsustainable and non-renewable materials is being developed [24]. This direction of development also includes medical implants. Although metallic implants have been greatly improved in terms of biocompatibility and corrosion resistance, their environmental impact is impossible to avoid [155]. Multiple studies have proven that sustainable biomaterials are not only suitable for the production of different implants, but their production process also has a significantly lower ecological footprint compared to any non-sustainable resources [27,29,156]. In the search for suitable biomaterials to be used in bone repair, wood has been one of the potential options studied. The earliest studies published in the last century were conducted using in vivo models. These studies proved the osteoconductive abilities of natural wood implants. Osteoconductivity is a passive attribute of implants, where they allow bone tissue ingrowth on the surface or inside pores of an implant [157–159]. This feature of wood implants has been proven in a few studies by obtaining microscopic pictures of new bone trabeculas inside wood pores [54,55,68]. Another important feature of bone implants is osseointegration—direct contact and anchorage between bone and implant, which is maintained over the long-term [160,161]. This feature was demonstrated with juniper prosthesis in rabbits. Animals were able to bear weight with no restrictions up to 3 years, without any implant failures [68]. Studies done in the 21st century are executed in vitro. Throughout the studies, not every species of wood had proved to be equally suitable for bone implants; e.g., lime and willow showed an acute inflammatory reaction [62]. It was concluded that the inflammatory reaction depends on the soluble substances of wood (extractives). Additionally, bamboo before treatment produced cytotoxicity in an in vitro study [162]. Other species such as birch, ash, and juniper have presented excellent biocompatibility [54,55,59,68]. To reduce a possible inflammatory reaction, pre-treatment of the wood is a crucial step. Not only for the sake of asepsis; additional components in untreated wood, e.g., fungi, have been found [163]. They alone can produce an inflammatory reaction [164]. Techniques for pre-treatment differ in various articles. Juniper for an in vivo study has been pre-treated only in boiling water for 10 min [68]. Other authors have tried ethanol pre-treatment that was not sufficient to avoid an acute inflammatory reaction [59]. In the following studies, pre-treatment with higher temperatures became more popular. Aho, Rekola et al. used heat as high as 220 °C for birch and ash implants [55]. This range of temperature neutralizes bacteria, fungi, and all organic extracts, such as essential oils. Thus, pre-treatment at high temperatures reduces the risk of toxic reaction to the minimum. Other authors considered wood as a scaffold for creating new biomaterials. In the oldest studies, charcoal was researched as a possible biomaterial. Although osteoconductive properties were preserved, poor mechanical properties limited further applications. Since the first articles on wood-derived ecoceramics were published, this biomaterial has gained considerable research interest, as ecoceramics combine wood's favorable properties such as porosity, mechanical and heat resistance. Aside from bone implant development, these characteristics raise interest in a wide range of industrial uses, i.e., filters, catalysts, electric sensors, etc. [165]. Despite the fact that only afew studies were published on biocompatibility of ecoceramics [166,167], the concept was advanced by combining ecoceramics with other biomaterials, such as HA and bioactive glass. Both concepts of wood as biomaterial—hybrid biocomposites and pure wood—have been highlighted in research as a potential biomaterial for repairing damaged bone [18–22,25–33]. It should be noted that other biomaterials are also available for this purpose, such as calcium phosphate bioceramics, bioactive glasses [168], and different composite materials combining bioactive inorganic materials with biodegradable polymers [169,170]. Wood is also a main source for cellulose, which has a wide application for medical devices and wound dressings as well. Although only a small number of studies involving cellulose scaffolds have been devoted to bone surgery, some promising results have been reported for bone tissue proliferation and local drug delivery to improve bone healing [171,172]. Orthopedic implants have demanding mechanical requirements to sustain long periods of mechanical loading. For this reason, thus far, metal implants are dominant, and only two groups of authors have processed wood with a goal of orthopedic implants for fracture fixation and joint arthroplasty [62,68,173].

5. Conclusions

Wood is a sustainable and renewable source suitable for the production of biomaterials. The processing of wood is more environmentally friendly, especially compared to titanium production, which emits carbon monoxide and other toxic by-products. Nowadays, there is still limited research on the use of wood in bone implants, despite the fact that its great potential has been demonstrated by available studies. The in vivo studies done in the 20th century's last decades show great insight into some species of wood's great biocompatibility and osteoconductivity. Based on the provided review, the continued development of wood implants for further incorporation in surgical practice is suggested.

Author Contributions: Conceptualization, L.A.; methodology, L.A. and V.N.; investigation, V.N.; writing—original draft preparation, V.N., M.K. and L.A.; writing—review and editing, M.A., I.F., R.T., L.V. and M.K.; visualization, V.N., L.A. and L.V.; supervision, L.A. and M.K.; project administration,

L.A., M.A.; funding acquisition, L.A. All authors have read and agreed to the published version of the manuscript.

Funding: The preparation of review article was funded by Bioeconomy grants of Latvian State Institute of Wood Chemistry 2022 "JunBon" and 2023 "WoodBone" 03-23.

Institutional Review Board Statement: The study was conducted in accordance with the Declaration of Helsinki.

Informed Consent Statement: Not applicable.

Data Availability Statement: Not applicable.

Conflicts of Interest: The authors declare no conflict of interest.

References

1. Borgel, S.; Latimer, B.; McDermott, Y.; Sarig, R.; Pokhojaev, A.; Abulafia, T.; Goder-Goldberger, M.; Barzilai, O.; May, H. Early Upper Paleolithic human foot bones from Manot Cave, Israel. *J. Hum. Evol.* **2021**, *160*, 102668. [CrossRef] [PubMed]
2. Afshar, A.; Steensma, D.P.; Kyle, R.A. Albin Lambotte: Pioneer of Osteosynthesis (Bone Fixation). *Mayo Clin. Proc.* **2021**, *96*, 2012–2013. [CrossRef] [PubMed]
3. Roberts, T.T.; Prummer, C.M.; Papaliodis, D.N.; Uhl, R.L.; Wagner, T.A. History of the orthopedic screw. *Orthopedics* **2013**, *36*, 12–14. [CrossRef]
4. Fairbank, J. The Evolution of Orthopaedic Surgery, by Leslie Klenerman. *Spine* **2002**, *27*, 2298. [CrossRef]
5. Steinbach, L.W., IV. On the Use of Fixation Plates in the Treatment of Fractures of the Leg. *Ann. Surg.* **1900**, *31*, 436–442. [CrossRef]
6. Lambotte, A. *Chirurgie Operatoire des Fractures*; Masson: Issy-les-Moulineaux, France, 1913.
7. Venable, C.S.; Stuck, W.G. *The Internal Fixation of Fractures*; Thomas: New York, NY, USA, 1947.
8. Zioupos, P.; Currey, J.D. Changes in the stiffness, strength, and toughness of human cortical bone with age. *Bone* **1998**, *22*, 57–66. [CrossRef] [PubMed]
9. Wu, J.J.; Shyr, H.S.; Chao, E.Y.; Kelly, P.J. Comparison of osteotomy healing under external fixation devices with different stiffness characteristics. *J. Bone Jt. Surg. Am.* **1984**, *66*, 1258–1264. [CrossRef]
10. Sansone, V.; Pagani, D.; Melato, M. The effects on bone cells of metal ions released from orthopaedic implants. A review. *Clin. Cases Miner. Bone Metab.* **2013**, *10*, 34–40. [CrossRef]
11. Cadosch, D.; Chan, E.; Gautschi, O.P.; Simmen, H.P.; Filgueira, L. Bio-corrosion of stainless steel by osteoclasts—In Vitro evidence. *J. Orthop. Res.* **2009**, *27*, 841–846. [CrossRef]
12. Cadosch, D.; Al-Mushaiqri, M.S.; Gautschi, O.P.; Meagher, J.; Simmen, H.P.; Filgueira, L. Biocorrosion and uptake of titanium by human osteoclasts. *J. Biomed. Mater. Res. A* **2010**, *95*, 1004–1010. [CrossRef] [PubMed]
13. Rekola, J.; Aho, A.J.; Gunn, J.; Matinlinna, J.; Hirvonen, J.; Viitaniemi, P.; Vallittu, P.K. The effect of heat treatment of wood on osteoconductivity. *Acta Biomater.* **2009**, *5*, 1596–1604. [CrossRef]
14. Zhang, X.; Wang, J.; Xing, L. Metal artifact reduction in x-ray computed tomography (CT) by constrained optimization. *Med. Phys.* **2011**, *38*, 701–711. [CrossRef]
15. Hargreaves, B.A.; Worters, P.W.; Pauly, K.B.; Pauly, J.M.; Koch, K.M.; Gold, G.E. Metal-induced artifacts in MRI. *AJR Am. J. Roentgenol.* **2011**, *197*, 547–555. [CrossRef]
16. Aciduman, A.; Belen, D. The earliest document regarding the history of cranioplasty from the Ottoman era. *Surg. Neurol.* **2007**, *68*, 349–352. [CrossRef]
17. Kiernan, C.; Knuth, C.; Farrell, E. Chapter 6—Endochondral Ossification: Recapitulating Bone Development for Bone Defect Repair. In *Developmental Biology and Musculoskeletal Tissue Engineering*; Stoddart, M.J., Craft, A.M., Pattappa, G., Gardner, O.F.W., Eds.; Academic Press: Boston, MA, USA, 2018; pp. 125–148. [CrossRef]
18. Sasaki, G.; Watanabe, Y.; Miyamoto, W.; Yasui, Y.; Morimoto, S.; Kawano, H. Induced membrane technique using beta-tricalcium phosphate for reconstruction of femoral and tibial segmental bone loss due to infection: Technical tips and preliminary clinical results. *Int. Orthop.* **2018**, *42*, 17–24. [CrossRef]
19. Yan, L.; Jiang, D.M. Study of bone-like hydroxyapatite/polyamino acid composite materials for their biological properties and effects on the reconstruction of long bone defects. *Drug. Des. Devel. Ther.* **2015**, *9*, 6497–6508. [CrossRef]
20. Roffi, A.; Krishnakumar, G.S.; Gostynska, N.; Kon, E.; Candrian, C.; Filardo, G. The Role of Three-Dimensional Scaffolds in Treating Long Bone Defects: Evidence from Preclinical and Clinical Literature-A Systematic Review. *Biomed. Res. Int.* **2017**, *2017*, 8074178. [CrossRef]
21. Nguyen, B.B.; Moriarty, R.A.; Kamalitdinov, T.; Etheridge, J.M.; Fisher, J.P. Collagen hydrogel scaffold promotes mesenchymal stem cell and endothelial cell coculture for bone tissue engineering. *J. Biomed. Mater. Res. A* **2017**, *105*, 1123–1131. [CrossRef]
22. Ruan, S.Q.; Deng, J.; Yan, L.; Huang, W.L. Composite scaffolds loaded with bone mesenchymal stem cells promote the repair of radial bone defects in rabbit model. *Biomed. Pharmacother.* **2018**, *97*, 600–606. [CrossRef]
23. Zhang, M.; Matinlinna, J.P.; Tsoi, J.K.H.; Liu, W.; Cui, X.; Lu, W.W.; Pan, H. Recent developments in biomaterials for long-bone segmental defect reconstruction: A narrative overview. *J. Orthop. Transl.* **2020**, *22*, 26–33. [CrossRef] [PubMed]

24. Directorate-General for Research and Innovation (European Commission); De Smet, M.; Linder, M.; Koopmans, R.; Doorsselaer, K.; Velis, C.; De Wilde, B.; Ritschkoff, A.; Crippa, M.; Leyssens, J.; et al. *A Circular Economy for Plastics: Insights from Research and Innovation to Inform Policy and Funding Decisions*; Publications Office: Independence, MO, USA, 2019. [CrossRef]
25. Widuto, A. *The European Green Deal and Cohesion Policy*; European Parliamentary Research Service: Brussels, Belgium, 2021.
26. Ramos-Rodriguez, D.H.; Pashneh-Tala, S.; Bains, A.K.; Moorehead, R.D.; Kassos, N.; Kelly, A.L.; Paterson, T.E.; Orozco-Diaz, C.A.; Gill, A.A.; Ortega Asencio, I. Demonstrating the Potential of Using Bio-Based Sustainable Polyester Blends for Bone Tissue Engineering Applications. *Bioengineering* **2022**, *9*, 163. [CrossRef] [PubMed]
27. De Bortoli, L.S.; Schabbach, L.M.; Fredel, M.C.; Hotza, D.; Henriques, B. Ecological footprint of biomaterials for implant dentistry: Is the metal-free practice an eco-friendly shift? *J. Clean. Prod.* **2019**, *213*, 723–732. [CrossRef]
28. Trimeche, D.M. Biomaterials for bone regeneration: An overview. *Biomater. Tissue Technol.* **2017**, *1*, 1–5.
29. Oladapo, B.I.; Zahedi, S.A.; Ismail, S.O.; Olawade, D.B. Recent advances in biopolymeric composite materials: Future sustainability of bone-implant. *Renew. Sustain. Energy Rev.* **2021**, *150*, 111505. [CrossRef]
30. Osuchukwu, O.A.; Salihi, A.; Abdullahi, I.; Abdulkareem, B.; Nwannenna, C.S. Synthesis techniques, characterization and mechanical properties of natural derived hydroxyapatite scaffolds for bone implants: A review. *SN Appl. Sci.* **2021**, *3*, 822. [CrossRef]
31. Agbeboh, N.I.; Oladele, I.O.; Daramola, O.O.; Adediran, A.A.; Olasukanmi, O.O.; Tanimola, M.O. Environmentally sustainable processes for the synthesis of hydroxyapatite. *Heliyon* **2020**, *6*, e03765. [CrossRef] [PubMed]
32. Yadav, D.; Garg, R.K.; Ahlawat, A.; Chhabra, D. 3D printable biomaterials for orthopedic implants: Solution for sustainable and circular economy. *Resour. Policy* **2020**, *68*, 101767. [CrossRef]
33. Anil, A.; Ali Serdar, V. Wood-Reinforced Polymer Composites. In *Wood in Civil Engineering*; Giovanna, C., Ed.; IntechOpen: Rijeka, Croatia, 2017. [CrossRef]
34. Zhao, S.; Zhao, J.X.; Han, G.Z. Advances in the study of mechanical properties and constitutive law in the field of wood research. *IOP Conf. Ser. Mater. Sci. Eng.* **2016**, *137*, 012036. [CrossRef]
35. Wiedenhoeft, A.; Miller, R. 2 Structure and Function of Wood. In *Handbook of Wood Chemistry and Wood Composites*; CRC Press: Boca Raton, Fl, USA, 2005.
36. Rosa, N.; Moura, M.F.S.F.; Olhero, S.; Simoes, R.; Magalhães, F.D.; Marques, A.T.; Ferreira, J.P.S.; Reis, A.R.; Carvalho, M.; Parente, M. Bone: An Outstanding Composite Material. *Appl. Sci.* **2022**, *12*, 3381. [CrossRef]
37. Vaz, M.F.; Canhão, H.; Fonseca, J.O. *Bone: A Composite Natural Material*; BioMed Central: London, UK, 2011. [CrossRef]
38. Jeong, J.; Kim, J.H.; Shim, J.H.; Hwang, N.S.; Heo, C.Y. Bioactive calcium phosphate materials and applications in bone regeneration. *Biomater. Res.* **2019**, *23*, 4. [CrossRef]
39. Sansalone, V.; Naili, S.; Bousson, V.; Bergot, C.; Peyrin, F.; Zarka, J.; Laredo, J.D.; Haïat, G. Determination of the heterogeneous anisotropic elastic properties of human femoral bone: From nanoscopic to organ scale. *J. Biomech.* **2010**, *43*, 1857–1863. [CrossRef]
40. Su, N.; Yang, J.; Xie, Y.; Du, X.; Chen, H.; Zhou, H.; Chen, L. Bone function, dysfunction and its role in diseases including critical illness. *Int. J. Biol. Sci.* **2019**, *15*, 776–787. [CrossRef]
41. Van Leeuwenhoeck, A. Several observations on the texture of bone in animals compared with that of wood: On the bark of. trees: On the little scales found in the cuticula. *Philos. Trans. R. Soc. Lond.* **1693**, *17*, 838–843.
42. Jeronimidis, G. Chapter 1—Structure-Property Relationships in Biological Materials. In *Pergamon Materials Series*; Elices, M., Ed.; JSTOR: Pergamon, Turkey, 2000; Volume 4, pp. 3–16.
43. Spatz, H.C.; Köhler, L.L.; Niklas, K. Mechanical behaviour of plant tissues: Composite materials or structures? *J. Exp. Biol.* **2000**, *202*, 3269–3272. [CrossRef]
44. Fratzl, P.; Weinkamer, R. Nature's hierarchical materials. *Prog. Mater. Sci.* **2007**, *52*, 1263–1334. [CrossRef]
45. Weinkamer, R.; Fratzl, P. Mechanical adaptation of biological materials—The examples of bone and wood. *Mater. Sci. Eng. C* **2011**, *31*, 1164–1173. [CrossRef]
46. Stanzl-Tschegg, S.E. Wood as a bioinspiring material. *Mater. Sci. Eng. C* **2011**, *31*, 1174–1183. [CrossRef]
47. McKittrick, J.; Chen, P.Y.; Tombolato, L.; Novitskaya, E.; Trim, M.; Hirata, G.; Olevsky, E.A.; Horstemeyer, M. Energy absorbent natural materials and bioinspired design strategies: A review. *Mater. Sci. Eng. C* **2010**, *30*, 331–342. [CrossRef]
48. Yi, K.; Ruffini, A.; Srivastava, S.; Srivastava, R. Bio-Inspired Technology Edited by Ruby Srivastava. SRIVASTAVA, Ruby. Introductory Chapter: DNA as Nanowires. In *Bio-Inspired Technology*; IntechOpen: Rijeka, Croatia, 2019. [CrossRef]
49. Murdoch, A.H.; Mathias, K.J.; Shepherd, D.E. Investigation into the material properties of beech wood and cortical bone. *Biomed. Mater. Eng.* **2004**, *14*, 1–4. [PubMed]
50. Kristen, H.; Bosch, P.; Bednar, H.; Plenk, H., Jr. The effects of dynamic loading on intracalcaneal wood implants and on the tissues surrounding them. *Arch. Orthop. Trauma Surg.* **1979**, *93*, 287–292. [CrossRef] [PubMed]
51. Beck, P.; Caudullo, G.; de Rigo, D.; Tinner, W. *Betula pendula, Betula pubescens and Other Birches in Europe: Distribution, Habitat, Usage and Threats*; Publication Office of the European Union: Luxembourg, 2016.
52. Rastogi, S.; Pandey, M.M.; Rawat, A.K.S. Medicinal plants of the genus Betula—Traditional uses and a phytochemical–pharmacological review. *J. Ethnopharmacol.* **2015**, *159*, 62–83. [CrossRef]
53. Laskowska, A. The Influence of Process Parameters on the Density Profile and Hardness of Surface-densified Birch Wood (Betula pendula Roth). *Bioresources* **2017**, *12*, 6011–6023. [CrossRef]

54. Kristen, H.; Bosch, P.; Bednar, H.; Plenk, H., Jr. Biocompatibility of wood in bone tissue (author's transl). *Arch. Orthop. Unf.* **1977**, *89*, 1–14. [CrossRef] [PubMed]
55. Aho, A.J.; Rekola, J.; Matinlinna, J.; Gunn, J.; Tirri, T.; Viitaniemi, P.; Vallittu, P. Natural composite of wood as replacement material for ostechondral bone defects. *J. Biomed. Mater. Res. B Appl. Biomater.* **2007**, *83*, 64–71. [CrossRef]
56. Rekola, J.; Lassila, L.V.; Hirvonen, J.; Lahdenpera, M.; Grenman, R.; Aho, A.J.; Vallittu, P.K. Effects of heat treatment of wood on hydroxylapatite type mineral precipitation and biomechanical properties in vitro. *J. Mater. Sci. Mater. Med.* **2010**, *21*, 2345–2354. [CrossRef]
57. Rekola, J.; Lassila, L.V.; Nganga, S.; Yla-Soininmaki, A.; Fleming, G.J.; Grenman, R.; Aho, A.J.; Vallittu, P.K. Effect of heat treatment of wood on the morphology, surface roughness and penetration of simulated and human blood. *Biomed. Mater. Eng.* **2014**, *24*, 1595–1607. [CrossRef]
58. Etiegni, L.; Campbell, A. Physical and chemical characteristics of wood ash. *Bioresour. Technol.* **1991**, *37*, 173–178 [CrossRef]
59. Bosch, P.; Kristen, H.; Braun, F.; Kovac, W. Reaction of connective tissue and striated muscle tissue to implanted ashwood (author's transl). *Wien. Med. Wochenschr.* **1979**, *129*, 419–423.
60. Yildiz, S.; Gezer, E.D.; Yildiz, U.C. Mechanical and chemical behavior of spruce wood modified by heat. *Build. Environ.* **2006**, *41*, 1762–1766. [CrossRef]
61. Hanley, S.J.; Karp, A. Genetic strategies for dissecting complex traits in biomass willows (*Salix* spp.). *Tree Physiol.* **2014**, *34*, 1167–1180. [CrossRef] [PubMed]
62. Horsky, I.; Huraj, E.; Paukovic, J. Utilization of wood in the manufacture of orthopedic implants. *Acta Chir. Orthop. Traumatol. Cech.* **1987**, *54*, 3–13. [PubMed]
63. Füchtner, S.; Brock-Nannestad, T.; Smeds, A.; Fredriksson, M.; Pilgård, A.; Thygesen, L.G. Hydrophobic and Hydrophilic Extractives in Norway Spruce and Kurile Larch and Their Role in Brown-Rot Degradation. *Front. Plant Sci.* **2020**, *11*, 855. [CrossRef] [PubMed]
64. Warmiński, K.; Stolarski, M.J.; Gil, Ł.; Krzyżaniak, M. Willow bark and wood as a source of bioactive compounds and bioenergy feedstock. *Ind. Crops Prod.* **2021**, *171*, 113976. [CrossRef]
65. Tyśkiewicz, K.; Konkol, M.; Kowalski, R.; Rój, E.; Warmiński, K.; Krzyżaniak, M.; Gil, Ł.; Stolarski, M. Characterization of bioactive compounds in the biomass of black locust, poplar and willow. *Trees* **2019**, *33*, 1235–1263. [CrossRef]
66. Enescu, C.; Durrant, T.; Caudullo, G.; de Rigo, D. *Juniperus Communis in Europe: Distribution, Habitat, Usage and Threats*; Publication Office of the European Union: Luxembourg, 2016.
67. Semerdjieva, I.; Zheljazkov, V.D.; Radoukova, T.; Dincheva, I.; Piperkova, N.; Maneva, V.; Astatkie, T.; Kačániová, M. Biological Activity of Essential Oils of Four Juniper Species and Their Potential as Biopesticides. *Molecules* **2021**, *26*, 6358. [CrossRef]
68. Gross, K.A.; Ezerietis, E. Juniper wood as a possible implant material. *J. Biomed. Mater. Res. A* **2003**, *64*, 672–683. [CrossRef] [PubMed]
69. Andze, L.; Andzs, M.; Skute, M.; Nefjodov, V.; Kapickis, M.; Tupciauskas, R. Preliminary Study of Chemically Pretreated Densification of Juniper Wood for Use in Bone Implants. *Mater. Sci. Forum* **2022**, *1071*, 101–108. [CrossRef]
70. Colville, J.B.P.; Hoikka, V.; Vainio, K. Wood anatomy and the use of carbonised wood as a matrix for bone regeneration in animals. *Int. Assoc. Wood Anat.* **1979**, *12*, 3–6.
71. Kosuwon, W.; Laupattarakasem, W.; Saengnipanthkul, S.; Mahaisavariya, B.; Therapongpakdee, S. Charcoal bamboo as a bone substitute: An animal study. *J. Med. Assoc. Thai.* **1994**, *77*, 496–500.
72. Gordic, M.; Bucevac, D.; Ružić, J.; Gavrilovic, S.; Hercigonja, R.; Stankovic, M.; Matovic, B. Biomimetic synthesis and properties of cellular SiC. *Ceram. Int.* **2014**, *40*, 3699–3705. [CrossRef]
73. Byrne, C.E.; Nagle, D.C. Carbonization of wood for advanced materials applications. *Carbon* **1997**, *35*, 259–266. [CrossRef]
74. Greil, P.; Lifka, T.; Kaindl, A. Biomorphic Cellular Silicon Carbide Ceramics from Wood: I. Processing and Microstructure. *J. Eur. Ceram. Soc.* **1998**, *18*, 1961–1973. [CrossRef]
75. Polozov, I.; Razumov, N.; Masaylo, D.; Silin, A.; Lebedeva, Y.; Popovich, A. Fabrication of Silicon Carbide Fiber-Reinforced Silicon Carbide Matrix Composites Using Binder Jetting Additive Manufacturing from Irregularly-Shaped and Spherical Powders. *Materials* **2020**, *13*, 1766. [CrossRef]
76. Varela-Feria, F.M.; Fernández, J.; Ramirez de Arellano Lopez, A.; Singh, M. Low Density Biomorphic Silicon Carbide: Microstructure and Mechanical Properties. *J. Eur. Ceram. Soc.* **2002**, *22*, 2719–2725. [CrossRef]
77. Sieber, H.; Vogli, E.; Greil, P. *Biomorphic SiC-Ceramic Manufactured by Gas-Phase Infiltration of Pine Wood*; John Wiley & Sons, Inc.: Hoboken, NJ, USA, 2008; Volume 22, pp. 109–116.
78. Singh, M.; Salem, J.A. Mechanical properties and microstructure of biomorphic silicon carbide ceramics fabricated from wood precursors. *J. Eur. Ceram. Soc.* **2002**, *22*, 2709–2717. [CrossRef]
79. Singh, M. Environment conscious ceramics (Ecoceramics). *Ceram. Eng. Sci. Proc.* **2000**, *21*, 39–44.
80. Parfen'eva, L.S.; Orlova, T.S.; Kartenko, N.F.; Sharenkova, N.V.; Smirnov, B.I.; Smirnov, I.A.; Misiorek, H.; Jezowski, A.; Varela-Feria, F.M.; Martinez-Fernandez, J.; et al. Thermal conductivity of the SiC/Si biomorphic composite, a new cellular ecoceramic. *Phys. Solid State* **2005**, *47*, 1216–1220. [CrossRef]
81. Kardashev, B.K.; Burenkov, Y.A.; Smirnov, B.I.; de Arellano-Lopez, A.R.; Martinez-Fernandez, J.; Varela-Feria, F.M. Elasticity and inelasticity of biomorphic silicon carbide ceramics. *Phys. Solid State* **2004**, *46*, 1873–1877. [CrossRef]

82. Naslain, R.R. SiC-Matrix Composites: Nonbrittle Ceramics for Thermo-Structural Application. *Int. J. Appl. Ceram. Technol.* **2005**, *2*, 75–84. [CrossRef]
83. Orlova, T.; Smirnov, B.; de Arellano-Lopez, A.; Martinez Fernandez, J.; Sepulveda, R. Anisotropy of electric resistivity of Sapele-based biomorphic SiC/Si composites. *Phys. Solid State* **2005**, *47*, 220–223. [CrossRef]
84. Filardo, G.; Kon, E.; Tampieri, A.; Cabezas-Rodriguez, R.; Di Martino, A.; Fini, M.; Giavaresi, G.; Lelli, M.; Martinez-Fernandez, J.; Martini, L.; et al. New bio-ceramization processes applied to vegetable hierarchical structures for bone regeneration: An experimental model in sheep. *Tissue Eng. Part A* **2014**, *20*, 763–773. [CrossRef]
85. de Carlos, A.; Borrajo, J.P.; Serra, J.; Gonzalez, P.; Leon, B. Behaviour of MG-63 osteoblast-like cells on wood-based biomorphic SiC ceramics coated with bioactive glass. *J. Mater. Sci. Mater. Med* **2006**, *17*, 523–529. [CrossRef]
86. Qian, J.; Kang, Y.; Zhang, W.; Li, Z. Fabrication, chemical composition change and phase evolution of biomorphic hydroxyapatite. *J. Mater. Sci. Mater. Med.* **2008**, *19*, 3373–3383. [CrossRef]
87. Tampieri, A.; Sprio, S.; Ruffini, A.; Celotti, G.; Lesci, G.; Roveri, N. From wood to bone: Multi-step process to convert wood hierarchical structures into biomimetic hydroxyapatite scaffolds for bone tissue engineering. *J. Mater. Chem.* **2009**, *19*, 4973–4980. [CrossRef]
88. Finardi, U.; Sprio, S. Human bone regeneration from wood: A novel hierarchically organised nanomaterial. *Int. J. Healthc. Technol. Manag.* **2012**, *13*, 171–183. [CrossRef]
89. Bayer, E.A.; Chanzy, H.; Lamed, R.; Shoham, Y. Cellulose, cellulases and cellulosomes. *Curr. Opin. Struct. Biol.* **1998**, *8*, 548–557. [CrossRef]
90. Marchessault, R.H. Cellulose Structure Modification and Hydrolysis; Raymond, A., Roger, Y., Rowell, M., Eds.; Wiley-Interscience: New York, NY, USA, 1986, p. 379. *J. Polym. Sci. Part C Polym. Lett.* **1987**, *25*, 139–140. [CrossRef]
91. Layton, M.; Roper, D. 12—Investigation of the Hereditary Haemolytic Anaemias: Membrane and Enzyme Abnormalities. In *Dacie and Lewis Practical Haematology*, 12th ed.; Bain, B.J., Bates, I., Laffan, M.A., Eds.; Elsevier: Amsterdam, The Netherlands, 2017; pp. 228–253. [CrossRef]
92. Di Giuseppe, E. Analogue Materials in Experimental Tectonics. In *Reference Module in Earth Systems and Environmental Sciences*; Elsevier: Amsterdam, The Netherlands, 2018. [CrossRef]
93. Gupta, B.S. 1—Manufacture, types and properties of biotextiles for medical applications. In *Biotextiles as Medical Implants*; King, M.W., Gupta, B.S., Guidoin, R., Eds.; Woodhead Publishing: Cambridge, UK, 2013; pp. 3–47. [CrossRef]
94. Viljanto, J. A Cellstick Device for Wound Healing Research. In *Wound Healing and Skin Physiology*; Springer: Berlin/Heidelberg, Germany, 1995; pp. 513–522.
95. Martson, M.; Viljanto, J.; Hurme, T.; Laippala, P.; Saukko, P. Is cellulose sponge degradable or stable as implantation material? An in vivo subcutaneous study in the rat. *Biomaterials* **1999**, *20*, 1989–1995. [CrossRef]
96. Martson, M.; Viljanto, J.; Laippala, P.; Saukko, P. Connective tissue formation in subcutaneous cellulose sponge implants in the rat. The effect of the size and cellulose content of the implant. *Eur. Surg. Res.* **1998**, *30*, 419–425. [CrossRef]
97. Martson, M.; Viljanto, J.; Hurme, T.; Saukko, P. Biocompatibility of cellulose sponge with bone. *Eur. Surg. Res.* **1998**, *30*, 426–432. [CrossRef]
98. Muller, F.A.; Muller, L.; Hofmann, I.; Greil, P.; Wenzel, M.M.; Staudenmaier, R. Cellulose-based scaffold materials for cartilage tissue engineering. *Biomaterials* **2006**, *27*, 3955–3963. [CrossRef]
99. Tommila, M.; Jokinen, J.; Wilson, T.; Forsback, A.P.; Saukko, P.; Penttinen, R.; Ekholm, E. Bioactive glass-derived hydroxyapatite-coating promotes granulation tissue growth in subcutaneous cellulose implants in rats. *Acta Biomater.* **2008**, *4*, 354–361. [CrossRef]
100. Ekholm, E.; Tommila, M.; Forsback, A.P.; Martson, M.; Holmbom, J.; Aaritalo, V.; Finnberg, C.; Kuusilehto, A.; Salonen, J.; Yli-Urpo, A.; et al. Hydroxyapatite coating of cellulose sponge does not improve its osteogenic potency in rat bone. *Acta Biomater.* **2005**, *1*, 535–544. [CrossRef]
101. Daugela, P.; Pranskunas, M.; Juodzbalys, G.; Liesiene, J.; Baniukaitiene, O.; Afonso, A.; Sousa Gomes, P. Novel cellulose/hydroxyapatite scaffolds for bone tissue regeneration: In vitro and in vivo study. *J. Tissue Eng. Regen. Med.* **2018**, *12*, 1195–1208. [CrossRef]
102. Priya, G.; Madhan, B.; Narendrakumar, U.; Suresh Kumar, R.V.; Manjubala, I. In Vitro and In Vivo Evaluation of Carboxymethyl Cellulose Scaffolds for Bone Tissue Engineering Applications. *ACS Omega* **2021**, *6*, 1246–1253. [CrossRef]
103. Janmohammadi, M.; Nazemi, Z.; Salehi, A.O.M.; Seyfoori, A.; John, J.V.; Nourbakhsh, M.S.; Akbari, M. Cellulose-based composite scaffolds for bone tissue engineering and localized drug delivery. *Bioact. Mater.* **2023**, *20*, 137–163. [CrossRef]
104. Chen, C.; Kuang, Y.; Zhu, S.; Burgert, I.; Keplinger, T.; Gong, A.; Li, T.; Berglund, L.; Eichhorn, S.J.; Hu, L. Structure–property–function relationships of natural and engineered wood. *Nat. Rev. Mat.* **2020**, *5*, 642–666. [CrossRef]
105. Filipova, I.; Grinfelds, U.; Andze, L.; Rozenberga, L.; Lazdina, D.; Zeps, M.; Jansons, A. Chemical composition and fi ber properties of fast-growing species in Latvia and its potential for forest bioindustry. *For. Stud.* **2017**, *66*, 27–32. [CrossRef]
106. Filipova, I.; Grinfelds, U.; Andze, L.; Rozenberga, L.; Zeps, M.; Neimane, U.; Jansons, A. Effect of Refining on the Properties of Fibres from Young Scots (*Pinus sylvestris*) and Lodgepole Pines (*Pinus contorta*). *Balt. For.* **2017**, *23*, 529–533.
107. Liting, C.; Peng, Z.; Yajing, D.; Jian, D.; Wei, W. Correlation of wood properties with chemical composition and microstructure of wood components. *Res. Sq.* **2023**, *1*, 24. [CrossRef]
108. Sorieul, M.; Dickson, A.; Hill, S.J.; Pearson, H. Plant Fibre: Molecular Structure and Biomechanical Properties, of a Complex Living Material, Influencing Its Deconstruction towards a Biobased Composite. *Materials* **2016**, *9*, 618. [CrossRef] [PubMed]

109. Xu, E.; Wang, D.; Lin, L. Chemical Structure and Mechanical Properties of Wood Cell Walls Treated with Acid and Alkali Solution. *Forests* **2020**, *11*, 87. [CrossRef]
110. Jang, E.-S.; Kang, C.-W. Porosity analysis of three types of balsa (*Ochroma pyramidale*) wood depending on density. *J. Wood Sci.* **2022**, *68*, 31. [CrossRef]
111. Esteves, B.; Pereira, H. Wood modification by heat treatment: A review. *BioResources* **2009**, *4*, 370–404. [CrossRef]
112. Peng, Q.; Ormondroyd, G.; Spear, M.; Chang, W.-S. The effect of the changes in chemical composition due to thermal treatment on the mechanical properties of Pinus densiflora. *Constr. Build. Mater.* **2022**, *358*, 129303. [CrossRef]
113. Zhang, Y.; Meng, Y.; Wu, Y.; Wang, S.; Du, G.-B.; Jiang, H.; Zhou, Z.-B. Effect of enzyme treatment on the mechanical properties of wood cell walls by nanoindentation. *Bioresources* **2012**, *7*, 2249–2259. [CrossRef]
114. Wagenfuehr, A.; Tech, S.; Unbehaun, H. Modification of wood properties through enzymes. *J. For. Suisse* **2005**, *156*, 420–426.
115. Plötze, M.; Niemz, P. Porosity and pore size distribution of different wood types as determined by mercury intrusion porosimetry. *Eur. J. Wood Wood Prod.* **2011**, *69*, 649–657. [CrossRef]
116. Abbasi, N.; Hamlet, S.; Love, R.M.; Nguyen, N.-T. Porous scaffolds for bone regeneration. *J. Sci. Adv. Mater. Devices* **2020**, *5*, 1–9. [CrossRef]
117. Kazemi, M.; Dickinson, D.; Murphy, R.J. Effects of Initial Moisture Content on Wood Decay at Different Levels of Gaseous Oxygen Concentrations. *J. Agric. Sci. Technol.* **2001**, *3*, 293–304.
118. Teng, T.-J.; Arip, M.N.M.; Sudesh, K.; Nemoikina, A.; Jalaludin, Z.; Ng, E.-P.; Lee, H.-L. Conventional technology and nanotechnology in wood preservation: A review. *BioResources* **2018**, *13*, 9220–9252. [CrossRef]
119. Lorenzo, I.; Serra-Prat, M.; Yébenes, J.C. The Role of Water Homeostasis in Muscle Function and Frailty: A Review. *Nutrients* **2019**, *11*, 1857. [CrossRef] [PubMed]
120. Böhm, M.; Kobetičová, K.; Procházka, J.; Černý, R. Moisture sorption and thickness swelling of wood-based materials intended for structural use in humid conditions and bonded with melamine resin. *IOP Conf. Ser. Mater. Sci. Eng.* **2019**, *549*, 012042. [CrossRef]
121. Eckelman, C. *The Shrinking and Swelling of Wood and Its Effect on Furniture*; Purdue University Cooperative Extension Service: West Lafayette, IN, USA, 2010.
122. Bikovens, O.; Lauberte, L.; Pranovich, A.; Reunanen, M.; Telysheva, G. Chemical Composition of Lipophilic Extractives from Grey Alder (*Alnus incana*). *Bioresources* **2013**, *8*, 350–357. [CrossRef]
123. Kebbi-Benkeder, Z.; Colin, F.; Dumarçay, S.; Gérardin, P. Quantification and characterization of knotwood extractives of 12 European softwood and hardwood species. *Ann. For. Sci.* **2014**, *72*, 277–284. [CrossRef]
124. Tao, M.; Ao, T.; Mao, X.; Yan, X.; Javed, R.; Hou, W.; Wang, Y.; Sun, C.; Lin, S.; Yu, T.; et al. Sterilization and disinfection methods for decellularized matrix materials: Review, consideration and proposal. *Bioact. Mater.* **2021**, *6*, 2927–2945. [CrossRef]
125. Manea, A.; Bran, S.; Baciut, M.; Armencea, G.; Pop, D.; Berce, P.; Vodnar, D.C.; Hedesiu, M.; Dinu, C.; Petrutiu, A.; et al. Sterilization protocol for porous dental implants made by Selective Laser Melting. *Clujul. Med.* **2018**, *91*, 452–457. [CrossRef]
126. Teacă, C.; Rosu, D.; Bodîrlău, R.; Rosu, L. Structural Changes in Wood under Artificial UV Light Irradiation Determined by FTIR Spectroscopy and Color Measurements–A Brief Review. *BioResources* **2013**, *8*, 1478–1507. [CrossRef]
127. Bargh, S.; Silindir-Gunay, M.; Ozer, A.Y.; Colak, S.; Kutlu, B.; Nohutcu, R. The effects of gamma and microwave sterilization on periodontological grafts. *Chem. Phys. Impact* **2021**, *3*, 100046. [CrossRef]
128. Despot, R.; Hasan, M.; Rapp, A.; Brischke, C.; Humar, M.; Welzbacher, C.; Raem, D.A. *Changes in Selected Properties of Wood Caused by Gamma Radiation*; InTech: Rijeka, Croatia, 2012. [CrossRef]
129. Novotny, M.; Skramlik, J.; Šuhajda, K.; Tichomirov, V. Sterilization of Biotic Pests by Microwave Radiation. *Procedia Eng.* **2013**, *57*, 1094–1099. [CrossRef]
130. Horakova, J.; Klicova, M.; Erben, J.; Klapstova, A.; Novotny, V.; Behalek, L.; Chvojka, J. Impact of Various Sterilization and Disinfection Techniques on Electrospun Poly-ε-caprolactone. *ACS Omega* **2020**, *5*, 8885–8892. [CrossRef] [PubMed]
131. Hodder, E.; Duin, S.; Kilian, D.; Ahlfeld, T.; Seidel, J.; Nachtigall, C.; Bush, P.; Covill, D.; Gelinsky, M.; Lode, A. Investigating the effect of sterilisation methods on the physical properties and cytocompatibility of methyl cellulose used in combination with alginate for 3D-bioplotting of chondrocytes. *J. Mater. Sci. Mater. Med.* **2019**, *30*, 10. [CrossRef] [PubMed]
132. Kretschmann, D.E. Mechanical properties of wood. *Environments* **2010**, *5*, 34.
133. Cai, Z.; Ross, R.J. Mechanical properties of wood-based composite materials. In *Wood Handbook-Wood as an Engineering Material. Centennial Edition. General Technical Report FPL-GTR-190*; Department of Agriculture, Forest Service, Forest Products Laboratory: Madison, WI, USA, 2010; p. 12.
134. Xie, Y.; Fu, Q.; Wang, Q.; Xiao, Z.; Militz, H. Effects of chemical modification on the mechanical properties of wood. *Eur. J. Wood Wood Prod.* **2013**, *71*, 401–416. [CrossRef]
135. Emrich, W. *Handbook of Charcoal Making: The Traditional and Industrial Methods*; Springer: Amsterdam, The Netherlands, 2010.
136. Kai, T.; Yamamoto, S.; Ishihara, S. The Effect of Carbonization Temperature on the Properties of Wood Charcoal. *Resour. Process.* **2000**, *47*, 132–139. [CrossRef]
137. Manivasagam, G.; Singh, A.K.; Rajamanickam, A.; Gogia, A. Ti based biomaterials, the ultimate choice for orthopaedic implants–A review. *Prog. Mater. Sci.* **2009**, *54*, 397–425. [CrossRef]
138. Smith, C.B.; Smith, D.A. Relations between age, mineral density and mechanical properties of human femoral compacta. *Acta Orthop. Scand.* **1976**, *47*, 496–502. [CrossRef]

139. Wall, J.C.; Chatterji, S.K.; Jeffery, J.W. Age-related changes in the density and tensile strength of human femoral cortical bone. *Calcif. Tissue Int.* **1979**, *27*, 105–108. [CrossRef]
140. Morgan, E.F.; Unnikrisnan, G.U.; Hussein, A.I. Bone Mechanical Properties in Healthy and Diseased States. *Annu. Rev. Biomed. Eng.* **2018**, *20*, 119–143. [CrossRef]
141. Singh, D.; Rana, A.; Jhajhria, S.; Garg, B.; Pandey, P.; Kalyanasundaram, D. Experimental assessment of biomechanical properties in human male elbow bone subjected to bending and compression loads. *J. Appl. Biomater. Funct. Mater.* **2018**, *17*, 2280800018793816. [CrossRef]
142. González-Bárcenas, L.; Trejo-Camacho, H.; Suárez-Estrella, I.; Heredia-Barbero, A.; Magaña, C.; Bucio, L.; Orozco, E. Three Point Bending Test of Human Femoral Tissue: An Essay in Ancient and Modern Bones. *Bone* **2003**, *682*, 173–178. [CrossRef]
143. Xie, S.; Manda, K.; Pankaj, P. Time-dependent behaviour of bone accentuates loosening in the fixation of fractures using bone-screw systems. *Bone Jt. Res.* **2018**, *7*, 580–586. [CrossRef] [PubMed]
144. Cadosch, D.; Chan, E.; Gautschi, O.P.; Filgueira, L. Metal is not inert: Role of metal ions released by biocorrosion in aseptic loosening–current concepts. *J. Biomed. Mater. Res. A* **2009**, *91*, 1252–1262. [CrossRef] [PubMed]
145. Cadosch, D.; Schlett, C.L.; Gautschi, O.P.; Frei, H.C.; Filgueira, L. Metal ions: Important co-players in aseptic loosening. *Z. Orthop. Unf.* **2010**, *148*, 393–397. [CrossRef] [PubMed]
146. Amirhosseini, M. *Aseptic Loosening of Orthopedic Implants: Osteoclastogenesis Regulation and Potential Therapeutics*; Linköping University Electronic Press: Linköping, Sweden, 2019. [CrossRef]
147. Feng, X.; Lin, G.; Fang, C.X.; Lu, W.W.; Chen, B.; Leung, F.K.L. Bone resorption triggered by high radial stress: The mechanism of screw loosening in plate fixation of long bone fractures. *J. Orthop. Res.* **2019**, *37*, 1498–1507. [CrossRef] [PubMed]
148. Huber, M.; Reinisch, G.; Trettenhahn, G.; Zweymüller, K.; Lintner, F. Presence of corrosion products and hypersensitivity-associated reactions in periprosthetic tissue after aseptic loosening of total hip replacements with metal bearing surfaces. *Acta Biomater.* **2009**, *5*, 172–180. [CrossRef] [PubMed]
149. Ghassemi, T.; Shahroodi, A.; Ebrahimzadeh, M.H.; Mousavian, A.; Movaffagh, J.; Moradi, A. Current Concepts in Scaffolding for Bone Tissue Engineering. *Arch. Bone Jt. Surg.* **2018**, *6*, 90–99.
150. Laasri, S.; Taha, M.; Hlil, E.K.; Laghzizil, A.; Hajjaji, A. Manufacturing and mechanical properties of calcium phosphate biomaterials. *Comptes Rendus Mécanique* **2012**, *340*, 715–720. [CrossRef]
151. Martin, R.I.; Brown, P.W. Mechanical properties of hydroxyapatite formed at physiological temperature. *J. Mater. Sci. Mater. Med.* **1995**, *6*, 138–143. [CrossRef]
152. Vyas, V.; Arepalli, S.; Singh, S.; Pyare, R. Effect of nickel oxide substitution on bioactivity and mechanical properties of bioactive glass. *Bull. Mater. Sci.* **2016**, *39*, 1355–1361. [CrossRef]
153. Manssor, N.A.; Radzi, Z.; Yahya, N.A.; Mohamad Yusof, L.; Hariri, F.; Khairuddin, N.H.; Abu Kasim, N.H.; Czernuszka, J.T. Characteristics and Young's Modulus of Collagen Fibrils from Expanded Skin Using Anisotropic Controlled Rate Self-Inflating Tissue Expander. *Ski. Pharm. Physiol.* **2016**, *29*, 55–62. [CrossRef] [PubMed]
154. Ren, J.; Liu, Y.; Kaplan, D.; Ling, S. Interplay of structure and mechanics in silk/carbon nanocomposites. *MRS Bull.* **2019**, *44*, 53–58. [CrossRef]
155. Guzzo, D.; Carvalho, M.M.; Balkenende, R.; Mascarenhas, J. Mascarenhas. Circular business models in the medical device industry: Paths towards sustainable healthcare. *Resour. Conserv. Recycl.* **2020**, *160*, 104904. [CrossRef]
156. Girón, J.; Kerstner, E.; Medeiros, T.; Oliveira, L.; Machado, G.M.; Malfatti, C.F.; Pranke, P. Biomaterials for bone regeneration: An orthopedic and dentistry overview. *Braz. J. Med. Biol. Res.* **2021**, *54*, e11055. [CrossRef] [PubMed]
157. Roberts, T.T.; Rosenbaum, A.J. Bone grafts, bone substitutes and orthobiologics: The bridge between basic science and clinical advancements in fracture healing. *Organogenesis* **2012**, *8*, 114–124. [CrossRef] [PubMed]
158. Albrektsson, T.; Johansson, C. Osteoinduction, osteoconduction and osseointegration. *Eur. Spine J.* **2001**, *10* (Suppl. S2), S96–S101. [CrossRef]
159. Cornell, C.N. Osteoconductive materials and their role as substitutes for autogenous bone grafts. *Orthop. Clin. N. Am.* **1999**, *30*, 591–598. [CrossRef]
160. Albrektsson, T.; Branemark, P.I.; Hansson, H.A.; Lindstrom, J. Osseointegrated titanium implants. Requirements for ensuring a long-lasting, direct bone-to-implant anchorage in man. *Acta Orthop. Scand.* **1981**, *52*, 155–170. [CrossRef]
161. Branemark, P.I.; Hansson, B.O.; Adell, R.; Breine, U.; Lindstrom, J.; Hallen, O.; Ohman, A. Osseointegrated implants in the treatment of the edentulous jaw. Experience from a 10-year period. *Scand. J. Plast Reconstr. Surg. Suppl.* **1977**, *16*, 1–132.
162. Li, S.H.; Liu, Q.; de Wijn, J.R.; Zhou, B.L.; de Groot, K. In vitro calcium phosphate formation on a natural composite material, bamboo. *Biomaterials* **1997**, *18*, 389–395. [CrossRef]
163. Coskey, R.J.; Mehregan, A.H.; Rippon, J.H. Wood implantation with phaeomycotic contamination. *J. Am. Acad. Dermatol.* **1983**, *8*, 428–429. [CrossRef] [PubMed]
164. Meyer, R.F.; Hood, C.I. Fungus implantation with wooden intraocular foreign bodies. *Ann. Ophthalmol.* **1977**, *9*, 271–278. [PubMed]
165. Smirnov, I.A.; Smirnov, B.I.; Orlova, T.S.; Wlosewicz, D.; Hackemer, A.; Misiorek, H.; Mucha, J.; Jezowski, A.; Ramirez-Rico, J.; Martinez-Fernandez, J. Heat capacity of Bio-SiC and SiC/Si ecoceramics prepared from white eucalyptus, beech, and sapele tree wood. *Phys. Solid State* **2013**, *55*, 454–460. [CrossRef]

166. De Arellano-López, A.R.; Martínez-Fernández, J.; González, P.; Domínguez, C.; Fernández-Quero, V.; Singh, M. Biomorphic SiC: A New Engineering Ceramic Material. *Int. J. Appl. Ceram. Technol.* **2004**, *1*, 56–67. [CrossRef]
167. Zeni, T.L.; Rambo, C.R.; Magalhães, W.L.E.; Hotza, D. Control of Porosity and Permeability of Wood by Fungi Action for Biotemplating of SiC. *J. Wood Chem. Technol.* **2013**, *33*, 33–43. [CrossRef]
168. Gerhardt, L.C.; Boccaccini, A.R. Bioactive Glass and Glass-Ceramic Scaffolds for Bone Tissue Engineering. *Materials* **2010**, *3*, 3867–3910. [CrossRef]
169. Müller, F. *Bone-Inspired Biomaterials*; Wiley: Hoboken, NJ, USA, 2018; pp. 337–350. [CrossRef]
170. Navarro, M.; Michiardi, A.; Castaño, O.; Planell, J.A. Biomaterials in orthopaedics. *J. R. Soc. Interface* **2008**, *5*, 1137–1158. [CrossRef]
171. Guo, L.; Liang, Z.; Yang, L.; Du, W.; Yu, T.; Tang, H.; Li, C.; Qiu, H. The role of natural polymers in bone tissue engineering. *J. Control Release* **2021**, *338*, 571–582. [CrossRef]
172. Hoenich, N. Cellulose for medical applications: Past, present, and future. *BioResources* **2006**, *1*, 270–280. [CrossRef]
173. Ezerietis, E.; Vetra, J.; Gardovskis, J.; Gross, K.; Jupatovs, R.; Skundra, M.; Krumalis, J.; Blaus, A. *Hip Joint Prosthesis*; National Library of Medicine: Bethesda, MD, USA, 1997.

Journal of *Functional Biomaterials*

Correction

Correction: Omori et al. Sinus Mucosal Damage Triggered by Synthetic or Xenogeneic Bone Substitutes: A Histological Analysis in Rabbits. *J. Funct. Biomater.* 2022, *13*, 257

Yuki Omori [1], Daniele Botticelli [2,*], Stefano Migani [2], Vitor Ferreira Balan [3], Eduardo Pires Godoy [4] and Samuel Porfirio Xavier [3]

1 Department of Oral Implantology, Osaka Dental University, 8-1 Kuzuhahanazonocho, Osaka 573-1121, Japan
2 ARDEC Academy, Viale Giovanni Pascoli 67, 47923 Rimini, Italy
3 Department of Oral and Maxillofacial Surgery and Periodontology, Faculty of Dentistry of Ribeirão Preto, University of São Paulo, Av. do Café-Subsetor Oeste-11 (N-11), Ribeirão Preto 14040-904, Brazil
4 Department of Oral Biology, Faculty of Dentistry of Ribeirão Preto, University of São Paulo, Ribeirão Preto 14040-904, Brazil
* Correspondence: daniele.botticelli@gmail.com

Citation: Omori, Y.; Botticelli, D.; Migani, S.; Ferreira Balan, V.; Pires Godoy, E.; Xavier, S.P. Correction: Omori et al. Sinus Mucosal Damage Triggered by Synthetic or Xenogeneic Bone Substitutes: A Histological Analysis in Rabbits. *J. Funct. Biomater.* 2022, *13*, 257. *J. Funct. Biomater.* **2024**, *15*, 121. https://doi.org/10.3390/jfb15050121

Received: 12 April 2024
Accepted: 22 April 2024
Published: 6 May 2024

Error in Figure

In the original publication [1], there was a mistake in Figure 1 with regards to what was published. Figure 1a is wrong, as it is just a repetition of Figure 2a. The correct version of Figure 1a appears below.

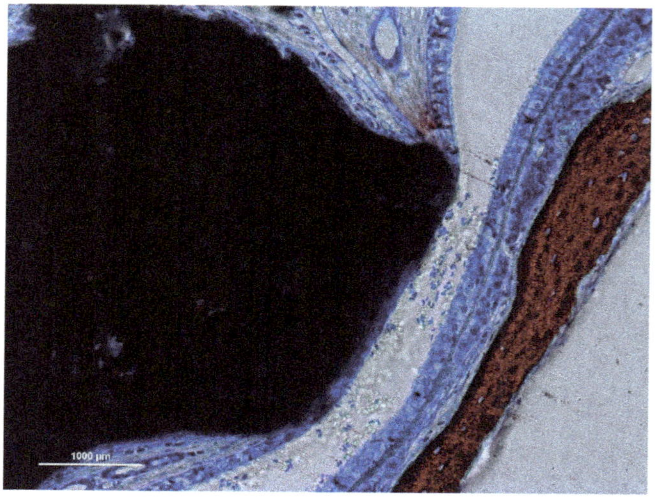

Figure Legend

In the original publication [1], there was a mistake in the legend for Figure 1. The legend of Figure 1a was wrong because it did not describe the correct figure that should have been used. The correct legend appears below.

Figure 1. (**a**) Synthetic site: Stevenel's blue and alizarin red stain. (**b**) Xenogeneic site: toluidine blue stain. Note the progressive decrease in width of both sinus mucosae and pseudostratified epithelia. A loss of cilia is evident in the thinnest sites on both biomaterials. While the process of resorption has a minimal impact on the xenogeneic graft, the synthetic graft has undergone a process already described as an interpenetrating bone network [18] characterized by concurrent bone formation within the biomaterial structure during its resorption.

The authors state that the scientific conclusions drawn in the paper are unaffected. This correction was approved by the Academic Editor. The original publication has also been updated.

Reference

1. Omori, Y.; Botticelli, D.; Migani, S.; Ferreira Balan, V.; Pires Godoy, E.; Xavier, S.P. Sinus Mucosal Damage Triggered by Synthetic or Xenogeneic Bone Substitutes: A Histological Analysis in Rabbits. *J. Funct. Biomater.* **2022**, *13*, 257. [CrossRef] [PubMed]

MDPI
St. Alban-Anlage 66
4052 Basel
Switzerland
www.mdpi.com

Journal of Functional Biomaterials Editorial Office
E-mail: jfb@mdpi.com
www.mdpi.com/journal/jfb

www.ingramcontent.com/pod-product-compliance
Lightning Source LLC
Chambersburg PA
CBHW061825210926
51850CB00006B/804
9783725811892